MOLECULAR BREEDING OF WOODY PLANTS

Progress in Biotechnology

Volume 1 New Approaches to Research on Cereal Carbohydrates (Hill and Munck, Editors)

Volume 2 Biology of Anaerobic Bacteria (Dubourguier et al., Editors)

Volume 3 Modifications and Applications of Industrial Polysaccharides (Yalpani, Editor)

Volume 4 Interbiotech '87. Enzyme Technologies (Blažej and Zemek, Editors)

Volume 5 In Vitro Immunization in Hybridoma Technology (Borrebaeck, Editor)

Volume 6 Interbiotech '89. Mathematical Modelling in Biotechnology (Blažej and Ottová, Editors)

Volume 7 Xylans and Xylanases (Visser et al., Editors)

Volume 8 Biocatalysis in Non-Conventional Media (Tramper et al., Editors)

Volume 9 ECB6: Proceedings of the 6th European Congress on Biotechnology (Alberghina et al., Editors)

Volume 10 Carbohydrate Bioengineering (Petersen et al., Editors)

Volume 11 Immobilized Cells: Basics and Applications (Wijffels et al., Editors)

Volume 12 Enzymes for Carbohydrate Engineering (Kwan-Hwa Park et al., Editors)

Volume 13 High Pressure Bioscience and Biotechnology (Hayashi and Balny, Editors)

Volume 14 Pectins and Pectinases (Visser and Voragen, Editors)

Volume 15 Stability and Stabilization of Biocatalysts (Ballesteros et al., Editors)

Volume 16 Bioseparation Engineering (Endo et al., Editors)

Volume 17 Food Biotechnology (Bielecki et al., Editors)

Volume 18 Molecular Breeding of Woody Plants (Morohoshi and Komamine, Editors)

Progress in Biotechnology 18

MOLECULAR BREEDING OF WOODY PLANTS

Proceedings of the
International Wood Biotechnology Symposium (IWBS)
held in Narita, Chiba, Japan, March 14–17, 2001

Edited by

Noriyuki Morohoshi
Professor,
Department of Environment Symbiotic Production System,
Graduate School of Bio-Applications and Systems Engineering,
Tokyo University of Agriculture and Technology,
Tokyo, Japan

Atsushi Komamine
Professor Emeritus of Tohoku University,
and Director of the Research Institute of Evolutional Biology,
Tokyo, Japan

2001

ELSEVIER

Amsterdam – London – New York – Oxford – Paris – Shannon – Tokyo

ELSEVIER SCIENCE B.V.
Sara Burgerhartstraat 25
P.O. Box 211, 1000 AE Amsterdam, The Netherlands

First edition 2001

Library of Congress Cataloging in Publication Data
A catalog record from the Library of Congress has been applied for.

ISBN: 0-444-50958-5

⊗ The paper used in this publication meets the requirements of ANSI/NISO Z39.48-1992 (Permanence of Paper).
Printed in The Netherlands.

Preface

We are facing several important problems on the environmental pollution and shortages of food, feed, biomass and natural resources, which will become more serious in the first quarter of this century. To solve the problems, we must be able to achieve a sufficient level of agriculture and forest productions to support an explosively increasing population, and all life on this small planet must be prevented from destruction of environments caused by pollution, global warming and desertification of lands. At present, plants and agricultural sciences are playing a leading role in the rescue of human beings from many problems in the biosphere; plant biotechnology may improve crop functions to rapidly promote food production. Plants and agricultural sciences may also produce plants tolerant to environmental stresses such as drought, salinity and coldness, and thus would expand land available for cultivation.

Of the plant species, woody plants are producing most abundant biomass resources, playing important roles in the suppression of carbon dioxide increase and supplying huge amounts of energy and resources to human beings in the biosphere. At present, we are also facing serious destruction in the tropical area caused by increasing population rapidly. It shows that 17 million hectares of the forests in the tropical area have been deforested or destroyed per year. There are extremely variant microbes, animals and plants in the tropical area. If the tropical forest is destructed, the treasury of such various genetic resources will be disappearing rapidly, without contributing to future human beings. We must stop the decrease of tropical forest and try to reforest.

On the other hand, the world demand of wood products is predicted to rise sharply in the next decade because we have to use wood biomass as essential energy and resource instead of fossil fuels in future. We must increase the productivity of woody plants to change the outlook for these problems. It is expected that "biotechnology of woody plants" is useful to the analysis, preservation and utilization of a gene resource, and can also contribute to the reproduction of the destroyed forest and increasing production of biomass.

The aim of this international conference and the Proceedings was to discuss the recent results of fundamental and applied researches for global resource and energy, biomass production and environmental problems from the aspect of woody science.

The conference encompassed the following topics:

1. Formation of the vascular bundle.
2. Biosynthesis of cellulose.

3. Lignin biosynthesis and transgenic woody plants.
4. Cell and tissue culture, and transformation in gymnosperms.
5. Micropropagation of woody plants.

It is my great pleasure to publish this book as the Proceedings of the International Wood Biotechnology Symposium, which contains 45 articles on the subjects mentioned. I trust that these proceedings may make a great contribution to develop wood biotechnology and finally to rescue human beings from global environmental, and energy and resources problems of the upcoming 21st century.

Noriyuki Morohoshi
Representative of the Organizing Committee of the
International Wood Biotechnology Symposium

Contents

Molecular Breeding of Woody Plants
N. Morohoshi and A. Komamine, editors.

REGULATION OF CELLULOSE BIOSYNTHESIS IN DEVELOPING XYLEM

Candace H. Haigler*, V. Michelle Babb, Sangjoon Hwang, and Vadim V. Salnikov

Department of Biological Sciences, Box 43131, Texas Tech University, Lubbock, Texas 79409-3131 USA

ABSTRACT

The advantages of using isolated mesophyll cells of *Zinnia elegans* as a model to study the regulation of cellulose synthesis will be discussed. These cells can be induced by different mechanisms to expand greatly *via* primary wall synthesis or to differentiate into tracheary elements with patterned secondary walls. Therefore, mechanisms of cellulose synthesis during primary and secondary wall deposition can be studied separately in cultured cells. Recent work discussed includes the activity and role of sucrose synthase and sucrose phosphate synthase during secondary wall cellulose synthesis and the diversity of cellulose synthase genes expressed during tracheary element differentiation. Evidence obtained during primary and secondary wall synthesis in cultured *Zinnia* cells is compared and contrasted. Data presented include immunolocalization of sucrose synthase and actin in cryogenically fixed cells, biochemical analysis of sucrose phosphate synthase activity during the time-course of tracheary element differentiation, and cloning and analysis of multiple cellulose synthase genes expressed differentially during secondary wall deposition in tracheary elements in culture. Differentiating cotton fibers and etiolated bean hypocotyls will be discussed as related systems. A composite cellular and metabolic model for cellulose synthesis will be presented.

KEYWORDS

Cellulose synthesis, cellulose synthase, primary wall, secondary wall, sucrose phosphate synthase, sucrose synthase, tracheary element, *Zinnia elegans*

INTRODUCTION

Isolated mesophyll cells of *Zinnia elegans* induced to form tracheary elements (TEs) semi-synchronously in culture are a valuable experimental system for analysis of xylem cell differentiation in general and cellulose synthesis in particular [1]. The TE secondary wall contains a large percentage of cellulose that is synthesized rapidly within a few hours before TE autolysis occurs to leave the water-conducting element. When *Zinnia* mesophyll cells are cultured in complex medium [2] in our laboratory, some cells divide, differentiation of small TEs begins at 56 – 60 h, the remaining undifferentiated cells continue to divide and expand, and large TEs differentiate in successive waves during the following days. In a simplified inductive medium [3], cell division is suppressed and later waves of TE differentiation are prevented, providing a system that is clearly interpretable in terms of one peak in TE differentiation. TEs differentiating in culture offer several other experimental advantages: (a) TE differentiation is largely separated from other developmental events; (b) many cells are at a similar developmental stage;

(c) cells are uniformly accessible to drugs in liquid medium; (d) secondary wall deposition occurs in a distinct pattern, facilitating analyses of associated cell biological phenomena; and (e) the single cells are amenable to cryogenic fixation to obtain accurate electron microscopic results. In addition, the isolated mesophyll cells can be stimulated to synthesize only primary walls, providing a valuable comparison to TEs differentiating *via* secondary wall deposition. For example, if the level of cytokinin in the medium is reduced, the mesophyll cells will divide and expand, forming a typical suspension culture but no TEs [2]. Alternatively, extensive expansion *via* primary wall synthesis can be induced before TE differentiation by stabilizing the pH of the medium to 5.5 – 6.0 [4].

In this paper, we will discuss three types of data focused on three aspects of cellulose synthesis in developing xylem cells: (a) cell biological analysis of a role for sucrose synthase (SuSy); (b) biochemical analysis of a role for sucrose phosphate synthase (SPS); and (c) molecular analysis of a role for expression of multiple cellulose synthase (CesA) genes. Analogies will be drawn with cotton fiber development and xylem differentiation in etiolated bean hypocotyls.

MATERIALS AND METHODS

Immunoelectron microscopy

A cryogenic method of sample preparation that should greatly hinder molecular movement was adapted from published methods [5]. Briefly, differentiating TEs or mesophyll cells induced to expand *via* primary wall synthesis were concentrated in simplified medium without centrifugation by slowing down the shaker for about 3 min. The stress of centrifugation was avoided because of the known lability of cellulose synthesis. The cells in about 0.5 ml medium were sucked out with a wide-bore pipet, frozen within 2 min by spraying through an artist's airbrush into re-solidifying liquid propane cooled by liquid nitrogen, freeze-substituted in acetone (acting as a mild aldehyde) 3 d at –80°C, infiltrated with Lowicryl resin 4 d at –80°C, and flat-embedded between two slides. The resin was polymerized by UV light for 4 d at –20°C (UV light turned on in a –40°C freezer) and 1 d at 4°C. Single differentiating TEs were selected in the light microscope, cut out in a square of thin resin, glued onto blank resin blocks, sectioned, and processed for immunoelectron microscopy by standard methods [6].

TE differentiation or cell expansion in culture

Mesophyll cells isolated from the first true leaves of *Zinnia elegans* were induced to differentiate into TEs in the dark as previously described on medium with sucrose as the carbon source [2]. The extent and timing of TE differentiation were manipulated by use of three kinds of medium. Two of the media, one complex [2] and one simplified [3], contained sufficient cytokinin to induce TE differentiation. The complex inductive medium supported cell division and successive waves of TE differentiation. The simplified inductive medium suppressed cell division and eliminated later waves of TE differentiation. The third medium did not induce TE differentiation; it differed from the complex inductive medium only by having a lower level of cytokinin [2]. This non-inductive medium also allowed cells to be maintained in culture until they lost visible starch grains about 7 d after culture (as detected by staining with I_2KI). Late TE differentiation was induced in starch-depleted cells by addition of cytokinin to equal the concentration in inductive medium.

Determining percent differentiation and percent live TEs in *Zinnia* cultures

Polarization (for early stage TEs) or bright-field (for late stage TEs) microscopy were used to count TEs among all cells in the culture over the time-course of differentiation. Sensitive polarization microscopy (Olympus BH-2 microscope) allowed the detection of cellulosic thickenings by their birefringence before they became visible in bright-field microscopy. Although differentiating TEs could have been detected about 2 h earlier by binding of the fluorescent brightener Tinopal LPW to their patterned cellulosic thickenings, it was not used because polarization microscopy was simpler and adequate to perceive the trends observed in these experiments. Percent TEs was calculated as [total TEs/(total TEs + other cells) x 100]. Evans Blue, a dye that permeates only dead cells, was used to quantify TE autolysis as previously described [7] over the time course of differentiation. Percent live TEs among all TEs was calculated as: [(total TEs – autolysed TEs)/total TEs x 100].

Growth and analysis of etiolated hypocotyls

Kidney beans (*Phaseolus vulgaris)* were purchased form the grocery store and germinated in the dark at 28-30°C in commercial potting soil. They grew into etiolated seedlings characterized by hyper-elongation and lack of chlorophyll and leaf development. However, the etiolated hypocotyl still contained differentiating TEs to support water conduction. Short (2-3 cm), medium (4-6 cm), and tall (7-8 cm) hypocotyls were analyzed for SPS activity. Hand sections were cut with a razor blade from the bottom, middle, and top of each length of hypocotyl, stained with safranin (1% w/v aq.), and examined in the light microscope to determine the relative amounts of xylem in each. Thirty-six hypocotyls of each size were stripped of roots and cotyledons, dehydrated in a 60°C oven for 3 d, and weighed.

SPS assay

Zinnia cells in medium were washed 3 x by low speed centrifugation in 0.2 M mannitol to remove exogenous sucrose, frozen in a concentrated suspension by drops in liquid nitrogen, and ground while frozen to a fine powder. Bean hypocotyl pieces were frozen and ground in liquid nitrogen. Ground tissue was thawed in 4°C extraction buffer [50 mM HEPES (pH. 7.4); 10 mM $MgCl_2$; 1 mM EDTA; 1 mM EGTA; 10% glycerol, 2% (w/v) polyvinylpolypyrrolidone, and 0.1% Triton X-100] and extracted by vortexing 5 sec, incubating on ice 5 min, and vortexing 20 sec. Cellular debris was pelleted (2 x 20 sec spin,14,000 rpm), and the supernatant was used to assay SPS. SPS assay proceeded in 70 μl reaction mixtures for 10 min at 34°C in: [50 mM HEPES (pH 7.4); 10 mM UDPG; 6 mM fructose 6-P; 20 mM glucose 6-P; 10 mM $MgCl_2$; 1 mM EDTA; 0.4 mM EGTA; 4% glycerol; 0.04% Triton X-100]. High substrate concentrations and the presence of the activator glucose-6-P define conditions for assay of Vmax SPS activity [8]. Three reaction tubes and 3 blanks (to normalize for possible different amounts of endogenous sucrose) were run for each sample. 1 N NaOH was added to the blanks before the plant extract. After 10 min, 1 N NaOH was added to stop the reaction, followed by boiling 10 min to destroy unreacted hexoses. 12 M HCL was added to hydrolyze sucrose into fructose and glucose, 0.1% (w/v in EtOH) resorcinol was added to react with fructose, and absorbance (A_{520}) of the pink reaction product was measured.

A sucrose standard curve was run in parallel, and protein concentration in the extracts was determined (BioRad protein assay kit).

Molecular methods

Using primers from conserved regions of several plant cellulose synthase genes near the U2 and U4 regions [9] , PCR was used to amplify gene fragments from total RNA isolated (Gibco BRL TRIzol system) from *Zinnia* cells differentiating into TEs at 60 h. Subsequent 3'-RACE PCR, cloning, and sequencing indicated that three distinct genes with different HVR2 (plant-specific, hypervariable) regions [9] and 3' untranslated regions were expressed during TE differentiation in culture. Northern analysis was performed according to standard methods. An unrooted cladogram was constructed according to methods described previously [10].

RESULTS & DISCUSSION

A role for SuSy in cellulose synthesis in tracheary elements

From research on cotton fibers, a particulate form of SuSy was proposed to cleave sucrose and channel UDP-glucose to the cellulose synthase in the plasma membrane [11]. SuSy was also demonstrated to have actin binding properties and to co-precipitate with actin in some cases [8, 12]. Patterns of cellulose synthesis are disturbed by actin antagonists in cotton fibers and TEs [13, 14]. Therefore, we hypothesized that SuSy also had a role in secondary wall cellulose synthesis in TEs and that actin might interact with it. We predicted that SuSy would be specifically enriched below the patterned sites of cellulose synthesis in differentiating *Zinnia* TEs [15]. The research summarized below will soon be published in complete form [16].

Antibodies against cotton SuSy [11] and chicken gizzard actin [17] were shown to recognize single bands in Western blots of *Zinnia* protein extracts. Immunofluorescence of SuSy yielded variable results—SuSy was only rarely observed over thickenings, and it sometimes appeared as dots over the whole cell surface or formed negative images of thickenings. Evidently the fixation or wall permeabilization steps in the immunofluorescence protocol led to artifactual rearrangement of SuSy, possibly due to effects such as those described on actin after processing for immunofluorescence [18]. Cryogenic electron microscopy methods followed by immunolocalization yielded consistent results. In these samples, quantitation of gold labeling relative to sites of secondary wall thickening showed that SuSy was preferentially localized near the plasma membrane under the thickenings of differentiating TEs. Rosettes, which have now been identified as organized cellulose synthases [19], are preferentially localized in the plasma membrane at the thickening sites [15]. No labeling of SuSy was observed in the cytoplasm or at the plasma membrane of cells induced to expand *via* primary wall synthesis (see Fig. 1; other data not shown). However, lower amounts of SuSy could be present at these locations, but not detected by immunolabeling. In differentiating TEs, actin was distributed over the whole cell cortex. Grazing sections of thickening sites showed that SuSy, actin, and microtubules lay close together in the TE cortex. However, actin lay above the microtubules but below SuSy, which was closer to the plasma membrane than actin or microtubules (data not shown). These results, when combined with other existing knowledge, are consistent with the regulation of secondary wall synthesis by a multi-protein complex including cellulose synthase, sucrose synthase, actin, and microtubules.

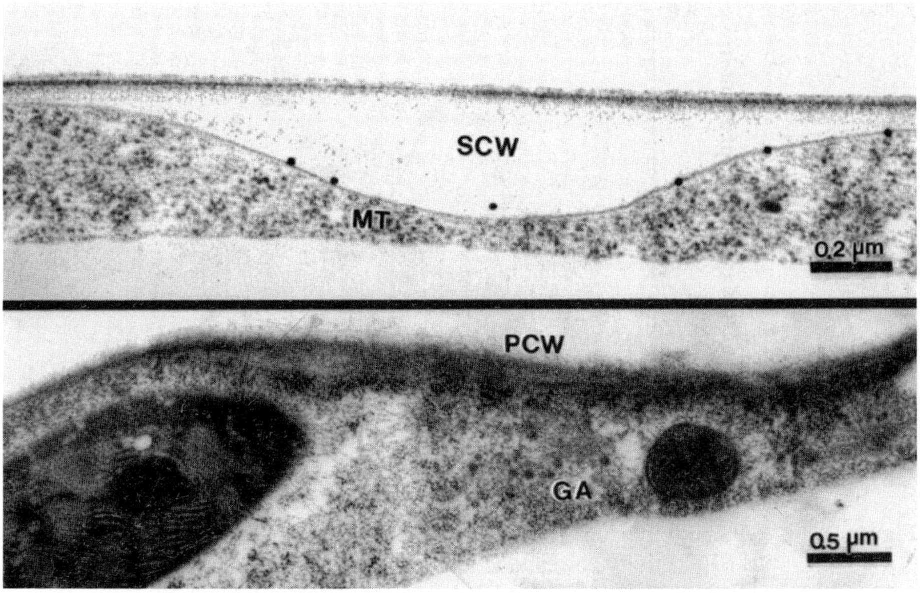

Figure 1. Immunolocalization of SuSy in *Zinnia* cells processed for electron microscopy by cryogenic methods. (Top) A differentiating TE with SuSy tagged with colloidal gold near the plasma membrane underneath a secondary wall thickening (SCW). The underlying microtubules (MT) are also visible. (Bottom) A mesophyll cell induced to expand *via* primary wall (PCW) synthesis in medium with stabilized pH [4]. The Golgi apparatus (GA) usually showed no SuSy labeling; rare label over the GA was interpreted as random background labeling.

A role for SPS in cellulose synthesis in tracheary elements

SuSy-mediated secondary wall cellulose synthesis implies that the availability of sucrose within the cell is critical. In leaves and sucrose-storing organs, SPS regulates the synthesis of sucrose-P from fructose-6-P and UDP-glucose [8]. (A phosphatase, which is generally not regulatory, removes the phosphate to form sucrose.) SPS in cellulose-sink cells could be important to recycle fructose released by SuSy (after phosphorylation to fructose-P) back to sucrose to support additional cellulose synthesis. SPS could assume an even more important role if translocated sucrose in plants or exogenous sucrose in cultured cells was cleaved by invertases before or immediately after entering the cellulose sink cells so that sucrose for secondary wall cellulose synthesis had to be resynthesized. These ideas have been more extensively discussed in a recent article [12]. We demonstrated that SPS activity increased about 5-fold as cotton fibers made the transition from primary to secondary wall synthesis [20]. We also obtained evidence in transgenic cotton plants that up-regulated SPS activity could increase the extent of fiber cellulose deposition when plants were growing under a 30/15°C day/night cycle [21].

We hypothesized that SPS activity would also rise during deposition of secondary cell walls in TEs, and we tested the hypothesis in the *Zinnia* cell culture system. In complex inductive medium, SPS activity rose from a low level at 24 h culture as the

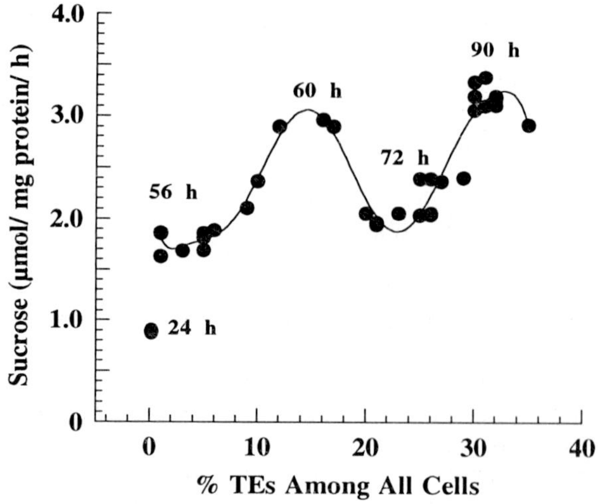

Figure 2. SPS activity vs. Percent TEs in Complex Medium. The two peaks of SPS activity correspond to two peaks in the percentage of living TEs in the culture. Time labels indicate hours after culture.

time of TE differentiation approached (Fig. 2). It peaked two times corresponding to two successive waves of TE differentiation (see Fig. 2; data on autolysis of many first-differentiated TEs by 72 h not shown). Other related data (Babb et al., in preparation) are summarized here. When only one wave of TE differentiation occurred in simplified inductive medium, SPS activity increased only one time. Detection of autolysed TEs by permeation with Evan's Blue showed that the single peak in SPS activity in simplified medium was paralleled by the maximum number of living TEs in the culture. SPS activity remained undetectable over several days in *Zinnia* mesophyll cells that were synthesizing only primary walls in non-inductive medium. However, when TE differentiation was induced late in these growing cells by addition of extra cytokinin, SPS activity rose in correlation with increasing numbers of TEs in the culture. The rise in SPS activity after late-induction occurred even when the cells were first allowed to deplete their internal starch stores completely over 7 days. This observation supports a role for SPS in TE differentiation beyond recycling carbon from starch into sucrose.

We obtained support for a role for SPS in differentiating xylem by analysis of etiolated hypocotyls of *Phaseolus vulgaris*. Their SPS activity peaked at medium height (4 – 6 cm) compared to short (2-3 cm) or tall (7-8 cm) height. Medium height was also the stage of maximum dry weight of the hypocotyls and maximum size of the vascular bundles (data not shown).

Expression of multiple *CesA* genes during xylem differentiation in culture

A PCR strategy using total RNA isolated from *Zinnia* cells actively differentiating into TEs revealed that at least three distinct *CesA* genes were expressed during TE differentiation in culture. The deduced amino acid sequences showed strong homology to other CesA proteins encoded by genes expressed in other secondary-wall-synthesizing cells. An unrooted cladogram made by D. Delmer according to published methods [10] confirmed that these *Zinnia* proteins were in a clade with other CesA proteins that have been associated with secondary wall stage cells of *Arabidopsis*, cotton, and poplar: *AtCesA04, AtCesA08, GhCesA02, GhCesA01, PtCesA01, and PtCesA02* (cladogram published elsewhere [12]). Partial sequences of these genes, *ZeCesA-01, -02, and –03* have been entered into the database (accession numbers AF323039, AF323040, AF323041). Northern analysis with gene-specific fragments including the HVR2 region with or without the 3' untranslated regions showed that these three genes were expressed at the onset of TE differentiation, but not prior to that point in TE-inductive medium and not in cultures induced to expand *via* primary wall synthesis (data not shown).

There are at least three reasons that multiple *CesA* genes might be expressed during TE differentiation in culture: (a) multiple CesA proteins might facilitate rapid synthesis of abundant cellulose; (b) two or more CesA proteins might be required to cooperate to synthesize cellulose; or (c) gene expression in the tissue culture system could be deregulated and not reflect any whole-plant phenomenon. Preliminary tissue prints indicate that *ZeCesA02* is expressed in the vascular bundles of leaves and stems in *Zinnia* plants (data not shown). Further work will determine whether there is differential or redundant expression of these three genes in *Zinnia* plants.

CONCLUSIONS

We have provided evidence that both SuSy and SPS have important roles in secondary wall synthesis in TEs. SuSy is close to other proteins in the cell cortex, including actin and microtubules, and to the cellulose synthases in the plasma membrane below secondary wall thickenings. Therefore, it appears that a multi-protein complex mediates secondary wall cellulose synthesis. The rising activity of SPS during secondary wall synthesis in three heterotrophic systems (cotton fibers, *Zinnia* TEs in culture, and etiolated bean hypocotyls) suggests that SPS is also important in facilitating high-rate cellulose synthesis. Our data cannot exclude that SuSy and SPS have the same roles during primary wall synthesis, but we have at least shown that SuSy is much more abundant and SPS is much more active during rapid cellulose synthesis for secondary wall deposition. We propose that these two enzymes work together to provide substrate to the cellulose synthase, and a cellular and metabolic model including both of them has recently been formulated (Haigler et al., in press). Further work will be required to determine whether multiple CesA genes expressed in differentiating *Zinnia* cultures have redundant or specialized functions and whether this class of CesA protein has particular domains to facilitate interaction with SuSy or its partners in the multi-protein complex.

ACKNOWLEDGEMENTS

We thank Mark Grimson for development of electron microscopy methods and Debby Delmer for construction of the cladogram including *Zinnia* CesA deduced protein sequences. This research was supported by NSF Plant Genomics grant DBI-9872627, a Howard Hughes Medical Institute grant through the Undergraduate Biological Sciences Education Program at Texas Tech University, and the Graduate School and Department of Biological Sciences, Texas Tech University.

REFERENCES

1. H. Fukuda, Xylogenesis: initiation, progression, and cell death, *Annu. Rev. Plant Physiol. Plant Mol. Biol.*, **1996**, 47, 299-325.
2. H. Fukuda & A. Komamine, Establishment of an experimental system for the study of tracheary element differentiation from single cells isolated from the mesophyll of *Zinnia elegans*, *Plant Physiol.,* **1980**, 65, 57-60.
3. A. W. Roberts, L. T. Koonce, & C. H. Haigler, A simplified medium for *in vitro* tracheary element differentiation in mesophyll cells from *Zinnia elegans*, *Plant Cell, Tissue, and Organ Culture*, **1992**, 28, 27-35.
4. A. W. Roberts & C. H. Haigler, Cell expansion and tracheary element differentiation are regulated by extracellular pH in mesophyll cultures of *Zinnia elegans* L., *Plant Physiol.,* **1994**, 105, 699-706.
5. T. N. Nicolas & J. M. Bassot, Freeze substitution after fast-freeze fixation in preparation for immunocytochemistry, *Microsc. Res. and Techn.*, **1993**, 24, 474-487.
6. G. Newman & J. Hobot, *Resin microscopy and on-section immunocytochemistry.* Berlin, Springer Verlag, **1993**.
7. A. W. Roberts & C. H. Haigler, Rise in chlorotetracycline fluorescence accompanies tracheary element differentiation in suspension cultures of *Zinnia*, *Protoplasma*, **1989**, 152, 37-45.
8. H. Winter & S. C. Huber, Regulation of sucrose metabolism in higher plants: Localization and regulation of activity of key enzymes, *Crit. Rev. Plant Sci.,* **2000**, 19, 31-67.
9. D. P. Delmer, Cellulose biosynthesis: Exciting times for a difficult field of study. *Ann. Rev. Plant Physiol. Mol. Biol.,* **1999**, 50: 245-276.
10. N. Holland, D. Holland, T. Helentjaris, K. S. Dhugga, B. Xoconostle-Cazares, & D. P. Delmer, A comparative analysis of the plant cellulose synthase (CesA) gene family, *Plant Physiol.*, **2000**, 123, 1313-1324.
11. Y. Amor, C. H. Haigler, S. Johnson, M. Wainscott, & D. P. Delmer, A membrane-associated form of sucrose synthase and its potential role in synthesis of cellulose and callose in plants, *Proc. Natl. Acad. Sci. USA,* **1995**, 92, 9353-9357.
12. C. H. Haigler, M. Ivanova-Datcheva, P. S. Hogan, V. V. Salnikov, S. Hwang, L. K. Martin, & D. P. Delmer, Carbon partitioning to cellulose synthesis, *Plant Mol. Biol.,* **in press**.
13. H. Kobayashi, H. Fukuda & H. Shiboka, Interrelation between the spatial disposition of actin filaments and microtubules during the differentiation of tracheary elements in cultured *Zinnia* cells, *Protoplasma,* **1988**, 143, 29-37.
14. R. W. Seagull, The effects of microtubule and microfilament disrupting agents on cytoskeletal arrays and wall deposition in developing cotton fibers, *Protoplasma,* **1990**, 159, 44-59.
15. C. H. Haigler & R. M. Brown, Jr., Transport of rosettes from the Golgi apparatus to the plasma membrane in isolated mesophyll cells of *Zinnia elegans* during differentiation to tracheary elements in suspension culture, *Protoplasma*, **1986**, 134, 111-120.
16. V. V. Salnikov, M. J. Grimson, D. P. Delmer, & C. H. Haigler, Sucrose synthase localizes to cellulose synthesis sites in tracheary elements, *Phytochem.,* **in press**.
17. J. L. Lessard, Two monoclonal antibodies to actin: one muscle selective and one generally reactive, *Cell Motility and the Cytoskeleton,* **1988**, 10, 349-362.

18. A. O. Frost & A. W. Roberts, Cortical actin filaments fragment and aggregate to form chloroplast-associated and free F-actin rings in mechanically isolated *Zinnia* mesophyll cells, *Protoplasma*, **1996**, 194, 195-207.
19. S. Kimura, W. Laosinchai, T. Itoh, X. Cui, R. Linder. & R. M. Brown Jr., Immunogold labeling of rosette terminal cellulose-synthesizing complexes in the vascular plant *Vigna angularis*. *The Plant Cell*, **1999**, 11, 2075-2085.
20. J. Tummala. Response of sucrose phosphate synthase activity to cool temperatures in cotton. M.S. thesis, **1996**, Texas Tech University, Lubbock, TX, U.S.A.
21. C. H. Haigler, A. S. Holaday, C. Wu, B. G. Wyatt, G. J. Jividen, J. G. Gannaway, W. X. Cai, E. F. Hequet, T. J. Jaradat., D. R. Krieg, L. K. Martin, R. E. Strauss, S. Nagarur, & J. Tummala. Transgenic cotton over-expressing sucrose phosphate synthase produces higher quality fibers with increased cellulose content and has enhanced seedcotton yield. Abstract 477. In: *Proc. Plant Biol.* **2000**, July 15 – 19, San Diego, CA. American Society of Plant Physiologists, Rockville, MD., [http://www.aspp.org/annual_meeting/pb-2000/2000.htm].

Molecular Breeding of Woody Plants
N. Morohoshi and A. Komamine, editors.

XYLEM FORMATION AND LIGNIFICATION IN TREES AND MODEL SPECIES

Edouard Pesquet[1], Magalie Pichon[1], Christophe Pineau[1], Philippe Ranocha[1], Catherine Digonnet[1], Alain Jauneau[1], Alain M. Boudet[1], Hiroo Fukuda[2], Taku Demura[2], & Deborah Goffner[1]*

[1] UMR –CNRS-UPS 5546 "Signaux et Messages Cellulaires chez le Végétaux"
Pôle de Biotechnologie Végétale, Chemin de Borde Rouge
31326 Castanet Tolosan, France

[2] Department of Biological Sciences, Graduate School of Sciences,
University of Tokyo, 7-3-1 Hongo, Japan

ABSTRACT

Laccases (EC 1.10.3.2) are blue copper oxidases that are found in a large variety of living organisms including bacteria, fungi, insects and plants [1-3]. To date their role in these organisms has not yet been clearly established. In higher plants, based on their capacity to oxidize monolignols *in vitro* and localization at the cell wall, laccases are considered candidate enzymes in the ultimate step in lignification. In order to provide functional evidence to support or refute this hypothesis, four lines of antisense poplars, each corresponding to a different gene, were obtained. Although none of the lines exhibited significant differences in either lignin content or composition, one line, *lac3*AS, is characterized by a two to three-fold increase in soluble phenolics and perturbations in cell adhesion of xylem fibers. The fact that several laccases from *Zinnia* (8 out of the 9 obtained) are heavily induced at the onset of lignification during the formation of tracheary elements (TEs) further suggest then involvement of laccases in secondary cell wall formation.

In order to make a quantitative leap in our understanding of lignification and vascular development, we are currently developing two strategies that will lead to the identification of new genes involved in these plant–specific processes. Firstly, we have constructed a "late xylogenesis" cDNA library by suppression subtractive hybridization (SSH) from differentiating TEs of *Zinnia*. Approximately 75% of the 800 clones obtained appear to be differentially expressed during TE formation. A limited number of differentially expressed clones were randomly chosen and sequenced. Among them, known molecular markers of late xylogenesis including a cysteine protease and an endonuclease were identified, demonstrating the quality of the library. Massive sequencing and the determination of detailed expression profiles of these cDNAs are now underway. Secondly, we have screened T-DNA tagged *Arabidopsis* mutants (Versailles collection) for atypical vascular patterns in floral stems. One of these mutants, *hca*, for high cambial activity, is characterized by the formation of a continuous ring of vascular tissue as opposed to the discrete vascular bundles typically observed in *Arabidopsis*. The identification of the gene responsible for this phenotype is now underway.

KEYWORDS

Lignin, laccase, antisense poplar, *Zinnia* subtractive library, *Arabidopsis* mutants

INTRODUCTION

The results of genetic engineering experiments using lignin biosynthetic genes have provided new insight into this complex metabolic pathway. The scientific community now acknowledges that this fundamental process is certainly more complex than previously imagined (for a recent review see [4]). If our understanding of the events leading to monolignol formation has been recently clarified, the terminal polymerization steps and the assembly of lignin subunits at the cell wall is still one of major outstanding enigmas in lignin biochemistry. For example, the nature of the enzyme(s) involved in the oxidative polymerization of lignin subunits is still a matter of controversy. "Peroxidases and /or laccases?; that is the question". One of the major difficulties encountered in addressing this issue arises from the fact that cell wall proteins are more often than not encoded by medium–to-large multigene families; peroxidases and laccases are no exceptions to this rule with approximately 70 different peroxidases and 15 laccases in the small genome of *Arabidopsis*. Therefore genetic engineering experiments may be difficult to interpret, especially when a phenotype is not detected (due to functional redundancy of other family members unaffected by transgene expression).

We originally characterized five different laccase genes in poplar (*lac1*, *lac2*, *lac3*, *lac90* and *lac110*) that are preferentially expressed in lignifying stem tissue [5]. Four independent lines of antisense poplars (each corresponding to a different gene) were generated; only one line, *lac3*AS, gave rise to a readily observable phenotype, the other three did not. Since elucidating laccase function is difficult *in planta*, we sought to obtain complementary information pertaining to laccase gene expression in the *Zinnia* TEs system, where secondary wall formation and lignification are tightly regulated [6]. If the expression of a given laccase gene or subset of genes is correlated with wall formation, it would then be considered a viable candidate for more labor-intensive characterization *in planta*. In this respect, the *Zinnia* system has already proven itself extremely valuable in establishing the role of a novel methyl transferase, caffeoyl-CoA *O*-methyltransferase (CCoAOMT), that had been exclusively associated with pathogenesis, in lignification [7].

In order to contribute to ongoing discoveries of novel molecular mechanisms of vascular development by isolating new genes involved in xylogenesis and wall formation, we have adopted strategies that rely on the one hand, on the exploitation of the *Zinnia* TE system, and on the other, on the search for novel vasculature *Arabidopsis* mutants. Since differentiation is tightly controlled and its stages semi-synchronous, the *Zinnia* system is particularly well-suited for obtaining differentially expressed genes. A handful of genes were obtained by Demura and Fukuda (1993) [8] and Ye and Warner (1993) [9]. More recently, Roberts and McCann (2000) have obtained hundreds of novel genes by cDNA AFLP [10]. In H. Fukuda's laboratory, a large-scale sequencing project of EST from differentiating TEs of *Zinnia* is underway. As a complementary approach, we have applied the recently-developed technique of suppression subtractive hybridization (SSH) to the *Zinnia* system. This powerful technique, based on the principles of subtractive hybridization and suppression PCR, has already been used with success, principally in animal systems, to select for differentially expressed, low abundance transcripts with a relatively low number of false positives [11].

In parallel, we have undertaken a genetic approach to obtain novel vascular development *Arabidopsis* mutants. Different types of vascular mutants have already been described. Some were the result of direct screening for aberrant vascular patterns in cotyledons and leaves [12-14] and stems [15] whereas others exhibit aberrant vasculature as a consequence of abnormal auxin perception and/or transport [16-19]. A range of mutants has also been obtained for abnormal xylem formation [20-22].

MATERIAL & METHODS

Poplar transformation and screening

Laccases are encoded by multigene families in all higher plants studied to date. In poplar, we have characterized five distinct laccases in poplar [5]. With the aim of elucidating the role of laccases, we have transformed poplar with four different genes (*lac1*, *lac3*, *lac90* and *lac110*) in the antisense orientation, under the control of a strong constitutive promoter, 35S CaMV. The four populations of antisense plants were screened by Northern blot analysis using specific 3'UTR as probe. Transformants with low residual levels of laccase transcript were selected for further analysis.

Extraction and analysis of total soluble phenolic compounds

Fresh stem tissue (0.5g) was ground to a fine powder. The powder was then homogenized three times in 50ml at 4°C in 80% ethanol. The crude extract was filtered and evaporated at 35°C under reduced pressure. The aqueous fraction was extracted twice with petroleum ether (40-60°) to remove lipids, freeze dried and stored at -20°C until further use. Total phenolic compounds were determined by the Folin-Ciocalteu method as described by Scalbert et al. (1989) [23].

Microscopic techniques

Hand sections from fresh poplar stems (fourth internode) were made with a razor blade and observed using an inverted microscope (Leitz DMIRBE, Leica) equipped with epifluorescence illumination (Excitation filter BP 340-380 nm, suppression filter LP 430 nm). Images were registered using a CCD camera (Colour Coolview, Photonic Science, UK) and treated by image analysis (Image Pro-Plus, Media Cybernetics, MD, USA).

Zinnia cultures

Zinnia elegans cv 'ENVY' TE cultures were performed according to Roberts *et al.* (1992) [24].

Construction of a subtractive library by suppression subtractive hybridization

To construct a late xylogenesis library, total RNA was extracted from TEs at the pre-cellulosic (72h), pre-lignification (96h) and pre-autolytic (120h) stages. One microgram of total RNA from each stage was pooled and used in subsequent experiments. For controls, RNA was extracted from the same time points indicated above without hormone or with auxin or cytokinin only. cDNA synthesis and library construction was performed using the SMART system for cDNA synthesis and PCR select kits respectively according to the manufacturers recommendations (Clontech). The resulting cDNAs were amplified by PCR using flanking oligonucleotides, blotted onto Nylon membranes, and hybridized with radiolabeled cDNAs from induced or control cultures.

Isolation of the *Arabidopsis* high cambial activity (*hca*) mutant

hca was isolated by screening approximately 5000 T2 lines from the T-DNA-tagged *Arabidopsis thaliana* ecotype Wassilevskija collection from INRA Versailles [25]. Inflorescence stems of 6-week-old greenhouse-grown plants were harvested and stored in fixative solution (60% ethanol, 5% acetic acid glacial, 10% formaldehyde). Free-hand sections were made from the stem base and examined microscopically under fluorescent illumination or stained for lignin with phloroglucinol-HCl.

RESULTS & DISCUSSION

Functional analysis of laccases in poplar and *Zinnia*

lac3 down-regulation in poplar results in an increase in soluble phenolic content and abnormal wall structure in xylem fibers

For all four independent lines of antisense laccase transformants, no differences were observed in overall growth and development between antisense and control poplars (*i.e.* height, stem diameter, phyllotaxy). In addition, none exhibited significant differences in lignin content or monomeric composition. These results are in agreement with previous preliminary data showing that individual laccase down-regulation had no effect on lignin profiles in *Liriodendron tulipifera* [26].

Interestingly, antisense suppression of *lac3* led to an increase in total soluble phenolic content. Ethanol-soluble phenolic compounds were quantified based on their reactivity *vis-à-vis* Folin's reagent (see Figure 1A). Two independent transformants *lac3.2*AS and *lac3.4*AS of *lac3* lines, exhibited a two to three-fold increase in soluble phenolic content. *lac90*AS and *lac110*AS poplars did not exhibit significantly different soluble phenolic content as compared to controls.

Microscopic observations of *lac3.2*AS, *lac3.4*AS revealed that the overall pattern of xylem tissue as seen in transverse sections was not dramatically disorganized as compared to control sections (see Figure 1B). All of the different cell types were easily recognizable: xylem vessels (v), fibers (f) ray parenchyma (rp) and phloem fiber (pf) cells. However, it was readily observed that the walls of *lac3.2*AS and *lac3.4*AS xylem fibers possessed a highly irregular cell contour as compared to controls (Figure 1B). Moreover, in antisense plants, the fluorescence emission was not homogeneous throughout the entire width of the wall; fluorescence was indeed negligible in the middle lamella/primary wall region between adjacent fibers. As a consequence, the cells appeared to be detached from one another. No differences in coloration were observed in transgenic stem sections stained with lignin-specific phloroglucinol and Maüle reagents as compared with controls. These results are in good agreement with the data indicating that there were no significant differences in lignin content and composition resulting from laccase suppression. We are currently analyzing these transformants for ultrastructural wall modifications, mechanical properties, and susceptibility to wall-degrading enzymes. These plants, in addition to *Arabidopsis* insertional mutants now available for laccase genes, provide excellent tools towards gaining a better understanding of laccase function in plants.

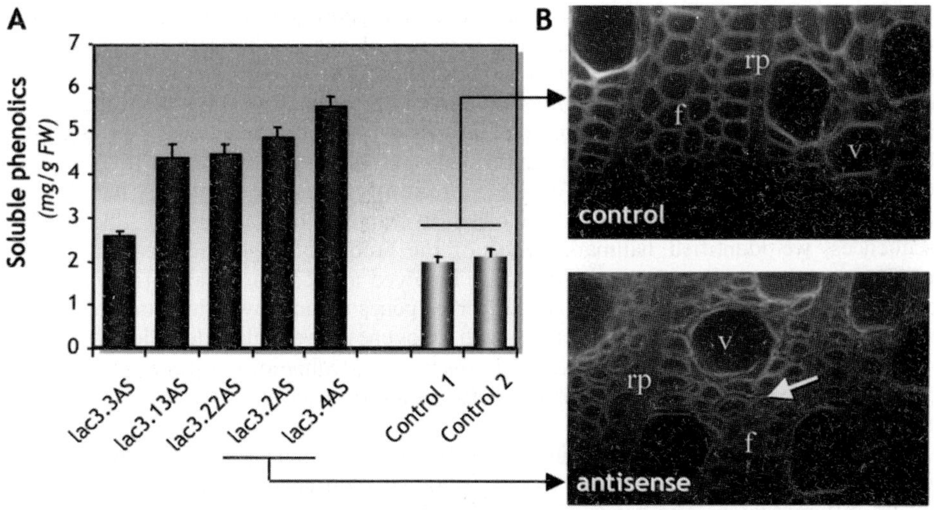

Figure 1. **A**. Soluble phenolic content of *lac3* antisense and control poplars. Five independent antisense transformants (*lac3.3*AS, *lac3.13*AS, *lac3.22*AS, and *lac3.2*AS, and *lac3.4*AS) and two different controls (C1 : untransformed poplar, C2: poplar transformed with an empty vector) were analyzed.

B. Cross sections of antisense *lac3.4*AS and control poplar stems visualized by UV fluorescence microscopy. v: xylem vessel, rp: ray parenchyma, f: xylem fiber.

Several laccase genes are induced at the onset of lignification in TEs of *Zinnia*

In order to complement our knowledge of laccase involvement in secondary wall synthesis and to determine which, if any, laccases are correlated with tracheary element formation, we obtained 9 partial laccase cDNAs of *Zinnia*. Based on sequence analyses of these and other known plant laccases, it is clear that they form three distinct phylogenetic groups. Both Northern blot analysis and RT-PCR indicated that laccases are differentially expressed in *Zinnia*. Eight out of nine laccases were tightly correlated with the onset of lignification, whereas one gene was abundantly expressed in mesophyll cells at the initiation of cell cultures and rapidly decreased thereafter during TE formation. These results suggests that 8 out of 9 *Zinnia* laccases are involved in late events of TE formation, most likely related to secondary wall formation. We are currently performing *in situ* hybridization experiments in order to determine the precise cellular location of laccase transcripts in *Zinnia*.

Gene discovery and vascular development in higher plants

Lots of new xylogenesis genes by SSH using the *Zinnia* system.

A 'late xylogenesis' SSH library was constructed from *Zinnia* TEs from the pre-cellulosic, pre-lignification and pre-autolytic stages. This library will likely be excellent source of cell-wall related genes. It contains approximately 800 clones ranging in size from 200-1300bp. Before sequencing, we screened clones to eliminate false positives and classified them according to their expression in TEs in comparison with control cells: those that were expressed exclusively during TE formation (37%), those that were strongly induced in TEs (42%) , those that did not hybridized with either TE or controls (12% - presumably low abundance transcripts), and those that hybridized to the same extent with TE and controls cDNAs (9% - presumably false positives). A handful of these clones were then sequenced in order to validate our approach. Among the sequences, we identified hallmark genes of the autolytic stage such as a cysteine protease[27] and an endonuclease [28]. Others are involved in secondary wall formation (cell wall proteins including a laccase). A number of genes of unknown function and those that, until now, have not been associated with xylogenesis were also identified. Massive cDNA sequencing is now underway. Using macroarray technology, we are establishing precise gene expression profiles during TE formation to determine which genes are co-regulated (gene clusters).

We are currently constructing an early-stage xylogenesis library that will be informative in identifying primary events in signal transduction networks leading to xylem differentiation. The genes obtained here by SSH is a complementary approach that will provide an additional source of new xylem-specific markers to these ongoing projects. As is the case for all global genomic approaches, one of the real challenges for the future will be the judicious selection of the most interesting genes/proteins to pursue by a gene-for-gene approach. High throughput techniques to study the function of large numbers of genes (*in situ* RT-PCR, modulation of gene function in transient expression assays in differentiating *Zinnia* cells) will undoubtedly help us in the selection process.

Identification of an *Arabidopsis* mutant with an atypical vascular pattern

In wild type *Arabidopsis* inflorescence stems, the primary vascular system is organised into 6 to 8 collateral vascular bundles which alternate with the interfascicular sclerenchyma fibers. Alternatively, the vascular system of an Arabidopsis named 'high cambial activity', *hca,* is characterized by a wide continuous ring of secondary xylem surrounded by numerous files of phloem. The sclerenchyma cells appear as small aggregates irregularly dispersed within xylem files. This atypical vascular organization and extensive secondary growth suggest an unusually high cambium activity. These anatomical alterations observed in *hca* were accompanied by pleiotropic effects such as a reduced growth habit and distorted leaves. Genetic analysis indicated that *hca* mutation is monogenic and recessive. Molecular cloning of *hca* is in progress. These data, together with an in-depth physiological characterization and transcriptome analysis will allow us to determine *hca* gene function in relation to cambial function.

CONCLUSIONS

A plenitude of functional genomic data is now becoming available using a variety of differential techniques applied to the *Zinnia* TE system. These investigations constitute the groundwork that will enable a quantitative leap in our knowledge of molecular mechanisms of plant-specific cell differentiation. Coupled to the use of mutants and transgenic plants, we can look forward to exciting times ahead in the field of xylem biology.

ACKNOWLEDGEMENTS

This work was supported, in part, by the European Commission (AIR programme : AIR2-CT93-1661) and Génoplante.

REFERENCES

1. P. Ranocha, D. Goffner, & A.M. Boudet, Plant laccases: are they involved in lignification?, In: *Cell and Molecular Biology of Wood Formation*, R. Savidge, J. Barnett & R. Napier (eds.), BIOS Scientific Publishers Ltd., Oxford, **2000**, pp397-410.
2. J.F.D. Dean & K.-E.L. Eriksson, Laccase and the deposition of lignin in vascular plants, *Holzforschung*, **1994**, 48, 21-33.
3. D.M. O'Malley, R. Whetten, W. Bao, C-L. Chen, & R.R. Sederoff, The role of laccases in lignification. *Plant J*, **1993**, 4, 751-757.
4. J. Grima-Pettenati & D. Goffner, Lignin genetic engineering revisited, *Plant Sci*, **1999**, 145, 51-65.
5. P. Ranocha, G. McDougall, S. Hawkins, R. Sterjiades, G. Borderies, D. Stewart, M. Cabanes-Macheteau, A.M. Boudet & D. Goffner, *Eur J Biochem*, **1999**, 259, 485-495.
6. H. Fukuda & A. Komamine, Establishment of an experimental system for the study of tracheary element differentiation from single cells isolated from the mesophyll of *Zinnia elegans, Plant Physiol,* **1980**, 65, 57-60.
7. Z-H. Ye, R.E. Kneusel, U. Matern & J.E. Warner, An alternative methylaton pathway in lignin biosynthesis in *Zinnia*, *Plant Cell*, **1994**, 6, 1427-1439.
8. T. Demura & H. Fukuda, Molecular cloning and characterization of cDNAs associated with tracheary element differentiation in cultured *Zinnia* cells, *Plant Physiol*, **1993**, 103, 815-821.
9. Z. H. Ye and J. E. Warner, Gene expression patterns associated with *in vitro* tracheary element formation in isolated single mesophyll cells of *Zinnia elegans*, *Plant Physiol*, **1993**, 103, 805-813.
10. K. Roberts & M. C. McCann, Xylogenesis : the birth of a corpse, *Curr Op Plant Biol,* **2000**, 517-522.
11. L. Diatchenko, Y-F.C. Lau, A.P. Campbell, A. Chenchik, F. Moqadam, B. Huang, S., Lukyanov, K. Lukyanov, N. Gurskaya, E.D. Sverdlov & P. D. Siebert, Suppression subtractive hybridization : A method for generating differentially regulated or tissue-specific cDNA probes and libraries, *Proc Natl. Acad. Sci, USA*, **1996**, 93, 6025-6030.
12. F. M. Carland, B. L. Berg, J. N. Fitzgerald, S. Jinamornphongs, T. Nelson, B. Keith, Genetic regulation of vascular tissue patterning in *Arabidopsis*. *Plant Cell* **1999**, 11, 2123-2137.

13. M. K. Deyholos, G. Cordner, D. Beebe & L. Sieburth, The *SCARFACE* gene is required for cotyledon and leaf vein patterning, *Development*, **2000**, 127, 3205-3213.
14. K. Koizumi, M. Sugiyama & H. Fukuda, A series of novel mutants of *Arabidopsis thaliana* that are defective in the formation of continuous vascular network : calling the auxin signal flow canalization hypothesis into question, *Development*, **2000**, 127, 3197-3204.
15. R. Zhong, J.J. Taylor & Z. Ye, Transformation of the collateral vascular bundles into amphivasal bundles in an *Arabidopsis* mutants, *Plant Physiol*, **1999**, 120, 53-64.
16. U. Mayer, G. Büttner & G. Jürgens, Apical-basal pattern formation in the *Arabidopsis* embryo : studies on the role of *GNOM* gene, *Development*, **1993**, 117, 149-162.
17. F. M. Carland & N. McHale. LOP1 : a gene involved in auxin transport and vascular patterning in *Arabidopsis*, *Development*, **1996**,122, 1811-1819.
18. C. S. Hardtke & T. Berleth, The *Arabidopsis* gene *MONOPTEROS* encodes a transcription factor mediating embryo axis formation and vascular development, *EMBO J*, **1998**, 17, 1405-1411.
19. L. Gälweiler, C. Guan, A. Müller, E. Wisman, K. Mendgen, A. Yephremov & K. Palme, Regulation of polar auxin transport by AtPIN1 in *Arabidopsis* vascular tissue, *Science*, **1998**, 2226-2230.
20. R. Zhong, J. J. Taylor & Z. Ye, Disruption of interfascicular fiber differentiation in an *Arabidopsis* mutant. *Plant Cell*, **1997**, 9, 2159-2170.
21. S. Turner & C. Somerville, Collapsed xylem phenotype of *Arabidopsis* identifies mutants deficient in cellulose deposition in the secondary cell wall, *Plant Cell*, **1997**, 9, 689-701.
22. R. Zhong, A. Ripperger & Z. Ye, Ectopic deposition of lignin in the pith of stems of two *Arabidopsis* mutants. *Plant Physiol*, **2000**, 123, 59-69.
23. A. Scalbert, B. Monties, & G. Janin, Tannins in wood : comparison of different estimation methods, *J Agri Food Chem*, **1989**, 37, 1324-1329.
24. A. W. Roberts, L. T. Koonce & C.H. Haigler, A simplified medium for *in vitro* tracheary element differentiation in mesophyll suspension cultures from *Zinnia elegans*, *Plant Cell, Tissue, and Organ Culture*, **1992**, 28, 27-35.
25. N, Bechtold, J. Ellis, & G. Pelletier, *C.R.Acad. Sci., Life Sciences,* **1993**, 316, 1194-1199.
26. J. F. D. Dean, P. R. Lafayette, C. Rugh, A. H. Tristram, J. T. Hoopes, K.E-L. Erikkson & S. A. Merkle, Laccases associated with lignifying vascular tissues, In: *Lignin and lignan biosynthesis*, N.G. Lewis & S. Sarkanen (eds.), Amercian Chemical Society, Washington, D.C., **1998**, pp.96-108.
27. A. Minami & H. Fukuda, Transient and specific expression of a cysteine endopeptidase during differentiation of *Zinnia* mesophyll cells into tracheary elements', *Plant Cell Physiol*, **1995**, 36, 1599-1606.
28. S. Aoyagi, M. Sugiyama & H. Fukuda, *BEN1* and *ZEN1* encoding S1-type DNases that are associated with programmed cell death in plants, *FEBS letters*, **1998**, 429, 134-138.

Molecular Breeding of Woody Plants
N. Morohoshi and A. Komamine, editors.

SPATIAL AND TEMPORAL REGULATION OF LIGNIFICATION DURING TRACHEARY ELEMENT DIFFERENTIATION

Yasushi Sato

Faculty of Science, Ehime University, Matsuyama, 790-7588, Japan

ABSTRACT

During tracheary element (TE) differentiation, lignin is deposited specifically to secondary cell walls. Spatial and temporal regulation of lignification during TE differentiation was investigated using an experimental system in which TEs are differentiated from isolated *Zinnia* mesophyll cells.

The mechanism how and whence monolignols are supplied to TEs undergoing programmed cell death was investigated. Analysis by HPLC and GC-MS showed that coniferyl alcohol, coniferaldehyde, and sinapyl alcohol were accumulated in cultured medium during differentiation inductive culture. The concentration of coniferyl alcohol peaked at the beginning of secondary wall thickening, decreased rapidly during secondary wall formation, then increased again. These results indicated that lignification of TEs progresses by supply of monolignols from not only TEs themselves but also surrounding xylem parenchyma-like cells through medium in vitro. Simultaneously, these results would suggest that lignification of TEs in vivo progresses by supply of monolignols from not only TEs themselves but also surrounding xylem parenchyma cells through apoplast.

For study of the final step of lignification, polymerization of monolignols, cell wall-bound peroxidase isoenzymes were analyzed, and a cationic isoenzyme P5 was shown to appear specifically for cells differentiating into TEs. Characterization of P5 strongly suggested that P5 is involved in lignin biosynthesis during TE differentiation. Furthermore, a peroxidase gene, ZPO-C, was isolated by PCR method. Transcripts of ZPO-C were accumulated transiently during thickening of secondary walls of TEs. By immunoelectron microscopy, the ZPO-C protein was shown to localize specifically in the lignified parts of secondary walls of TEs. On the other hand, basic laccases appeared specifically in differentiation inductive culture, too.

In conclusion, it was shown that the monolignols would be supplied to TEs from other cells, and polymerized to lignin by the peroxidases and/or laccases localized specifically in secondary cell walls of TEs during TE differentiation of cultured *Zinnia* mesophyll cells.

KEYWORDS

Cell differentiation, coniferyl alcohol, lignin synthesis, peroxidase, tracheary element, *Zinnia elegans*

INTRODUCTION

Lignins, complex phenolic heteropolymers characteristic for TEs, are synthesized through three steps, that is monolignol synthesis in cytoplasm, transportation of monolignols to the part of lignification in the cell walls, and polymerization of monolignols to the macromolecules, lignins at the part of lignification.

The step of monolignol synthesis involves three pathways, known as the shikimate, the general phenylpropanoid, and the specific lignin pathway. The enzymes involved in general phenylpropanoid and the specific lignin pathway have been investigated in detail for understanding of lignin biosynthesis [1]. Although the monolignols are thought to be transported to the cell walls by vesicles those are derived from the Golgi apparatus or the endoplasmic reticulum [2], the process of transportation of monolignols to the part of lignification is not elucidated so detailed. In the cell walls, they are polymerized into lignin by wall-bound peroxidases and/or laccases. It is not clear fully which isoenzymes spatially and temporally regulate the step of polymerization to lignin in many peroxidases and laccases.

For study of these subjects, in vitro experimental systems for TE differentiation are very useful. In the experimental system of TE differentiation of *Zinnia*, single cells isolated from first leaves of *Zinnia* differentiate to TEs with high frequency and synchrony. Furthermore, lignin is deposited to the secondary cell walls of TEs same as in vivo [3]. Therefore, this system is suitable for the study of transportation of monolignols and the enzymes involved in lignin polymerization. In this study, transportation of monolignols and polymerization of monolignols to lignin were investigated using the experimental system of TE differentiation of *Zinnia*..

TRANSPORTATION OF MONOLIGNOLS TO THE PART OF LIGNIFICATION DURING TE DIFFERENTIATION

(1) Lignification and PCD during TE differentiation

Differentiation into TEs is a typical example of programmed cell death (PCD) in higher plants, and mature TEs are completed by the loss of all cell contents. During TE differentiation of isolated *Zinnia* mesophyll cells, the disruption of the central vacuole, the irreversible step toward cell death, would occur just after the completion of secondary cell wall thickenings [4, 5]. Nevertheless, lignification of secondary cell walls progresses continuously until the completion of mature TEs as shown by Fukuda & Komamine (1982) [6]. This may indicate that TEs that have undergone PCD are lignified by receiving monolignols from outside, namely from other undifferentiated cells through the culture medium.

(2) Effects on lignification of TE by various modifications of culture conditions

It was hypothesized that higher cell culture density would result in heavier lignification by dint of the higher accumulation of lignin precursors secreted into the medium.

Isolated mesophyll cells were cultured at the initial cell density of 0.8×10^5 cells/ml in differentiation inductive medium (D medium). After 57 h of culture, the cells were resuspended in fresh D medium at each cell density (0.05, 0.2, and 0.8×10^5 cells/ml) and cultured for additional 48 h. No significant inhibitory effect of cell dilution and exchange of medium on the ratio of TE differentiation was detected. On the other hand, the cells cultured in the higher cell density had the higher lignin content.

Furthermore, we investigated the effect of used medium on lignification of TEs. After isolated mesophyll cells were cultured for 53 h in D medium, the cells were resuspended at low cell density (0.05 x 10^5 cells/ml) in the conditioned media which had been used for cultures at each cell density (0.05, 0.2, and 0.8 x 10^5 cells/ml) for 96 h, then cultured for further 48 h. As expected, lignin content was higher when conditioned medium of higher cell density was used.

It was expected that lignification should be suppressed by exchange of the medium for removal of lignin precursors, and addition of exogenous lignin precursors might overcome its inhibition. After isolated mesophyll cells were cultured for 53 h, the cells were collected and resuspended in fresh D medium or fresh D medium containing each concentration (0.3, 3, or 30 μM) of coniferyl alcohol (CA) at intervals of three hours for 48 h. After that, the lignin content in the cells cultured under each condition was determined. Lignification was almost perfectly inhibited by continuous exchanging of medium. On the other hand, treatment with higher concentrations of CA resulted in more lignin contents.

(3) Analysis of lignin precursors in medium

Differentiation inductive medium (D medium) and control medium (CN medium) after 120 h of culture were analyzed by HPLC. The peaks corresponding to CA and sinapyl alcohol (SA) at 270 nm and coniferaldehyde (CD) at 340 nm were detected in D medium by elution with authentic compounds. However, none of these peaks were detected in CN medium. The medium after 120 h of D culture was analyzed by GC-MS. The MS results of GC peaks of CA and CD in D culture were identical to that of authentic CA and CD. Although mass spectrograph of SA showed that SA certainly exist in the fraction, the MS pattern had a little noise because of insufficiency of amounts. The changes in concentrations of three lignin precursors, CA, SA, and CD in medium during culture were followed (Fig. 1). The concentration of CA was the highest in these three precursors. The concentration of CA in D medium increased and reached about 5 μM at 48 h of culture when secondary wall thickening start. Between 48 h and 60 h of culture, CA concentration suddenly decreased to 0.2 μM. Thereafter, CA concentration increased steadily again. The concentrations of CD and SA kept very low levels.

The results obtained here were summarized. Before secondary wall thickening of TEs, monolignols may be secreted from all cultured cells. After start of secondary cell wall thickening of TEs, monolignols are incorporated by TEs meanwhile they are secreted from immature TEs and xylem parenchyma-like cells. After PCD of TEs, monolignols secreted from xylem parenchyma-like cells are continuously incorporated by TEs. Simultaneously, xylem parenchyma cells would be suggested to supply monolignols to vessels through apoplast in vivo.

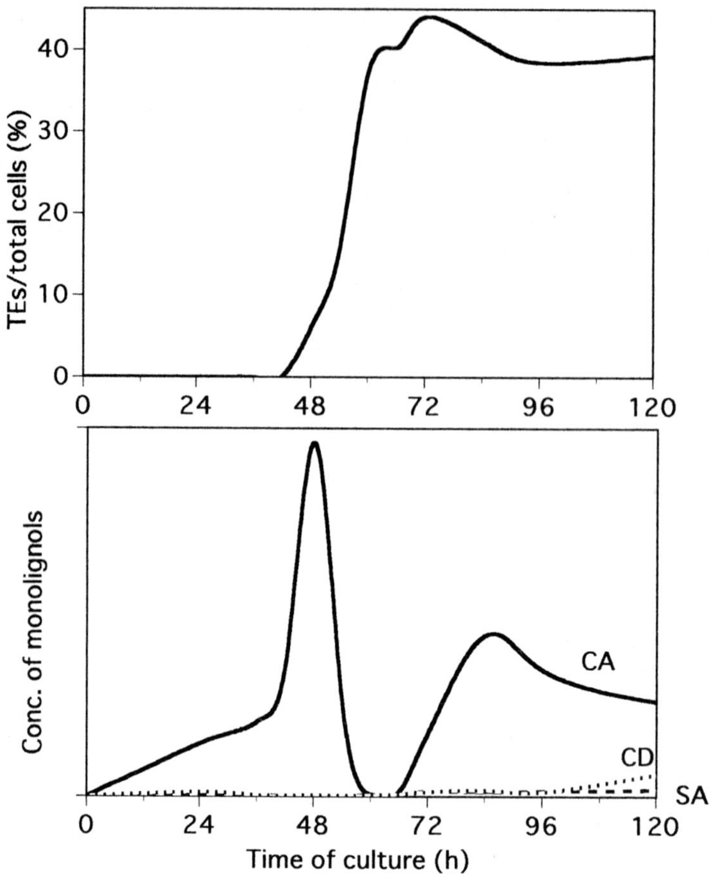

Figure 1.
Upper graph: Time course of TE differentiation during culture of isolated mesophyll cells of *Zinnia*. **Lower graph:** Changes in the concentration of each lignin precursor (CA; coniferyl alcohol, CD; coniferaldehyde, SA; sinapyl alcohol) in D medium during culture of isolated *Zinnia* mesophyll cells.

POLYMERIZATION OF MONOLIGNOLS TO LIGNIN

(1) Changes in wall-bound peroxidase isoenzymes during TE differentiation

Monolignols supplied are thought to be polymerized into lignin by wall-bound peroxidases and/or laccases. However, there are many isoenzymes of peroxidases and laccases in the cell walls. In order to determine which isoenzymes of peroxidase or laccase catalyze lignin synthesis, it is necessary to examine substrate specificity, subcellular localization, and temporal and spatial correlation with active lignification.

By activity staining after native polyacrylamide gel electrophoresis, five peroxidase isoenzymes, P1-P5, bound ionically to the cell walls were detected during TE differentiation of *Zinnia.* Among these isoenzymes, P4 and P5 appeared specifically in TE inductive culture [7]. Fractionation of *Zinnia* cells by centrifugation in Percoll solutions revealed that P1, P2, and P5 were present in TEs [8]. These peroxidase isoenzymes were separated by several column chromatographies. During these steps, P5 activity was separated into P5A and P5B activities. Finally, enzymatically pure preparations of P1, P3, P5A and P5B were obtained and used for characterization of each isoenzyme. All isoenzymes tested oxidized coniferyl alcohol efficiently, whereas *p*-coumaryl alcohol and sinapyl alcohol were poor substrates for all isoenzymes. Therefore, P5 was the peroxidase isoenzyme specific for TEs and having affinity to coniferyl alcohol [9]. These support hypothetical involvement of P5 (P5A and P5B) in lignification.

(2) Changes in wall-bound laccase isoenzymes during TE differentiation

Laccase is another candidate of the enzyme involved in lignin synthesis. We analyzed changes in the activites of laccase isoenzymes during TE differentiation of *Zinnia.* Flat type native PAGE was carried out for active staining of laccase isoenzymes using diaminofluorene (DAF) as a substrate. Basic laccase activities appeared in differentiation specific manner.

(3) Isolation and characterization of a gene for peroxidase involved in lignin synthesis

Isolation and characterization of differentiation specific peroxidase gene were attempted by PCR amplification of cDNA derived from the mRNA from *Zinnia* mesphyll cells of 48 h of D culture. As a result, a differentiation specific clone for peroxidase was isolated and designated ZPO-C. The ZPO-C cDNA contained an open reading frame of 1116 bp, and the deduced polypeptide sequence contained 317 amino acids with a signal peptide at N-terminal region and a pI of 8.59 of cationic isoenzyme. The transcripts of ZPO-C were expressed specifically and transiently between 48 and 60 h in D culture. This timing was coincident with thickening of secondary walls of TE and the onset of lignification.

(4) Characterization of the localization of the peroxidase encoded in ZPO-C

A polyclonal antiserum against ZPO-C fusion protein was raised. Western blotting using the anti-ZPO-C protein antiserum showed that ZPO-C protein was detected in ion-extracted fraction of cell walls from D culture. Anti-ZPO-C protein IgG was purified from antiserum by affinity chromatography.

By immuno-histochemistry using anti-ZPO-C protein IgG as first antibody and alkaline phosphatase labeled anti rabbit IgG as secondary antibody, specific signals were shown to localize to vessels in stem from 30-d-old plants and TEs of 67 h of D culture. By immuno-electron microscopy, the signals of gold particles were already localized in developing secondary walls of immature TEs. Signals were observed in developed secondary walls of mature TEs, too (Fig. 2). From these results, ZPO-C was appeared to be a gene for a peroxidase isoenzyme involved in polymerization of lignin.

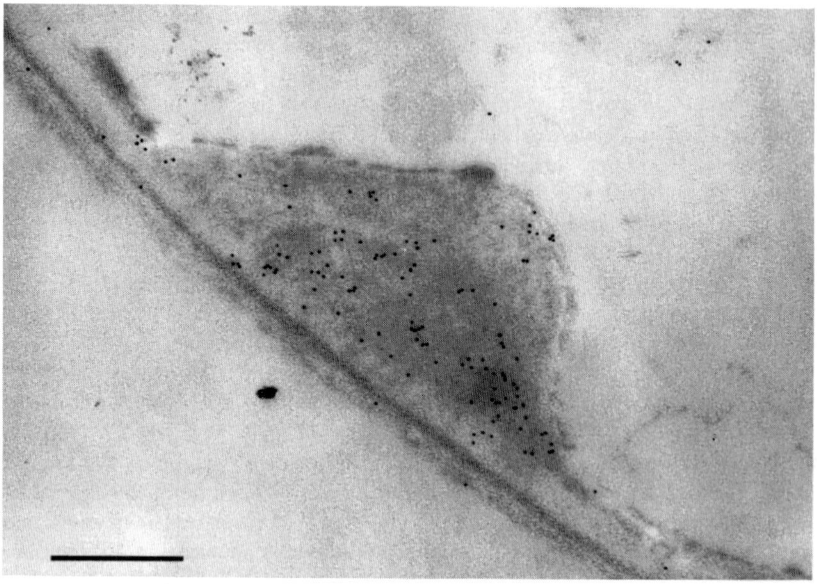

Figure 2.
Immunolocalization of ZPO-C protein in a TE of *Zinnia*. Rabbit anti-ZPO-C fusion protein was used as first antibody. Goat anti-rabbit IgG conjugated to colloidal gold (10 nm particle size) was used as second antibody. Bar=500 nm.

CONCLUDING REMARKES

In this study, analysis of spatial and temporal regulation of lignification was tried using an experimental system of TE differentiation of *Zinnia*. The results obtained here were summarized to an illustration (Fig. 3). Monolignols, mainly CA, are released outside from all cells previous to secondary cell wall formation in *Zinnia* system. Monolignols are accumulated in the medium at the highest level in this period. In developing TEs before PCD, the peroxidase corresponded to ZPO-C is produced and incorporated to thickening secondary walls. The laccases specific for TE differentiation may be incorporated to the secondary cell walls of TEs, too. After PCD of TE, monolignols are supplied from xylem parenchyma-like cells and polymerized to lignin by the peroxidase corresponded to ZPO-C and laccases localized in secondary walls of TEs.

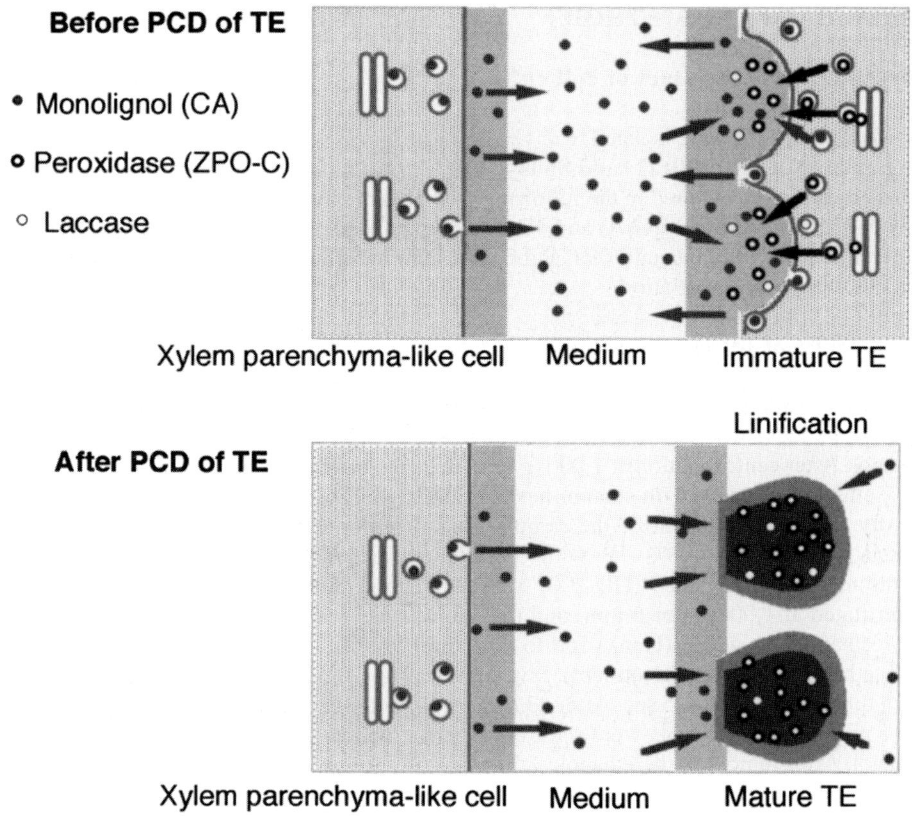

Figure 3.

Model of lignification of TEs differentiated from isolated *Zinnia* mesophyll cells. Before PCD of TEs, monolignols (mainly CA, coniferyl alcohol) are secreted from all cells. Peroxidases (including the peroxidase corresponded to ZPO-C) and laccases are incorporated to thickening secondary walls. After PCD of TEs, stop of secretion of monolignols from TEs undergone PCD and increase of secretion of monolignols from xylem parenchyma-like cells occur. Peroxidases and laccases localized in secondary walls of TEs polymerize monolignols to lignin.

MATERIALS AND METHODS

Plant material and start of cell culture

Mesophyll cells were isolated from the first true leaves of 14-d-old seedlings of *Zinnia elegans* L. cv. Canary Bird (Takii Shubyo Co., Kyoto, Japan) as described previously [9]. Isolated cells were cultured in the following media: D medium, which contained 0.1 mg/L 1-naphthaleneacetic acid (NAA) and 0.2 mg/L benzyladenine (BA) and which induced the differentiation of the cells into TEs; and CN medium, which contained 0.1 mg/L NAA and did not support differentiation.

Measurement of lignin content

For measurement of lignin content, cells (approximately 3.2 x 10^6 cells) cultured for various periods were ultrasonically homogenized (UD-200; TOMY, Tokyo) in 95% ethanol. After centrifugation at 1,000xg for 5 min, the pellet was washed three times with 95% ethanol and twice with ethanol-hexane (1:2, v/v). The washed pellet was allowed to air-dry. The lignin content of the samples was determined according to the method of Garcia and Latge (1987) [10] with some modifications. The dried samples were ultrasonically resuspended (UD-200) in 90% ethanol, divided into 4 micro-tubes and centrifuged at 1,000xg for 5 min, and the ethanol was removed. To one tube 0.5 ml of 90% ethanol was added (blank) and to the other three 0.5 ml of 2% phloroglucinol in 90% ethanol was added (treatment). Five minutes later, 0.5 ml of 5.6N HCl was added to each tube and mixed for 5 sec. After 13 min incubation, tubes were centrifuged at 1,000 g for 5 min, the supernatants were removed, and the pellets were washed in 1 ml of 90% ethanol. After removing the ethanol, 0.5 ml of 25% acetyl bromide in glacial acetic acid was added to each tube, and the samples were shaken by hand for 5 sec. Glacial acetic acid (0.5 ml) was then added and the tubes were mixed at 10 min intervals. Twenty-five minutes later, samples were centrifuged at 10,000xg for 5 min and the absorbance of the supernatant was measured at 545 nm to determine the lignin content.

Determination of lignin precursors in cultured media by HPLC

Thirty-five ml of cultured media were acidified by addition of 35 μl of acetic acid, and filtered through a 0.2 μm filter (PTFE; Millipore). Filtrates were applied to Sep Pak C18 cartridges (Waters) that had been pre-wetted with 10 ml of ethanol and equilibrated with 10 ml of water containing 0.1% acetic acid. The cartridges were washed with 10 ml of water containing 0.1% acetic acid. The lignin precursors were eluted by 2 ml of a 60% ethanol solution.

Aliquots (20 μl) of samples were fractionated by gradient HPLC on C-18 column (Lichrosorb PR18-5, 4.0 x 250 mm, GC Science Inc.) using solvent A: 2% (v/v) acetic acid; solvent B: 2% (v/v) acetic acid in acetonitrile; gradient conditions: 5-20% B over 45 min, 20-25% B over 5 min, 25-100% B over 4 min, 100-0% B over 1 min, 0 % B for 5 min; solvent flow: 1 ml/min. The eluate was monitored at 270 nm and 340 nm, and the peak areas of the eluting peaks were determined by integration.

GC-MS analysis

Two hundred and seventy ml of the medium of D culture for 96 h were concentrated and fractionated repeatedly by HPLC as described above, and respective peaks of CA, SA, and CD were collected. The collected samples were lyophilized and the resulting residues were extracted individually with methanol. Each methanol solution was dried in vacuo and subjected to GC-MS analysis.

Gas chromatography-mass spectrometry was performed on a JMS-DX303HF mass spectrometer (JEOL Ltd.) equipped with a Hewlett-Packard 5890J gas chromatograph and a JMA-DA5000 mass data system [electron impact mode, 70 eV; gas-chromatographic column, Shimadzu Hicap CBP-10M25-025 (5 m x 0.22 mm); temperature, 40°C at t=0 to 2 min, then to 240°C at 30°C/min; carrier gas, He; splitless injection]. The samples for GC-MS were dissolved in N, O-bis(trimethylsilyl)acetamide and left standing at 60°C for 45 min; then an aliquot of the solution was subjected to GC-MS analysis.

Production of anti-ZPO-C protein antibody

The QIAexpress pQE31 vector (QIAGEN) was used for expression vector. The construct was designed to produce the ZPO-C fusion protein, whose sequence had 6 His tag at N-terminus and the amino acid sequence of 49-317 of ZPO-C protein. The construct was transformed into XL1-blue, and.the fusion protein was indeuced by addition of 2 mM IPTG. The fusion protein was purified by Ni-NTA resin (QIAGEN) according to instructions, separated by SDS-PAGE and subsequent cutting out from the gels. Fusion protein was injected with ground polyacrylamide gel in a rabbit three times for raising antiserum against ZPO-C fusion protein. Anti-ZPO-C protein IgG was purified from antiserum against ZPO-C fusion protein using affinity column chromatography. Affinity column conjugated ZPO-C fusion protein was prepared by conjugation of ZPO-C fusion protein to CNBr-activated Sepharose 4B (Pharmacia) and used for purification of anti-ZPO-C protein IgG according to instructions.

Electron microscopy (immunogold labeling)

Cultured *Zinnia* cells were fixed in 4% (w/v) paraformaldehyde and 0.1% (v/v) glutaraldehyde in 30mM sodium phosphate buffer (pH 7.2) at 4°C for 2 h. Cells were dehydrated through alcohol series and embedded in hydrophilic resin, LR White (London Resin Co.). Ultrathin sections were first immersed in 20 μL of 1% BSA in PBS-T for 1h at room temperature, then replaced with 20 μL of antibody solution [purified anti-ZPO-C protein IgG at 1:50 (20 μg/mL IgG final concentration); or rabbit IgG (Jackson ImmnoResearch, USA) at 1:1400 (20 μg/mL IgG final concentration) in 1 % BSA in PBS-T] for 14 h at room temperature. Sections were then washed thoroughly for 5min with PBS-T. Sections were immersed in 20 μL gold (10 nm) conjugated goat anti-rabbit IgG (1:40 working solution in 1% BSA in PBS-T) (Sigma) for 1h. Sections were then washed thoroughly with PBS-T and distilled water, and dried. Sections were stained with saturated uranyl acetate in 20% butylalcohol for 50 min in darkness, washed with distilled water, and examined in an electron microscope (JEM-2000, JEOL Ltd.) at 80 kV.

ACKNOWLEDGEMENTS

The author is very grateful to Professor A. Komamine of the Research Institute of evolutionary Biology, Professor. H. Fukuda of University of Tokyo, Dr. M. Sugiyama of University of Tokyo, Professor R.J. Gorecki of University of Agriculture and Technology Poland, Professor T. Takagi of Tohoku University, Professor T. Umezawa of Kyoto University, Dr. S. Suzuki of Kyoto University, M.Sc. M. Hosokawa of Ehime University for collaboration of this work. This work was supported in part by Grants-in-Aid from the Ministry of Education, Science and Culture of Japan (No. 09740599) and from the Japan Society for the Promotion of Science (JSPS-RFTF).

REFERENCES

1. A. M. Boudet, C. Lapierre, & J. Grima-Pettenati, 'Tansley review No. 80. Biochemistry and molecular biology of lignification', *New Phytol.*, **1995**, 129, 203-236.
2. N. G. Lewis, & E. Yamamoto, 'Lignin: occurrence, biogenesis and biodegradation', *Annu. Rev. Plant Physiol. Plant Mol. Biol.*, **1990**, 41, 455-496.
3. H. Fukuda, 'Tracheary element differentiation', *Plant Cell,* **1997**, 9, 1147-1156.
4. A. Groover, N. DeWitt, A. Heidel, & A. Jones, 'Programmed cell death of plant tracheary elements differentiating in vitro', *Protoplasma,* **1997**, 196, 197-211.
5. H. Kuriyama, 'Loss of tonoplast integrity programmed in tracheary element differentiation', *Plant Physiol.*, **1999**, 121, 763-774.
6. H. Fukuda , & A. Komamine, 'Lignin synthesis and its related enzymes as markers of tracheary-element differentiation in single cells isolated from the mesophyll of *Zinnia elegans*', *Planta,* **1982**, 155, 423-430.
7. Y. Sato, M. Sugiyama, R. J. Gorecki, H. Fukuda, A. Komamine, 'Interrelationship between lignin deposition and the activities of peroxidase isoenzymes in differentiating tracheary elements of *Zinnia*', *Planta,* **1993**, 189, 584-589.
8. Y. Sato, M. Sugiyama, A. Komamine, & H. Fukuda, 'Separation and characterization of the isoenzymes of wall-bound peroxidase from cultured *Zinnia* cells during tracheary element differentiation', *Planta,* **1995**, 196, 141-147.
9. H. Fukuda, & A. Komamine, 'Establishment of an experimental system for the study of tracheary element differentiation from single cells isolated from the mesophyll of *Zinnia elegans*', *Plant Physiology,* **1980**, 65, 57-60.
10. S. Garcia, & J. P. Latge, 'A new colorimetric method for dosage of lignin', *Biothechnol. Techniques*, **1987**, 1, 63-68.

Molecular Breeding of Woody Plants
N. Morohoshi and A. Komamine, editors.

FINAL AND FATAL STEP OF TRACHEARY ELEMENT DIFFERENTIATION

Alan M. Jones, Andrew Groover, Xiaohong Yu, & Tony Perdue

Department of Biology, The University of North Carolina at Chapel Hill, Chapel Hill, NC, 27599, USA.

ABSTRACT

The process of terminal differentiation which produces the hollow tracheary element cell corpse requires strict coordination of two developmental events, construction of a rigid, persistent secondary cell wall and programmed cell death. We show that tracheary element programmed cell death involves an influx of Ca^{2+} into the cell which may play a causative role in executing cell death. Ca^{2+} influx is triggered by an extracellular signal, and leads to the rapid cessation of cytoplasmic streaming and collapse of the large hydrolytic vacuole. This specific means of effecting cell death is a necessary prerequisite for postmortem developmental events including autolysis and chromatin degradation. A protease ("trigger protease) is secreted during secondary cell wall synthesis which may be the primary trigger of cell death, because specific proteolysis of the extracellular matrix is necessary and sufficient to trigger Ca^{2+} influx, vacuole collapse, cell death, and chromatin degradation. We propose a model in which secondary cell wall synthesis and cell death are coordinated by the concomitant secretion of the trigger protease with secondary cell wall precursors. Subsequent cell death is triggered upon realization of a critical extracellular activity of protease corresponding with completion of a functional secondary cell wall. Because increased Ca^{2+} levels are associated with cell death involving what has been termed the mitochondrial pathway in animals, we investigated the possible role of released mitochondrial factors in the death mechanism in tracheary elements. We find that some cytochrome c is released to the cytosol at a time when death occurs and induced by calcium influx. Concomitant with this release are changes in the inner membrane voltage potential and the morphology of the mitochondria. However, cytochrome c release is insufficient to induce death in these cells. This suggests that the events triggered by the extracellular "trigger" protease may set in motion events shared by the mitochondrial pathway for apoptosis in animal cells.

KEYWORDS

Mitochondrial pathway, programmed cell death, regulatory protease, tracheary elements, Zinnia

INTRODUCTION

Most terminally differentiated cells fulfill specialized functions until they die, but for some cell types, function does not begin until after death. The developmental programs producing such functional cell corpses involve the coordination of cell differentiation with PCD. The classic example of terminal differentiation in plants is the tracheary element (TE), a functional cell corpse that forms a single unit of the water-conducting vessels of the xylem. We previously used a cell-culture system in which mechanically isolated mesophyll cells differentiate as TEs in vitro to characterize morphological changes during PCD of TEs[1]. During differentiation the living TE constructs a rigid, interlacing secondary cell wall between the primary cell wall and the plasma membrane. Secondary cell wall synthesis is accompanied by the synthesis of nucleases and proteases[2-6], and influx of Ca^{2+} [7,8]. An average of 6 h after secondary cell wall thickenings become visible, the large central vacuole collapses rapidly, cytoplasmic streaming ceases abruptly, and the contents of the hydrolytic vacuole mix with the cytoplasm[1]. Enzymatic degradation of the cell contents ensues and nDNA degradation can be detected in single cells with TUNEL both in vitro[1,9] and in vivo[10,11].

As in animal systems, there are indications that the signals initiating PCD in plants vary among cell types. Developmental programs culminating in cell death are initiated by ethylene in aerenchyma formation[12], by GA3 in aleurone cells[13], and by auxin[14] and brassinolides[15] in TEs, although it is not clear if these hormones modulate PCD directly or if they initiate developmental programs in which PCD is a subroutine. The extracellular matrix is an important component of at least some types of plant PCD.

For a cell to "commit suicide," catabolic processes must overwhelm the metabolic processes that normally sustain it. Although it is not known how this is regulated by plant cells, most if not all animal cells irreversibly commit to (execute) PCD through the action of the caspase family of Cys proteases[16]. Although protease activity in plants has been correlated with developmental events culminating in PCD, including the hypersensitive response[17] and TE cell autolysis[3-5], it is not known if proteolysis plays a role in regulating or executing cell death. With the Arabidopsis genome sequenced, no prototypical caspases are found, however recently a family of caspase homologs designated metacaspases have been identified[18]. However, no functional data is yet available to indicate that these metacaspases have caspase activity.

We present evidence that cell death during TE differentiation is controlled by a signaling mechanism coordinated with secondary cell wall synthesis. We correlate cell death with the secretion of a trigger protease and provide data implicating this protease as a primary trigger of cell death. Execution of cell death requires an influx of Ca^{2+}, and is morphologically marked by collapse of the hydrolytic vacuole and the mixing of the vacuole with the cytoplasm. We propose a model in which execution of cell death is coordinated with completion of a functional secondary cell wall by the requirement of either a critical extracellular concentration of protease or the arrival of a substrate whose proteolytic cleavage produces a signaling product.

MATERIALS & METHODS

Plants, cell culture, and chemicals

Seedlings of zinnia (Zinnia elegans L. cv Green Envy; Stokes Seed, Buffalo, NY) were grown in a growth chamber at 25°C and 60% RH with 14 h of light (110 μmol photons m^2 s^2) per day. Cells were isolated by the method described by Fukuda and Komamine[19] using modifications described by Groover and Jones[9].

Protein extraction

Intracellular proteins were isolated by homogenizing cells in extraction buffer (50 mM Tris-HCl, pH 7.5, 2 mM DTT, 250 mM sucrose) at 4°C, followed by centrifugation at 12,000 X g at 4°C for 15 min to pellet cell debris. For concentration of proteins from the medium, cultures were centrifuged twice to remove cells, and the supernatant was passed through a 2-µm filter. The proteins in the filtered supernatant were concentrated at 4°C using a pressure cell concentrator (Amicon, Beverly, MA) with a 10-kD cutoff filter (YM10, Amicon). Protein samples were mixed with an equal volume of the sample buffer described by Ye and Droste[3] without heating and loaded onto 0.75-mm-thick 12% SDS acrylamide gels containing heat-denatured gelatin (0.1 mg mL1). For expression of protease activity, gels were incubated overnight at room temperature in 50 mL of 50 mM sodium citrate, pH 5.0, 5 mM DTT, 5 mM $CaCl_2$, and 1 mM $ZnCl_2$.

Transmission Electron Microscopy

Cells were collected at various time points and prepared for TEM (transmission electron microscopy) by fixation in a phosphate buffered solution (Sorensen's phosphate, pH 5.8) of 2.5% glutaraldehyde with 0.15% sucrose and 2% mannitol to maintain proper osmoticum for 12-24 hr at 4°C. Cells were post-fixed in 2% osmium tetroxide in the same buffer for 1hr. After rinsing, the cells were exposed to 2% uranyl acetate (aqueous) and subsequently embedded in 2% agar. Samples were dehydrated in a graded ethanol series, infiltrated with Spurr's resin, embedded, and cured at 70°C for 24 hr. Ultra-thin sections were cut and stained with uranyl acetate and Reynold's lead citrate prior to observation in a Zeiss EM10 microscope.

Detection of mitochondria depolarization

Mitotracker dye JC-1 was purchased from Molecular Probes (Eugene, OR, USA). 24h and 72h zinnia cell cultures were incubated with 10 µg/ml JC-1 for 20 minutes at room temperature, then cells were washed with fresh medium for 3 times. Cells were imaged using a Zeiss LSM410 confocal scanning microscope equipped with an Argon-Kryptor laser. Images were collected using an excitation of 488 nm to observe green fluorescence (emission BP 515-540) and an excitation of 568 nm to observe red fluorescence (emission BP 575-640). The resulting two images were combined and overlaid with a brightfield image of the cell to demonstrate the degree of secondary cell wall formation in that particular cell.

RESULTS & DISCUSSION

Cell Death Is Marked by the Rapid Collapse of the Vacuole and Leads to Autolysis and nDNA Fragmentation

The first morphological manifestation of differentiation occurs approximately 72 h after cell isolation, when nascent TEs synthesize an elaborate secondary cell wall between their primary cell wall and the plasma membrane. Approximately 6 h after the appearance of visible cell wall thickenings, the large central vacuole collapses rapidly and cytoplasmic streaming ceases simultaneously[1], marking the irreversible termination of normal metabolism and providing a distinct morphological marker of a critical event during PCD, the execution of cell death (video microscopy of vacuole collapse can be viewed at http://www.unc.edu/depts/biology/joneslhp/pcd/). The contents of the hydrolytic vacuole mix with the cytoplasm, leading to active degradation of organelles by hydrolytic enzymes synthesized during differentiation. nDNA is degraded and can be assayed in individual cells using TUNEL[1], an in situ labeling method.

The Process Executing Cell Death Influences Postmortem Development and Is Distinct from Necrosis

The immediate question centers on the significance of cell death during PCD. Specifically, does the endogenous mechanism used to end normal metabolism (i.e. to execute cell death) significantly influence postmortem developmental events, including autolysis? A related question is whether vacuole collapse and DNA fragmentation (assayed by TUNEL) discern PCD from necrotic death under our experimental conditions. We reasoned that these questions could be addressed directly by treating cultures containing nascent TEs (before the onset of cell death during PCD) with drugs that modulate specific components of cell signaling or metabolic pathways and assaying for premature collapse of the vacuole and TUNEL.

Among the various drugs tested, only mastoparan induced significant numbers of cells to prematurely fragment nDNA. Concentrations of other drugs tested included lethal doses, but did not induce DNA fragmentation detectable with TUNEL, immediately suggesting that cell death must be executed in a specific fashion for postmortem DNA fragmentation to occur, and showing that TUNEL is a robust marker of PCD in this system. Mastoparan is an activator of heterotrimeric G-proteins that stimulate enzymes or ion channels in response to ligand-mediated receptor activation.

Mastoparan activates an endogenous process required for the rapid collapse of the vacuole, leading to autolysis and fragmentation of DNA. Low levels of Mas 7, an active synthetic analog of mastoparan, and mastoparan-induced cell death and DNA fragmentation occur in a dose-dependent manner, whereas Mas 17, an inactive synthetic analog, had no effect above control levels, showing that the effects of mastoparan were specific and not attributable to contaminating substances. Other agents that killed cells with similar kinetics and efficacy as mastoparan did not induce DNA fragmentation. Furthermore, as observed with time-lapse videomicroscopy, 83% of cells ($n = 12$) dying in response to mastoparan treatment displayed the rapid vacuole collapse characteristic of TE cell death within minutes of treatment, with cytoplasmic streaming ending instantaneously with collapse of the vacuole. Cells dying from hydrogen peroxide treatment (10 mM; $n = 6$) gradually slowed cytoplasmic streaming without collapse of the vacuole; cells dying from sodium azide treatment (40 μM; $n = 14$) rapidly stopped cytoplasmic streaming but did not display vacuole collapse; cells dying from Triton X-100 treatment (0.02%; $n = 13$) stopped streaming gradually, plasmolyzed, then showed dissolution of chloroplast membranes.

Mastoparan did not cause DNA fragmentation directly, and only cells differentiating as TEs fragmented DNA in response to mastoparan treatment. Cells cultured in medium without exogenous hormones did not differentiate as TE, undergo PCD, or fragment DNA in response to mastoparan treatment. Cells induced to differentiate with hormones fragmented DNA in response to mastoparan treatment only after reaching a developmental stage within approximately 6 h before the appearance of secondary cell wall thickenings visible with light microscopy. Mastoparan induced a high rate of cell death in all of the cultures, but the percentage of dying cells fragmenting DNA in response to mastoparan treatment was correlated with the percentage of cells differentiating as TEs, suggesting that mastoparan treatment leads to DNA fragmentation only in cells differentiating as TEs. The ability of mastoparan to trigger premature vacuole collapse and DNA fragmentation suggests that it activates part of the endogenous mechanism that executes cell death. Because cell death was also induced in cells not differentiating as TEs, mastoparan must activate cellular components used during PCD that are not unique to differentiating TEs.

Execution of Cell Death Requires Ca^{2+} Influx

The rapid collapse of the vacuole and the cessation of cytoplasmic streaming that occur during PCD of TEs and in response to mastoparan treatment likely represent changes in cell turgor and membrane potential that might be explained by ion flux across the plasma membrane. Consistent with this notion, pretreatment of cultures containing nascent TEs with either EGTA (to chelate extracellular Ca^{2+}) or La^{3+} or ruthenium red (to inhibit Ca^{2+} influx) reduced both cell death and DNA fragmentation resulting from mastoparan treatment. The antagonistic effect on cell death by inhibiting Ca^{2+} influx was limited, although the level of DNA fragmentation was reduced to near control levels. This may indicate that DNA fragmentation has a more stringent requirement for Ca^{2+} influx than cell death during PCD. Regardless, these results indicate that mastoparan prematurely induces cell death during PCD by a mechanism requiring an influx of Ca^{2+} into the cell, probably through plasma membrane channels.

Imposing Ca^{2+} influx directly is sufficient to prematurely initiate vacuole collapse leading to DNA fragmentation. Cultures containing nascent TEs were treated with the Ca^{2+} ionophore A23187. Cells in medium containing 1 mM $CaCl_2$ treated with A23187 died (approximately 55%) and fragmented DNA (approximately 20%), whereas about one-half as many cells treated with A23187 in medium lacking supplemental $CaCl_2$ died and fragmented DNA. A23187 caused vacuole collapse in 57% of dying cells (n = 28) cultured in 1 mM $CaCl_2$ (videomicroscopy not shown).

TE Cell Death Can Be Manipulated by Extracellular Proteolysis

We envisioned that extracellular changes could coordinate cell wall synthesis and PCD. For example, the synthesis of a secondary cell wall between the primary wall and the plasma membrane could sever connections between the cytoskeleton and the extracellular matrix, which triggers cell death. Alternatively, the hydrolysis of the primary cell wall during TE differentiation could release a signal molecule triggering cell death, as during cell death in response to wall-derived elicitor molecules during the hypersensitive response. To test these possibilities, cultures containing nascent TEs were treated with exogenous hydrolytic enzymes targeting specific components of the extracellular matrix and assayed for cell death and DNA fragmentation. Although several of the hydrolases tested caused an increase in the percentage of dead cells, only trypsin caused cell death leading to DNA fragmentation (Table I). Moreover, trypsin (0.5%) caused vacuole collapse in 87% of killed cells (n = 15) observed with time-lapse videomicroscopy (not shown). The observation that other proteases did not trigger DNA fragmentation suggests that specific proteolysis of the extracellular matrix is required to trigger cell death mimicking PCD of TEs.

Table I. *Cell death and DNA fragmentation induced by hydrolytic enzymes. Cells were treated with 1% w/v of each hydrolase 67 h after isolation and scored for the percentage of dead cells 6 h later. At least 200 cells were scored for each treatment. Data taken from Groover and Jones[9].*

Hydrolase	% Dead	% TUNEL
Control	25	2.7
Macerozyme	40	4.8
Pectinase	40	0.5
Cellulase	58	0
Proteinase K	48	0
Protease XIV	31	0
Protease XXIV	54	0
Protease XVII-B	21	0
Papain	61	3.2
Chymotrypsin	28	0
Trypsin	96	**21.9**

Trypsin initiated cell death via an influx of Ca^{2+}, which is consistent with the activation of the endogenous mechanism executing cell death. Trypsin-induced death and DNA fragmentation were inhibited by chelating extracellular Ca^{2+} with EGTA or by blocking Ca^{2+} channels with La^{3+} or ruthenium red. Trypsin-induced death was also inhibited by soybean trypsin inhibitor, indicating that cell death resulted from the proteolytic activity of trypsin, not from contaminating substances.

Selective inhibition of extracellular proteolysis specifically inhibited PCD. Cells at different points in development were treated with soybean trypsin inhibitor. When present between 24 and 70 h of culture, soybean trypsin inhibitor did not cause necrosis or inhibit cell division, indicating that the inhibitor had negligible toxicity in this system. In contrast, soybean trypsin inhibitor present between 48 and 96 h effectively inhibited TE differentiation and PCD in a dose-dependent fashion. The 21-kD soybean trypsin inhibitor would not be expected to cross the plasma membrane, suggesting that its inhibitory effects on TE cell death were exerted in the extracellular matrix.

A Protease Is Secreted Coincident with PCD

A secreted protease whose properties implicated it as an activator of cell death was identified with substrate-activity gels. As shown in Figure 1, several intracellular proteases were recognized in protein preparations from cells, as in previous reports[4,6], whereas the activity of a unique protease of approximately 40 kD (Fig. 1A) increased in the medium of cultures as PCD progressed. Although several strong protease activities were detected in intracellular protein samples, the 40-kD activity did not accumulate intracellularly, which is consistent with secretion. Leakage of intracellular proteases could be detected in the culture supernatants at later time points. However, leakage of protease from dying cells was not responsible for the 40-kD activity, because the abundant intracellular proteases showed little activity in the medium (Fig. 1A).

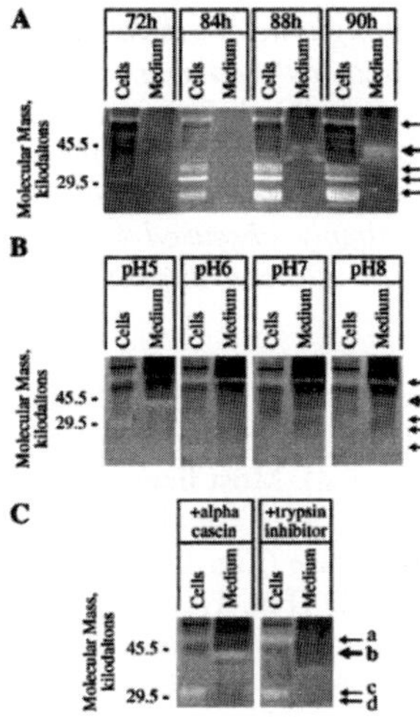

Figure 1. *Timing of expression and characteristics of proteases expressed by differentiating TEs. A, Intracellular proteins (Cells) and proteins concentrated from media (Medium) of the same culture at the indicated times after culture initiation were assayed on protease activity gels. After development, protease activities are recognized as clear bands resulting from hydrolysis of the gelatin substrate. At the time of harvest, the percentages of dead TEs were 0%, 20%, 49%, and 78% for the 72-, 84-, 88-, and 90-h cultures, respectively. Several intracellular protease activities can be seen (arrows a, c, d, and e), to previous findings of Beers and Freeman (1997) and Ye and Droste (1996). Protease activity is visible at approximately 40 kD (arrow b) in media after 84 h. The exact time during development that protease secretion commences cannot be determined directly from this technique, and accumulation of detectable protease activity in the medium may significantly lag behind the onset of secretion. Approximately 0.015 μg of medium protein and 0.5 μg of intracellular protein were loaded per sample. B, Aliquots of the same preparations of intracellular proteins (90-h culture) and medium proteins (88-h culture) were run on the same protease activity gel. After fractionation the gel was sliced into four pieces, and each piece was*

incubated in an activity buffer with the indicated pH overnight before development. The 40-kD activity in medium proteins (arrow b) is detected only at pH 5. C, of the same preparations of intracellular proteins (90-h culture) and medium proteins (88.5-h culture) were run on the same protease activity gel. The gel was divided in half, and one-half was immersed in ice-cold activity buffer containing 10 mg/mL soybean trypsin inhibitor (21 kD) and the other half was immersed in ice-cold activity buffer containing 10 mg/mL dephosphorylated -casein (23 kD) for 45 min to allow the proteins to diffuse into the gels. Gels were then incubated at room temperature overnight before development. -Casein has no protease inhibitory property, so it was used as a control for increasing background staining attributable to protein infusion into the gel. The soybean-trypsin-infused gel does not show the 40-kD activity in the medium, whereas the -casein-infused gel does show the 40-kD activity (arrow b), indicating that the activity was not simply obscured by the infused proteins, but was specifically inhibited by soybean trypsin inhibitor. Figure from Groover and Jones, 1999 [9].

The 40-kD protease was active at pH 5.0 but not at a more basic pH (Fig. 1B), which is consistent with the wall pH in planta and in vitro (the culture medium pH was 5.5 at the time of culture initiation). Most importantly, the 40-kD protease was inhibited by soybean trypsin inhibitor (Fig. 9C). The observations that (a) the 40-kD protease was the only detectable secreted protease (Fig. 1A); (b) the appearance of the 40-kD protease activity was coincident with PCD (Fig. 1A); and (c) soybean trypsin inhibitor inhibited both the endogenous TE PCD mechanism and the secreted protease (Fig. 1C) provide strong indirect evidence that the 40-kD protease triggers TE cell death.

We have addressed two fundamental questions concerning TE differentiation: How is the synthesis of the secondary cell wall coordinated with PCD? And how does the cell execute cell death? We found that a principal part of the mechanism executing cell death is a regulated influx of Ca^{2+}, probably through plasma membrane channels. Death is morphologically manifest by rapid collapse of the hydrolytic vacuole, mixing of the vacuole and the cytoplasm, and immediate cessation of cytoplasmic streaming. This mechanism does not simply terminate normal metabolism, but also creates an environment necessary for postmortem developmental events, including autolysis, to proceed. Vacuole collapse results from either a transition from the gradual Ca^{2+} influx shown to occur during secondary cell wall synthesis[7,8] to a rapid influx, or the activation of additional ion channels upon exceeding a threshold level of intracellular Ca^{2+}.

The coordination of secondary cell wall synthesis and PCD begins well in advance of the execution of cell death, with the approximately concurrent commencement of secondary cell wall synthesis and the production of hydrolytic enzymes. All of the inhibitors shown to block PCD also block secondary cell wall synthesis, suggesting that these developmental programs are not only concurrent, but also molecularly interdependent. However, we were able to implicate a protease as a key coordinating factor by exploiting PCD-specific markers that report cell death independently of cell wall synthesis. We designate this protease as the "trigger" protease for its role in triggering the collapse of the vacuole. The protease was secreted by cells coincident with PCD, and the protease and cell death were both inhibited by soybean trypsin inhibitor. Execution of cell death can be triggered prematurely by exogenous application of another protease, trypsin, which presumably mimics the action of the endogenous protease.

Mitochondrion change morphologically during TE PCD

The "mitochondria pathway" is regarded as a central component of some types of programmed cell death in animal cells where specific signals cause the release of cytochrome c from mitochondria into cytoplasm to trigger a calcium-initiated proteolytic cascade involving caspases. However, plant cells lack prototypical caspases, therefore a role for the mitochondria in PCD in plant cells is not obvious. Thus, under the conditions we have described for TE formation above, we have examined the mitochondria to determine if and when changes correlating to cell death occur. Figure 2 shows that just prior to execution of cellular autolysis initiated by the rupture of the large central vacuole to release sequestered hydrolases, mitochondria adopt a definable morphology distinct from mitochondria found in necrotic cells. The matrix condenses, fine ultrastructure is lost and the outer membrane looses integrity, but this morphology is clearly different than that observed during necrotic death (c.f. Fig 2B vs. 2C). We have also observed that the inner membrane voltage collapses at a time commensurate with the onset of cell death (data not shown). The mitochondrial membrane potential was monitored with the voltage-sensitive dye, JC-1, over time of transdifferentiation. Mitochondria from newly isolated cells fluoresced bright red indicating a negative membrane potential, however by 72 hours in culture when cell death is measurable, many mitochondria are fluorescing green. At a later time point, all mitochondrial are green and dead cells do not bind the dye. This indicates a progressive loss of mitochondrial membrane potential during the period when cells are dying by PCD.

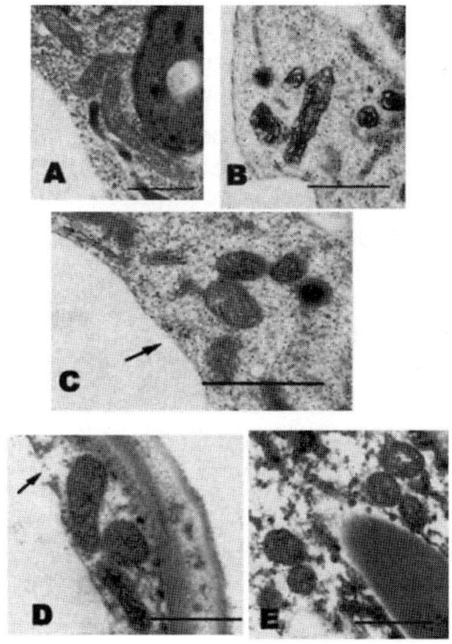

Figure 2. *Electron microscopy of mitochondria in non-differentiating cells and differentiating cells. A. Healthy 24 h cell, note that the mitochondria have a well organized ultrastructure. B a necrotic cell. Note the electron-dense mitochondrial matrix which clearly distinguishes these necrotic cells from non-necrotic mitochondria or mitochondria in late differentiating cells (c.f. C-E). C, Cell with a secondary cell wall but intact tonoplast (arrow), thus this cell has not yet triggered death. Note that some mitochondria exhibit unaltered, well organized ultrastructure , while some exhibit electron-dense mitochondrial matrix. D. Tonoplast shows signs of breakdown (arrow), increasing electron-dense mitochondrial matrix (F). In the late stage of TEs formation, the tonoplast was broken (arrow in G) and the cytoplasm shows signs of degradation. E. After autolysis begins the ultrastructure of the mitochondria is severely altered with the outer mitochondrial membrane broken. chl: chloroplast; SCW: secondary cell wall. Standard bar = is 1 μm*

Cytochrome c is released during differentiation but is not sufficient to induce death

The available antisera to cytochrome c does not recognize native cytochrome c in plants therefore we addressed the question of cytochrome c released using fractionation of microsomal and cytosolic compartments of cells and measuring the relative amount of cytochrome c in each compartment by immunoblot analysis. Figure 3 shows that the % of cytochrome c in the cytosol increases slightly during the time that cell death is occurring suggesting that cytochrome c release to the cytosol is triggering cell death in plant cells as it is proposed to do in animals.

To investigate this further, we induced a collapse in the membrane potential using betulinic acid, which induces a permeability transition pore in animal mitochondria, and cytochrome c release. We found that Bet A induces the release of cytochrome c and mitochondrial membrane depolarization. In animal cells, cyclosporin A blocks the effect of BetA and , in many cases, prevents PCD. Therefore we examined the effect of cyclosporin A on cytochrome c release, TUNEL, and death. While cyclosporin A was effective at blocking cell death induced by BetA, it had much less effect on cytochrome c release. These results suggest a role for the mitochondria in TE PCD but do not support the current animal paradigm for a causative role on PCD by cytochrome c release.

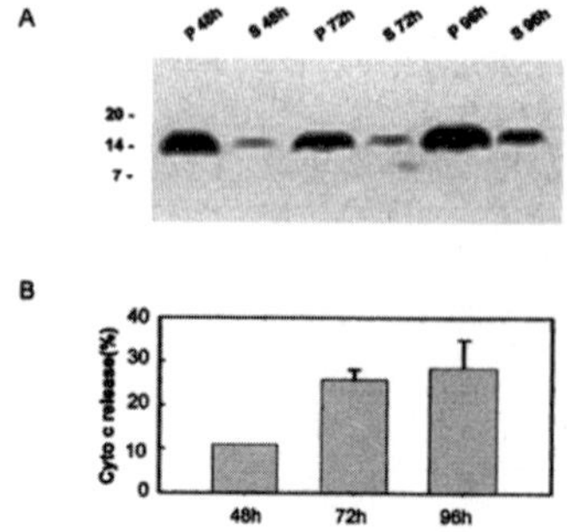

Figure. 3 *Immuno-blot of cytochrome c release during TEs formation 48, 72 and 96 h old cultures were collected, protein from cytosol and mitochondria were separated and analyzed by immunoblotting with a monoclonal cytochrome c antibody (A), IMAGEQUANT program quantified the signal and showed the ratio of cytochrome c release (B). Cytochrome c release is reported as cytochrome c in cytosol versus the total amount of cytochrome c in mitochondria and in cytosol. P: pellet protein represents for mitochondrial protein; S: supernatant protein represents for cytosol protein. Error bar represents twice independent experiments.*

A model for the trigger of death

The extracellular matrix is of fundamental importance for the PCD of at least some animal cell types, and can be a primary regulator of apoptosis[20-22]. Disruption of the extracellular matrix is involved in PCD during normal development in mammals [23-25] and during Xenopus laevis metamorphosis[26].

Extracellular proteases and their inhibitors have been shown to be vital components of fundamental developmental processes in animals. For example, a secreted Ser protease in Drosophila melanogaster encoded by the Easter gene proteolytically releases a ligand (derived from the product of the Spatzle gene) that activates the receptor encoded by Toll[27-28]. This pathway is responsible for establishing the dorsal-ventral asymmetry of the embryo. Our results indicate that secreted proteases may play important roles during plant development.

A simple model describes the coordination of cell death with secondary cell wall synthesis. The secretion of secondary cell wall precursors during differentiation is accompanied by secretion of the trigger protease, leading to increasing protease activity in the extracellular matrix as secondary cell wall synthesis proceeds. The secreted protease activates Ca^{2+} influx, and upon realization of a critical extracellular activity of protease or the arrival of signal substrate, cell death is executed via Ca^{2+} influx. The accumulation of protease in the extracellular matrix would thus act to measure the progression of secondary cell wall synthesis, and activates cell death only after a critical amount of secondary cell wall synthesis is achieved.

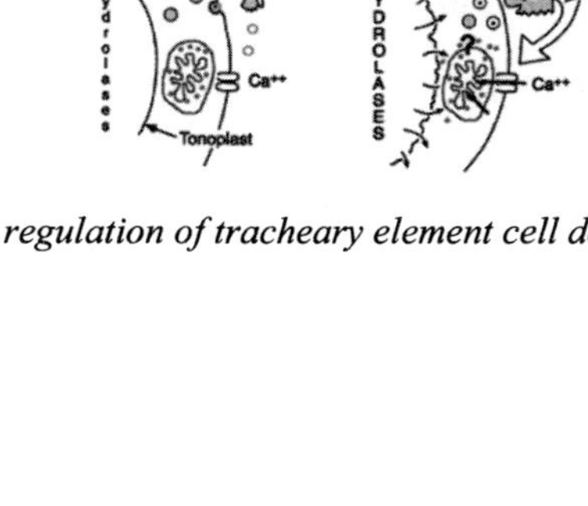

Figure 4. *A model of the regulation of tracheary element cell death*

What is the mechanism of activation? One possibility is that the trigger protease causes calcium mobilization. Since both plasma membrane (La^{3+}) and mitochondrial (RR) calcium channel inhibitors, blocked PCD, we speculate that mobilization from extracellular or matrix stores of calcium positively reinforces the each other. An increase in cytoplasmic calcium at low ATP levels is known to initiate PTP formation. Analogous to animal mitochondria, the formation of PTPs releases apoptotic factors. However, unlike in animals, cytochrome c is not a factor in TE PCD or it is insufficient to initiate a cascade of events that irreversibly lead to death. We do not exclude the release of other plant specific factors. It is also possible that PTP formation or simply the collapse of the mitochondrial membrane potential is sufficient to trigger the collapse of the vacuole. Mitochondrial dysfunction is expected to cause an increase in reactive oxygen species (ROS) and there is evidence that ROS is a central regulator of death in other PCDs.

The model incorporating the speculations above is shown in Figure 4. As discussed, when the level of proteolytic activity reaches a threshold, either by its activation, accumulation, or the loss of an inhibitor, calcium mobilization occurs to increase the cytosolic concentration. Presumably this change induces a collapse in the mitochondrial membrane potential and the release of an unknown apoptotic factor, possibly a ROS.

REFERENCES

1. A. Groover, N. DeWitt, A. Heidel, & A. Jones Programmed cell death of plant tracheary elements differentiating in vitro. *Protoplasma,* **1997,** 196, 197-211.

2. M. Thelen & D. Northcote Identification and purification of a nuclease from Zinnia elegans: a potential molecular marker for xylogenesis. *Planta,* **1989**, 179, 181-195.

3. Z. Ye & D. Droste Isolation and characterization of cDNAs encoding xylogenesis-associated and wounding-induced ribonucleases in Zinnia elegans. *Plant Mol Biol,* **1996** 30, 697-709.

4. Z. Ye & J. Varner Induction of cysteine and serine proteases during xylogenesis in Zinnia elegans. *Plant Mol Biol,* **1996** 30, 1233-1246.

5. A. Minami & H. Fukuda Transient and specific expression of a cysteine endopeptidase associated with autolysis during the differentiation of Zinnia mesophyll cells into tracheary elements. *Plant Cell Physiol,* **1995,** 36, 1599-1606.

6. E. Beers & T. Freeman Proteinase activity during tracheary element differentiation in Zinnia mesophyll cultures. *Plant Physiol.,* **1997**, 113, 873-880.

7. A. Roberts & C. Haigler Rise in chlorotetracycline fluorescence accompanies tracheary element differentiation in suspension cultures of Zinnia. *Protoplasma*, **1989**, 152, 37-45.

8. A. Roberts & C. Haigler Tracheary-element differentiation in suspension-cultured cells of Zinnia requires uptake of extracellular Ca2+. *Planta*, **1990**, 180, 502-509.

9. A. Groover, & A. JonesProgrammed cell death in transdifferentiating tracheary elements is triggered by a secreted protease. *Plant Physiology,* **1999** 119,375-384.

10. R. Mittler & E. Lam Identification, characterization, and purification of a tobacco endonuclease activity induced upon hypersensitive response cell death. *Plant Cell* **1995**, 7, 1951-1962.

11. R. Mittler & E. Lam In situ detection of nDNA fragmentation during the differentiation of tracheary elements in higher plants. *Plant Physiol.*, **1995**, 108, 489-493.

12. C. He, P. Morgan & M. Drew Transduction of an ethylene signal is required for cell death and lysis in the root cortex of maize during aerenchyma formation induced by hypoxia. *Plant Physiol.*, **1996** 112, 463-472.

13. M. Wang, B. Oppedijk, X. Lu, B. Van Duijn & R. Schilperoort Apoptosis in barley aleurone during germination and its inhibition by abscisic acid. Plant Mol Biol., **1996**, 32, 1125-1134.

14. G. Dalessandro & L. Roberts Induction of xylogenesis in pith parenchyma explants of Lactuca. *Am J Bot.*, **1971**, 58, 378-385.

15. R. Yamamoto, S. Fujioka, T. Demura, S. Takatsuto, S. Yoshida, H. Fukuda Brassinosteroid levels increase drastically prior to mrphogenesis of tracheary elements. 2001, *Plant Physiology,* 2001, **125**, 556-563.

16. D. Nicholson & N. Thornberry Caspases: killer proteases. *Trends Biochem Sci.,* **1997**, 22, 299-306.

17. A. Levine, R. Pennell, M. Alvarez, R. Palmer, C. Lamb Calcium-mediated apoptosis in a plant hypersensitive disease resistance response. *Curr Biol.*, **1996** 6, 427-437.

18. A. Uren, K. Orourke, L. Aravind, M. Pisabarro, S. Seshagiri, E. Koonin, V. Dixit Identification of paracaspases and metacaspases: two ancient families of caspase-like proteins, one of which plays a key role in MALT lyphoma. *Molecular Cell* **2000**, 6, 961-967.

19. H. Fukuda H & A. Komamine Establishment of an experimental system for the study of tracheary element differentiation from single cells isolated from the mesophyll of Zinnia elegans. *Plant Physiol* **1980**, 65, 57-60.

20. J. Meredith, B. Fazeli, M. Schwartz The extracellular matrix as a cell survival factor. *Mol Biol Cell*, **1993**, 4, 953-961.

21. S. Frisch & H. Francis Disruption of epithelial cell-matrix interactions induce apoptosis. *J Cell Biol.,* 1994, 124, 619-626.

22. E. Ruoslahti & J. Reed Anchorage dependence, integrins, and apoptosis. *Cell* **1994** 77, 477-478.

23. R. Talhouk, M. Bissel, & Z. Werb Coordinated expression of extracellular matrix-degrading proteinases and their inhibitors regulates mammary epithelial function during involution. *J Cell Biol.*, **1992** 118, 1271-1282.

24. N. Bourdreau, C. Sympson, Z. Werb, & M. Bissell Suppression of ICE in mammary epithelial cells by extracellular matrix. *Science* **1995**, 267, 891-893.

25. E. Coucouvanis & G. Martin Signals for death and survival: a two-step mechanism for cavitation in the vertebrate embryo. *Cell,* **1995**, 83, 279-287.

26. D. Patterson, W. Hayes, & Y. Shi Transcriptional activation of the matrix metalloproteinase gene stromelysin-3 coincides with thyroid hormone-induced cell death during frog metamorphosis. *Dev Biol.*, **1995**, 167: 252-262.

27. D. Morisato & K. Anderson Signaling pathways that establish the dorsal-ventral pattern of Drosophila melanogaster. *Annu Rev Genet.*, **1995**, 29, 371-399.

28. S. Misra, P. Hecht, R. Maeda, & K. Anderson Positive and negative regulation of Easter, a member of the serine protease family that controls dorsal-ventral patterning in the Drosophila embryo. *Development*, **1998**, 125, 1261-1267.

Molecular Breeding of Woody Plants
N. Morohoshi and A. Komamine, editors.

ARABIDOPSIS AS A MODEL FOR INVESTIGATING GENE ACTIVITY AND FUNCTION IN VASCULAR TISSUES

Eric P. Beers* and Chengsong Zhao

Department of Horticulture,
Virginia Polytechnic Institute and State University,
Blacksburg, Virginia 24061, USA

ABSTRACT

Using the zinnia mesophyll cell system for tracheary element differentiation, members of at least three families of mechanistically distinct (serine, cysteine and threonine) proteases have been implicated in the regulation of tracheary element differentiation. We are currently using Arabidopsis to facilitate genetic analysis of the roles of proteolytic enzymes during vascular tissue differentiation. Although not typically considered as a model for secondary growth, Arabidopsis forms a true cambium and produces a relatively large amount of secondary xylem and phloem within the root and hypocotyl. This potential for secondary growth is best realized under conditions that delay senescence, i.e., inflorescence removal and growth at low population density. After growing Arabidopsis under these conditions for at least eight weeks it is possible to isolate biochemical quantities of xylem and phloem for analysis of tissue-specific protease activities. After dissecting approximately 300 root-hypocotyl segments into xylem and bark fractions we isolated RNA for the construction of the first xylem and bark cDNA libraries from Arabidopsis. Using gene-specific primers and degenerate primers we screened the libraries by PCR and amplified fragments from seven protease genes including members of serine, cysteine, and aspartic acid protease families. Expression for three of these genes (*XCP1*, *XCP2*, and *XSP1*) is xylem-specific. *XCP1* and *XCP2* are predicted to encode papain-like cysteine proteases and *XSP1* is predicted to encode a subtilisin-like serine protease. To identify additional genes with potential importance to vascular tissue differentiation and physiology, we analyzed 1,000 (500 from xylem and 500 from bark) ESTs.

KEYWORDS

Arabidopsis, protease, peptidase, papain, subtilisin, xylem, phloem, cambium, cell wall, tracheary element, zinnia, programmed cell death

INTRODUCTION

Plant vascular tissues represent exceptional opportunities for novel discoveries of great fundamental significance. For example, despite the highly specialized and distinctive nature of xylem and phloem, both tissues are derived from a common precursor, the procambium (for primary tissues) or the cambium (for secondary tissues), indicating that the cambium is a rich resource for the discovery of novel cell fate regulators. Gene expression in the cambium is not the only subject of interest, however. The tracheary element (TE) secondary cell wall modification and death programs have also attracted a high level of attention over the past decade. Tracheary element research has benefited enormously from the availability of the zinnia mesophyll cell culture system for TE differentiation [1,2]. The principal advantages of the zinnia system are the semi-synchronous transdifferentiation of a high percentage (up to 70%) of live mesophyll cells and the ability to uniformly expose cultured cells to signaling molecules, growth regulators and inhibitors.

We are interested in defining the roles played by proteolytic enzymes during the differentiation of TEs where proteases may act as inducers, transducers or effectors of processes essential to differentiation and programmed cell death (pcd). We previously provided a partial characterization of proteases isolated from differentiating zinnia TEs [3], reporting that activities of both serine and cysteine proteases increased during TE differentiation, consistent with data from others studying zinnia TEs [4,5]. It was further demonstrated that in the presence of SDS, proteases present in TEs were capable of complete lysis of protein extracted from cultured TEs [3]. Similar proteolytic activity was not detected in extracts from non-induced control cultures. It was also shown, by single-cell assay, that increased protease activity was associated specifically with TEs and not with other cell types present in the same culture [3]. Activity of the cysteine proteases present in zinnia TEs can be blocked by carbobenzoxy-leucinyl-leucinyl-leucinal (MG132), a peptide-aldehyde inhibitor of calpain. In the presence of MG132, 85% of TEs (versus 15% for solvent control TEs) were unable to complete the autolytic phase of pcd by 96 hours in culture [6]. The proteasome has also been implicated as a regulator of TE differentiation, as differentiation, but not death, of zinnia TEs can be blocked by the proteasome-specific inhibitor lactacystin [7] (as the *claso*-lactacystin β-lactone) [6].

Despite its obvious utility as a model for the biochemistry and molecular cell biology of TE differentiation, zinnia has not emerged as a tractable genetic model. Consequently, the ability to address zinnia gene function through reverse genetic approaches has been limited to experiments in heterologous systems [8]. Moreover, while the zinnia system is a powerful tool for investigating TE gene expression, vascular tissue research would benefit from the development of models that allow for profiling of gene expression in all vascular cell types. To some extent such models have been provided, as work with economically important tree species loblolly pine [9] and poplar [10] has yielded EST databases for these species. However, as with the zinnia system, trees present some limitations as models. None of these models for vascular tissue studies possesses the combined attributes of the genetic model Arabidopsis, i.e., a fully sequenced and annotated genome, short life span, small size, high fecundity, ease of stable genetic transformation and support from a wealth of resources that facilitate research. Yet, while Arabidopsis develops a true cambium and produces secondary xylem and phloem [11-15], as recently as 1998 it was believed that it was not possible to isolate vascular tissues from Arabidopsis [10].

Realizing the benefits to be derived from developing Arabidopsis as a model for vascular tissue research, we succeeded in demonstrating that xylem and phloem can be isolated from Arabidopsis. We subsequently produced the first xylem and bark cDNA libraries from Arabidopsis--the first libraries from these tissues for any monocarpic plant--and we have made rapid progress in the identification and partial characterization of xylem-specific peptidases [15]. In addition, we have produced a modest EST database from these libraries. In this article we describe the method for isolation of xylem, phloem and bark from Arabidopsis and summarize our results concerning vascular tissue proteases and the discovery of other genes expressed in vascular tissues of Arabidopsis.

MATERIALS AND METHODS

Plant growth and isolation of xylem, phloem/cambium and bark

Arabidopsis thaliana ecotype Columbia was grown in Sunsine Mix 1 (Wetsel Seed Co., Harrisonburg, VA) under continuous light, at a planting density of 4 to 6 plants per 10-cm pot. It is important to firmly tamp the potting medium before sowing seed. Plants were watered with nutrient solution. Under these conditions inflorescences were typically visible 3 weeks after germination. For the next 5 weeks, inflorescences were routinely removed as they emerged. Eight-week-old plants were harvested and potting medium was washed from the roots with a strong stream of cool tap water. Approximately 1 cm of root-hypocotyl was excised from just below the cotyledons and lateral roots were trimmed from the primary root with a razor blade. Prior to dissection, root-hypocotyl segments were washed with dd H_2O, blotted dry and placed on ice. Separation of root-hypocotyl segments into xylem, phloem and nonvascular fractions or xylem and bark fractions was done under the dissecting microscope. Briefly, using a double-edged razor blade, a longitudinal cut was made along the entire length of the root-hypocotyl segment passing through the nonvascular tissue and secondary phloem but not into the xylem. Using a dissecting probe and forceps, the nonvascular tissue was peeled from the phloem and placed in liquid N_2. Xylem and phloem were then separated and placed in liquid N_2. It is assumed that the cambium remains associated with the phloem or bark. When xylem and bark samples were isolated, the nonvascular tissue and phloem were separated from the xylem as a unit. It is possible to isolate approximately 1 μg poly$(A)^+$ RNA from 60 xylem or bark segments prepared as described in this report. Additional information concerning methods for cDNA library construction and screening, RNA gel blots and quantitative RT-PCR can be found in Zhao et al. [15].

RESULTS AND DISCUSSION

Vascular proteases

Secondary growth is evident in the root-hypocotyl of 8-day-old Arabidopsis seedlings [11]. By 14 days, the procambium-derived vascular cambium is producing secondary xylem internally and secondary phloem externally and the pericycle-derived cambium is also active. By 6 to 8 weeks of age, the cambium in the root-hypocotyl of Arabidopsis, grown as described here and originally by Lev-Yadun [14], is a continuous lateral meristem producing secondary xylem internally and secondary phloem externally [14, 15].

From cDNA libraries constructed from xylem and bark isolated from the root-hypocotyl of 8-week-old plants we cloned two full-length cDNAs predicted to code for two closely related papain-like cysteine endopeptidases (*XCP1* and *XCP2*) and one full-length cDNA predicted to code for a subtilisin-like serine endopeptidase (*XSP1*) [15]. An additional papain-like enzyme possessing a granulin-like C-terminal extension, *XBCP3*, was also cloned. Using *XSP1* as a marker for TE differentiation, competitive RT-PCR was conducted using RNA from 2-, 4-, 6- and 8-week-old Arabidopsis roots. The

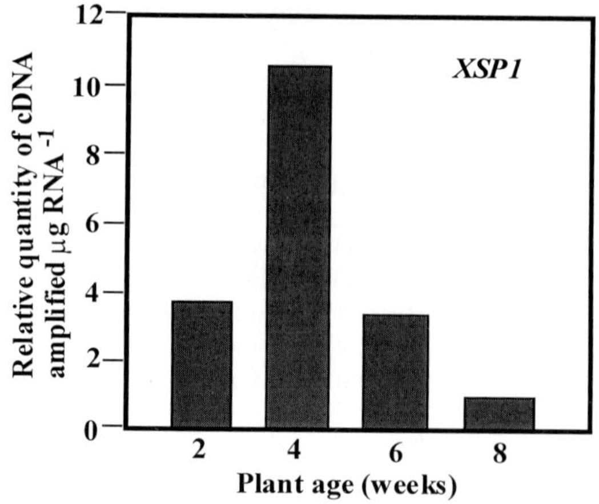

Figure 1. Quantitative RT-PCR for *XSP1* expression in roots from 2-, 4-, 6- and 8-week-old Arabidopsis. Levels of cDNA, relative to that for week-8 set at one unit, obtained from RNA isolated at the weeks indicated are shown. Quantitative RT-PCR was performed as described in Zhao et al. [15].

results shown in Figure 1 indicate that the highest level of gene expression associated with TE differentiation occurs in 4-week-old roots and is nearly 11-fold greater than that observed for 8-week-old roots. These results indicate that 4-week-old roots may be better subjects for evaluation of TE-associated gene expression than the 8-week-old organs used to construct xylem and bark cDNA libraries [15].

Quantitative RT-PCR for various tissues and organs indicates that the expression levels for *XCP2* are 10 to 20-fold higher than those observed for *XCP1* [15]. This is consistent with the observation that *XCP2* promoter-*GUS* plants show GUS activity that is predictive (i.e., detectable prior to visible thickening of secondary cell walls of TEs) of tertiary vein positioning, while *XCP1* promoter-*GUS* plants show activity only in late stage TEs (Table 1). In addition to TEs, both *XCP1* and *XCP2* promoter-*GUS* plants show GUS activity at the base of trichomes on young expanding leaves. Immunofluorescence confocal microscopy indicates that XCP1 localizes to TEs (E. Beers, unpublished observation), consistent with the localization of GUS activity for *XCP1* promoter-*GUS* plants.

The papain-like cysteine peptidases described here (XCP1 and XCP2) are typical three-domain zymogens (recently reviewed by Beers et al. [16]), that exhibit 70% identity at the amino acid level. XCP1 is currently the only papain-like enzyme from among the 28 predicted papain-like enzymes encoded by the Arabidopsis genome for which there is experimental evidence for proteolytic activity. Under acid (pH 5.5) conditions, inactive polyhistidine-tagged proXCP1 is apparently autocatalytically processed to yield the active mature form of XCP1 [15]. When expressed ectopically in transgenic Arabidopsis, XCP1 is detectable by immunoblot as a 29 kD polypeptide that comigrates with proteolytic activity not detected in control plants (E. Beers, unpublished observation). Independent *35S::XCP1* transformants exhibit phenotypes ranging from severely stunted plants to those without obvious abnormalities. Some stunted plants

Table 1. Summary of GUS activity specified by putative promoters for the indicated peptidases isolated from Arabidopsis xylem and bark cDNA libraries. H, hydathodes; T, trichomes; PTE, protoxylem tracheary elements; MTE, metaxylem tracheary elements, STE, secondary xylem tracheary elements; C/P, cambium/phloem; XP, xylem parenchyma.

Promoter-*GUS* fusion	Cell or tissue type						
	H	T	PTE	MTE	STE	C/P	XP
XCP1	-	+	+	+	+	-	-
XCP2	-	+	+	+	+	-	-
XSP1	-	-	+	+	+	-	-
XBCP3	+	-	-	-	-	+	+

produce curled leaves or leaves that senesce prematurely. High XCP1 levels correlate with phenotype severity. XCP1 has been localized to isolated vacuoles purified from protoplasts prepared from *35S::XCP1* Arabidopsis (E. Beers, unpublished observation).

Xylem and bark ESTs

To generate a partial profile of gene expression for xylem and bark from Arabidopsis we have produced a modest EST database consisting of 390 high quality sequences out of 500 randomly selected clones from xylem and 397 high quality sequences from 500 randomly selected clones from bark. Sequences were analyzed if they contained no ambiguities for at least 125 nucleotides. Selected ESTs coding for enzymes involved in cell wall modification are shown in Table 2. The distribution of these ESTs between the bark and xylem libraries is consistent with frequencies reported for homologous genes from poplar cambium and developing xylem cDNA libraries [10]. These data support the conclusion that xylem and bark libraries prepared from Arabidopsis are valuable resources for investigating gene expression with relevance to wood formation.

The development of Arabidopsis as a model for vascular tissue research also provides new opportunities for the identification and functional analysis of novel genes expressed in cambium, xylem and phloem cells. As Arabidopsis xylem and bark libraries are vastly enriched for vascular tissue-specific transcripts relative to all other existing Arabidopsis cDNA libraries, it is reasonable to investigate unique ESTs from these libraries as potential genes with vascular tissue-specific roles. Similarly, xylem or bark ESTs with only one corresponding EST in the AtEST database may also be worth investigating. For example, within the AtEST data set, the TE peptidases *XCP1* has no corresponding EST and the TE peptidase *XSP1* has only one corresponding EST. Both of these peptidases were previously uncharacterized and were cloned from the Arabidopsis xylem cDNA library. This does not mean, of course, that genes with two or more ESTs cannot exhibit vascular expression patterns. Numerous ESTs represent the peptidase XCP2 and yet *XCP2* promoter-*GUS* data indicate TE expression for this gene. Tables 3 lists predicted functions for selected Arabidopsis xylem and bark ESTs with no identical AtESTs reported.

Table 2. Selected ESTs from xylem and bark predicted to code for enzymes involved in cell wall formation. Approximately 1,000 clones (500 from xylem, 500 from bark) were selected at random, amplified by PCR using T3 and T7 vector primers and sequenced from the 5' end. A single copy was found for each cDNA listed, except for S-adenosylmethionine synthetase, which is present as two copies from the same gene in the current xylem EST data set. The number in parentheses indicates that two genes predicted to code for cinnamyl alcohol dehydrogenase are represented in the xylem EST data set.

Library	Predicted identity
xylem	Blue copper protein
xylem	S-adenosylmethionine:
	2-demethylmenaquinone methyltransferase
xylem	S-adenosylmethionine synthetase
xylem	Cinnamate-4-hydroxylase
xylem	Cinnamoyl-CoA reductase
xylem	Cinnamyl alcohol dehydrogenase (2)
xylem	Endo-1,3-1,4-β-D-glucanase
xylem	O-methyltransferase
xylem	Phenylalanine ammonia lyase
xylem	Polygalacturonase
bark	β-galactosidase
bark	β-D-glucan exohydrolase
bark	β-glucosidase
bark	Pectate lyase

Among those ESTs shown in Table 3 are cDNAs that are predicted to code for proteins with potential roles in signal transduction, gene activation, cell wall modification and disease resistance. As further confirmation of the value of the Arabidopsis xylem and bark libraries for discovery of genes important to vascular cell fate, within 500 clones selected at random we identified a cDNA that codes for Athb-8 (Table 3). Baima et al. [17] used degenerate PCR primers to clone *Athb-8*, and other homeobox genes, and demonstrated that *Athb-8* is expressed in procambial cells. *Athb-8* apparently has not been sequenced for an EST project prior to the Arabidopsis xylem data set reported here. Xylem and bark ESTs with only one identical AtEST (Table 4) also represent an interesting set of cDNAs that includes proteins likely to be involved in signal transduction, regulation of gene expression and secretory pathway trafficking.

Figure 2 illustrates that 34% of xylem ESTs are predicted to code for proteins of unknown function. This percentage is very close to the 37% noted for unknowns from developing xylem ESTs from poplar [10]. Arabidopsis bark EST percentages for the classifications shown in Figure 2 are nearly identical to those for xylem (data not

Table 3. Xylem and bark ESTs with no identical AtESTs reported. Shown are the predicted identities for genes corresponding to xylem or bark ESTs for which no identical EST was present among the AtEST data set using BLAST [18] via TAIR, http://www.arabidopsis.org, with our EST as a query.

Library	Predicted identity
xylem	AMP binding protein
xylem	Disease resistance protein, RPP8-like
xylem	FK506 binding protein
xylem	Homeobox gene *Athb-8*, expressed in procambial cells [17]
xylem	MAP kinase
xylem	Microtubule-associated EB1-like protein
xylem	Nucleotide repair protein
xylem	PhyA signal transduction 1 protein (GRAS regulatory protein family) [19]
xylem	Protein kinase C
xylem	RING zinc finger protein
xylem	RNA-binding protein
xylem	Serine/threonine-specific protein kinase
bark	Glucosyl transferase
bark	GTP-binding protein
bark	Phospholipase D
bark	Polygalacturonase
bark	PRM1 homolog

A	B	C	D	E	F	G	H	I	J
7	15	7	5.5	5.5	34	4	4	13	6

Figure 2. Classification of 397 ESTs from the xylem cDNA library. Protein identity predictions were determined as for Table 3. **A**, signal transduction/hormone early response genes; **B**, protein synthesis (ribosomal proteins, RNA-binding, elongation factors, chaperones); **C**, protein/lipid/nucleic acid degradation; **D**, stress response (oxidative, pathogen, salt); **E**, cytoskeletal/cell wall; **F**, unknown; **G**, DNA-binding; **H**, protein kinase/phosphatase; **I**, metabolism/photosynthesis; **J**, ion or sugar transport and protein trafficking. Numbers indicate the percentage of total xylem ESTs within each class.

Table 4. Xylem and bark ESTs with a single identical AtEST reported. ESTs were obtained and analyzed as described for Table 3.

Library	Predicted identity
xylem	ABC transporter
xylem	Jasmonic acid regulatory protein (transcriptional activator)
xylem	Receptor-protein kinase-like
xylem	Signal recognition particle, 68kD protein-like
bark	AtFP3 gene (isoprenylated metal-binding protein)
bark	Protein phosphatase-2c
bark	Transport protein particle component Bet3p-like protein
bark	WD-40 repeat protein

shown). Twenty-three percent of the total xylem and bark ESTs coding for unknown Arabidopsis proteins have no corresponding ESTs among the AtEST dataset. When Arabidopsis xylem and bark ESTs coding for unknown proteins were compared with poplar xylem and cambium ESTs, 27 ESTs (16% of combined Arabidopsis xylem and bark ESTs) were found to share identity with poplar sequences resulting in BLAST [17] scores greater than 100.

CONCLUSIONS

By exploiting the potential of Arabidopsis for secondary growth it is possible to produce cDNA libraries from xylem and bark. These libraries represent valuable resources that enable the rapid isolation of cDNAs from genes with expression limited to vascular tissue cell types. Novel vascular tissue promoters are easily identified from these cDNAs via the annotated Arabidopsis genome. With this development of Arabidopsis as a tool for the discovery of genes that function in vascular tissues, it is reasonable to predict that Arabidopsis will soon become an important source for genes and promoters useful in the modification of wood in economically important tree species. In addition, the utility of Arabidopsis as a genetic model combined with its ability to produce woody tissue argues for its increased use as a model for rapid testing of strategies aimed at introducing quantitative and qualitative changes to secondary vascular tissues in many other species.

ACKNOWLEDGEMENTS

Support for this research was provided by the United States Department of Agriculture-National Research Initiative Competitive Grants Program (9801401), the National Science Foundation (MCB-9418377) and the College of Agriculture and Life Sciences, Virginia Polytechnic Institute and State University. We thank Drs. Bonnie Woffenden and Greg Welbaum for critical reading of the manuscript.

REFERENCES

1. H. Fukuda & A. Komamine, Establishment of an experimental system for the study of tracheary element differentiation from single cells isolated from the mesophyll of *Zinnia elegans*, *Plant Physiol.*, **1980**, 65, 57-60.
2. H.W. Kohlenbach & B. Schmidt, Cytodifferenzierung in form einer direkten umwandlung isolierter mesophyllzellen zu tracheiden, *Z. Pflanzenphysiol.*, **1975**, 75, 369-374.
3. E.P. Beers & T.B. Freeman, Proteinase activity during tracheary element differentiation in *Zinnia* mesophyll cultures, *Plant Physiol.,* **1997**, 113, 873-880.
4. A. Minami & H. Fukuda, Transient and specific expression of a cysteine endoproteinase associated with autolysis during differentiation of *Zinnia* mesophyll cells into tracheary elements, *Plant Cell Physiol.*, **1995**, 36, 1599-1606.
5. Z.-H. Ye & J.E. Varner, Induction of cysteine and serine proteinases during xylogenesis in *Zinnia elegans*, *Plant Mol. Biol.*, **1996**, 30, 1233-1246.
6. B.J. Woffenden, T.B. Freeman & E.P. Beers, Proteasome inhibitors prevent tracheary element differentiation in Zinnia mesophyll cell cultures, *Plant Physiol.*, **1998**, 118, 419-430.
7. S. Omura, T. Fujimoto, K. Otoguro, K. Matsuzaki, R. Moriguchi, H. Tanaka, & Y. Sasaki, Lactacystin, a novel microbial metabolite, induces neuritogenesis of neuroblastoma cells, *J. Antibiot.*, **1991**, 44, 113-116.
8. M. Igarashi, T. Demura, & H. Fukuda, Expression of the Zinnia TED3 promoter in developing tracheary elements of transgenic Arabidopsis, *Plant Mol. Biol.*, **1998**, 36, 917-927.
9. I. Allona, M. Quinn, E. Shoop, K. Swope, S.S. Cyr, J. Carlis, J. Riedl, E. Retzel, M.M. Campbell, R. Sederoff & R.W. Whetten, Analysis of xylem formation in pine by cDNA sequencing, *Proc. Natl. Acad. Sci. USA*, **1998**, 95, 9693-9698.
10. F. Sterky, S. Regan, J. Karlsson, M. Hertzberg, A. Rohde, A., Holmberg, B. Amini, R. Bhalerao, M. Larsson, R. Villarroel, M. Van Montagu, G. Sandberg, O. Olsson, T.T. Teeri, W. Boerjan, P. Gustafsson, M. Uhlen, B. Sundberg, & J. Lundeberg, Gene discovery in the wood-forming tissue of poplar: Analysis of 5,692 expressed sequence tags, *Proc. Natl. Acad. Sci. USA*, **1998**, 95, 13330-13335.
11. J.S. Busse & R.F. Evert, Pattern of differentiation of the first vascular elements in the embryo and seedling of *Arabidopsis thaliana, Int. J. Plant Sci.*, **1999**, 160, 1-13.
12. L. Dolan & K. Roberts, Secondary thickening in roots of *Arabidopsis thaliana*: anatomy and cell surfaces, *New Phytol.,* **1995**, 131, 121-128.
13. E.A. Kondratieva-Melville & L.E. Vodolazsky, Morphological and anatomical structure of *Arabidopsis thaliana* (Brassicaceae) in ontogenesis, *Bot. J.*, **1982**, 67, 1060-1069.
14. S. Lev-Yadun, Induction of schlereid differentiation in the pith of *Arabidopsis thaliana* (L.) Heynh, *J. Exp. Bot.*, **1994**, 45, 1845-1849.
15. C. Zhao, B.J. Johnson, B. Kositsup & E.P. Beers, Exploiting secondary growth in Arabidopsis. Construction of xylem and bark cDNA libraries and cloning of three xylem endopeptidases, *Plant Physiol.*, **2000**, 123, 1185-1196.
16. E.P. Beers, B.J. Woffenden & C. Zhao, Plant proteolytic enzymes: possible roles during programmed cell death, *Plant Mol. Biol.,* **2000**, 44, 399-415.

17. S. Baima, F. Nobili, G. Sessa, S. Lucchetti, I. Ruberti & G. Morelli, The expression of the *Athb-8* homeobox gene is restricted to provascular cells in *Arabidopsis thaliana*, *Development*, **1995**, 121, 4171-4182
18. S.F. Altschul, W. Gish, W. Miller, E.W. Myers, & D.J. Lipman, Basic local alignment search tool, *J. Mol. Biol.*, **1990**, 215, 403-10.
19. C. Bolle, C. Koncz & N.H. Chua, PAT1, a new member of the GRAS family, is involved in phytochrome A signal transduction, Genes Dev., **2000**, 14, 1269-78.

Molecular Breeding of Woody Plants
N. Morohoshi and A. Komamine, editors.

MOLECULAR MECHANISMS OF VASCULAR PATTERN FORMATION

Hiroo Fukuda[1,3], Koji Koizumi[2], Kenji Motomatsu[1], Hiroyasu Motose[1], Munetaka Sugiyama[2]

[1]*Department of Biological Sciences, Graduate School of Science, University of Tokyo, Tokyo 113-0033, Japan*

[2]*Botanical Gardens, Graduate School of Science, University of Tokyo, Tokyo 112-0001, Japan*

[3]*Plant Morphology Group, Plant Science Center, RIKEN, Wako 351-0198, Japan*

ABSTRACT

Vascular plants have developed a complex network of vascular systems through the plant body, allowing efficient transport of water, nutrients and signals. To understand molecular mechanisms of vascular pattern formation, we have made two approaches. First we have isolated *Arabidopsis* mutants with defects in vascular pattern formation. Microscopic and genetic examination of the cotyledonary venation of 3400 M3 lines led to the identification of 8 mutant lines whose abnormality was caused by mutations in 7 genetic loci designated *VAN1-VAN7*. Morphological analysis of *van* mutants indicated that *van1-van6* mutations caused fragmentation of lateral veins in cotyledons and of tertiary veins in rosette leaves, but did less injurious effects on the formation of their main veins or of vasculatures in hypocotyls and roots. *van* mutants were further characterized using *pAthb8::GUS* and *pTED3::GUS* as molecular markers of provascular cells and tracheary element precursor cells, respectively. As a result, it was revealed that most of *van* mutants lacked provascular cells at the disconnection points of the vascular network even at walking stick stage of embryogenesis. These results suggest that *VAN* genes are involved in the spatial control of provascular tissue differentiation, which realizes a continuous network of the vascular system. Second, we have analyzed regulation of cell-cell communication that may be involved in continuous formation of the vascular system using *Zinnia* cell culture. For this purpose, we developed two culture methods, thin-sheet culture and microbead culture. These culture methods indicated the presence of a high-molecular weight proteinaceous substance that promotes tracheary element differentiation. An improved microbead culture method brought about the partial purification of the substance, revealing that it is an arabinogalactan protein. Based on these results, we will discuss molecular mechanism of vascular pattern formation.

KEYWORDS
Arabidopsis, Arabinogalactan protein, Tracheary element, Vascular pattern, *Zinnia*

INTRODUCTION

The vascular tissues of plants, which are composed of specialized conducting tissues, xylem and phloem, form continuous systems through the plant body and provide transport pathways for water, nutrients, and signaling molecules and support a plant body against mechanical stresses. These functions of the vascular system are realized through fine regulation of the timing and position of vascular differentiation to form continuous files of each kind of vascular cells. However, the molecular mechanisms controlling vascular differentiation remain to be elucidated.

To identify machineries responsible for spatial regulation of vascular tissue formation, genetic analysis using *Arabidopsis*, which has played an increasingly important role in recent studies of plant development, is a promising approach. Several kinds of mutants showing aberrant vein pattern formation have been reported in *Arabidopsis* to date (*monopteros*[1]; *gnom*[2]; *fackel*, *fass*[3,4]). These mutants were identified originally as being impaired in body organization of the seedling. Such morphological abnormalities were traced back to altered embryogenesis. Thus, it is possible that the aberrant venation in these mutants may not be due to the primary effects of mutations but the secondary effects caused by a disorganized body plan. Therefore, it is important to find out mutations that primarily and directly affect the vascular network formation and to genetically identify the components of the regulatory system. To search for such mutations, cotyledonary venation provides a good selection trait because it is determined at the very early stage of plant development. The cotyledonary venation also offers the simplest pattern of venation that is advantageous in detecting genetic defects in the vascular patterning. Thus, we screened EMS-mutagenized populations of *Arabidopsis* for mutants that exhibit some abnormalities in cotyledonary venation.

Continuous formation of vascular strands suggested a possible involvement of local intercellular communication guiding neighboring cells into the same fate in vascular formation, which had never been characterized. Xylem differentiation can be induced *in vitro* from various parenchymatous cells. Such xylogenesis *in vitro* provides a unique opportunity for direct detection of local intercellular communication and isolation of local communicators, which are difficult or substantially impossible *in planta*. Among various *in-vitro* xylogenic cultures, *Zinnia* xylogenic culture, in which about half of the isolated mesophyll cells transdifferentiate into tracheary elements (TEs) in a synthetic medium supplemented with adequate concentrations of auxin and cytokinin[5], is most advantageous to the study of local intercellular communication controlling xylogenesis. In this culture system, intercellular relationships can be manipulated experimentally, and positional information pre-existing in leaves is canceled by isolation of mesophyll cells and dispersion of isolated cells into the culture medium. Using these advantages of the *Zinnia* system, we analyzed the cell-cell interaction in xylogenesis. Newly developed culture methods, thin-sheet culture and microbead culture, were successfully applied to provide the first evidence for involvement of local intercellular communication in xylogenesis. Furthermore, the results demonstrated that a proteinaceous macromolecule of larger than 25 kDa in molecular weight mediates such local intercellular communication. This factor was named as "xylogen" with reference to its xylogenic activity.

MATERIALS & METHODS

Isolation of mutants

Ler seeds of *Arabidopsis thaliana* were mutagenized by treatment with 0.3% EMS solution for 16-20 hours at room temperature. After two cycles of self-fertilization, M3 lines were constructed, so that each M3 line consists of seeds harvested from only a single M2 plant. Screening was carried out with a portion of seeds from each M3 line. The resultant seedlings were decolored in 99% ethanol and stained with 1% phloroglucinol in 20% hydrochloric acid. The vessels of the cotyledons were examined under a light microscope.

Linkage analysis

Ler plants heterozygous for one of *van* mutations were crossed with wild-type Col plants. Seeds were collected from self-pollinated F1 plants heterozygous for the *van* mutation, yielding a polymorphic F2 population. The genotype of the *VAN* locus and SSLP loci or CAPS loci were determined for each F2 plant using its F3 progeny. For SSLP and CAPS analysis DNA was isolated from each F3 plant with a IsoPlant Kit. PCR was performed with a thermal cycler by repeating 40 times the following cycle: heat-denaturation at 94°C for 30 seconds; annealing at 55°C for 30 seconds; and polymerization at 72°C for 1 minute. PCR products were resolved by electrophoresis on a Nusieve 3:1 agarose gel and recombination frequencies between *VAN* loci and SSLP or CAPS loci were scored.

Histochemical localization of GUS activity

Transgenic *Arabidopsis* (WS) plants, which carried the chimeric gene *pAthb8::GUS* consisting of the *Athb-8* promoter and the GUS structural gene[6], or the chimeric gene *pTED3::GUS* consisting of the *TED3* promoter and the GUS structural gene[7] were used to distinguish vascular tissues. Seeds of transgenic *Arabidopsis* carrying *pAthb8::GUS* were kindly provided by Dr. Baima and Dr. Morelli, Unità di Nutrizione Sperimentale. The *van3* mutation was introduced into each transgenic line by crossing artificially between heterozygous *van3* plants and the transgenic plants. Histochemical GUS staining was performed with seedlings of the F2 generation or embryos of the F3 generation.

Thin-sheet culture

Seeds of *Zinnia elegans* L. cv. Canary bird were purchased from Takii Shubyo (Kyoto, Japan). Zinnia seedlings were grown on vermiculite at 25°C under a cycle of 14 h of light and 10 h of darkness. The first true leaves of 14-d-old seedlings were used as the source material for isolation of mesophyll cells. Mesophyll cells were isolated mechanically by homogenization of surface-sterilized leaves in culture medium according to the procedure of Sugiyama and Fukuda[8]. The culture medium was a slightly modified version of that described by Fukuda and Komamine[5] and contained 0.1 mg l^{-1} (0.54 μM) 1-napthaleneacetic acid and 0.2 mg l^{-1} (0.89 μM) benzyladenine as phytohormones. In order to increase the percentage of single cells in the population of obtained cells, the leaf homogenate was filtered through a 72-μm nylon mesh and subsequently through a 42-μm mesh (this two-step filtration increased the percentage of single cells up to 80%). Mesophyll cells were precipitated by centrifugation of the filtrate at 150 x g for 1 min, rinsed with the culture medium, and suspended in the same

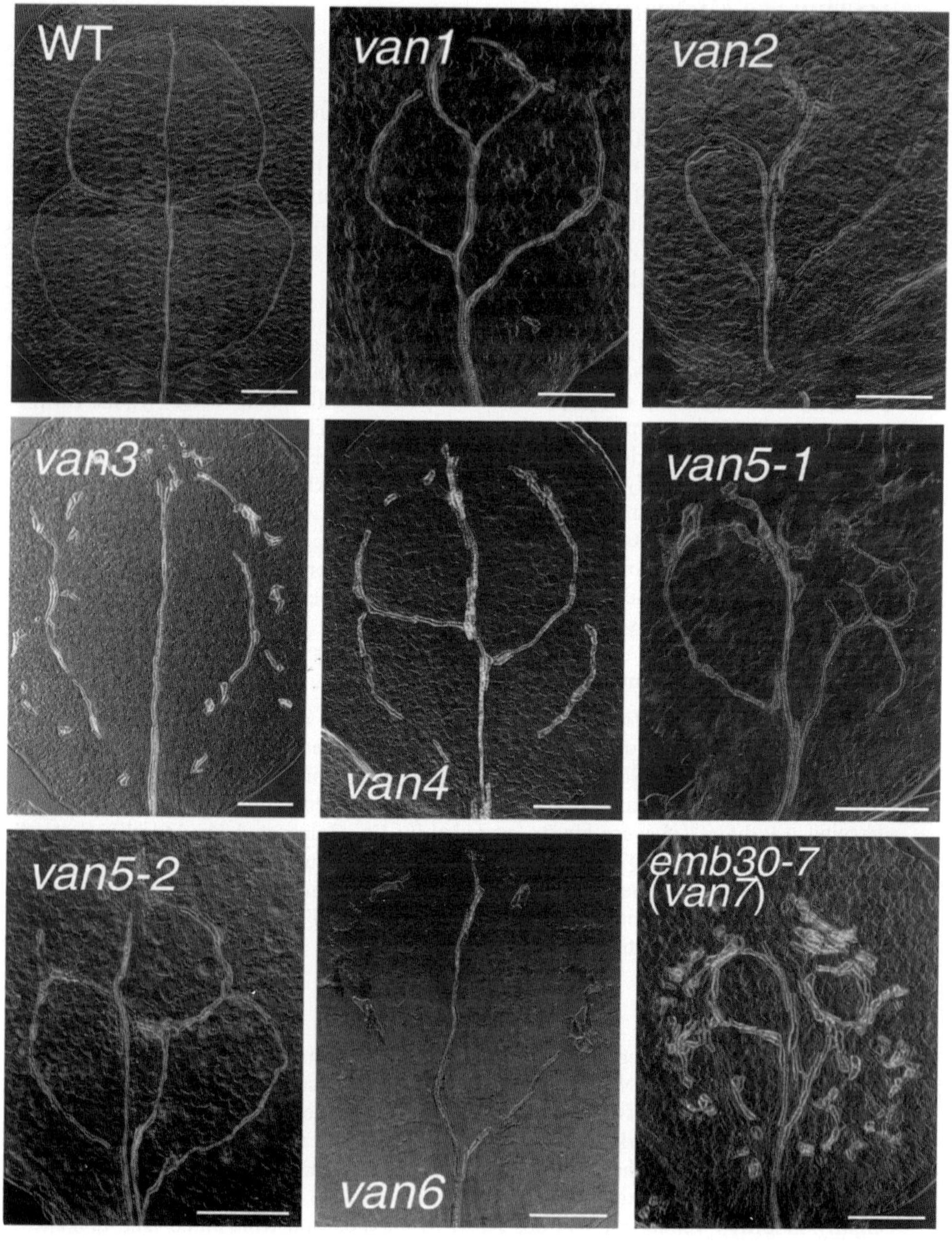

Fig.1 Vascular patterns in the cotyledon.

culture medium at twice the final cell density. The cell suspension was warmed to 30°C and mixed with an equal volume of medium containing 5.0 to 5.6% of low-melting-temperature agarose, which had been heated to melt agarose and then cooled to 30°C. The mixture was dropped onto the groove of a glass mold that was specially designed for making gel sheets, overlaid with a coverslip, and cooled down to 18°C. This produced an even sheet of agarose gel of 9 x 10 mm^2 in size and 200 μm in thickness. The sheet was transferred onto culture medium gelled with 0.25% gellan gum in a plastic dish and cultured in the dark at 27°C.

Microbead culture

Cell suspension prepared as described above was mixed with an equal volume of culture medium containing 4.0% of low-melting-temperature agarose at 30°C. Ten-microliter aliquots of the mixture were dropped onto siliconized glass slides. Each drop was solidified into a lens-shaped microbead of 3 mm in diameter by cooling the slides to 18°C. The microbeads were transferred into the liquid medium in a test tube cultured in the dark at 27°C while being rotated at 10 rpm on a revolving drum.

Determination of the frequencies of TE differentiation and cell division, and cell viability

For quantitative evaluation of TE differentiation, cell division, and cell viability, TEs, divided cells (cells that divided during culture), and dead cells (non-TE cells that died at cell isolation or during culture), which could be distinguished morphologically under a microscope, were counted for each culture. Here, a cell clump formed through cell division from an initially single cell was scored as one divided cell. A single-cell-derived clump containing TE(s) was scored as one divided cell and also as one TE. The numbers of TEs and divided cells are indicated as percentages of the number of initially living cells, which equals the initial cell number minus the initial number of dead cells. Cell viability is defined as the ratio of initial cell number minus dead cell number to the initial cell number.

RESULTS AND DISCUSSION

***van* mutants**

As the first step for the genetic analysis of regulatory mechanisms underlying vein pattern formation, we isolated mutants of *Arabidopsis thaliana* impaired in vascular formation. A microscopic examination of the cotyledonary venation of 3,400 M3 lines led to the identification of 14 mutant lines. These mutant lines could be categorized into two types: the type I mutants with abnormal patterning of lateral veins (12 lines) and the type II mutants with thickened veins (2 lines). Genetic analysis indicated that all of these mutations were monogenic and recessive. Out of 14 mutants, 8 mutants belonging to the type I were subjected to further analysis. A complementation test of these 8 mutants showed that their abnormalities in the vascular system are caused by mutations in 7 genetic loci. We designated these loci *VAN1~7* (vascular network defective). Map positions of *VAN* loci were determined utilizing DNA polymorphism between Ler and Col strains.

Phenotypic characterization of *van* mutants of *Arabidopsis* was conducted with a

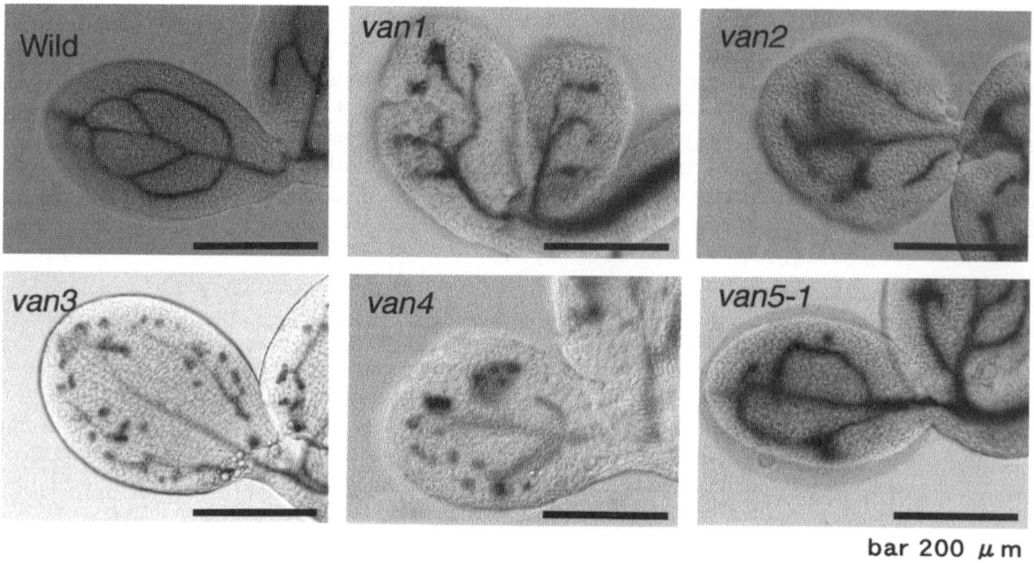

Fig. 2. *pAthb8::GUS* gene expression in the wild-type and *van* mutants at the late stage of embryogenesis

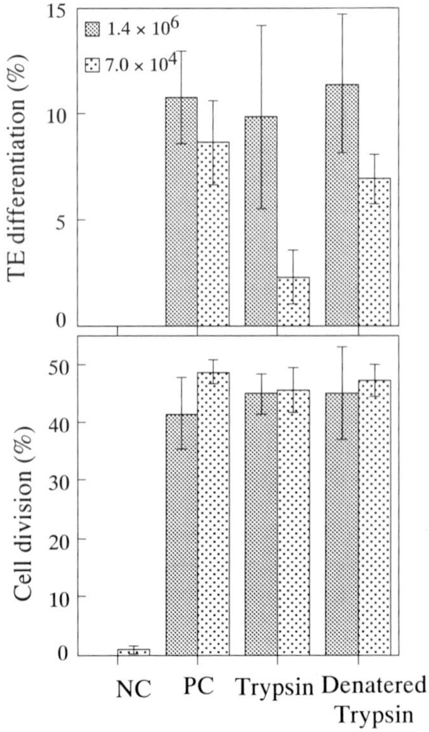

Fig. 3. Effects of insertion of an agarose gel sheet containing immobilized trypsin on TE differentiation, cell division, and viability in sandwich cultures. Low-density sheets of agarose gel containing mesophyll cells at the cell density of 7.0 x 104 cells ml-1 were laid on high-density sheets containing cells at 1.4 x 106 cells ml-1 with insertion of agarose sheets containing native or denatured trypsin. NC represents negative control, in which the low-density sheets were cultured separately from the high-density sheets. PC represents positive control, in which the low-density sheets were laid on the high-density sheets with insertion of cell-free, enzyme-free sheets of agarose. The frequencies of TE differentiation, and the frequencies of cell division were determined after 96 h in culture for the low-density (stippled bars) and high-density (shaded bars) sheets. Data are mean values of three replicates ± SD.

predominant focus on vascular network patterning. Histological analysis revealed that *van1-van6* mutations caused fragmentation of lateral veins in the cotyledon and that of tertiary veins in the rosette leaf, but did less injurious effects on the formation of major veins (Fig. 1). Despite the disconnection of minor vein networks, most of the *van* mutants had a relatively normal architecture of venation as a whole. Morphological analysis at the whole-plant level showed that seedlings of all the *van* mutants exhibited a relatively normal appearance in cotyledon expansion, greening, and hypocotyl elongation.

With respect to the possible mechanisms determining vein patterns, two hypotheses, i.e., auxin signal flow canalization and diffusion-reaction prepattern, have been proposed and discussed. The auxin canalization hypothesis assumes that a promotive effect of auxin flux on the capacity of auxin polar transport, forming a positive feedback loop, leads to canalization of the flow of auxin, which directs the vascular differentiation[9,10]. The diffusion-reaction prepattern hypothesis is based on diffusion-reaction wave theory, derived from the pioneering model of Turing[11] which postulates interaction among at least two diffusible substances with different diffusion rates, resulting in autonomous formation of patterns[12,13]. The features of the *van* mutants seem to support the diffusion-reaction hypothesis rather than the auxin canalization hypothesis.

The *van1-van5* mutants were further characterized, using *pAthb8::GUS* and *pTED3::GUS* as molecular markers for the early stages of vascular tissue formation. Results indicated that provascular cells as well as mature vascular cells were absent at the disconnection points of the vascular network in the *van* mutants and never formed a complete network during its development (Fig. 2). This suggests that the most of *VAN* genes are involved in the spatial regulation of provascular tissue differentiation, which realizes a continuous network of the vascular system. Moreover toward the isolation of the *VAN3* gene, the *VAN3* locus was mapped to a sub-cM region in chromosome 5.

Intercellular communication in the *Zinnia* xylogenic culture

To investigate intercellular communication in the *Zinnia* xylogenic culture, two types of culture method were developed, in which mesophyll cells were embedded in a thin sheet of agarose gel and cultured on solid medium, or embedded in microbeads of agarose gel and cultured in liquid medium. A statistical analysis of the two-dimensional distribution of TEs in the thin-sheet cultures demonstrated a positive intercellular communication between TEs. In the microbead cultures, the frequency of TE differentiation was shown to depend on the local cell densitiy (the cell density in each microbead): TE differentiation required local cell densities of more than 10^5 cells ml^{-1}. These results suggest that TE differentiation involves intercellular communication mediated by a locally-acting diffusible factor. This presumptive factor was characterized by applying a modified version of the sheet culture, which used two sheets of different cell densities, a low-density sheet and a high-density sheet. TE differentiation in the low-density sheet could be induced only by bringing it into contact with the high-density sheet. Insertion of a 25-kDa-cutoff membrane between the high-density and low-density sheets severely suppressed such induction of TEs in the low-density sheet while a 300-kDa-cutoff membrane did only slightly. Insertion of agarose sheets containing immobilized pronase E or trypsin also interfered with the induction of TEs in the low-density sheets (Fig. 3). Thus, a proteinaceous macromolecule of 25 kDa to 300 kDa in molecular weight was assumed to mediate the local intercellular communication required for TE differentiation. This substance was designated "xylogen" with reference to its xylogenic activity. The time of requirement for xylogen during TE differentiation was assessed by experiments in which cells in the low-density sheet were separated from xylogen produced in the high-density sheet at various times by insertion of a 25-kDa-cutoff membrane between the two sheets, and was estimated to be from the 36th hour to the 60th hour of culture (12 – 36 h before visible thickening of secondary cell walls of TEs).

To characterize and isolate xylogen, a bioassay system to monitor the activity of xylogen was developed, in which mesophyll cells were embedded in microbeads of agarose gel at a low (2.0 to 4.3 x 10^4 cells ml^{-1}) or high density (8.0 to 9.0 x 10^5 cells ml^{-1}) and microbeads of different cell densities were cultured together in a liquid medium to give a total density of 2.1 to 2.5 x 10^4 cells ml^{-1}. Without any additives, the frequency of TE differentiation was much lower in the low-density microbeads than in the high-density microbeads. This low level of TE differentiation in the low-density microbead was attributable to the shortage of xylogen. When cultures were supplemented with conditioned medium (CM) prepared from *Zinnia* cell suspensions undergoing TE differentiation, the frequency of TE differentiation in the low-density microbeads increased remarkably, indicating the activity of xylogen in the CM. The xylogen activity in CM was sensitive to protease treatments. Xylogen was bound to galactose-specific lectins such as *Ricinus communis* agglutinin and peanut agglutinin, and precipitated by β-glucosyl Yariv reagent. These results indicate that xylogen is a kind of arabinogalactan protein (AGP).

The diffusion-reaction prepattern hypothesis is based on diffusion-reaction wave theory, derived from the pioneering model of Turing[11]. which postulates interaction among at least two diffusible substances with different diffusion rates, resulting in autonomous formation of patterns[12,13]. In the simplest case, an autocatalytic, local activator and a long-range inhibitor are necessary and sufficient for pattern formation. In the light of the above argument, one of the fascinating possibilities for the *in-planta* function of xylogen is that xylogen may participate as an autocatalytic, local activator in the generation of a diffusion-reaction wave, which directs the position of vascular differentiation. Now this is not a red herring, not only because the molecular weight of xylogen is enough large (75 kDa to 300 kDa) to diffuse and act locally as an activator, but also because the positive feedback loop implicated in xylogen signalling *in vitro* hints at autocatalytic self-activation of xylogen.

These results have recently been published[14, 15], or are in press[16].

ACKNOWLEDGEMENTS

This work was supported in part by Grants-in-Aid from the Ministry of Education, Science, Sports and Culture of Japan (No. 10304063, No. 10219201, No. 10182101), from the Japan Society for the Promotion of Science (JSPS-RFTF96L00605), and from the Ministry of Agriculture, Forestry and Fisheries (Gene discovery and elucidation of functions of useful genes in rice genome by gene-expression monitoring system).

REFERENCES

1. T. Berleth, & G. Jürgens, The role of the *monopteros* gene in organising the basal body region of the *Arabidopsis* embryo. *Development,* **1993,** 118, 575-587.
2. G. Jürgens, U. Mayer, R. A. Torres Ruiz, T. Berleth & S. Miséra, Genetic analysis of pattern formation in the *Arabidopsis* embryo. *Development* Supplement, **1991,** 1, 27-38.

3. U. Mayer, R. A. Torres Ruiz, T. Berleth, S. Miséra, & G. Jürgens, Mutations affecting body organization in the *Arabidopsis* embryo. *Nature*, **1991**, 353, 402-407.
4. R. A. Torres Ruiz & G. Jürgens, Mutations in the *FASS* gene uncouple pattern formation and morphogenesis in *Arabidopsis* development. *Development,* **1994**, 120, 2967-2978.
5. H. Fukuda, & A. Komamine, Establishment of an experimental system for the study of tracheary element differentiation from a single cell isolated from the mesophyll of *Zinnia elegans*. *Plant Physiol,* **1980**, 65, 57-60
6. S. Baima, F. Nobili, G. Sessa, S. Lucchetti, I. Ruberti & G. Morelli, The expression of the *Athb-8* homeobox gene is restricted to provascualr cells in *Arabidopsis thaliana*. *Development,* **1995**, 121, 4171-4182.
7. M. Igarashi, T. Demura & H. Fukuda, Expression of the *Zinnia TED3* promoter in developing tracheary elements of transgenic *Arabidopsis*. *Plant Mol. Biol.* **1998**, 36, 917-927.
8. M.Sugiyama, & H. Fukuda, *Zinnia* mesophyll culture system to study xylogenesis. In "Plant Tissue Culture Manual Edited by Lindsey K Supplement 5" **1995**, pp H2 1-15 Kluwer Academic Publishers, Dordrecht
9. T. Sachs, Cell polarity and tissue patterning in plants. *Development Supplement*, **1991**, 1, 83-93.
10. T. Sachs, Integrating cellular and organismic aspects of vascular differentiation. *Plant Cell Physiol,* **2000**, 41, 649-656
11. A.M. Turing, The chemical basis of morphogenesis. *Philos Trans R Soc London Ser B*, **1952**, 237, 37-72.
12. H. Meinhardt, *Models of biological pattern formation.* **1982**, London: Academic Press.
13. A. J. Koch & H. Meinhardt, Biological pattern formation: from basic mechanisms to complex structures. *Rev. Mod. Phys,* **1994**, 66, 1481-1507.
14. K. Koizumi, M. Sugiyama & H. Fukuda, A series of novel mutants of *Arabidopsis thaliana* that are defective in the formation of continuous vascular network: Calling the auxin signal flow canalization hypothesis into question. *Development*, **2000**, 217, 3197-3204.
15. H. Motose, M. Sugiyama & H. Fukuda, An arabinogalactan protein(s) is a key component of a fraction that mediates local intercellular communication involved in tracheary element differentiation of zinnia mesophyll cells. *Plant Cell Physiol.,* **2001**, 42, 129-137.
16. H. Motose, H. Fukuda & M. Sugiyama, Involvement of local intercellular communication in the differentiation of zinnia mesophyll cells into tracheary elements. *Planta*, **2001**, in press.

Molecular Breeding of Woody Plants
N. Morohoshi and A. Komamine, editors.

THE *ASYMMETRIC LEAVES2* (*AS2*) GENE OF *ARABIDOPSIS THALIANA* REGULATES LAMINA FORMATION AND IS REQUIRED FOR PATTERNING OF LEAF VENATION

Endang Semiarti[1,2], Yoshihisa Ueno[1], Hidekazu Iwakawa[1], Hirokazu Tsukaya[3], Chiyoko Machida[1*], and Yasunori Machida[1*]

[1] *Division of Biological Science, Graduate School of Science, Nagoya University, Chikusa-ku, Nagoya 464-8602, Japan*

[2] *Permanent address: Faculty of Biology, Gadjah Mada University, Sekip Utara, Yogyakarta 55281, Indonesia*

[3] *National Institute for Basic Biology/Center for Integrated Bioscience, Myodaiji-cho, Okazaki 444-8585, Japan, and Form and Function, PRESTO, Japan Science and Technology Corporation, Kawaguchi, Japan*

ABSTRACT

To understand the molecular mechanisms behind symmetrical development of leaf, we have analyzed the *asymmetric leaves2* (*as2)* mutant of *A. thaliana*, which generated leaf lobes and leaflet-like structures from the petioles of leaves in a bilaterally asymmetric manner. The delayed formation of the primary vein and the asymmetric formation of secondary veins were apparent in leaf primordia of *as2* plants. A distinct midvein, which is the thickest vein and is located in the longitudinal center of the leaf lamina of wild-type plants, was often rudimentary even in mature *as2* leaves. However, several parallel veins of very similar thickness were evident in such leaves. The malformed veins were visible prior to the development of asymmetry of the leaf lamina, and were maintained in the mature *as2* leaves. Culture *in vitro* on phytohormone-free medium of leaf sections from the *as2* mutants and from the *asymmetric leaves1* (*as1*) mutant, which has a phenotype similar to that of *as2*, revealed an elevated potential in both cases for regeneration of shoots from leaf cells. Analysis by the reverse transcription-polymerase chain reaction showed that *AS2* and *AS1* negatively regulates the homeobox genes *KNAT1, KNAT2* and *KNAT6* in leaves. Taken together, our results suggest that *AS2* and *AS1* are involved in establishment of leaf venation and the formation of symmetric leaf lamina, which might be related to repression of expression of the homeobox genes in leaves.

KEY WORDS

Arabidopsis thaliana, *asymmetric leaves1*, *asymmetric leaves2*, *KNOX* homeobox genes, leaf morphology, venation pattern, midvein, shoot

INTRODUCTION

The establishment of left-right symmetry is one of the most important factor for the leaf morphogenesis of plants. It is generally accepted that leaves of many angiosperms exhibit obvious but approximate left-right symmetry with the rachis as the axis (Hickey, 1979; Sinha, 1999), except *Begonia* spp. (Lieu and Sattler, 1976) and *Tropaeolum* (Whaley and

---

* To whom correspondence should be addressed

Whaley, 1942). Regardless of the complexity of leaf shape (e.g., a simple leaf or a compound leaf), the two sides of each leaf are nearly mirror images of one another (Ogura, 1994). However, our understanding of the way in which the nearly mirror-image architecture arises during leaf development remains at a descriptive level (see below) and the molecular and genetic basis for this phenomenon remains to be analyzed.

Previous studies using *A. thaliana* have focused on two aspects of leaf symmetry. It has been demonstrated that the number and the positions of the serration on the margin of a leaf lamina are bilaterally symmetric (Tsukaya and Uchimiya, 1997). It has also been demonstrated that venation patterns in the leaf laminas of Arabidopsis exhibit bilateral symmetry (Candela et al., 1999). A single primary vein, the midvein, is the thickest vein and is located at the center of the leaf lamina (Hickey, 1979; Nelson and Dengler, 1997).

It has previously been pointed that ectopic expression of a SAM (Shoot Apical Meristem)-related homeobox gene might affect the symmetrical pattern of serration on the leaf margin (Tsukaya and Uchimiya, 1997). The SAM retains stem cells in its central zone, which is required for self-regeneration and maintenance of undifferentiated state, but the SAM can also generate leaf primordia from its peripheral zone (Steeves and Sussex, 1989; Howell, 1998). The *SHOOT MERISTEMLESS* (*STM*) gene, a member of the family of class 1 *knox* homeobox genes, is required for the development of the SAM, as well as for the maintenance of stem cell identity throughout the life of the plant (Long et al., 1996). *KNAT1* and *KNAT2* are other members of the knox class I genes. The transcripts of these genes are localized primarily in the region around the SAM and the floral meristem, with down-regulation of expression in the presumptive region of a new leaf primordium (Long and Barton, 2000). The studies of the ectopic overexpression of *KNAT1* in Arabidopsis have shown that leaf cells can be converted from the meristematic indeterminate state to the determinate state, and back again, and that their levels of expression are closely related to the extent of leaf serration or formation of lobes (Sinha et al., 1993; Lincoln et al., 1994; Chuck et al., 1996).

To understand the development of symmetrical leaves, we analyzed the *asymmetric leaves2* (*as2*) mutant of *A. thaliana*, which was originally isolated by Rédei (ABRC, OH, Machida et al., 1997), and another similar mutant, *asymmetric leaves1* (*as1*) that was reported to show the distorted bilateral symmetry of leaves (Rédei & Hirono, 1964; Tsukaya and Uchimiya, 1997). In the present study, we analyzed the phenotype of the *as2* mutant, focusing on patterns of serration, formation of lobes, and venation in the leaves and leaf-like organs, and found that leaf serration in *as2* was asymmetric, with generation of leaflet-like structures from petioles and malformed midvein. The relationship between such an abnormality and misexpression of the class I *knox* genes in the *as2* leaves will also be discussed.

MATERIALS & METHODS

Plant materials and growth conditions

Arabidopsis thaliana ecotypes Col-0 (CS1092) and Arabidopsis mutant *as2-1* (CS3117), and *as1-1* (CS3374) were obtained from the Arabidopsis Biological Resource Center (Columbus, OH, USA; ABRC). Seeds were sown on a soil and cultured under the light for 16 hours and darkness for 8 hours after 4°C for 4 days in the darkness.

Analysis of vasculature and numbers of branching points of leaf veins

Specimens were prepared as described below. Leaves were taken from 23-day-old plants and fixed in 14% glacial acetic acid, plus 84% ethanol overnight. Samples were dehydrated twice in 70% ethanol and twice in 99.5% ethanol, and they were cleared in a solution of chloral hydrate [trichloroacetaldehyde monohydrate, 200 g; glycerol, 20 g; distilled water, 50 g]. Photographs of leaf veins were observed under the dark-field microscope (Axiophot or Stemi2000; Zeiss, Germany) and number of branching points were counted.

Reverse transcription-polymerase chain reaction (RT-PCR)

Total RNA was extracted from leaves and shoot apices of 19-day-old plants by using the RNeasy Plant Minikit (QIAGEN GmbH, Hilden, Germany) and poly(A)+ RNA was purified by two rounds of treatment with Dynabeads (Dynal). The first strand of cDNA was synthesized by using the kit of You-Prime First-Strand Beads (Amersham Pharmacia) according to the manufacturer's instruction. PCR was performed as described by Hamada *et al.* (2000). Sequences of primers for the PCR analysis will be provided on request.

Culture of leaf sections in vitro

Leaves of 19- to 21-day-old plants were halved and incubated on plates of Murashige and Skoog (MS) basic medium (Onouchi et al., 1995) at 22°C under continuous white light.

RESULTS AND DISCUSSION

Prominent leaf lobes, leaflet-like structures, and reduced development of vein systems in *as2* plants

Fig. 1 shows typical leaf phenotypes. In terms of overall shape, the lamina of the *as2* leaf was often plump and humped at its base; the leaf surface was wavy (Fig. 1a, 1b); the leaf often had many deep and irregularly split lobes; and plants had leaflet-like structures (ls) on petioles which were relatively shorter (Fig.1c-1e).

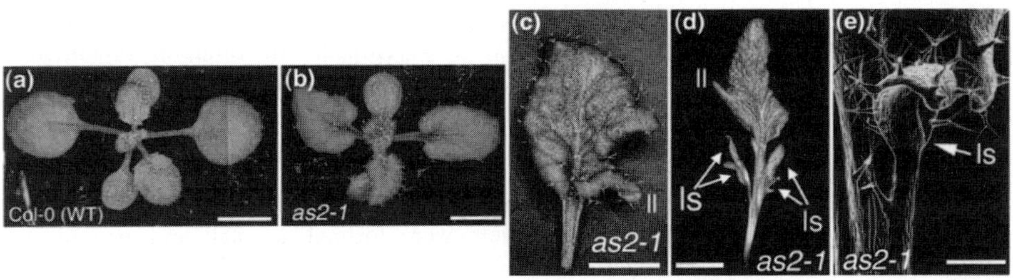

Figure 1. Phenotype of the *as2* mutant. The overall morphology of 18-day-old wild-type (a) and *as2* plants. (c-e) Fifth leaves of *as2* mutant at 23 day-old (c,e) and 60 day- old. ll, leaf lobe. ls, leaflet-like structure. Bars: 5 mm in a–d; 500 μm in e.

We analyzed the venation in rosette leaves of *as2* and wild-type plants by dark-field microscopy. In the wild type, there was a single, distinct and maximally thick midvein in the center of each leaf lamina and a number of thinner secondary veins were connected to the midvein (Fig. 2a). The severity of the effect of the *as2* mutation on venation varied. In extreme cases, no midvein was obvious, and several veins of similar thickness to one another were evident with a proximo-distal orientation (Fig. 2b). In both mild (Fig. 2a, 2c) and extreme cases, several secondary veins failed to connect with the midvein in the leaf lamina and sometimes they ran separately through the petiole. We also found that the complexity of vein network was reduced in the mutant (Fig. 2b, 2c).

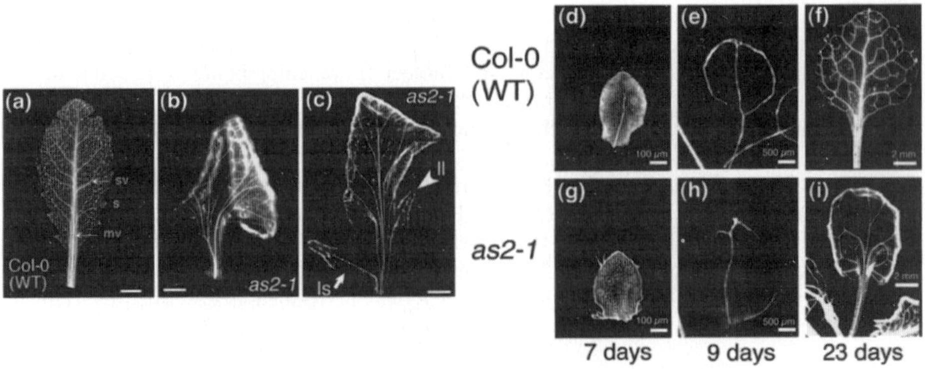

Figure 2. Patterns of venation in wild-type and *as2* leaves. (a) Seventh leaf of wild type, (b) second and (c) fifth leaf of *as2*. (d-i) Vascular development of the first leaf of wild type (d-f) and *as2* (g-i). sv, secondary vein; s, serra; mv, midvein; ll, leaf lobe; ls, leaflet-like structure. Bars: 2 mm in a-c.

We investigated the earliest stages of vein development in the first leaves of both wild-type and *as2*. Fig. 4A shows the delayed development of the primary vein in *as2*. In leaf primordia of most wild-type plants seven days after sowing, primary veins were visible (Fig. 2d), but such was not the case in those of *as2* plants, even though the *as2* leaf primordia were normal in shape (Fig. 2g). On day 8, a primary vein appeared for the first time in the *as2* leaf primordium (data not shown). In the wild type, the primary vein bifurcated at its distal end, and forming two strands (secondary veins) that extended basipetally and connected to middle positions on the primary vein (Fig. 2e). By contrast, in almost all the primordia of *as2*, the primary veins bifurcated irregularly and asymmetrically (Fig. 2h). The secondary veins of *as2* were developed with bilateral asymmetry, approached the primary vein at a more acute angle than in the wild type, in some cases, did not connect with the primary vein in the leaf lamina (Fig.4a-c,f,j,l). The development of higher-ordered veins had ceased at very earlier stage in *as2* leaves by day 15.

We similarly analyzed vein development in the third rosette leaves of both wild-type and *as2* plants. The shape of *as2* lamina was normal prior to the formation of the primary vein. At the stage during which the secondary vein were formed, a small and asymmetric protrusion which seemed to become larger to generate asymmetric deep lobes, was often visible at the marginal region of the lamina. (data not shown). It suggests that the *AS2* gene is involved in the establishment of the prominent midvein and the network of lateral veins. And the establishment of such vein systems is related to the lamina symmetry. Chracterization of the *AS2* gene may provide a clue to answer this question.

Ectopic expression of meristem-related homeobox genes

The morphology of *as2* and *as1* leaves was similar to that of leaves of transgenic Arabidopsis that ectopically express the class 1 *KNOX* homeobox gene *KNAT1*. W e examined the expression of *KNAT1* and other meristem-related homeobox genes, namely *KNAT2*, *KNAT6* and *STM*. in leaves of wild type, *as1* and *as2* using RT-PCR. As shown in Fig. 3a, the products of PCR correspond to KNAT1 KNAT2, and KNAT6 cDNA were detected in all the leaves and shoot apices of *as1* and *as2* but not of wild-type plants. Transcripts of the *STM* gene accumulated in first pair of leaves of *as1* plants as well as their shoot apices, although the relative levels were lower than those of the *KNAT1* (Fig. 3a).

In Arabidopsis, overexpression of the *KNAT1* gene results in the formation of ectopic

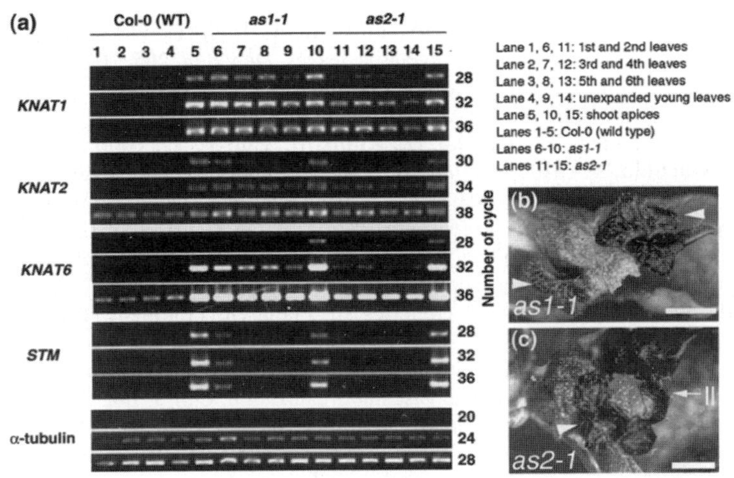

Figure 3. Misexpression of the class 1 *KNOX* homeobox genes in *as1* and *as2* leaves (a) and the autonomous shoots in *as1*(b) and *as2*(c) tissue culture (See text for detail).

shoots on rosette leaves. However, no ectopic shoots appeared on rosette leaves of *as1* and *as2* mutant. We examined whether or not *as1* and *as2* leaves could regenerate shoots during culture *in vitro* on MS medium without exogenous phytohormones. The shoots could regenerated from 2-3% of leaf sections from *as1* and *as2* plants, but not from those of wild type (Fig. 3b,c), suggesting that *as1* and *as2* leaves had a higher potential for regeneration of shoots in vitro than did wild type. And, the morphological changes in *as2* mutants might be explained by the ectopic expression of *KNAT1*, *KNAT2* and *KNAT6* in the mutant leaves.

CONCLUSION

The *as2* mutation in Arabidopsis generated lobes in leaf margins and leaflet-like structures from petioles in a bilaterally asymmetric manner. The *as2* mutants failed to produce rosette leaves with a thickened and prominent midvein, and exhibited the asymmetrical patterns of secondary veins and the reduced complexity of higher-order veins. The *as2* mutation enhanced the ability of leaf cells to regenerate shoots in vitro and increased the accumulation of transcripts of the *KNOX* genes in leaves. These observations suggest that *AS2* plays roles in the establishment of the entire vein systems including the thickened midvein, which is the structural axis of left-right symmetry in the leaf, as well as the formation of lamina symmetry. *AS2* also functions in maintaining a developmentally determinate state of leaf cells and repressing expression of the class 1 *knox* genes. Although these morphological, physiological and molecular events might be related to each other, such relationships must be investigated by further experimentation.

AKCNOWLEDGEMENTS

The authors acknowledge the Ministry of Education, Science, and Culture and Sports of Japan for General Scientific Research (no. 12640598) and to the Ministry of Agriculture, Forestry, and Fisheries of Japan for the grant for Pioneering Research Projects in Biotechnology. E. S. was supported by a scholarship from the Ministry of Education, Science, and Culture and Sports of Japan.

REFERENCES

1. **Candela, H., Martinez-Laborda, A. and Micol, J. L.** (1999). Venation pattern formation in *Arabidopsis thaliana* vegetative leaves. *Dev. Biol.* **205,** 205-16.
2. **Chuck, G., Lincoln, C. and Hake, S.** (1996). *KNAT1* induces lobed leaves with ectopic meristem when overexpressed in Arabidopsis. *Plant Cell* **8,** 1277-1289.
3. **Hamada, S., Onouchi, H., Tanaka, H., Kudo, M., Liu, Y. -G., Shibata, D., Machida, C. and Machida, Y.** (2000). Mutations in the *WUSCHEL* gene of *Arabidopsis thaliana* result in the development of shoots without juvenile leaves. *Plant J.* **24,** 91-101.
4. **Hickey, L. J.** (1979). A revised classification of the architecture of dicotyledonous leaves. In: Anatomy of the Dicotyledons (ed. C.R. Metcalfe and L. Chalk), pp. 25-39. New York: Oxford University Press.
5. **Howell, S.** (1998). Molecular Genetics of Plant Development. Cambridge, UK: Cambridge University Press
6. **Lieu, S. M. and Sattler, R.** (1976). Leaf development in *Begonia hispida* var. *cucullifera* with special reference to vascular organization. *Can. J. Bot.* **54,** 2108-2121.
7. **Lincoln, C., Long, J., Yamaguchi, J., Serikawa, K. and Hake, S.** (1994). A *knotted1-like* homeobox gene in *Arabidopsis* is expressed in the vegetative meristem and dramatically alters leaf morphology when overexpressed in transgenic plants. *Plant Cell* **6,** 1859-1876.
8. **Long, J. A., Moan, E. I., Medford, J. I. and Barton, M. K.** (1996). A member of the KNOTTED class of homeodomain proteins encoded by the *STM* gene of *Arabidopsis. Nature* **379,** 66-69.
9. **Long, J. A. and Barton, M. K.** (2000). Initiation of axillary and floral meristems in *Arabidopsis. Dev. Biol.* **218**: 341-353.
10. **Machida, C., Onouchi, H., Koizumi, J., Hamada, S., Semiarti, E., Torikai, S. and Machida, Y.** (1997). Characterization of the transposition pattern of the *Ac* transposable element in *Arabidopsis thaliana* using endonuclease I-*Sce*I. *Proc. Natl. Acad. Sci. U.S.A.* ***94,*** 8675-8680.
11. **Nakashima, M., Hirano, K., Nakashima, S., Banno, H., Nishihama, R. and Machida, Y.** (1998). The Expression Pattern of the Gene for *NPK1* Protein Kinase Related to Mitogen-Activated Protein Kinase Kinase Kinase (MAPKKK) in a Tobacco Plant: Correlation with Cell Proliferation. *Plant Cell Physiol.* **39,** 690-700.
12. **Nelson, T. and Dengler, N.** (1997). Leaf Vascular Pattern Formation. *Plant Cell* **9,** 1121-1135
13. **Onouchi, H., Nishihama, R., Kudo, M., Machida, Y. and Machida, C.** (1995). Visualization of site-specific recombination catalyzed by a recombinase from *Zygosaccharomyces rouxii* in *Arabidopsis thaliana. Mol. Gen. Genet.* **247,** 653-660.
14. **Ori, N., Eshed, Y., Chuck, G., Bowman, J. L. and Hake, S.** (2000). Mechanisms that control *knox* gene expression in the *Arabidopsis* shoot. *Development* **127,** 5523-5532.
15. **Rédei, G. P. and Hirono, Y.** (1964). Linkage studies. *Arabidopsis Inf. Serv.* **1,** 9.
16. **Sinha, N.** (1999). Leaf development in angiosperms. *Annu. Rev. Plant Physiol. Plant Mol. Biol.* **50,** 419-446.
17. **Sinha, N. R., Williams, R. E. and Hake, S.** (1993). Overexpression of the maize homeobox gene, *KNOTTED-1,* causes a switch from determinate to indeterminate cell fates. *Genes Dev.* **7,** 787-795.
18. **Steeves T. A. and Sussex, I. M.** (1989) Patterns in Plant Development. Cambridge: Cambridge University Press.
19. **Tsukaya, H. and Uchimiya, H.** (1997). Genetic analyses of the formation of the serrated margin of leaf blades in *Arabidopsis*: combination of a mutational analysis of leaf morphogenesis with the characterization of a specific marker gene expressed in hydathodes and stipules. *Mol. Gen. Genet.* **256,** 231-238.
20. **Whaley, W.G. and Whaley, C.Y.** (1942). A developmental analysis of inherited leaf patterns in *Tropaeolum. Am. J. Bot.* **29,** 105-194.

Molecular Breeding of Woody Plants
N. Morohoshi and A. Komamine, editors.

BIOSYNTHESIS OF CELLULOSE

Inder M. Saxena* & R. M. Brown Jr.

Section of Molecular Genetics and Microbiology, School of Biological Sciences, University of Texas at Austin, Austin, TX 78712, U.S.A.

ABSTRACT

Cellulose is synthesized by a large number of living organisms ranging from the bacterium *Acetobacter xylinum* to forest trees. *A. xylinum* produces abundant amounts of cellulose and this bacterium has been used as a model system for studies on cellulose biosynthesis and structure of the cellulose product. Cellulose is synthesized by the enzyme cellulose synthase, a membrane protein that catalyzes the direct polymerization of glucose from the substrate UDP-glucose into a cellulose product. Genes for cellulose synthases have been identified from many bacteria, *Dictyostelium discoideum*, and higher plants. Analysis of the predicted protein sequences has allowed identification of conserved residues in cellulose synthases from different organisms. The conserved residues are found in the globular region of the cellulose synthases. Using site-directed mutagenesis experiments we have shown that the conserved amino acid residues are required for cellulose synthase activity in *A. xylinum*. Although cellulose synthase activity can be monitored *in vitro* using membrane fractions from *A. xylinum*, it is not easy to monitor this activity when membrane fractions from plants are used. We have initiated experiments to analyze cellulose synthases from plants in *A. xylinum* in an effort to characterize the different cellulose synthases, for example the ones involved in cellulose biosynthesis during primary cell wall formation and those that are active during secondary wall synthesis. A general model describing the possible sequence of events in the cellulose synthase catalytic site will be presented to provide sufficient details not only into the biosynthesis of cellulose but also other polysaccharides.

KEYWORDS

Cellulose, biosynthesis, cellulose synthase, *Acetobacter xylinum*

INTRODUCTION

Cellulose is a major polysaccharide produced mainly by plants. A number of other organisms also synthesize cellulose and the capacity to synthesize this polysaccharide may be present in a larger variety of organisms than presently known. For example, sequences similar to cellulose synthase and other proteins identified in the cellulose-synthesizing operon of *A. xylinum* have been identified in *E. coli* and other bacteria even though no cellulose production has been reported in these bacteria [1]. Although plants produce cellulose as a major product, genes for cellulose synthase were identified only in the last few years [2,3]. More recently, the identification of a large gene family of cellulose synthases and related proteins in plants [4] has allowed a number of interesting observations to be made in terms of the expression of different genes in different tissues and in the requirement of different genes for the synthesis of cellulose in the primary and secondary cell wall [5-7]. Even before multiple cellulose synthase genes were identified in plants, two cellulose synthases genes were identified in *A. xylinum* [8]. Whether the enzymes coded by the different cellulose synthase genes differ in their catalytic activity, their regulation, and their association with similar or different catalytic subunits and accessory protein remains to be understood.

CELLULOSE SYNTHASES

Cellulose is synthesized by the enzyme cellulose synthase, and in all cases this enzyme is predicted to be a membrane protein that utilizes UDP-glucose as the sugar donor in a direct transfer reaction [9]. The glucan chain is elongated from the non-reducing end [10] processively and although suggestions for the requirement of a primer by cellulose synthase have been made [11], no primer has yet been identified. The cellulose synthases from plants show similarity to the *A. xylinum* cellulose synthase in a globular region that contains the putative catalytic region and the conserved amino acid residues. Cellulose synthase activity from *A. xylinum* can be assayed *in vitro* and the enzyme has been partially purified. On the other hand, cellulose synthase has not been sufficiently purified from plant membranes and the analysis of the cellulose product is complicated because of the synthesis of other polysaccharides, especially callose. In terms of regulation of the enzyme activity, the *A. xylinum* enzyme is specifically activated by c-di-GMP and does not seem to require any additional factors. In biochemical studies with membrane preparations from plants it has not been possible to identify any specific activator of cellulose synthase; however, cellobiose is required for increased uptake of glucose from UDP-glucose into cellulose (characterized as the ANR-insoluble product) in *in vitro* reactions. Since the mechanism by which cellulose synthase performs glycosyl transfer is predicted to be essentially the same in bacteria and plants, we are interested in understanding this mechanism and the various modes of regulation involved in the activity of cellulose synthase from different sources.

The process of cellulose biosynthesis can be viewed as a number of simultaneous or sequential events that requires an active site in the cellulose synthase and may involve accessory factors and proteins. We have investigated the role of different regions of the *A. xylinum* cellulose synthase in a number of experiments to determine the amino acid residues essential for enzyme activity (by site-directed mutagenesis experiments) [12,13] and regions required for enzyme activity (by transposon insertion mutagenesis and deletion mutagenesis) [14].

The different regions and residues of cellulose synthase that may be essential for the enzyme activity include:

(a) The catalytic region, which includes the substrate (UDP-glucose) binding residues and the catalytic residue (that functions as a base). We investigated the role of the conserved residues in the D, D, D, QXXRW motif, identified in β-glycosyltransferases, by site-directed mutagenesis. Replacement of the conserved aspartic acid residues and the conserved residues in the QXXRW motif led to a loss of cellulose synthase activity in *A. xylinum*.

(b) Region(s) for binding of the growing glucan chain (and containing the glucose residue in the growing chain that functions as an acceptor). This region may be essential for the processivity of cellulose synthase and other processive glycosyltransferases.

(c) Transmembrane regions for membrane insertion (so far no cellulose synthase activity has been observed in the globular region that has been expressed as a cytosolic protein).

(d) Activator-binding site – in *A. xylinum*, the c-di-GMP- binding site may be present in either the same polypeptide chain that carries the catalytic region or in a separate polypeptide chain. A deletion of amino acid residues 707 – 1108 of the AcsAB protein, where this region corresponds to the c-di-GMP-binding region, shows no activity and no reaction with the antibody against the 93-kDa polypeptide (Saxena and Brown, unpublished observations).

(e) Protein-protein interaction sites for binding one subunit with homologous or/and heterologous subunit(s). In *A. xylinum*, cellulose synthase activity is still observed even when one of the two cellulose synthase genes is mutagenized by insertion [8]. Moreover insertion in the *acsAII* gene does not result in a phenotypic change (the cells are still able to make a ribbon of cellulose) suggesting that the AcsAB and the AcsAII proteins are not required together for forming the cellulose synthesizing complex in *A. xylinum*. When insertions take place in the *acs* operon genes, the *acsAII* gene and operon is not able to assemble a cellulose synthesizing complex. However, in plants multiple cellulose synthase catalytic subunits may interact to give rise to homomeric or heteromeric structures (see Figure 1). The interaction of these subunits has been suggested based on mutant and protein interaction analysis in *A. thaliana* [6,7].

(f) Residues that bind to Mg^{2+} in the catalytic region.

(g) Residues that function in a ratchet-like mechanism for movement of the growing polymer chain from the active site.

CELLULOSE SYNTHESIZING COMPLEXES

Is the smallest unit in a cell responsible for producing a cellulose I microfibril a single cellulose synthase enzyme or an aggregate of cellulose synthase catalytic subunits? The production of non-crystalline cellulose and cellulose II probably can take place from single catalytic subunits of cellulose synthase as observed in *in vitro* reactions and in mutants where the organization of the subunits has been disturbed. However, assembly of cellulose I microfibrils probably require other proteins that allow assemblage of the catalytic subunits for efficient synthesis and export of the cellulose product. In *A. xylinum*, these proteins have been identified by mutant analysis; however, they have not been characterized biochemically. In higher plants, a larger number of proteins may be required for cellulose I biosynthesis, including more than one kind of cellulose synthase catalytic subunit (Figure 1). Apart from the catalytic subunits, other proteins may be required for the processing of the proteins before they are exported to the plasma membrane. In all cases, cellulose I microfibrils have been found to be associated with organized structures observed on the cell membrane. These structures are believed to be the cellulose synthesizing complexes. They are referred to as terminal complexes (TCs) and are visualized as rosettes or linear structures by freeze-fracture electron microscopy.

The nature of the components in these structures has so far only been inferred based on the attachment of cellulose microfibril with these structures. Recently, techniques that allow labeling of freeze-fracture replicas by antibodies have led to the localization of some of the components in these complexes. In plants, the cellulose synthase catalytic subunit has been localized to the rosette structure [15] and an activator (c-di-GMP) binding protein has been localized to the linear complexes observed in *A. xylinum*. So far, no other protein has been found to be associated with the cellulose synthases in the rosette complex of higher plants; however suggestions have been made for the role of a number of proteins that may associate with the catalytic subunits in the rosette complex. It will certainly be interesting to determine the nature of interaction between the different catalytic subunits in the organization of the rosette. Do the different subunits interact directly with each other or do they require the assistance of other proteins? Certainly the variable regions in cellulose synthases from plants may provide sites for interaction with other proteins. Since the amino acid sequence in the variable regions vary amongst the different cellulose synthases, these regions may provide sites for specific interactions with other proteins. Modeling of the globular region of cellulose synthase from cotton suggests that the variable regions observed in higher plant cellulose synthases are present as loops on the surface where they are accessible for interaction with other proteins [13].

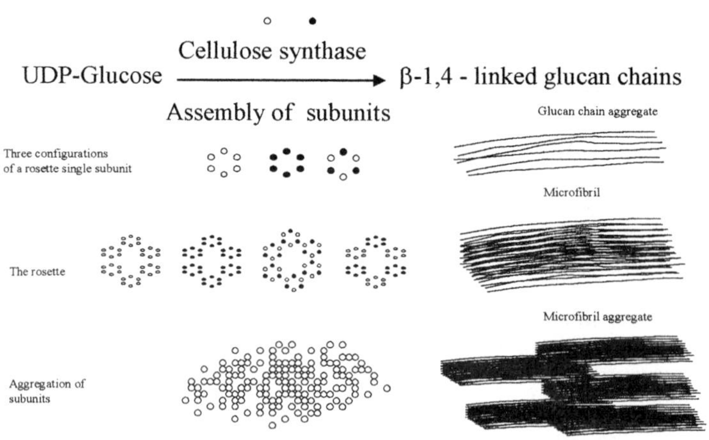

FIGURE 1. Similar or different cellulose synthase catalytic subunits may associate to form rosettes that give rise to a cellulose microfibril. In some cases, during the formation of secondary cell wall, subunits may aggregate without forming rosettes and these aggregates can form microfibril bundles.

FUNCTIONAL ANALYSIS OF CELLULOSE SYNTHASES AND CELLULOSE SYNTHASE-LIKE PROTEINS FROM HIGHER PLANTS

The globular region of cellulose synthases and cellulose synthase-like proteins contain the conserved amino acid residues involved in catalysis and this region is homologous to the globular region of the cellulose synthase from *A. xylinum*. In order to understand the function of the globular region of the plant proteins and develop a system where specific residues or regions can be altered for understanding their function, we have developed a system for substituting the globular region of the bacterial cellulose synthase with the homologous region from the plant cellulose synthases and cellulose synthase-like proteins. The goal is to produce chimeric proteins that have the transmembrane and regulatory (c-di-GMP-binding) regions of the *A. xylinum* cellulose synthase and the catalytic region from plant proteins. Expression of these chimeric proteins in *A. xylinum* would allow for their systematic analysis. We have chosen to express the globular region from two different cellulose synthases and a cellulose synthase-like protein using this system at present (Figure 2). Expression of the chimeric protein using the globular region of the cellulose synthase-like protein has been observed in *A. xylinum*; however, this chimeric protein was not found to produce any cellulose in the *in vitro* assay validating the suggestion that these proteins are probably involved in the synthesis of non-cellulosic polysaccharides.

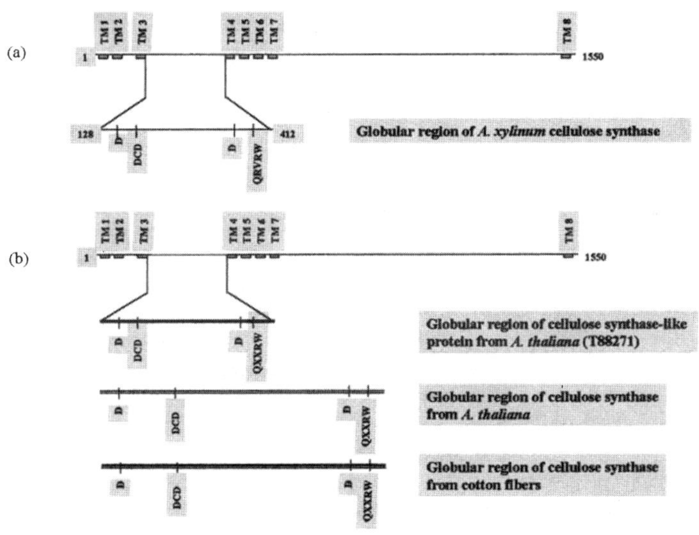

FIGURE 2 Substitution of the (a) globular region of cellulose synthase of *A. xylinum* with (b) globular region of cellulose synthases and cellulose synthase-like proteins from plants.

STRUCTURE AND FUNCTION OF CELLULOSE SYNTHASES AND OTHER GLYCOSYLTRANSFERASES

Interpretations of recently determined structures of glycosyltransferases (of unknown function or those that attach a single sugar residue to an acceptor molecule) suggest a single catalytic center in these enzymes [16]. This proposal is also extended to enzymes for the synthesis of cellulose, hyaluronan, and other β-linked polysaccharides. Although the single active site model may be useful in understanding non-processive addition of sugar residues, it does not explain the 180° rotation of glucose residues in the glucan chains of cellulose or the addition of two different sugars with two different linkages in hyaluronan. A Class II hyaluronan synthase from *Pasteurella multocida* has been shown to have duplication of domain A and the third conserved aspartic acid providing it with two catalytic centers in a single polypeptide chain [17]. Duplicated domains have not been identified by sequence analysis in other processive β-glycosyltransferases and this has led to the thinking that these enzymes have a single catalytic center. However, the two catalytic centers do not have to be generated from duplicated domains. The addition of two sugar residues in a sequential or simultaneous fashion probably requires two catalytic centers that may be present in the same polypeptide chain or two polypeptide chains. So far, no crystal structure has been obtained for any processive β-glycosyltransferase. In our model of the globular region of cellulose synthase, an extended catalytic cavity has been observed. This cavity can accommodate two UDP-glucose residues as well as hold the growing end of the glucan chain [13]. Whether this cavity has two catalytic centers is not known at present.

CONCLUSIONS

In the last few years remarkable progress has been made in the identification of genes coding for cellulose synthases and cellulose synthase-like proteins in plants. Progress in the identification of these genes has certainly allowed determination of the function of some of these proteins by mutant analysis; however, the nature of these proteins from biochemical analysis remains to be understood. A number of approaches will have to be utilized for the functional analysis of the cellulose synthases and cellulose synthase-like proteins identified in higher plants, including expression of complete proteins or parts of these proteins in non-plant hosts where their function can be analyzed. The recent structure determinations of a number of glycosyltransferases has provided an insight into the catalytic centers of enzymes involved in glycosyl transfer and although these structures have been very useful, they fail to explain fully the mechanisms of biosynthesis of cellulose and other polysaccharides. Purification and crystallization of native cellulose synthase or regions of this enzyme will be important in obtaining an understanding of not only the mechanism of glycosyl transfer but also the mode of processivity and the manner in which alternate residues are inverted in the glucan chains of cellulose. Certainly all these studies will aid in understanding the conditions that regulate biosynthesis of cellulose and allow for production of cellulose with desirable properties.

ACKNOWLEDGEMENTS

The authors acknowledge grant support from the U. S. Department of Energy (DE-FG03-94-ER20145).

REFERENCES

1. H. J. Sofia, V. Burland, D. L. Daniels, G. Plunkett III & F. R. Blattner, Analysis of the *Escherichia coli* genome. V. DNA sequence of the region from 76.0 to 81.5 minutes, *Nucleic Acids Res.*, **1994**, 22, 2576-2586.
2. J. R. Pear, Y. Kawagoe, W. E. Schreckengost, D. P. Delmer & D. M. Stalker, Higher plants contain homologs of the bacterial celA genes encoding the catalytic subunit of cellulose synthase, *Proc. Natl. Acad. Sci. USA*, **1996**, 93, 12637-12642.
3. T. Arioli, L. Peng, A. S. Betzner, J. Burn, W. Wittke, W. Herth, C. Camilleri, H. Höfte, J. Plazinski, R. Birch, A. Cork, J. Clover. J. Redmond & R. E. Williamson, Molecular analysis of cellulose biosynthesis in *Arabidopsis*, *Science*, **1998**, 279, 717-720.
4. T. A. Richmond & C. R. Somerville, The cellulose synthase superfamily, *Plant Physiology*, **2000**, 124, 495-1324.
5. N. Holland, D. Holland, T. Helentjaris, K. Dhugga, B. Xoconostle-Cazares & D. P. Delmer, A comparative analysis of the plant cellulose synthase (CesA) gene family, *Plant Physiology*, **2000**, 123, 1313-498.
6. M. Fagard, T. Desnos, T. Desprez, F. Goubet, G. Refregier, G. Mouille, M. McCann, C. Rayon, S. Vernhettes & H. Höfte, *PROCUSTE1* encodes a cellulose synthase required for normal cell elongation specifically in roots and dark-grown hypocotyls of Arabidopsis, *Plant Cell*, **2000**, 12, 2409-2423.
7. N. G. Taylor, S. Laurie & S. R. Turner, Multiple cellulose synthase catalytic subunits are required for cellulose synthesis in Arabidopsis, *Plant Cell*, **2000**, 12, 2529-2539.
8. I. M. Saxena & R. M. Brown, Jr., Identification of a second cellulose synthase gene (*acsAII*) in *Acetobacter xylinum*, *J. Bacteriol.*, **1995**, 177, 5276-5283.
9. I. M. Saxena & R. M. Brown, Jr., Cellulose synthases and related enzymes, *Curr. Op. Pl. Biol.*, **2000**, 3, 523-531.
10. M. Koyama, W. Helbert, T. Imai, J. Sugiyama & B. Henrissat, Parallel-up structure evidences the molecular directionality during biosynthesis of bacterial cellulose, *Proc. Natl. Acad. Sci. USA*, **1997**, 94, 9091-9095.
11. W. Lukowitz, T. C. Nickle, D. W. Meinke, R. L. Last, P. L. Conklin & C. R. Somerville, *Arabidopsis cyt1* mutants are deficient in a mannose-1-phosphate guanylyltransferase and point to a requirement of N-linked glycosylation for cellulose biosynthesis, *Proc. Natl. Acad. Sci. USA*, **2001**, 98, 2262-2267.
12. I. M. Saxena & R. M. Brown, Jr., Identification of a cellulose synthase(s) in higher plants: Sequence analysis of processive β-glycosyltransferases with the common motif 'D,D,D35Q(R,Q)XRW,' *Cellulose*, **1997**, 4, 33-49.

13. I. M. Saxena, R. M. Brown, Jr. & T. Dandekar, Structure-function characterization of cellulose synthase: Relationship to other glycosyltransferases, *Phytochemistry*, **2001** (in press).
14. I. M. Saxena, K. Kudlicka, K. Okuda & R. M. Brown, Jr., Characterization of genes in the cellulose-synthesizing operon (*acs* operon) of *Acetobacter. xylinum*: Implications for cellulose crystallization, *J. Bacteriol.*, **1994**, 176, 5735-5752.
15. S. Kimura, W. Laosinchai, T. Itoh, X. Cui, C. R. Linder & R. M. Brown, Jr., Immunogold labeling of rosette terminal cellulose-synthesizing complexes in the vascular plant *Vigna angularis*, *Plant Cell*, **1999**, 11, 2075-2085.
16. S. J. Charnock, B. Henrissat & G. J. Davies, Three-dimensional structures of UDP-sugar glycosyltransferases illuminate the biosynthesis of plant polysaccharides, *Plant Physiology*, **2001**, 125, 527-531.
17. W. Jing & P. L. DeAngelis, Dissection of the two transferase activities of the *Pasteurella multocida* hyaluronan synthase: two active sites exist in one polypeptide, *Glycobiology*, **2000**, 10, 883-889.

Molecular Breeding of Woody Plants
N. Morohoshi and A. Komamine, editors.

FUNCTIONAL ANALYSIS OF POLYSACCHARIDE SYNTHASES RESPONSIBLE FOR CELL WALL SYNTHESIS IN HIGHER PLANTS

Rachel A. Burton, David M. Gibeaut & Geoffrey B. Fincher*

Department of Plant Science, University of Adelaide, Waite Campus, Glen Osmond, South Australia 5064, Australia

ABSTRACT

Traditional biochemical approaches have rarely been successful in the isolation and characterization of polysaccharide synthase enzymes that are responsible for cell wall biosynthesis in higher plants. However, emerging genetic technologies are now being applied in functional genomics programs to identify candidate genes that might encode these enzymes. While sequence comparisons between candidate genes and related genes in various databases has been used in the past to assign a specific function to candidate genes, in many cases it is not possible to simply use sequence similarities to identify the gene. Thus, an important component of a functional genomics program is the development of a number of systems to identify gene function. Here we briefly review both loss-of-function and gain-of-function systems that can be used to define the identify putative polysaccharide synthase genes involved in cell wall deposition in plants.

KEYWORDS

Gain-of-function, gene knockout, heterologous expression, RNA interference, transformation, virus-induced gene silencing.

INTRODUCTION

Technical difficulties associated with the purification of enzymes responsible for cell wall synthesis in plants have frustrated efforts to characterize these enzymes through normal biochemical routes. Indeed, there are no examples of purified plant enzymes that can be linked unequivocally with the synthesis of the polymeric backbone of cell wall polysaccharides and, as a result, no genes encoding this group of polysaccharide synthases have been identified from the amino acid sequence of a purified enzyme. Difficulties associated with the purification of membrane-bound enzymes, the inherent instability of many polysaccharide synthases, and the likely requirements for crucial cofactors or for other protein constituents of a multi-enzyme complex have all contributed to our poor progress in this area.

A defining breakthrough came from the laboratory of Delmer, who identified candidate genes for cellulose synthases in cotton[1]. Evidence was subsequently accumulated to show that these genes do encode cellulose synthases[2], but examination of the sequence databases generated by many large scale functional genomics programs around the world reveals that most plant species have a large number of putative cellulose synthase or closely related genes[3]. Furthermore, it has become clear that it is not possible to identify a newly discovered gene as a cellulose synthase gene solely on the basis of sequence similarity with confirmed cellulose synthase genes[4]. In the case of cell wall biosynthesis in cereals, polysaccharide products generated by a cellulose synthase, a xylan synthase, a (1,3;1,4)-β-glucan synthase or a xyloglucan synthase have common biochemical and structural features, and one might anticipate that genes encoding such a group of synthase enzymes would share sequence similarities. For example, the backbone (1,4)-β-glucan chain of a xyloglucan molecule would presumably be synthesized by an enzyme very similar in structure with one that synthesizes the unsubstituted (1,4)-β-glucan of cellulose. Either a single enzyme or two closely related enzymes could conceivably participate in the biosynthesis of these two types of polysaccharide. In another example, the (1,4)-β-xylan backbone of cereal glucuronoarabinoxylans is structurally very similar to the (1,4)-β-glucan molecule that makes up cellulose microfibrils. Again the xylan and cellulose synthases would be expected to share structural similarities.

As a result, the functions of individual genes that are believed to encode polysaccharide synthases involved in cell wall synthesis need to be analyzed carefully before conclusions are drawn regarding their biological roles in wall synthesis. Indeed, it is not usually sufficient to base the identification of genes on sequence similarity with other better-characterized genes. A number of functional analysis systems have now been developed and the potential of these methods will be discussed below in relation to the unequivocal identification of genes that mediate plant cell wall biosynthesis.

CELL WALLS IN HIGHER PLANTS

Primary cell walls of higher plants are extracellular structures consisting predominantly of polysaccharides, but also containing proteins, glycoproteins and phenolic acids, that are deposited by growing cells. They are ultimately responsible for the strength and flexibility of the plant but are sufficiently porous to allow the passage of water, nutrients, phytohormones and other small molecules. When cell growth ceases, secondary wall deposition may lead to wall thickening and lignin is often deposited to further strengthen the wall.

The composition of walls varies considerably between plant species, but walls are generally characterized by the presence of cellulosic microfibrils, usually associated with xyloglucans, glucomannans and heteroxylans, embedded in a matrix of pectins, (1,3;1,4)-β-glucans and other components[5,6].

CELLULOSE SYNTHASE AND RELATED GENES

As mentioned above, the cellulose synthase genes of higher plants were the first genes for which a function in cell wall synthesis was firmly established[1,2,4]. However, since that time numerous putative cellulose synthase genes (*CesA*) and cellulose synthase-like genes (*Csl*) have been identified[7]. In addition, other potential β-glucan synthase genes, including callose synthase genes, have been placed in the *Gsl* group[7]. Thus, in *Arabidopsis thaliana* there are more than 12 known *CesA* genes, although the classification of genes into this group is often based on sequence similarity and not all have been subjected to rigorous functional analysis. The *Csl* genes of *Arabidopsis* have been divided into six subfamilies and the cellulose synthase and cellulose synthase-like gene superfamily of *Arabidopsis*, which consists of all the *CesA* and *Csl* genes, has at least 40 members[7].

Members of the superfamily share a number of common features. They encode integral membrane proteins that have similar topographies with respect to the number and disposition of transmembrane domains. Sequence similarities are greatest in the large cytoplasmic region and D,D,D,QxxRW amino acid sequences are believed to represent the nucleotide sugar-binding region of the catalytic site[7].

LOSS-OF-FUNCTION ANALYSES

Early attempts to assign functions to unknown plant genes often involved transformation of the plant with the gene of interest in such a way that expression levels of the endogenous gene would be decreased. Thus, antisense gene constructs were introduced through transformation in the expectation that decreased expression of the gene of interest would lead to a phenotype that could be correlated with the suspected function of the gene. However, results were often variable and difficult to interpret. In particular, both sense and antisense constructs could lower mRNA abundance of the target gene in a process now referred to as co-suppression[8]. Furthermore, transformation in some species was, and remains, problematical; many research groups still find it hard to transform important cereal species such as wheat and barley at high efficiency. Copy number and position effects of the transgene(s) can have unexpected effects on expression, and if loss of function of the target gene proves to be lethal, no transgenic plants can be recovered for analysis. For example, if all cellulose synthesis were blocked by transformation with a single cellulose synthase gene fragment, it seems highly unlikely that a transgenic plant could be regenerated. Despite these difficulties, a number of loss-of-function systems have proved to be useful in the analysis of gene function, and selected examples of these are briefly discussed below.

Virus-induced gene silencing (VIGS)

Virus-induced gene silencing (VIGS) has been used to examine gene function in *Nicotiana benthamiana*[9]. Genes or gene fragments of interest have been inserted into a modified potato virus X (PVX) cDNA and RNA transcripts run off from the cDNA are used to infect young seedlings. Viral particles spread through the plant, although they are excluded from meristematic tissue. Post-transcriptional gene silencing at the mRNA level leads to the destruction of mRNA transcribed from the transgene, as well as mRNA of endogenous genes that share about 80% or more sequence identity with the introduced gene fragment[9]. The precise mechanism of the silencing has not yet been defined, but is likely to involve the formation of double-stranded RNA (dsRNA)[10]. Advantages of the VIGS system are that full-length cDNAs or genes are not required, silencing can be detected much more quickly than in antisense or co-suppression approaches, and the method can be used to silence genes for which loss of activity would be potentially lethal, because seedlings become established before the knockout construct is introduced.

The functions of two putative cellulose synthase genes from *Nicotiana tabacum*, *NtCesA1* and *NtCesA2*, have been tested in the VIGS system[4]. Plants infected with the *NtCesA1* constructs had a dwarf phenotype, short internodes and small leaves. Abnormally large cells ballooned from the epidermal layer of the undersurfaces of leaves, consistent with a weakening of the cell wall. Methylation analyses showed that the cellulose content of the walls had decreased by about 25% in the infected, dwarf plants, and mRNA levels corresponding to the introduced gene were less abundant than in control plants[4]. It was concluded therefore that the introduced *NtCesA1* gene encoded a cellulose synthase, and silenced that gene. In contrast, the *NtCesA2* gene, which had more than 80% sequence identity with the *NtCesA1* gene, had no major effects on plant phenotype or wall composition[4]. This was not to say that the *NtCesA2* gene did not encode a cellulose synthase. It might have encoded a cellulose synthase that was expressed in a different tissue at a different time. Furthermore, the observation that the cellulose content of walls in the dwarf plants only decreased by 25% suggested that other cellulose synthase genes were being expressed and were unaffected by the VIGS construct used. Thus, some interpretative difficulties are encountered when members of a multigene family are used in the VIGS system, and this problem is likely to apply to other gene silencing procedures. In addition, the VIGS system currently works best in *Nicotiana benthamiana* and genes from other species might not silence the endogenous genes because their nucleotide sequences do not exhibit 80% or more identity with the endogenous genes. It therefore becomes difficult to examine potential functions of homologous genes, such as those from wheat or barley, in the *Nicotiana benthamiana* system.

One particularly important result of the VIGS experiments described above was that they demonstrated the presence of interconnecting feedback loops between the cellular pathways that mediate cell wall synthesis. This conclusion was based on the observation that the 25% decrease in wall cellulose content in the VIGS plants was offset by an increase in homogalacturonan content of the walls[4]. The content of other wall polysaccharides was not affected. Furthermore, the degree of esterification of the homogalacturonan decreased from about 50% to about 33%, and this might be expected to enhance wall strength through the formation of extended junction zones in pectic polysaccharides[4]. Similar effects had been noted previously in tissue-cultured cells grown in the presence of the herbicide 2,6-dichlorobenzonitrile (DCB). Walls of DCB-adapted cells had reduced cellulose contents that were offset by large increases in pectic polysaccharides with lower than usual degrees of esterification[11].

Double-stranded RNA interference (RNAi)

Another form of post-transcriptional gene silencing that is finding applications in studies of plant-pathogen interactions and elsewhere is referred to as double-stranded RNA interference, or simply RNAi[12]. In this procedure, individual epidermal cells are bombarded with either dsRNA or cDNA encoding a 'hairpin' RNA that will spontaneously form a double-stranded structure in the cell. Silencing of endogenous genes related to the introduced dsRNA is observed through microscopic examination of single bombarded cells. The mechanism for gene silencing is likely to be the same as for co-suppression and for the VIGS system, where dsRNA is believed to be formed after highly abundant mRNAs accumulate in plant cells and cellular responses that lead to the destruction of the dsRNA will silence both the introduced gene and homologous, endogenous genes.

The method has been used to investigate, quickly and easily, the role of the *Mlo* gene in the resistance of barley to the causal agent of powdery mildew, *Blumeria graminis*. Furthermore, it has been used to define the functions of genes in pigment biosynthesis pathways, again in maize and barley epidermal cells[12]. Advantages of RNAi include the speed and ease with which analyses of gene function can be effected. Gene constructs that encode intron-spliced RNA with a hairpin structure can induce post-transcriptional gene silencing with almost 100% efficiency[13]. In addition, the procedure has been developed for several important plant species, including the cereals wheat, maize and barley[12], and one would anticipate that it will become a commonly used tool for reverse genetics in the future.

Catalytic RNAs (Ribozymes) and DNAs (DNAzymes)

Catalytic RNAs, or ribozymes, can hydrolyse RNA in a sequence-dependent manner and therefore have the potential to silence the expression of specific genes. Ribozymes are often susceptible to hydrolysis by nucleases, but more stable synthetic ribozymes have recently been designed[14]. Catalytic DNAs, or DNAzymes, that catalyse the sequence-specific hydrolysis of RNA have also been identified[15]. Neither catalytic RNAs nor DNAs have been used extensively in functional analyses to silence plant genes, but their possible application in therapeutics is under investigation[14,15].

Transposon-tagged mutant libraries

Genes encoding polysaccharide synthases involved in cell wall biosynthesis could be identified through the analysis of mutants in which obvious lesions have occurred in the process of cell wall deposition. Naturally-occurring mutants or mutants generated by chemical mutagens or by bombardment with atomic particles can be screened by eye at the phenotypic level, or through histochemical or chemical analysis of walls. A number of such mutant collections exist, although their availability is often limited, particularly if they have been generated in the private sector. Moreover, the screening procedure is unlikely to be straightforward and even when it becomes apparent that there is an alteration in a gene that plays a critical role in cell wall biosynthesis, isolation of the affected gene can be a difficult and protracted exercise.

Many of these problems can be overcome through the generation of libraries of mutants in which genes are inactivated by insertion of a specific DNA 'tag'. Upon recognition of the lesion in cell wall synthesis, for example, the inactivated gene can be easily cloned because it is tagged with a DNA fragment with a known sequence. Plant transposable elements have been used as efficient insertional mutagens and are increasingly used in functional genomics programs to identify gene candidates for processes such as cell wall biosynthesis[16]. The maize *Activator-Dissociation* (*Ac/Ds*) system is one of the most common transposon systems for generating tagged libraries of insertional mutants and has recently been applied to produce transposon-tagged libraries of barley[17]. Provided a species can be transformed at a reasonable efficiency, stable transformants carrying the *Ds* element can be generated, together with transformants carrying the *Ac* transposase gene. An important objective is to have numerous transformants in which the *Ds* elements are dispersed widely across the genome. The *Ds* elements can subsequently be activated by crossing the *Ac* and *Ds* transgenic plants. In barley, about 75% of transposition events lead to the movement of the *Ds* elements to linked sites on the genome, and about 25% to unlinked sites[17]. If the *Ds* elements in the stable transformants can be mapped, it might be possible to speed up the process by selecting a line in which the *Ds* element is close to the gene of interest or close to a position known to influence a trait of interest, such as the level of a particular cell wall polysaccharide. In this way, the possibility of the *Ds* element will be transposed into the linked gene is higher than if the gene were unlinked, elsewhere on the genome.

Analysis of cell wall polysaccharide synthesis through mutant libraries could be further enhanced by complementation experiments. Thus, if a mutant had lost its ability to generate a particular wall polysaccharide and a transferred gene restored the ability to produce that wall polysaccharide, the function of the gene in wall synthesis would be confirmed. Indeed, complementation was used to show that the *rsw1* gene of *Arabidopsis* encodes a cellulose synthase[2].

GAIN-OF-FUNCTION ANALYSES

Interpretative difficulties might be experienced in certain loss-of-function systems because these systems often provide only indirect evidence for the role of a particular gene in cellular metabolism. Transformation can be accompanied by genetic rearrangements that could interfere with expression patterns of unrelated genes or that could indirectly silence expression of the target structural gene through changes in genes encoding transcription factors. Gain-of-function systems might therefore find applications in the identification of genes involved in cell wall biosynthesis. Transfer of cereal genes into *Arabidopsis*, tobacco and yeast might ultimately prove to be very useful in analyzing cell wall biosynthesis in cereals. For example, identification of genes encoding cereal cell wall polysaccharide synthases could be effected in yeasts or in dicotyledonous plants, where ancillary proteins or other components of eukaryotic wall synthesis are likely to be present. If a barley (1,3;1,4)-β-glucan synthase gene was inserted into easily transformable species such as *Arabidopsis* and tobacco, or even into yeast, the barley polysaccharide might be expected to be synthesized and deposited in the wall. None of these species normally synthesize (1,3;1,4)-β-glucans, which are found only in the Poaceae family[5,6]. The (1,3;1,4)-β-glucan would be easy to detect in the transgenic plants or in yeast using monoclonal antibodies or histochemical procedures, and could later be analysed more thoroughly by enzymic and chemical methods.

DIRECT ANALYSIS OF EXPRESSED ENZYME ACTIVITY

An even more direct way to define the function of a putative cell wall polysaccharide synthase gene would be to express that gene, or portion of it, in heterologous expression systems in bacteria, yeast or *Pichia pastoris*, or in the baculovirus/insect cell system. Such systems have been used to express mammalian Golgi-derived glycosyl transferases involved in polysaccharide synthesis. The expressed polysaccharide synthases can be easily purified, and enzyme activity and specificity can be analyzed in simple *in vitro* assays. However, the procedures are not without their difficulties with respect to the identification of plant polysaccharide synthases. These enzymes usually have very high molecular masses, they are usually integral proteins of the plasma membrane or Golgi apparatus and have multiple transmembrane domains. Furthermore, many may simply represent single components of a multi-enzyme or multi-protein complex that will exhibit no activity in isolation. For these reasons, together with associated difficulties encountered in obtaining correct folding of proteins in heterologous systems, the usefulness of this approach might be limited. Attempts to overcome some of these problems by expressing only the predicted cytoplasmic, soluble region of the enzyme might well be successful, but one would not expect this to be so in all cases.

CONCLUSIONS

As large functional genomics and genome sequencing programs generate lists of candidate genes for polysaccharide synthases that are required for the deposition and modification of cell wall polysaccharides in higher plants, the availability of a range of rapid, high-throughput functional analysis systems will become increasingly important. The functions of individual genes will need to be defined unequivocally and it is likely that a combination of methods, generating corroborating evidence, will be required to provide proof-of-function for the genes. Robust functional analysis systems, particularly those that are easily transferable for use in different species, will be in demand and will be linked to the detailed structural analysis of cell wall polysaccharides during plant growth and development. Understanding how these genes mediate wall deposition and modification will undoubtedly suggest ways in which plant productivity might be enhanced, it will offer solutions to certain processing difficulties, and will present opportunities to improve the quality of plant products.

ACKNOWLEDGMENTS

This work was supported by grants from the Australian Research Council and from the Grains Research and Development Corporation of Australia.

REFERENCES

1. J.R. Pear, Y. Kawagoe, W.E. Schreckengost, D. P. Delmer & D.M. Stalker, 'Higher plants contain homologs of the bacterial *cel*A genes encoding the

catalytic subunit of cellulose synthase'. *Proc. Natl. Acad. Sci. USA*, **1996**, 93, 12637-12642.

2. T. Arioli et al. 'Molecular analysis of cellulose biosynthesis in Arabidopsis'. *Science,* **1998**, 279, 717-720.
3. S. Cutler & C.R. Somerville, 'Cellulose synthesis: cloning in silico' *Curr. Biol.*, **1997**, 7, 108-111.
4. R.A. Burton, D.M. Gibeaut, A.Bacic, K. Findlay, K. Roberts, A. Hamilton, D.C. Baulcombe & G.B. Fincher, 'Virus-induced silencing of a plant cellulose synthase gene'. *Plant Cell*, **2000**, 12, 691-705.
5. A. Bacic, P.J. Harris & B.A. Stone, 'Structure and function of plant cell walls'. In: *The Biochemistry of Plants: A Comprehensive Treatise, Vol. 14, Carbohydrates*, J. Preiss, (ed.), Academic Press, New York, **1988**, pp. 297-371.
6. N.J. Carpita & D.M. Gibeaut, 'Structural models of primary cell walls in flowering plants: Consistency of molecular structure with the physical properties of the walls during growth', *Plant J.*, **1993**, 3, 1-30.
7. T.A. Richmond & C.R. Somerville, 'The Cellulose Synthase Superfamily'. *Plant Physiol.* **2000**, 124(2): 495-498.
8. J.M. Kooter, M.A. Matzke & P. Meyer, 'Listening to the silent genes: transgene silencing, gene regulation and pathogen control', *Trends Plant Sci.*, **1999**, 4, 340-347.
9. M.T. Ruiz, O. Voinnet & D.C. Baulcombe, 'Initiation and maintenance of virus-induced gene silencing' *Plant Cell*, **1998**, 10, 937-946.
10. D.C. Baulcombe, 'Fast forward genetics based on virus-induced gene silencing' *Curr. Opin. Plant Biol.*, **1999**, 2, 109-113.
11. E. Shedletzky, M. Schmuel, T. Trainin, S. Kalman & D.P. Delmer, 'Cell wall structure in cells adapted to growth on the cellulose synthesis inhibitor 2,6-dichlorobenzonitrile', *Plant Physiol.*, **1992**, 100, 120-130.
12. P. Schweizer, J. Pokorny, P. Schulze-Lefert & R. Dudler, 'Double-stranded RNA interferes with gene function at the single-cell level in cereals' *Plant J.*, **2000**, 24, 895-903.
13. N.A. Smith, S.P. Singh, M-B. Wang, P.A. Stoutjesdijk, A.G. Green & P.M. Waterhouse, 'Total silencing by intron-spliced hairpin RNAs' *Nature*, **2000**, 407, 319-320.
14. N. Usman & L.M. Blatt, 'Nuclease-resistant synthetic ribozymes: developing a new class of therapeutics' *J. Clin. Invest.*, **2000**, 106, 1197-1202.
15. L.M. Khachigian, 'Catalytic DNAs as potential therapeutic agents and sequence-specific tools to dissect biological function' *J. Clin. Invest.*, **2000**, 106, 1189-1195.
16. D. Bouchez & H. Hoefte, 'Functional genomics in plants' *Plant Physiol.*, **1998**, 118, 725-732.
17. T. Koprek, D. McElroy, J. Louwerse, R. Williams-Carrier & P.G. Lemaux, 'An efficient method for dispersing *Ds* elements in the barley genome as a tool for determining gene function', *Plant J.*, **2000**, 24, 253-263.

Molecular Breeding of Woody Plants
N. Morohoshi and A. Komamine, editors.

ANALYSIS OF SECONDARY CELL WALL FORMATION IN *ARABIDOPSIS*

Simon R. Turner, Neil G. Taylor and Louise Jones

University of Manchester, School of Biological Science, 3.614 Stopford Building, Oxford Road, Manchester M13 9PT, United Kingdom

ABSTRACT

A genetic approach has been taken to study secondary cell wall formation in *Arabidopsis*. Xylem elements that fail to form a secondary cell wall are unable to withstand the negative pressure generated during water transport and collapse inwards. We have named plants that exhibit these collapsed xylem phenotype irregular xylem (irx) mutants. To date, of the 5 *irx* complementation groups identified, one (*irx4*) is deficient in lignin deposition and the remaining 4 are deficient in cellulose deposition. Although the secondary cell walls of *irx4* plants have wild type levels of cellulose and xylan, they are greatly expanded, demonstrating the importance of lignin in cross-linking components of the secondary cell wall. The gene defective in *irx4* is one of 10 genes identified in *Arabidopsis* that appear to encode Cinnamoyl CoA Reductase. We have demonstrated that *irx3* is caused by a mutation in a member of the CesA gene family (AtCesA7). Furthermore, it has been shown that mutations in *irx1* are caused by a defect in another member of the same gene family. We have shown that both of these gene products are essential for cellulose synthesis in the same cell types and that IRX1 and IRX3 interact directly as components of the same cellulose synthase complex. More recently, we have demonstrated that *irx5* is caused by a defect in AtCesA4. We are currently studying the reasons why IRX1, IRX3 and IRX5 are all essential for cellulose synthesis and how different CesA family members are organised within the cellulose synthase complex.

KEY WORDS

Arabidopsis, mutant, xylem, cellulose, lignin, protein complex.

INTRODUCTION

Plant cell walls may be classified as primary or secondary cell walls. Primary cell walls are synthesised while the cell is still expanding. The cellulose-xyloglucan network is considered the main load-bearing network and is thought to be essential in controlling cell expansion. In addition, the orientation of cellulose microfibrils within the wall controls the direction of cell expansion. Consequently, cellulose within the plant cell wall has a key role in controlling cell shape and hence plant morphology. In contrast, secondary cell walls are laid down once the cell has attained its final shape. These secondary cell walls are often responsible for the mechanical strength of plant material. The essential role of cellulose in secondary cell wall formation is well documented. Plants that exhibit a decrease in the cellulose content of the secondary cell wall have dramatically altered physical properties[1].

Until relatively recently, no genes for any of the subunits of the higher plant cellulose synthase subunit had been cloned. This situation changed when Pear *et al.*[2] described a clone from cotton (now described as a member of the CesA gene family) which showed homology to the catalytic subunit of bacterial cellulose synthases and contained several conserved sequences indicative of a processive glucosyl transferase. Conclusive proof that a member of this family of genes represents the higher plant cellulose synthase was provided by studies on a temperature sensitive mutant of *Arabidopsis* (*rsw1*)[3]. At the restrictive temperature, *rsw1* plants die at an early stage and have only half the cellulose content of the wild type. *rsw1* has a mutation in a member of the CesA family of genes[3]. We have also shown that the *irx3* mutation, which has a specific defect in secondary cell wall cellulose synthesis[1], is caused by a mutation in a gene corresponding to a different member of the CesA family[4]. Analysis of the completed *Arabidopsis* genome suggests that it contains a superfamily containing more than 40 genes showing homology to bacterial cellulose synthases (http://cellwall.stanford.edu/cellwall/index.html). The CesA genes form a clear subfamily. There are at least 10 members of the CesA gene family in *Arabidopsis*. The role of these different CesA family members is an area of intense interest.

Much of the information on the structure of the higher plant cellulose synthase complex has come from scanning electron microscopy of freeze fractured plasma membranes. Such studies have revealed the existence of rosettes made up of six 'globules' embedded in the plasma membrane. These rosettes are considered to be the cellulose synthase complex[5]. Unequivocal confirmation that these rosettes are the sites of cellulose synthesis has come from genetic studies on the temperature sensitive mutant *rsw1*. At the restrictive temperature *rsw1* plants exhibit reduced cellulose in the primary cell wall, the breakdown of the organisation of rosettes to give disorganised globules and the synthesis of β(1-4) glucose chains not organised into crystalline microfibrils[3]. The fact that a mutation in a CesA gene causes the rosettes to become disorganised clearly indicates an essential role for these genes in both the catalysis of β(1-4) linked glucose and the organisation of the cellulose synthase complex.

Several models have suggested a very complex structure for the cellulose synthase complex. For example Delmer and Amor[5] have suggested that each globule of the rosette contain six subunits of each polypeptide required to synthesise cellulose. Consequently, according to this model each rosette would be a '36mer', simultaneously synthesising 36 chains of cellulose, the number required to make a microfibril. Using solid state NMR, however, Ha *et al.*[6] have suggested that cellulose is synthesised initially as an 'elementary fibril', which is composed of approximately 18 chains. Larger microfibrils are constructed from these elementary fibrils. To date, however, it is unclear exactly how many β(1-4) cellulose chains are synthesised by a single rosette and whether a rosette synthesises one or more elementary fibrils. A proper understanding of how the different CesA proteins are organised within the rosette and how many chains of cellulose are made by each rosette is clearly a prerequisite to understanding how higher plants synthesise cellulose.

Lignin is the second most abundant polymer in the secondary cell wall. Whilst many of the steps involved in the lignin biosynthesis pathway have been identified and characterised in a variety of different plant species, many questions remain. For example, how are lignin monomers transported out of the cell into the wall and how are they polymerised within the wall. How different secondary cell wall polymers such as lignin and cellulose are assembled together within the wall also remains unclear.

The many advantages of *Arabidopsis* as a model for molecular genetic analysis are well documented. The availability of the complete genome sequence that can be used in conjunction with mass spectrometry (MS) analysis for protein identification, and large populations of insertional mutants for reverse genetics are invaluable tools. Secondly, rates of secondary cell wall synthesis are high in stems of the appropriate age.

Whilst cellulose constitutes only a small percentage of seedlings, up to 35% of the ethanol-insoluble fraction of mature stems is cellulose[1]. Consequently developing stems are an excellent source of starting material for any biochemical analysis. Most importantly it is possible to isolate very severe mutations. We have previously isolated *Arabidopsis* irregular xylem (*irx*) mutants that synthesise little or no cellulose in the secondary cell wall[1]. For example, the cellulose content in stem segments of *irx3* plants is decreased more than 5-fold (the reduction in secondary cell wall cellulose is even greater), but despite this the plants remain relatively healthy[1].

MATERIALS AND METHODS

Tissue Prints

Inflorescence stems were cut cleanly from plants that had recently bolted using a razor blade and the cut surface pressed onto Immuno blot PVDF membrane (Biorad) which had been wetted in methanol and then equilibrated in water. After 4 seconds of gentle pressure, a second print was made of the same cut surface on another piece of PVDF. A section was then taken by hand using a razor blade and stained with Toluidine blue in order to visualise the distribution of tissues within the printed section. The tissue prints were then blocked in 5% skimmed milk powder in TBS 0.1% Tween 20 (TBS-T) for 60 minutes followed by incubation in either 1/5000 anti IRX3 antisera or 1/1000 anti IRX1 antisera[7] diluted in 10% skimmed milk powder in TBS for 60 minutes. After three ten minute washes in TBS-T, the blots were incubated in 1/1000 alkaline phosphatase conjugated anti sheep secondary antibody in 10% skimmed milk powder in TBS. After 60 minutes, the blots were again washed three times in TBS-T and the signal detected using BCIP/NBT. When a reasonable signal was observed the reaction was stopped by washing in a large excess of water before drying the blots for visualisation under a microscope.

Construction of epitope tagged IRX3

An 8.3 kb XhoI-MunI genomic DNA fragment carrying the entire *IRX3* coding region and 1.7 kb of promoter sequence[4] cloned into pCB2300 was cut with NheI and a double stranded oligonucleotide (the product of annealing His1 (5'-CTAGGGGATCCCATCACCATCACCATCACC-3') and His2 (5'-CTAGGGTCATGGTGATGGTCATCGGATCCC-3') ligated to insert the epitope. The correct insertion of this epitope was confirmed by sequencing the relevant area of the gene. This construct was transformed into *irx3* plants by vacuum infiltration.

Purification of epitope tagged IRX3

1 g of stems from transformed plants were ground well in lysis buffer (50 mM NaH_2PO_4 pH 8.0, 300 mM NaCl) containing 10 mM imidazole to reduce non-specific interactions. After clarification by centrifugation at 13000 rpm in a microcentrifuge, Triton X-100 was added to a final concentration of 2%. 100 μl of NiNTA Superflow (Qiagen) was added to these solubilised extracts, which were mixed end over end for 60 minutes. After centrifugation as before, the resin was washed 3 times with 250 μl of lysis buffer containing 20 mM imidazole. Proteins were eluted from the resin twice with 30 μl lysis buffer containing 250 mM imidazole. The entire purification procedure was carried out at 4 °C in the presence of protease inhibitors (Protease Inhibitor cocktail for mammalian cell extracts, Sigma, Poole, Dorset). 10 μl aliquots were denatured in loading buffer for 60 minutes at 37 °C before electrophoresis through 7.5% SDS

polyacrylamide gels. After transfer to Immuno-blot PVDF membrane (Biorad), protein gel blots were carried out following standard protocols. Epitope tagged IRX3 was detected using an anti-RGSHis monoclonal antibody (Qiagen) at a dilution of 1/1000 and IRX1 was detected using anti IRX1 antisera at a dilution of 1/1000. Secondary antibodies conjugated to alkaline phosphatase were used followed by colormetric development using BCIP/NBT.

Neutral Sugar Content

The neutral sugar content of wild type and *irx4* cell walls was assessed by gas chromatography (GC). Mature stems, harvested from 6 week old wild-type and *irx4* plants, were divided into four equal parts (designated tip, upper middle, lower middle and base) and crude cell wall preparations were isolated from each section of stem material. Crude cell wall fractions were obtained following the extraction of soluble material in 70% (v/v) ethanol at 70°C for 1 hour[8]. The dry weight of the cell wall material was recorded prior to the analysis of neutral sugar content. The cell wall material was initially hydrolysed in 2 M H_2SO_4 for 1 hour at 121°C and the alditol acetate derivatives of these sugars analysed by GC, as previously described[1]. All measurements were carried out on at least 6 replicates for each developmental stage.

Phenolic measurements

The lignin content of wild-type and *irx4* cell wall material, isolated as described above, was determined by thioglycolic acid (TGA) analysis[9]. The crude cell wall preparations were treated with 1 M NaOH prior to extraction with TGA for 3 hours at 80°C. Following centrifugation, the insoluble material was washed with distilled water and incubated overnight in 1 M NaOH on a rotating shaker at room temperature. The supernatant was collected and transferred to a fresh tube and 200 μl of concentrated HCl added. The precipitate was collected by centrifugation and resuspended in 1 M NaOH. All samples were diluted 10-fold and the absorbance measured at 280 nm. All measurements were again carried out on at least 6 replicates.

RESULTS

Genetic analysis of secondary cell wall formation

The secondary cell walls of the tracheary elements are specialised to withstand the negative pressures generated during the transport of water and solutes. Bean seedlings grown in the presence of the PAL inhibitor AOPP fail to synthesise and deposit normal

Complementation group	No. of alleles	Defect
irx1	4	Reduced Cellulose
irx2	2	Reduced Cellulose
irx3	2	Reduced Cellulose
irx5	3	Reduced Cellulose
irx4	1	Reduced Lignin
Unassigned	3	Unknown

Table 1. Summary of known *irx* complementation groups.

levels of lignin in the secondary cell wall[10]. Consequently, the tracheary elements of these plants collapse inwards. A number of *Arabidopsis* mutants have been isolated that exhibit a similar phenotype. These mutants have been named *irregular xylem* (*irx*) due to the collapsed appearance of their tracheary elements. To date, 12 mutants have been isolated from at least 5 different complementation groups (Table 1). Four of these complementation groups correspond to plants that exhibit decreased cellulose deposition in the secondary cell wall. The remaining complementation group appears to exhibit decreased lignin deposition in the secondary cell wall.

All of the *irx* mutants characterised to date appear to act as recessive Mendelian loci and segregate in a 3:1 manner. In addition, the plants all appear to grow quite normally and are fertile. Other than the irregular xylem, the only other phenotype caused by the collapsed xylem vessels is a slight decrease in stature and a slightly darker green coloration.

Characterisation of *irx4*

Examination of the secondary cell walls of *irx4* plants using both light and electron microscopy show the walls to be much thicker than in wild type. In mature plants the secondary wall may expand to fill almost the entire cell[11]. Furthermore, in contrast to wild type secondary cell walls, which stain blue with toluidine blue, the walls of *irx4* plants stain very poorly. In addition, an abnormal staining pattern, with light and dark staining areas, is also revealed using TEM.

The phenolic content of mature stems from *irx4* plants is approximately 50% that of wild type[11]. This figure is in agreement with solid state NMR data that demonstrates a 50% reduction in lignin in the mutant. The accumulation of phenolics in *irx4* appears to occur later in secondary cell wall formation than it does in the wild type. This lag in lignin accumulation, in addition to the overall decrease, may contribute to the alterations in cell wall morphology observed in *irx4* plants.

The effect of *irx4* on other secondary cell wall components has been examined by measuring cellulose and neutral sugar composition from developing stems. *irx4* plants have similar levels to the wild type, they accumulate slightly less cellulose than the corresponding wild type plants throughout development. It is unclear, however, whether these small differences are important in view of the alterations in growth rate and stature observed for *irx4* plants. Similarly, there are little differences between neutral sugar composition between wild type and *irx4* plants. Whilst there is a increase in xylose during stem development, correlated with increased secondary wall deposition, and a decrease in arabinose, the pattern is very similar for both *irx4* and wild type plants[11]. These results demonstrate that it is possible to specifically decrease lignin deposition without substantially affecting the other major secondary cell wall components.

Ultrastructure of *irx* mutant cell walls

Comparison of *irx3*, the most severe cellulose deficient mutant, with the lignin deficient mutant *irx4*, demonstrates that these two polymers appear to have opposite effects on cell wall morphology. Both light and electron microscopy show that *irx3* plants have thin, uneven, darkly staining secondary walls. In contrast, *irx4* plants exhibit walls that are much thicker than the wild type and expand to fill almost the entire cell[11]. Since there appears to be no increase in other secondary cell wall components, such as xylan and cellulose[11], the increase in secondary cell wall thickness in *irx4* plants is due to an expansion of the existing cellulose-xylan network. These observations demonstrate the importance of lignin in the structure of the cell wall and in particular the way it appears to be the 'glue' that holds other secondary cell wall components together.

Cloning *IRX1* and *IRX3*

The *irx3* mutant was initially mapped to the top arm of chromosome V. Analysis of a large number of ESTs showing homology to bacterial cellulose synthases[12] revealed that one of these ESTs (75G11) mapped to a region close to *irx3*. Subsequent complementation analysis demonstrated that the *irx3* mutation was indeed caused by a mutation in the gene corresponding to 75G11[4]. Using the systematic nomenclature suggested by Delmer[13] this gene corresponds to AtCesA7.

Initial analysis of *irx1* showed that it mapped to the top arm of chromosome 4. Subsequent completion of the genome sequence in this region revealed the presence of another member of the CesA gene family. Complementation analysis confirmed that *irx1* was indeed caused by a mutation in the AtCesA8 gene.

Careful examination of the tracheary elements of the xylem using light and transmission electron microscopy indicated that both the *irx1* and *irx3* mutations appear to give rise to an identical phenotype. The secondary cell walls of the tracheary elements from both of these mutants have characteristic even, thin, dark-staining secondary cell walls.

Analysis of the interaction between IRX1 and IRX3

Since we have both the *irx3* mutation and the gene that complements the mutation, we are able to insert epitopes into the *IRX3* gene and ensure that this does not disrupt the way in which the protein functions by demonstrating that the recombinant protein still complements the *irx3* mutation. Initial experiments have utilised an RGSHis tag, which contains a run of 6 histidines for use in immobilised metal affinity chromatography as well as a recognition site for a monoclonal antibody. Insertion of this tag close to the N-terminus results in a fully functional protein, which may be recognised using the monoclonal antibody (fig. 1). Whilst the RGSHis tag is comparatively small at only 9 amino acids, we have recently shown that it is possible to add GFP at the same site and still retain normal activity. In addition, we have raised highly specific antibodies to both variable region 1 and the constant regions of IRX3 (fig. 1).

The epitope-tagged IRX3 protein was solubilised in Triton X100 and incubated with a metal affinity resin. A substantial proportion of the protein bound to the resin when they were spun down. Using an IRX1 specific polyclonal antibody it was possible to demonstrate that a similar proportion of IRX1 was also co-precipitated with the IRX3 protein[7]. Precipitation of IRX3 with the affinity resin is completely dependent upon the hexa-histidine tag and other plasma membrane markers did not co-precipitate. Taken together these results demonstrate that there is a specific interaction between IRX1 and IRX3 and that they are likely to be part of the same complex.

Localisation of IRX3 and IRX1

We have used tissue printing as a convenient means of examining the localisation of IRX1 and IRX3. Using successive prints from the cut surface of a mature inflorescence stem it is clear that IRX3 and IRX1 have a very similar distribution. Both proteins localise to the xylem and to the cells of the interfascicular region. This is in agreement with the phenotype of *irx3* mutant plants that show dramatic alterations in cellulose content in both the xylem and interfascicular region. In contrast, the phenotype of *irx1* plants exhibit a much less dramatic affect on the interfascicular cells. It is unclear at

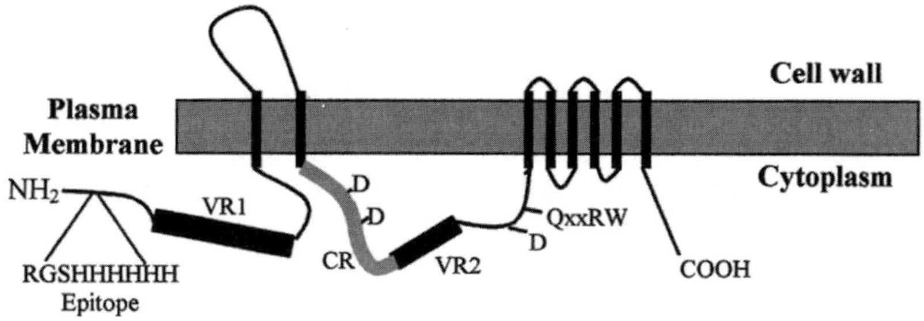

Figure 1. Schematic diagram showing the predicated organisation and membrane topology of the epitope tagged IRX3 protein. The constant region (CR), variable regions 1 and 2 (VR1, VR2) are indicated together with three aspartate residues (D) and QxxRW motif conserved in all processive glucosyl transferases. Antibodies were raised against variable region 1 and the constant region. The position of the epitope tag close to the amino terminus is also indicated.

present why the phenotype of *irx1* plants is less pronounced in the interfascicular region. However, there is the possibility that some functional redundancy exists and another member of the gene family may be able to compensate for the absence of IRX1 function.

Characterisation of *irx5*

Our preliminary analysis suggests that, like *irx1* and *irx3*, *irx5* is caused by a mutation in another member of the *Arabidopsis* CesA family. Using a similar approach to that described for the interaction between IRX3 and IRX1, IRX5 appears to be part of a complex containing IRX1 and IRX3.

CONCLUSIONS

Our data suggests that at least three members of the *Arabidopsis* CesA gene family

Gene name	Mutant	Reference
AtCesA1	(*radial swelling1*) *rsw1*	3
AtCesA2		
AtCesA3	(*isoxaben resistant*) *ixr1*	Sheible and Somerville unpub.
AtCesA4	(*irregular xylem5*) *irx5*	Taylor *et al.* unpub.
AtCesA5		
AtCesA6	(*isoxaben resistant*) *ixr2*	Fagard and Hofte unpub.
	(*procuste*) *prc1*	14
	Quill	15
AtCesA7	(*irregular xylem3*) *irx3*	4
AtCesA8	(*irregular xylem1*) *irx1*	7
AtCesA9		
AtCesA10		

Table 2. Summary of known mutations in Arabidopsis CesA gene family

are required to make cellulose (Table 2). In addition, IRX1, IRX3 and IRX5 are all apparently specific for secondary cell wall cellulose biosynthesis. Mutations in *RSW1* (AtCesA1) or *PRC1* (AtCesA6) appear to affect the primary cell wall. In addition, *ISOXABEN RESISTANCE1* (AtCesA3) may also be a cellulose synthase involved in primary cell wall biosynthesis. Consequently, it is possible that two non redundant groups of three CesA genes are required to make cellulose in the primary (AtCesA1,3 and 6) or secondary cell wall (AtCesA4, 7 and 8). Many question about how these rosettes are organised and why so many different family members are required awaits further study.

REFERENCES

1. S.R. Turner and C.R. Somerville,Collapsed xylem phenotype of Arabidopsis identifies mutants deficient in cellulose deposition in the secondary cell wall. Plant Cell **1997**, 9, 689-701.
2. J.P. Pear, Y. Kawagoe, W.E. Schrenkengost, D.P. Delmer and D.M. Stalker, 'Higher plants contain homologs of the bacterial CelA genes encoding the catalytic subunit of cellulose synthase', Proc. Natl. Acad. Sci, USA **1996,** 93, 12637-12642.
3. T. Arioli, L. Peng, A.S. Betzner, J. Burn, W. Wittke, W. Herth, C. Camilleri H. Hofte J.Plazinski, R. Birch and R Williamson, 'Molecular analysis of cellulose synthesis in Arabidopsis', Science **1998,** 279 717-720.
4. N.G. Taylor W.-R. Shieble, S. Cutler, S., C.R. Somerville and S.R. Turner, 'The irregular xylem3 locus of Arabidopsis encodes a cellulose synthase required for secondary cell wall deposition'. Plant Cell, **1999**, 9, 689-701.
5. D.P. Delmer and Y.Amor, 'Cellulose biosynthesis'. Plant Cell **1995,** 7, 987-1000.
6. M.A. Ha, D.C. Apperley, B.W. Evans, M. Huxham, W.G. Jardine, R.J. Vietor, D. Reis, B. Vian and M.C. Jarvis. Fine structure in cellulose microfibrils: NMR evidence from onion and quince. Plant Journal **1998**,16, 183-190.
7. N.G. Taylor, S. Laurie and S.R. Turner, 'Multiple Cellulose Synthase Catalytic Subunits are required for cellulose synthesis in Arabidopsis'. Plant Cell **2000**, 12, 2529-2539.
8. W.D. Reiter, C.Chapple and C.R. Somerville, 'Altered growth and development in a fucose deficient cell wall mutant of Arabidopsis'. Science, **1993**, 261, 1032-1035.
9. M.M. Campbell and B.E. Ellis, 'Fungal elicitor-mediated responses in pine cell cultures: Cell wall-bound phenolics'. Phytochemistry, **1992**, 31, 737-742.
10. C.C. Smart and N. Amrhein, 'The influence of lignification on the development of vascular tissue in Vigna radiata L.' Protoplasma, **1985**, 124, 87-95.
11. L. Jones, A.R. Ennos and S.R. Turner, 'Cloning and characterisation of *irx4*: a severe lignin deficient mutant of Arabidopsis'. Plant Journal **2001** in press.
12. S. Cutler and C. Somerville, 'Cellulose synthesis: cloning *in silico*'. Curr. Biol., **1997**, 7, R108-R111.
13. D.P. Delmer, 'Cellulose biosynthesis: Exciting times for a difficult field of study'. Annu. Rev. Plant Physiol. Plant Mol. Biol. **1999,** 50, 245-276.
14. M. Fagard. T. Desnos, T. Desprez, F. Goubet, G. Refegier, G. Mouille, M. McCann, C. Rayon, S. Vernhettes and H. Hofte, 'PROCUSTE1 encodes a cellulose synthase required for normal cell elongation specifically in roots and dark-grown hypocotyls of Arabidopsis'. Plant Cell, **2000**, 12, 2409-2423.
15. M.-T. Hauser, A. Morikami, and P.N. Benfey, 'Conditional root expansion mutants of Arabidopsis'. Development **1995,** 121, 1237-1252.

Molecular Breeding of Woody Plants
N. Morohoshi and A. Komamine, editors.

ORGANIZATION OF CELLULOSE-SYNTHESIZING TERMINAL COMPLEXES

Kazuo Okuda & Satoko Sekida

Department of Natural Environmental Science, Faculty of Science, Kochi University, 2-5-1 Akebono-cho, Kochi 780-8520, Japan.

ABSTRACT

A marine dinoflagellate *Scrippsiella hexapraecingula* Horiguchi et Chihara produces a cell covering called a pellicle in the non-motile phase. We found cellulose microfibrils present in the pellicle and new cellulose-synthesizing terminal complexes (TCs) that synthesize those microfibrils. Treatment of pellicle with acetic-nitric reagent yielded 2 to 14-nm microfibrils, which were shown to be cellulosic by electron diffraction. Negatively stained preparations revealed these microfibrils to be composed of 2-nm fibrils. The TCs were linear particle arrays that are associated with the tip of microfibril impressions in the PF of the plasma membrane. Five to forty particles, each with a diameter of 5-15 nm, were arranged in two rows, but the positions of the particles were almost random. The length and the width of the TCs ranged 62-290 nm and 15-31 nm, respectively. Several TCs up to 7 were often associated laterally with each other and consolidated as a cluster. Such a TC cluster synthesized bands of microfibrils. Thus, the TCs of *Scrippsiella hexapraecingula* are quite distinctive from those found so far both in the particle arrangement of TCs and in the structure of microfibrils synthesized.

KEYWORDS

cellulose microfibril, cellulose-synthesizing terminal complex, dinoflagellate, freeze fracture, pellicle, *Scrippsiella hexapraecingula*

INTRODUCTION

Cellulose microfibrils are synthesized generally by plasma membrane-bound enzymes[1,2]. Freeze-fracture investigations have demonstrated particle aggregates in association with the tip of microfibril impression in the fractured plasma membrane[3,4]. The particle aggregates are called terminal complexes (TCs), which are assumed to function as cellulose-synthesizing enzyme complexes (see Brown 1985[5] for review). In this concept, TCs are assumed to passively move on the plasma membrane under a force that is generated when cellulose microfibrils crystallize[6]. Recently, the catalytic subunit of cellulose synthase has been shown to be associated with TCs in vascular plants[7]. Several distinct TCs have been found in various organisms, although all TCs have the common function of synthesizing cellulose microfibrils[8]. It has been pointed out that distinct TCs synthesize microfibrils with characteristic morphologies[9]. This suggests the occurrence of a relationship between the organization of TCs and microfibril assembly. However, there is also an example that distinct TCs synthesize similar microfibrils[10].

Cellulose is found widely in different phylogenetic groups such as prokaryotes, slime molds, glaucophytes, chlorophytes including land plants, rhodophytes, haptophytes, chromophytes, fungi and invertebrates[11]. The presently known variations in TC organization and microfibril structure reflect a divergent evolution for cellulose synthesis and their regulation. This suggests that some distinct origins that had acquired and evolved the organized enzyme structure essential for cellulose microfibril assembly occurred independently in different evolutionary lines. Searching a new TC and microfibrils that the TC synthesizes is, therefore, one of clues for understanding the origin and evolution of cellulose biogenesis. In the present study, the structures of TCs and cellulose microfibrils in the dinoflagellate *Scrippsiella hexapraecingula* are demonstrated and compared with those already known in other organisms in order to consider cellulose microfibril assembly in an evolutionary aspect.

MATERIALS & METHODS

Culture

Scrippsiella hexapraecingula Horiguchi et Chihara was collected from tide pools on the Turugizaki coast, Miura Peninsula, Kanagawa Prefecture, Japan, on 30 August, 1992. Unialgal cultures were maintained in PES medium[12] at 22 °C in a 14:10 h LD cycle at a photon flux density of 90 $\mu E/m^2/s$. Under these conditions, motile cells swam during the light period and settled on the substrate to become non-motile cells at the beginning of the dark period[13]. A cell covering called pellicle formed a cell periphery in the non-motile phase. One or two daughter cells were produced internally, and they escaped through the pellicle and became motile cells after the beginning of the next light period.

For preparation of non-motile cells, the motile-cell suspension was harvested 1 h before the beginning of the dark period. A drop of the cell suspension was put on a small (4x4 mm) piece of membrane filter (Millipore JAWPO 4700), which was directly placed on a piece of filter paper (Whatman GF/C) to absorb excess cell culture medium. This procedure transformed motile cells into non-motile cells settling on the membrane filter and thus induced pellicle formation. The membrane filters were transferred into petri-dishes containing fresh medium and cultured for 1-2 hours. These non-motile cells were used for freeze-etching experiments.

Negative staining and electron diffraction

Non-motile cells were sonicated in distilled water for 1 min and treated with 5 % sodium hypochlorite solution for 2 days. After washing with distilled water, the suspended material was extracted with acetic-nitric reagent in a boiling-water bath for 30 min. Crystalline cellulose is insoluble in the acetic-nitric reagent used for the determination of cellulose[14]. The acid-insoluble residue was rinsed with water, mounted on formvar-coated grids or carbon-coated grids, negatively stained with 1 % uranyl acetate, and then observed with a JEOL JEM 1010-T electron microscope. For electron diffraction, cellulose microfibrils isolated from the tunic of the tunicate *Halocynthia* as a standard were used. Electron diffraction was carried out with the electron microscope operating at 100 kV, and a camera length of 15 cm. Diffraction patterns were recorded on Mitsubishi MEM electron microscope films.

Freeze fracture

The procedures of freeze fracture electron microscopy were similar to those described by Okuda *et al.* (1994)[10]. Before freeze fixation, non-motile cells were scratched with a razor from the surface of the membrane filter. The cells were mounted on the holes of 3-mm double replica aluminum supports, immediately frozen without prior chemical fixation in liquid propane cooled with liquid nitrogen, and then stored in liquid nitrogen until fracture. Freeze fracture and metal shadowing were performed with a Baltec BAF 060 apparatus at –106 °C and 1×10^{-6} mbar. Specimens were shadowed unidirectionally at an angle of 60 degrees with platinum-carbon and subsequently coated with carbon. Replicas were cleaned by placing them in a 2.5% sodium dichromate-50% sulfuric acid mixture overnight, washed with distilled water, and mounted on Formvar-coated grids for examination with the electron microscope. Non-motile cells were fixed 1-2 h after the cells settled on the membrane filter.

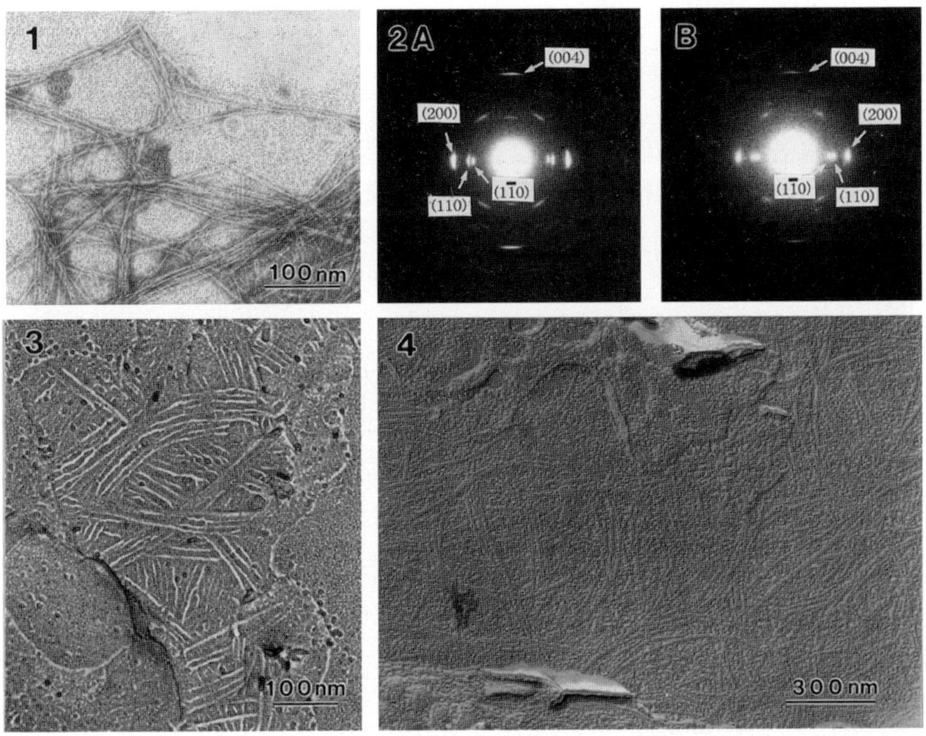

Figure 1. Negative staining of microfibrils from pellicle of *Scrippsiella*.
Figure 2. Electron diffraction patterns of tunicate cellulose (A) and *Scrippsiella* pellicle microfibrils (B).
Figure 3. Inner surface of pellicle showing microfibril bands.
Figure 4. EF of the plasma membrane in *Scrippsiella*. Random and curved microfibrils were deposited. Note the fracture membrane etched.

RESULTS

Microfibril structure

Treatment of the pellicle with acetic-nitric reagent yielded microfibrils with a variable width in the range of 2-14 nm (Fig. 1). The microfibrils consisted of very fine fibrils, about 2 nm in diameter. The microfibrils were deposited with random orientations and often curved (Fig. 4). Some microfibrils were associated laterally with each other (Figs. 4, 5). Spaces between abreast microfibrils were not always constant.

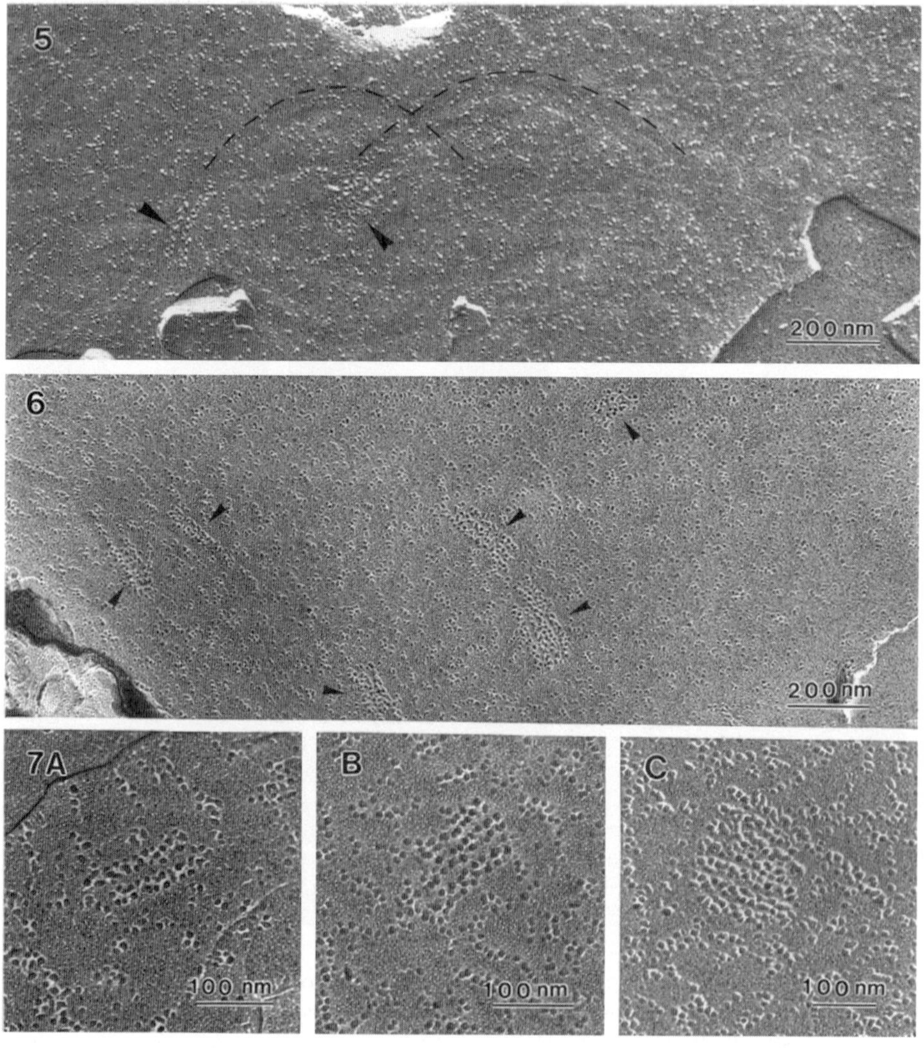

Figures 5-7. PF of the plasma membrane in *Scrippsiella*. **5.** TCs (arrowheads) and microfibril impressions (broken lines). **6.** Various TCs (arrowheads). **7.** Two (A), Three (B), Seven (C) TCs consolidated.

Electron diffraction of microfibrils

The diffraction diagram obtained from tunicate cellulose microfibrils was used a standard and showed a typical pattern of cellulose I with four spots, $1\bar{1}0$, 110, 200, 004 (Fig. 2A). Figure 2B shows the diffraction diagram from the acetic-nitric reagent insoluble residues, where four reflection spots corresponded to those in tunicate cellulose microfibrils. This indicates that the microfibrils isolated from the pellicle are of crystalline cellulose I.

TCs in *Scrippsiella*

The plasmatic fracture face (PF) of the plasma membrane in *Scrippsiella* non-motile cells clearly revealed many particle complexes (Figs. 5-7). Since these particle complexes were often associated with the tips of microfibril impressions (Fig. 5), they were regarded as TCs. The TCs consisted of two rows of particle subunits, but the particle subunits were not arranged at regular intervals (Figs. 7). The number of the subunits varied from 5 to 40 among the TCs. The diameter of the subunits also varied between 5 and 15 nm. The length and the width of the TCs ranged 62-290 nm and 15-31 nm, respectively. Several TCs up to 7 were often associated laterally with each other and consolidated as a cluster (Figs. 7). In the exoplasmic fracture face (EF) of the plasma membrane, no particle complex was observed at the tips of microfibril impressions.

DISCUSSION

In dinoflagellates thecal plates and cyst wall have been known to contain cellulose[15,16]. The present study showed that cellulose microfibrils are present in the pellicle of the dinoflagellate *Scrippsiella hexapraecingula*. Further, freeze fracture electron microscopy revealed a new TC that synthesizes the cellulose microfibrils.

TCs have been categorized into two types, rosette and linear TCs[17]. However, some linear TCs synthesize extremely large microfibrils as in *Valonia*[18], whereas others synthesize thin, ribbon-like microfibrils like in *Vaucheria*[19]. We classify TCs into 4 groups by means of structures of microfibrils that TCs synthesize (Fig. 8). (1) rosette TCs synthesizing 3.5-nm-microfibrils in the Chlorophyta including charophycean green algae[20,21] and land plants[4]. (2) linear TCs synthesizing large microfibrils in the Glaucophyta[22], the Chlorophyta including Chlorophycean[3] and Ulvophycean green algae[18], and invertebrates such as tunicates[23]. (3) linear TCs synthesizing thin, ribbon-like microfibrils in prokaryotes such as *Acetobacter*[24], the Rhodophyta[25], the Chromophyta including Xanthophycean[19] and Phaeophycean[26] heterokonts, and slime molds[27]. (4) linear TCs synthesizing bundles of 2-nm-microfibrils in the dinophyte *Scrippsiella hexapraecingula* (the present study) and possibly in some members of the Haptophyta[28]. According to the endosymbiont hypothesis based on gene sequences of *rbcL*, the Rhodophyta, Glaucophyta and Chlorophyta arose through the endosymbiosis of a photoautotrophic bacterium with a heterotrophic flagellate[29]. The phylogenetic tree based on gene sequences of 18S rRNA suggests that the other algal phyla arose form different heterotrophic flagellate ancestors, through the incorporation of a primeval photosynthetic, eukaryotic alga[30,31]. The Chromophyta is suggested to have evolved when primeval heterokonts incorporated rhodophyte-like algae as chloroplasts into the cells. The ability to synthesize cellulose in the Chromophyta might have brought from such rhodophyte-like algae, since the TCs of the Rhodophyte and Chromophyte assemble similar thin, ribbon-like microfibrils. The TCs of the dinoflagellate *Scrippsiella* described in the present study seem to be primitive, because the microfibrils synthesized are aggregates or bundles consisting of fine 2-nm-fibrils. If the TCs acquire regulatory mechanisms by which the 2-nm-fibrils coalesce laterally to each other and crystallize, they could assemble thin, ribbon-like microfibrils with a thickness of 2 nm. However, no dinoflagellate species that synthesizes ribbon-like microfibrils has yet found. On the other hand, there is possibility that the dinoflagellate-type TCs may occur also in other phylogenetic groups, especially in the Rhodophyta and Chromophyta.

One interesting phenomenon to be noticed is consolidation of TCs. In the dinoflagellate *Scrippsiella*, a cluster of several TCs synthesizes bands of microfibrils, indicating that the TC cluster functions as a single TC by the consolidation of TCs. This is consistent with the case where individual rosette TCs form hexagonal arrays during secondary wall formation in zygnematalean algae belonging to the Charophyceae[20], and with the formation of multiple linear TCs in a slime mold[27]. The consolidation of TCs may be a result of parallel evolution in distinct phylogenetic groups.

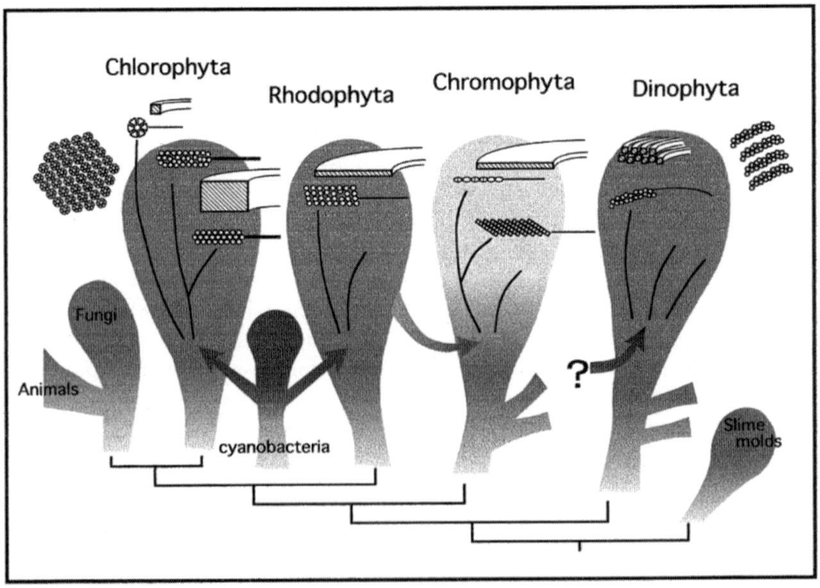

Figure 8. TC organization and microfibril structure in an evolutionary aspect.

REFERENCES

1. R. M. Jr. Brown, The biosynthesis of cellulose, *J. Macromal. Sci.-Pure Appl. Chem.*, **1996**, A33, 1345-73.
2. D. P. Delmer, Cellulose biosynthesis: Exciting times for a difficult field of study, In: *Annual Review of Plant Physiology and Plant Molecular Biology*, R. L. Jones, H. J. Bohnert & V. Walbot (eds.), **1999**, pp 245-76.
3. R. M. Jr. Brown & D. Montezinos, cellulose microfibrils: visualization of biosynthetic and orienting complexes in association with the plasma membrane, *Proc. Natl. Acad. Sci. USA*, **1976**, 73, 143-7.
4. S. C. Mueller & R. M. Jr. Brown, Evidence for an intramembrane component associated with a cellulose microfibril-synthesizing complex in higher plants, *J. Cell Biol.*, **1980**, 84, 315-26.
5. R. M. Jr. Brown, Cellulose microfibril assembly and orientation: recent developments, *J. Cell Sci. Suppl.*, **1985**, 2, 13-32.
6. D. Montezinos, A cytological model of cellulose biogenesis in the alga *Oocystis apiculata*, In: *Cellulose and other natural polymer systems: biogenesis, structure and degradation*, R. M. Jr. Brown (ed.), Plenum Press, New York, **1982**, pp 3-21.
7. S. Kimura, W. Laosinchai, T. Itoh, X. Cui, C. R. Linder & R. M. Jr. Brown, Immunogold labeling of rosette terminal cellulose-synthesizing complexes in the vascular plant *Vigna angularis*, *Plant Cell*, **1999**, 11, 2075-85.
8. I. Tsekos, The sites cellulose synthesis in algae: diversity and evolution of cellulose-synthesizing enzyme complexes, *J. Phycol.* **1999**, 35, 635-55.
9. S. Kuga & R. M. Jr. Brown, Correlation between structure and the biogenic mechanisms f cellulose: new insights based on recent electron microscopic findings, In: *Cellulose and Wood-Chemistry and Technology*, C. Schuerch (ed.), Wiley, New York, **1989**, pp 677-88.
10. K. Okuda, I. Tsekos & R. M. Jr. Brown, Cellulose microfibril assembly *Erythrocladia subintegra* Rosenv.: an ideal system for understanding the relationship between synthesizing complexes (TCs) and microfibril crystallization, *Protoplasma*, **1994**, 180, 49-58
11. P. A. Richmond, Occurrence and function of native cellulose, In: *Biosynthesis and Biodegradation of Cellulose*, C. H. Haigler & P. J. Weimer (eds.),Marcel Dekker, New York, **1991**, pp 5-23.
12. L. Provasoli, Media & prospects for the cultivation of marine algae, In: *Culture and collections of algae*, A. Watanabe & A. Hattori (eds.), Japanese Society of Plant Physiologists, Tokyo, **1966**, pp 63-75.
13. T. Horiguchi & M. Chihara, *Scrippsiella hexapraecingula* sp. nov. (Dinophyceae), a tide pool dinoflagellate from the Northwest Pacific, *Bot. Mag.* Tokyo, **1983,** 96, 351-8.
14. D. M. Updegraff, Semimicro determination of cellulose in biological materials, *Anal. Biochem.*, **1969**, 32, 420-4.
15. S. Sekida, T. Horiguchi & K. Okuda, Direct evidence for cellulose microfibrils present in thecal plates of the dinoflagellate *Scrippsiella hexapraecingula*, *Hikobia*, **1999**, 13, 65-9.
16. E. Swift & C. C. Remsen, The cell wall of *Pyrocystis* spp. (Dinococcales), *J. Phycol.* **1970**, 6, 79-86.
17. R. M. Jr. Brown, C. H. Haigler, J. Suttie, A. R. White, E. Roberts, C. Smith, T. Itoh

& K. Cooper, The biosynthesis and degradation of cellulose, *J. Appl. Polym. Sci. Appl. Polym. Symp.*, **1983**, 37, 33-78.

18. T. Itoh & R. M. Jr. Brown, The assembly of cellulose microfibrils *Valonia macrophysa* Kütz, *Planta*, **1984**, 160, 160-9.
19. S. Mizuta, E. M. Roberts & R. M. Jr. Brown, A new cellulose synthesizing complex in *Vaucheria hamata* and it relation to microfibril assembly, In: *Cellulose and Wood-Chemistry and Technology*, C. Schuerch (ed.), Wiley, New York, **1989**, pp 659-76.
20. T. H. Giddings, Jr., D. L. Brower & L. A. Staehelin, Visualization of particle complexes in the plasma membrane of *Micrasterias denticulate* associated with the formation of cellulose fibrils in primary and secondary cell walls, *J. Cell Biol.*, **1980**, 84, 327-39.
21. K. Okuda & R. M. Jr. Brown, A new putative cellulose-synthesizing complex of *Coleochaete scutata*, *Protoplasma*, **1992**, 168, 51-63.
22. J. H. M. Willison & R. M. Jr. Brown, Cell wall structure and deposition *Glaucocystis*, *J. Cell Biol.*, **1978**, 77, 103-19.
23. S. Kimura & T. Itoh, New cellulose synthesizing complexes (terminal complexes) involved in animal cellulose biosynthesis in the tunicate *Metandrocarpa uedai*, *Protoplasma*, **1996**, 194, 151-63.
24. R. M. Jr. Brown, J. H. M. Willison & C. L. Richardson, Cellulose biosynthesis in *Acetobacter xylinum*: visualization of the site of synthesis and direct measurement of the in vivo process, *Proc. Natl. Acad. Sci. USA*, **1976**, 73, 4565-9.
25. I. Tsekos & H. D. Reiss, Occurrence of the putative microfibril-synthesizing complexes (linear terminal complexes) in the plasma membrane of the epiphytic marine red alga *Erythrocladia subintegra* Rosenv., *Protoplasma*, **1992**, 169, 57-67.
26. H. Tamura, I. Mine & K. Okuda, Cellulose-synthesizing terminal complexes and microfibril structure in the brown alga *Sphacelaria rigidura* (Sphacelariales, Phaeophyceae), *Phycol. Res.*, **1996**, 44, 63-68.
27. M. J. Grimson, C. H. Haigler & R. L. Blanton, Cellulose microfibrils, cell motility, and plasma membrane protein organization change in parallel during culmination in *Dictyostelium discoideum*, *J. Cell Sci.*, **1996**, 109, 3079-87.
28. D. K. Romanovics & R. M. Jr. Brown, Biogenesis and structure of Golgi-derived cellulosic scales in *Pleurochrysis*. II. Scale composition and supramolecular structure, *Appl. Polymer Symp.*, **1976**, 28, 587.
29. G. I. McFadden, P. R. Gilson & R. Waller, Molecular phylogeny of chlorarachniophytes based on plastid rRNA and *rbcL* sequences, *Arch. Protistenkd*, **1995**, 145, 231-9.
30. G. I. McFadden, P. R. Gilson & D. R. A. Hill, *Goniomonas*: rRNA sequences indicate that this phagotrophic flagellate is a close relative of the host component, *Eu. J. Phycol.*, **1994**, 145, 29-32.
31. G. I. McFadden, P. R. Gilson, C. J. B. Hofmann & U. G. Mairer, Evidence that an amoeba acquired a chloroplast by retaining part of an engulfed eukaryotic alga, *Proc. Natl. Acad. Sci. USA*, **1994**, 91, 3690-4.

Molecular Breeding of Woody Plants
N. Morohoshi and A. Komamine, editors.

REGULATION OF DYNAMIC CHANGES IN CELL WALL POLYSACCHARIDES

Naoki Sakurai and Naoki Nakagawa

Faculty of Integrated Arts and Sciences
Hiroshima University, Higashi Hiroshima
739-8521, Japan

ABSTRACT

When tobacco BY-2 suspension cells are cultured in the presence of a cellulose synthesis inhibitor, 1 μM 2,4-dichrolobenzonitrile (DCB), the cells swell, grow slowly but survive. The cell walls consist of 66% pectin, 30% hemicellulose and only 3% cellulose. DCB treatment did not affect the mRNA level of CesA (cellulose synthase). Electron micrography with immunogold labeling technique demonstrated that DCB-habituated cells accumulated CesA protein in cytoplasm more than normal BY-2 cells. DCB probably inhibits the cellulose synthesis after transcription of CesA. The DCB treatment, however, decreased the mRNA level of EXGT (endoxyloglucan transferase). Treatment of BY-2 cells with cellulase also decreased the mRNA level of EXGT, suggesting that defect or weakening of cell walls regulates EXGT expression. Cellulase treatment did not affect mRNA level of CesA. The cellulase-induced decrease in EXGT expression was partly overcome by the addition of osmoticum (0.45M mannitol) to the medium, implying that turgor change or stretch of plasmamembrane is involved in the regulation mechanism. Pretreatment of actinomycin D cancelled the decrease in mRNA level of EXGT by cellulase treatment, indicating that the regulation of EXGT expression needs function of another gene. These results suggest that (1) plant cells can monitor alternations in cell wall architecture probably by sensing the stretch of plasmamembrane as demonstrated in yeast and bacteria, (2) EXGT expression is regulated by some gene involved in the sensing mechanism, (3) expression of CesA gene is rather insensitive to the alternations in cell wall architecture in BY-2 cells.

KEYWORDS

Calcium, Cellulase, Cellulose, CesA, DCB (2,6-dichlorobenzonitrile), EXGT, turgor, plasmamembrane, tobacco BY-2

INTRODUCTION

Higher plants contain a membrane-bound glucan synthase responsible for the synthesis of cellulose. However, almost all the attempts at in vitro synthesis of cellulose have resulted in the formation of only 1,3-β-glucans or only a very limited synthesis of 1,4-β-glucans[1, 2]. Plant cDNAs with similarity in terms of amino acid sequence to the bacterial cellulose synthase were found from cotton fibers[3] and named CesA[4]. Mutants of *Arabidopsis* impaired in cellulose production were selected with the use of radial swelling phenotype (*rsw*)[5], which mimics the responses to cellulose synthesis inhibitors such as DCB (2,6-dichlorobenzonitrile)[6]. Tomato cultured-cells that were habituated in the presence of DCB lacked a cellulose-xyloglucan network[7]. In the present work, we examined the response of tobacco BY-2 cells to DCB to know the regulation mechanism of cellulose synthesis in higher plants.

MATERIALS & METHODS[8, 9]

Plant material

A suspension culture of tobacco BY-2 cells derived from *Nicotiana tabacum* L. cv. Bright Yellow 2 was kindly donated by Dr. T. Asada (Osaka University) and was propagated in a rotary shaker at 125 rpm at 27°C. The cells were maintained by weekly subculture. DCB-habituated cells were propagated by exposing BY-2 cells to medium containing 1 μM DCB throughout subculture routines for several months. DCB-habituated cells were cultured in a rotary shaker at 80 rpm at 27°C and maintained by subculture at intervals of 21 days. DCB-habituated cells were used for experiments after at least 3 months from the beginning of habituation.

Analysis of cell wall polysaccharides

Cell wall polysaccharides of BY-2 cells and DCB-habituated cells were fractionated into pectin, hemicellulose and cellulose. Cultured cells (ca. 1 g fr wt) were boiled for 15 min in 15 ml of methanol and centrifuged for 10 min at 1,000 x g. The residue was homogenized in deionized water with a mortar and pestle. The homogenate was boiled for 10 min and then centrifuged at 1,000 x g. The residue (cell wall fraction) was treated with porcine pancreatic α-amylase (Type I-A; Sigma, St. Louis, MO, U.S.A.) in 50 mM sodium acetate buffer (pH 6.5) for 2 h at 37°C. The EDTA-soluble substances were extracted three times, for 15 min each, from the cell walls with 50 mM EDTA in 50 mM sodium phosphate buffer (pH 6.8) at 95°C. Next, hemicellulosic substances were extracted for 18 h at 25°C with 17.5% NaOH that contained 0.02% $NaBH_4$. The residue was washed three times with 0.03 M acetic acid and with a mixture of ethanol and ether (1 : 1, v/v) and dried for one day at 25°C and for two days at 40°C. The dried materials were designated the cellulose fraction. Sugar contents of each fraction were determined by the phenol-sulfuric acid method.

Preparation of an antibody against CesA1 protein

Synthesis of antigen peptide fragment combined with MAP (multiple antigen peptide) resin and rabbit antiserum production were prepared by Sawady Technology Co. (Tokyo, Japan). The amino acid sequence of the antigen peptide was KEAIHVISCGYEDKS. This sequence was based upon the analysis of cotton celA1 cDNA sequence[4] for antigenicity and surface probability and comparison of the CesA1 cDNA sequences of cotton, Arabidopsis and rice. The antiserum was affinity-purified with the immobilized antigen peptide to increase.

Preparation of microsomal fractions

The BY-2 cells (4 days or 12 days after subculture) and DCB-habituated cells (12 days after subculture) were used for this experiment. The normal cells were in exponential-growth phase on day 4. Since the growth of DCB-habituated cells was slow, they were in exponential-growth phase on day 12. For short-term DCB treatment, 1/1000 volumes of stock DCB solution (1 mM or 10 mM, dissolved in DMSO) was added to BY-2 cells 3 days after subculture. As a control, the same volume of DMSO without DCB was added to another culture. After 24 h, they were harvested for the preparation of microsomal fractions. Cells (about 1.5 g) were homogenized with a mortar and pestle in 8 ml of extraction buffer (50 mM MOPS-KOH pH 7.5 containing 0.25 M sucrose, 10 mM dithiothreitol, 14 mM 2-mercaptoethanol, 1 mM *p*-APMSF and 5 mM EDTA) and centrifuged at 6,000 x g for 20 min at 4°C to remove heavy particles and organella. The supernatants were then ultracentrifuged at 100,000 x g for 90 min at 4°C (Beckman XL-90, Ti90 rotor, 40,000 rpm). After the precipitates were resuspended in 10 ml of 50 mM MOPS-KOH pH 7.5 containing 0.25 M sucrose, the samples were ultracentrifuged again for 60 min. The precipitates were suspended in 50 mM MOPS-KOH pH 7.5 containing 0.25 M sucrose, 14 mM 2-mercaptoethanol, 1 mM *p*-APMSF, and 10% glycerol and brought to a protein concentration of 2 mg/ml. These samples were stored at −70°C until use.

Immunoblot analysis of microsomal fraction

Microsomal fractions prepared as described above were solubilized by slowly adding one seventh volume of 4% (w/v) CHAPS (final concentration is 0.5%) and kept on ice for 30 min followed by ultracentrifugation. The supernatants were mixed with a half volume of 3 x SDS-sample buffer. The pellets were resuspended in the same volume of 1 x SDS-sample buffer as that of supernatant. The samples (20 μl/lane, containing about 15 μg of protein) were subjected to immunoblot analysis. A portion of the antibody was passed through a sepharose 4B column of immobilized antigen peptide twice and then used as a "absorbed antibody" to visualize non-specific binding. Binding of the antibody was visualized by Western Blot Chemiluminescence Reagent Plus (New England Biolabs, NEL103).

Cellulase, Aphidicolin and Actinomycin D treatment

Three ml of BY2 cell cultures (7 days after subculture) were transferred to a 30 ml of fresh culture medium containing 0 to 0.5 % (w/v) of "CELLULASE ONOZUKA" R-10 (Yakult, Tokyo) followed by incubation on a rotary shaker for the indicated times. For DCB treatment, 1/1000 volume of stock DCB solution (5 mM, dissolved in DMSO) was added to fresh medium instead of cellulase. As a control, the same volume of DMSO without DCB was added to another culture. Aphidicolin solution (5 mg/ml) was dissolved in DMSO and 1/1000 volume was added to fresh medium. Actinomycin D was dissolved in 40 % ethanol to make the stock solution (3 mg/ml) and 1/100 volume was added to each culture. As a control, the same volume of 40 % ethanol was added.

Northern blot analysis

Total RNA was extracted by using an Isogen RNA extraction system (Nippon Gene, Tokyo). First-strand cDNA synthesis and PCR for probe preparation was performed with the Takara RNA LA PCR Kit (Takara, Kyoto). Primers used in the reactions for the amplification of EXGT were designed based on the EXGT-N1 sequence which has been demonstrated to be abundant in BY2 cells[10]. The sequences of primers for EXGT-N1 were 5'-AGTCACCACATC AAGTTACCTCA-3' and 5'-CTCCACCAATGA TACACTCAAA-3', respectively. Primers for the amplification of cellulose synthase were designed based on the sequences of the highly conserved regions of the enzyme. The sequences of the primers for cellulose synthase were 5'-GAAGGTTGGACT ATGCAAGA(CT)GG-3' and 5'-ATAGATCCATCCAATCTCT-TTTCCCCA-3', respectively. The primer set for cellulose synthase produced a 950 bp long cDNA fragment covering between H-2 (homologous region 2, Pear et al. 1996) and H-3 of the mRNA, and its identity was ensured by sequencing from both ends. Total RNA (20 μg for each lane) was separated on a 1.5% agarose gel containing formaldehyde. After blotting and fixation to nylon membrane, RNA was stained by methylene blue to check RNA integrity. The hybridisation, washing and autoradiography were carried out as described[11].

RESULTS & DISCUSSION

Accumulation of CesA protein in DCB-habituated cells[8]

When BY-2 cells are habituated to 1 μM DCB, the cells become swollen and rounded. Normal BY-2 cells are rectangular. Although the cells can grow in the presence of DCB, the growth is much slower than in the absence of DCB. We analyzed the content of cell wall polysaccharides at the three growth stages, lag phase, log phase and stationary phase. Normal BY-2 cells produced cellulose as they proliferated and matured. At the stationary phase, cellulose accounted for 60% of total cell wall polysaccharides of normal cells. But DCB-habituated cells produced much less cellulose (7 to 11%) throughout the growth stages. Interestingly, DCB-habituated cells produced more pectin than normal cells. It was assumed that, because of the severe inhibition of cellulose synthesis, DCB-habituated cells pay much effort to produce normal cellulose level. Therefore, we expected that DCB-habituated cells try to increase CesA transcripts and protein. Antibody was raised against CesA protein to estimate the cellulose synthase level in DCB cells. An oligopeptide of 15 amino acids (KEAIHVISCGYEDKS) was selected for the antigen that corresponds to the region just after the so-called hyper valuable region of CesA protein. After homogenized BY-2 cells and centrifuged the cytoplasmic fraction, the microsomal fraction was treated with or without 0.5% CHAPS and load on SDS-PAGE. The antibody detected several intense bands only for DCB-

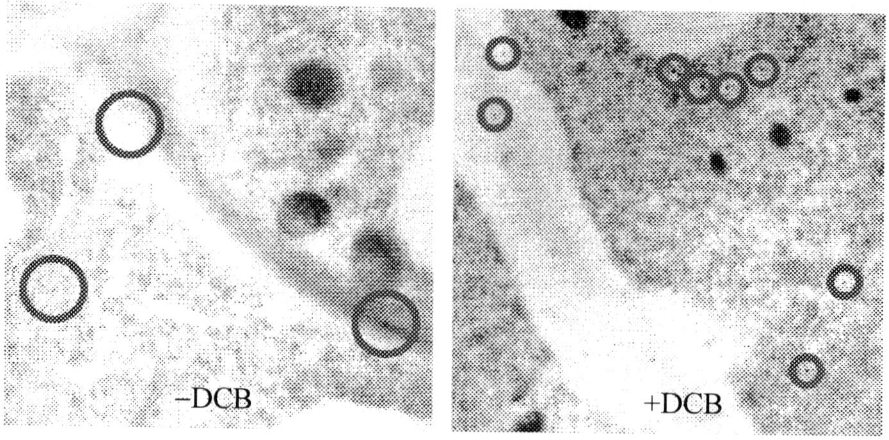

Figure 1. Immunogold lebeling of section of BY-2 cells. −DCB, normal BY-2 cells; +DCB, DCB-habituated BY-2 cells.

habituated cells. They were 117, 107, and 94 kDa as well as a much higher band about 164 kDa. These lower three bands correspond to those of the reported CesA protein between 100 to 120 kDa. More than 50% of these proteins are solubilized by CHAPS. Normal cells did not have such a positive band even on both day 4 and 12. When fresh normal BY-2 cells are treated with 1 μM of DCB for 24 hr, the positive band around 110 kDa was detected. However a higher concentration of DCB 10 μM did not show such accumulation of CesA protein. To confirm the accumulation of CesA protein in the DCB-habituated cells, section of BY-2 cells are labeled by an immunogold technique (Fig. 1). Gold particles shown by circles were observed sparsely in the normal cells without DCB. There were two dots on the cell walls. There were many gold particles that were localized in the cytoplasm of DCB-habituated cells. The results suggest that DCB inhibits the turnover of CesA protein, but it may not affect the gene expression of CesA. To confirm this possibility, we analyzed the CesA expression in the presence of DCB.

DCB did not affect CesA expression but EXGT[9]

CesA gene is amplified between motif 1 and 2 using BY-2 mRNA. Northern analysis revealed that at any growth stage, DCB did not inhibit CesA expression. It follows that even when DCB-habituated cell has produced less cellulose, no compensatory promotion of CesA gene expression takes place. But surprisingly, endo-xyloglucan transferase (EXGT) gene was suppressed by DCB after log phase. There are two EXGT genes in tobacco BY-2 cells, N1 and N2, but N1 is predominantly expressed. We further studied why EXGT-N1 expression was reduced by DCB in more detail.

If DCB inhibits EXGT expression, it means that condition of less cellulose in the cell walls inhibits EXGT. To manipulate the cellulose content in the cell walls, we treated the normal cells with cellulase solution. When the normal cells are inoculated into the fresh medium, EXGT expressed more by 24 hr, but when they are inoculated into the fresh medium containing 0.02 or 0.1% of Cellulase Onozuka R-10 for 10 and 24 hour, EXGT expression was reduced. Cellulase treatment mimics DCB treatment in EXGT expression. Under the same condition, CesA expression did not change.

To confirm that the inhibition of expression is caused by cellulase activity and not by the other factors, cellulase solution was heated at 120°C for 5 min. The heating treatment completely abolished the inhibitory effect of cellulase on EXGT expression. Akashi and Shibaoka (1991)[12] reported that apoplastic extracellular protease can influence the cortical microtubules. To check the possibility that cellulase preparation may contain protease and that protease affect the EXGT expression, the normal cells were treated with 0.02% pronase, but pronase did not inhibit the expression. These results demonstrate that cellulase activity reduces EXGT expression in BY-2 cells.

There is a possibility that DCB has a non-specific growth inhibition or chemical stress. To check this possibility, the normal BY-2 cells were inoculated into the fresh medium containing DCB or Aphidicolin that specifically inhibits cell division by the inhibition of DNA polymerase. Aphidicolin specifically repressed M-phase specific cyclin, while it did not affect EXGT. On the other hand, DCB repressed EXGT but did not affect cyclin-M. These results supported the assumption that the expression of EXGT was repressed as a result of a cell wall defect or weakening.

If cell walls are treated with cellulase, the cell walls become thinner. If the cell walls are thinner, the cell must swell. We assumed that this swelling cause stretching of cell membrane that might lead to changes in the expression of EXGT. To examine this hypothesis, we added 0.45 M of mannitol into the ambient medium to prevent swelling by cellulase treatment. This concentration is isotonic for BY-2 cells in the present condition. When we added the isotonic concentration of 0.45 M mannitol in addition to cellulase, cellulase-induced reduction of EXGT expression was partly recovered. Hyper or hypotonic concentration of mannitol did not restore the inhibition. These results suggest that membrane stretch by weakening of cell walls interferes the expression of EXGT. This is a new sensing mechanism of turgor pressure that regulates cell wall-associated genes.

Next we examined the mechanism of reduction of EXGT expression by cell wall defect or weakening. We analyzed how fast cellulase reduces EXGT expression. The higher concentration of cellulase shortened the time required for the reduction of EXGT expression, but the very high concentration 0.5% of cellulase still needs 1 hr to inhibit EXGT expression. Therefore, we assumed that another gene or factor mediates the control mechanism. To check this possibility, BY-2 cells were pretreated with Actinomycin D. Cellulase (0.02%) suppressed EXGT expression after 2 hour. When the BY-2 cells were pre-treated with Actinomycin D, cellulase effect was eliminated. Even in the presence of cellulase, there was no reduction of EXGT expression when the incubation medium included Actinomycin D. It clearly indicates that EXGT mRNA is degraded more in the presence of cellulase, and the degradation is regulated by another gene expression.

Fig. 2 summarizes the regulation of cell wall-associated genes by cell wall detect or weakening. Our results clearly show that DCB did not suppress the mRNA level of CesA. The immunogold labeling experiment demonstrated that DCB-habituated cells accumulated CesA protein in the cytoplasm. The less production of cellulose may cause cell wall defect or weakening. There are two possible ways to transmit this signal to gene expression. One is pressure change. If the wall becomes weakened, turgor pressure exceeds cell wall pressure. It leads to the cell swelling which we actually observed in DCB-habituated cells. Cell swelling, in turn, stretches plasma membrane. In yeast or bacteria, stretching signal is coupled with calcium channel. Calcium is a second messenger of stretching signal. We, so far, do not know whether or not this signal is also coupled with calcium channel in plants. The other possible way is an oligo-saccharide that is released from cell walls induces another gene X expression that regulates EXGT degradation. EXGT mRNA has AUUUA sequence in the 3' un-translated region. This sequence is known to control mRNA stability[13]. Through calcium or direct oligosaccharide signal, the gene X product may bind to EXGT 3' UTR and accelerate the degradation. Note that the cell wall defect or weakening neither reduces nor accelerates CesA gene expression. Why should cell wall defect or weakening lead to the inhibition of EXGT expression? If the BY-2 EXGT contributes to making cross-links between xyloglucan and cellulose microfibrils, DCB cell should save the energy to produce EXGT, because very few cellulose microfibrils exist in the cell walls.

The present finding also suggests the mechanism of tension wood. In angiosperm, stretching of the stem or trunk results in tension wood. Tension wood consists of more cellulose and less lignin. If the tree bent, gravity force stretches each cell of one side, and may result in the use of a different recipe of the cell wall synthesis. In our

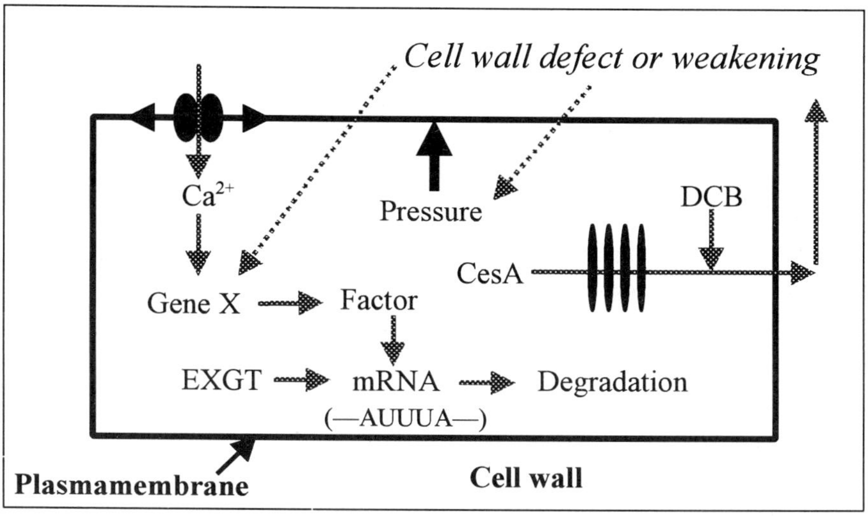

Figure 2. Schematic diagram of sensing mechanism of cell wall defect or weakening that controls EXGT mRNA stability.

experiment, stretching signal seems not to affect CesA expression in BY-2 cells. Therefore if the stretching signal reduces only the gene expression associated with lignin synthesis, the new wall consists of more cellulose and less lignin. Thus stretching signal can be used for the alteration of cell wall synthesis.

ACKNOWLEDGEMENT

We are grateful to Dr. Kimura and Itoh of Kyoto University for taking electron micrographs with immunogold labeling.

REFERENCES

1. D. P. Delmer, Cellulose biosynthesis, Annu. Rev. Plant Physiol., **1987**, 38: 259-290.
2. K. Okuda,L. Li., K. Kudicka, S. Kuga & R. M. Brown,β-glucan synthesis in the cotton fiber. I. Identification of β–1,4– and β–1,3–glucan synthesized in vitro. Plant Physiol., **1993**, 101, 1131-1142.
3. J. R. Pear, Y. Kawagoe, W. E. Schekengost, D. P. Delmer & D. M. Stalker, Higher plants contain homologuew of the bacterial celA genes encoding the catalytic subunit

of cellulose synthase, Proc. Natl. Acad. Sci. USA, **1996**, 93, 12637-12642.

4. D. P. Delmer, Cellulose biosynthesis: Exciting times for a difficult field of study. *Annu. Rev. Plant Physiol. Plant Mol. Biol.*, **1999**, 50, 245-276.
5. T. I. Baskin, A. S. Betzner, R. Hoggart, A. Cork & R. E. Williamson, Root morphology mutants in *Arabidopsis thaliana*, Aust. J. Plant Physiol., **1992**, 19, 427-437.
6. T. Hogetsu, H. Shibaoka & M. Simokoriyama, Involvement of cellulose synthesis in actions of gibberellin and kinetin on cell expansion. 2,6-dichlorobenzonitrile as a new cellulose synthesis inhibitor, Plant Cell Physiol., **1974**, 15, 389-393.
7. E. Shedletzky, M. Schmuel, D. P. Delmer & D. T. A. Lamport, Adaptation and growth of tomato cells on the herbicide 2,6-dichlorobenzonitrile leads to production of unique cell walls virtually lacking a cellulose-xyloglucan network, Plant Physiol., **1990**, 94, 980-987.
8. N. Nakagawa & N. Sakurai, Increase in the amount of celA protein in tobacco BY-2 cells by a cellulose biosynthesis inhibitor, 2,6-dichlorobenzonitrile, Plant Cell Physiol., **1998**, 39, 779-785.
9. N. Nakagawa & N. Sakurai, Cell wall integrity countrols expression of endoxyloglucan transferase in tobacco BY2 cells, Plant Cell Physiol., **2001**, 42, 240-244.
10. H. Ito & K. Nishitani, Visualization of EXGT-mediated molecular grafting activity by meanse of a fluorescent-labeled xyloglucan oligomer. Plant Cell Physiol. **1999**, 40: 1172-1176.
11. N. Nakagawa & H. Mori, K. Yamazaki & H. Imaseki, Cloning of a comlementaryDNA for auxin-induced 1-aminocyclopropane-1-carboxylate synthase and differential expression of the gene by auxin and wounding. Plant Cell Physiol. **1991**, 32: 1153-1163
12. T. Akashi & H. Shibaoka, Involvement of transmembrane proteins in the association of cortical microtubules with the plasma membrane in tobacco BY-2 cells, J. Cell Science, **1991**, 98, 169-174.
13. M. A. Johnson, E. J. Baker, J. T. Colbert & P. J. Green, Determinants of mRNA stability in plants, In: A look beyound transcription, J. Bailey-Serres & S. R. Gallie (eds.), American Society of Plant Physiologists, **1998**, pp. 40-53.

Molecular Breeding of Woody Plants
N. Morohoshi and A. Komamine, editors.

MICROFIBRILS BUILD ARCHITECTURE: A GEOMETRICAL MODEL

A.M.C. Emons*[1], B. M. Mulder[2]

[1]*Laboratory of Plant Cell Biology, Department of Plant Sciences, Wageningen University, Arboretumlaan 4, 6703BD Wageningen, The Netherlands*
[2]*FOM Institute for Atomic and Molecular Physics (AMOLF), Kruislaan 407, 1098 SJ Amsterdam, The Netherlands*

ABSTRACT

Several mechanisms for cellulose microfibril (CMF) orientation have been proposed, of which guidance by cortical microtubules is widely accepted for growing cells, but in many non-growing cells CMF orientations are not correlated to microtubule arrangement. We have formulated a theory for wall deposition consistent with present day experimental data on walls and cellular processes. It appeals to a very generic origin, geometrical constraints, as the underlying cause of cell wall architecture. This is a fully mathematical model, allowing for specific predictions of qualitative and quantitative nature. It was formulated on the basis of extensive observations on the helicoidal wall texture of *Equisetum hyemale* root hairs. The key point of the geometrical theory is the coupling of the CMF synthase, i.e. rosette, trajectories to the density of rosettes. This provides the cell with a route to manipulate the wall structure by creating controlled local variations of the number of active rosettes. We discuss how the model can describe known wall textures, underlining the flexibility of the proposed mechanism.

KEYWORDS

Cell wall texture, cellulose microfibril, geometrical model, microfibril angle, role of microtubules

INTRODUCTION

Plant cellulose is produced in the form of microfibrils, crystalline aggregates of linear polymers of D-glucopyranosyl residues[1,2], linked in the β-(1→4) conformation. The pattern in which these cellulose microfibrils occur in walls, the wall texture, is cell type and developmental stage specific, and is a major determinant of wall properties. These cellulose microfibrils (CMFs) are embedded in a matrix of polysaccharides, proteins and often phenolics[3].

The CMFs lie perpendicularly to the direction of elongation. The cortical microtubules in most expanding walls lie perpendicularly to the direction of cell elongation as well. This correlation in the orientations of CMFs and cortical microtubules, together with experiments with microtubule depolymerizing drugs, have inspired the hypothesis that the movement of cellulose synthases, though driven by cellulose crystallization[4] is guided by cortical microtubules[5].

In a growing number of cells, mainly non-elongating cells or cell parts, in which the secondary wall, the wall laid down after cell expansion, is deposited, the co-alignment does not occur (reviews[6,7]).

The co-alignment is seen in expanding cells, but in only very few of these cases have the CMFs been studied reliably. The polarizing microscope gives an overall view only and staining with uranyl acetate-lead citrate of glutaraldehyde-osmiumtetroxide fixed thin sections does not reveal the wall texture[8].

A helicoidal cell wall is made of lamellae which are one CMF thick. The orientation of the CMFs in every subsequent lamella changes with a constant angle according to the orientation in the previous lamella. Therefore, this wall type is ideal to study the supposed paralellism of CMFs and cortical microtubules. Such studies have been carried out for the *Equisetum hyemale* root hair, and the cortical microtubules next to the wall were net-axial in stead of rotating regularly[6].

Because many data are known from this root hair cell wall, a tip-growing cell with helicoidal secondary wall deposition away from the tip in the non-expanding tube, a geometrical model for CMF deposition could be formulated on the basis of these data[9,10,11,12,13]. In the present paper, we show that the geometrical model can easily explain crossed polylamellate, helical and axial) cell wall textures as well, the basic textures that constitute most known cell walls.

THE GEOMETRICAL MODEL IN BIOLOGICAL TERMS

During the process of exocytosis, Golgi vesicles carrying wall matrix material as cargo and cellulose synthases in their membranes, deposit these into the wall and the plasma membrane, respectively. Freeze fractured plasma membranes show these cellulose synthases, which consist of six particles arranged in a ring[14], a so-called particle rosette. These rosettes spin out the cellulose polymers[15,16], which crystallize as bundles, the cellulose microfibrils (CMFs). The polymerization forces cause the synthases to move in the plasma membrane. At the growing CMF-tip is the particle rosette/terminal globule, the CMF synthase[17], which transports and polymerizes UDP-glucoses producing cellulose (ß-1,4-D-linked glucose), while moving in the fluid plane of the plasma membane. It is driven by the crystallization of the developing CMF[3,16]. The CMF synthase is brought into the membrane by a Golgi vesicle from one or more Golgi complexes situated near the plasma membrane.

Our geometrical model supposes that the cellulose synthases are inserted into the membrane or activated by insertion domains (IDs), groups of Golgi bodies moving or being activated adjacent to the plasma membrane. At the one end of this plasma membrane domain new CMF synthases are constantly added/activated in the plasma membrane, while at the other end the complexes disintegrate after having completed the formation of a CMF.

Cellulose microfibrils are long structures, and no reliable estimate of their length is currently available, because hardly any microfibril ends are seen in wall preparations. Therefore, an individual cellulose microfibril can contribute to several wall lamellae. Freeze fracture images show that the cell wall is appressed against the plasma membrane. Therefore, wall deposition occurs under space-limiting conditions.

THE GEOMETRICAL MODEL IN MATHEMATICAL TERMS

The model for cellulose microfibril deposition[9,10,11,12,13] quantitatively relates the deposition angle of cellulose microfibrils with respect to the cell axis to (1) the density of active synthases in the plasma membrane, (2) the distance between individual microfibrils within a wall lamella and (3) the geometry of the cell.

The model considers that CMF deposition is a tip-growing process at the plasma membrane and that deposition occurs in the limited space between plasma membrane and cell wall. This makes CMF deposition a process under space limiting conditions. The laws of geometry then determine the orientation of deposited CMFs. The path that a CMF synthase follows is a geodesic, the shortest distance between two points on a surface. At any time point the deposition orientation is determined by:

$$\text{sine of helical angle} = \frac{\text{number of strings x string width}}{\text{cylinder diameter x } \pi}$$

in which, α = helical angle, the angle between the CMFs and the transverse direction
N = number of strings, (one string is one CMF including its adhering wall matrix)
D = cylinder circumference, the cell dimensions
d = string width

In the geometrical model the number of cellulose synthases active locally, determines the orientation of the CMFs at the plasma membrane/cell wall interface, comparable to the wrapping of strings around a cylinder. One string in this context represents one CMF together with its adherent matrix material. String insertion occurs at the inside of the existing cell wall: the outer circumference of the complete structure does not increase, but the inner circumference decreases slightly with deposition of each new CMF lamella. Further, string insertion occurs via a mobile 'insertion domain' which moves up and down in the cell and there might be more of such domains active at the same time.

Nature can exploit the geometrical constraints imposed by the shape of the cell and the limited space in which microfibrils are being deposited to let the wall textures essentially 'build themselves'. This does not imply that the cell can not control its wall texture. On the contrary, the cell has ample regulatory mechanisms to control wall texture formation, i.e. by controlling synthase insertion and the distance between individual microfibrils within a wall lamella.

Apart from the helicoidal wall texture, the following cell wall textures are known: crossed polylamellate, helical and axial. Other types of texture derive from these basic types by successive deposition of different textures. In the present paper we show that the geometrical model can describe the development of all these textures. We will deal with the random wall texture in another publication.

SOLUTIONS OF THE MODEL

In mathematical terms our model is a partial differential equation, whose solutions describe the evolution of the density of active rosettes both in space and in time. From these solutions the tracks of individual rosettes can be reconstructed and hence the texture of the deposited CMFs. The model has only four parameters, roughly speaking 1) the size of the putative insertion domains, 2) their speed of motion, 3) the lifetime of an active rosette and

4) a parameter describing the temporal characteristics of rosette creation and activation within the insertion domains (Mulder and Emons, 2001, in press, Journal of Mathematical Biology). By studying the solutions of the model under variations of these parameters different wall structures are obtained. Moreover, one is able to predict quantitatively how changes in the parameters are reflected in the details of the deposited walls, opening up the way for experimental verification of the model.

Helicoidal cell wall texture

In a helicoidal cell wall, the CMF orientation changes by a constant angle going from one lamella to another producing a staircase-like structure (hence the name helicoid). To obtain such a structure in which the deposition angle progresses from transversal (α=0°) through axial (α=90°) and back to transversal (α=180°), after which the cycle can repeat itself, the parameters of the model need to be precisely tuned. In a purely transverse cross-section of the wall the orientation of the CMFs steadily increases from the transverse direction eventually crossing the axial direction and continuing on to complete a full turn, after which the whole sequence is repeated.

Crossed polylamellate cell wall texture

In this wall successive lamellae have CMF orientations at right angles to each other. Our model readily describes such a structure. In this case the rosette creation rate is so high that the maximum density is achieved almost instantaneously. This yields an initially purely axial deposition. By tuning the size and speed of the insertion domain to the lifetime of the rosettes, the density can suddenly drop dramatically yielding a stretch of purely transverse deposition. This analysis shows that the crossed polylamellate texture is in fact a special case of a helicoidal texture.

Helical cell wall texture

In a helical wall texture, all lamellae make a more or less constant angle with the cell axis. Many parts of wood cell walls have such a texture. This texture is obtained in our model when the lifetime of the rosettes is large compared to the time it takes the insertion domain to pass through a given location along the cell. In this case the rosette density builds up to a specific value, which determines the helical angle, and thereafter remains essentially constant.

Axial cell wall texture

In the axial texture CMFs run parallel to the cell axis in all lamellae, a common texture in many wood cells. This texture is also often seen in cells in which cell diameter has shrunk considerably, so in a sense it is a finite size effect. It is in fact an extreme case of the helical texture where the rosette density builds up all the way to the maximum density, yielding purely axial deposition. This is readily understood in the case of thin cells, where the maximum number of rosettes N_{max} = cell circumference / string width that can be accommodated is smaller in an absolute sense.

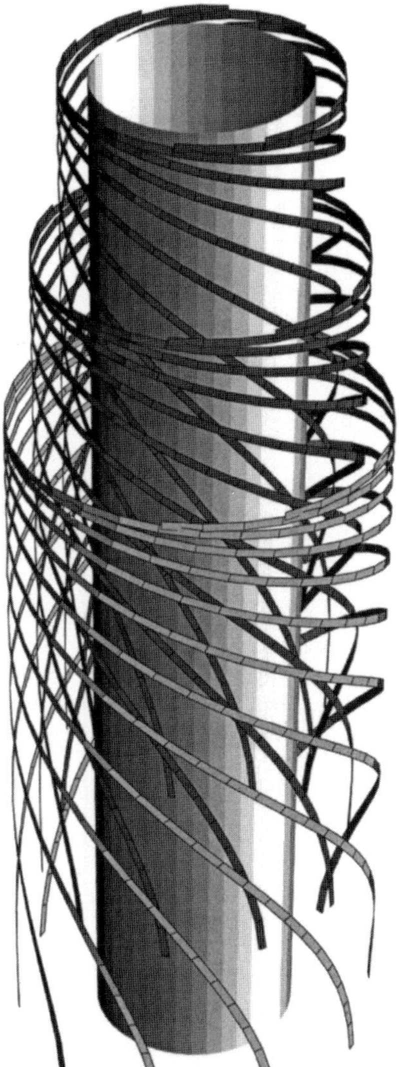

A. Helicoidal wall texture

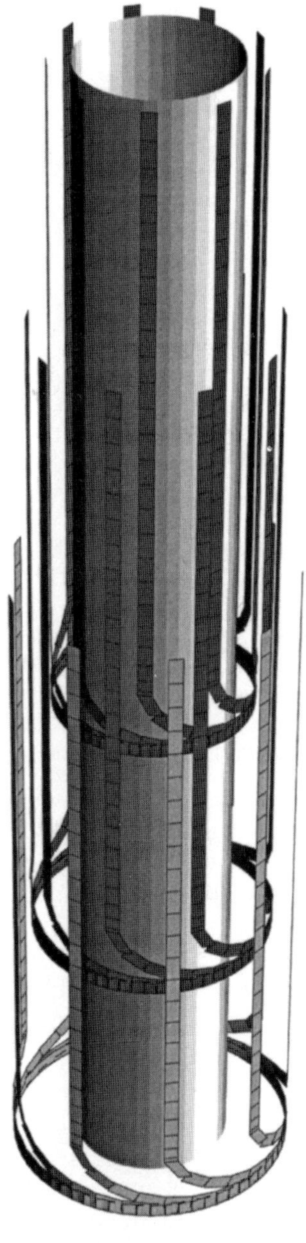

B. Crossed polylamellate wall texture

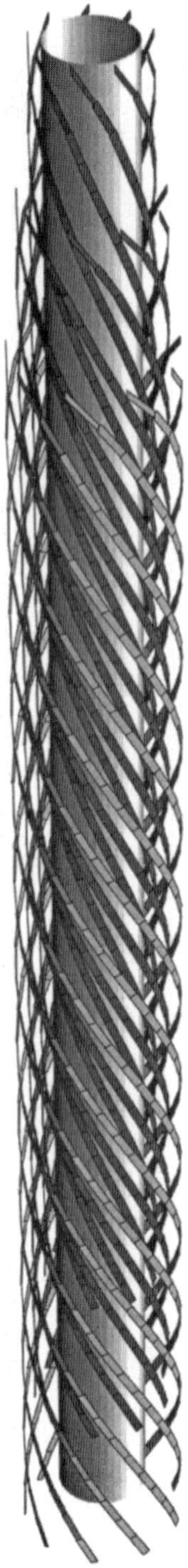

C. Helical wall texture

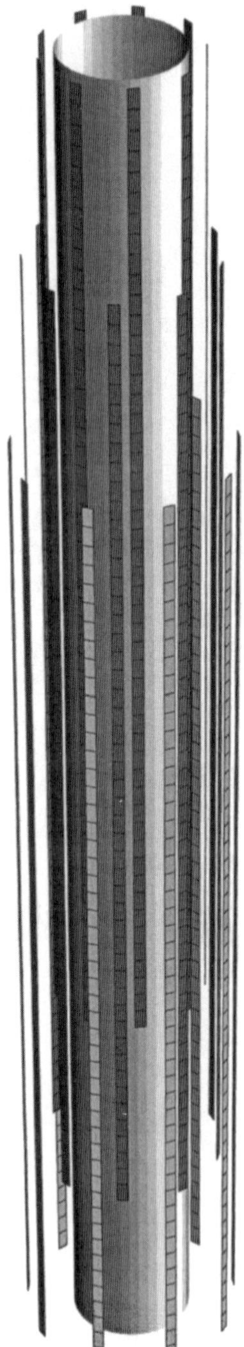

D. Axial wall texture

THE ROLE OF CORTICAL MICROTUBULES

In elongating plant cells, the cortical microtubules as well as the CMFs that are being deposited are both seen transverse to the elongation direction of the cells[18]. This correlation has led to the hypothesis that cortical microtubules determine the orientation of the CMFs[19] by stiffening the plasma membrane, locally forming corridors for the cellulose synthases[20]. From such data it has been concluded that cortical microtubule orientation determines the deposition orientation of the CMFs, reviewed in[21]. However, in most concentrations used, these drugs also stop unidirectional cell elongation, a parameter that could be involved in CMF alignment. Therefore, the effect that microtubule drugs have on nascent CMFs should preferably be studied in fully expanded cells. In the fully expanded cells studied, cortical microtubules and most recently deposited CMFs are not generally in parallel alignment[6].

The hypothesis that microtubules orient wall CMFs, though accepted in many text books, is still controversial. Even the fact that cortical microtubules are parallel to nascent CMFs is unproven in many cells because cell wall texture was not thoroughly studied in combination with the cortical microtubules. Uranyl acetate/lead citrate staining of glutaraldehyde/osmium tetroxide fixed plant material often does not reveal the actual CMF alignment[8] and polarization microscopy gives an overall image of the whole wall only. Because the cases in which no parallelism is found often constitute non-growing cells or cell parts, it may be stated that in expanding cells microtubules might orient CMFs by being appressed to the plasma membrane more than in non-expanding walls[6]. However, studies on mung bean hypocotyl[22] are among the few of elongating cells, in which cortical microtubule and CMF orientation are both investigated reliably. These studies show that growing coleoptile cells do not have a cell wall with CMFs transverse to the elongation direction, but a helicoidal wall, which lies adjacent to cortical microtubules in other directions. Also in cortical cells of explants of *Nicotiana tabacum* L. flower stalks, which were brought under tissue culture conditions, neither colchicine nor cremart, which depolymerized the microtubules, influenced the changes in CMF deposition[23]. Recently, Sugimoto et al.[24] determined that cortical microtubules and CMFs are both transversely aligned in the region of the *Arabidopsis* roots where relative growth rate is rising and this agrees with data from roots of other species. However, these authors also show that cortical microtubules are less universely transverse as CMFs in the cell division-elongation zone and have even more contrasting orientations in the distal elongation zone. Such lack of paralellism in post elongation root zones with CMFs transverse next to oblique to longitudinal cortical microtubules, has been reported before[25,26,27].

Cortical microtubules do not seem to be the guiding principle of CMF orientation. Mutants disturbed in either cortical microtubule or CMF alignment will be important for elucidating the microtubule microfibril paradigm.

REFERENCES

1. A. Bacic & B.A. Stone, Structure and function of plant cell walls, In: *The Biochemistry of Plants: A Comprehensive Treatise,* Vol. 14, P.K. Stumpf, & E.E. Conn, E.E. (eds.), **1988** pp. 297-371
2. W-D Reiter, W-D The molecular analysis of cell wall components, *Trends in Plant Science* **1998**, 3, 27-32

3. N.C. Carpita & D.M. Gibeaut, Structural models of primary cell walls in flowering plants: consistency of molecular structure with the physical properties of the walls during growth, *Plant J.* **1993,** 3, 1-30
4. W. Herth, Calcofluor white and congo red inhibit chitin microfibril assembly of *Poteriochromonas*: evidence for a gap between polymerization and microfibril formation, *J. Cell Biol.* **1980,** 87, 442-450
5. T.H. Giddings & L.A. Staehelin, Microtubule-mediated control of microfibril deposition: a re-examination of the hypothesis. In: *The cytoskeletal basis of plant growth and form.* C.W. Lloyd, (ed.) **1991,** pp.85-100
6. A.M.C. Emons A.M.C., J.H.M. Derksen J.H.M. & M.M.A. Sassen M.M.A. Do microtubules control plant cell wall microfibrils? *Physiologia Plantarum* **1992,** 84, 486-493
7. T. Itoh Biogenesis of cellulose microfibrils and the role of microtubules in green algae. ACS Symposium series no. 399: In: *Plant Cell wall polymers: Biogenesis and biodegradation* N.G. Lewis & M.G. Paice, (eds.) **1989**
8. A.M.C. Emons, A comparison of methods for visualization of the cell wall microfibrils. *Acta Botanica Neerlandica,* **1988,** 37, 31-38
9. A.M.C. Emons, Winding threads around plant cells: a geometrical model for microfibril deposition, *Plant Cell & Evironment.* **1994**, 17, 3-14
10. A.M.C. Emons & H. Kieft, Winding cells around plant cells: applications of the geometrical model for microfibril deposition, *Protoplasma* **1994**,180, 59-69
11. A.M.C.Emons & B.M. Mulder, Plant cell wall architecture, *Comments Theor. Biol.* **1997**, 4, 115-131
12. A.M.C. Emons, & B.M. Mulder, The making of the architecture of the plant cell wall: how cells exploit geometry, *Proc. Natl. Acad. Sci.* USA **1998**, 95, 7215-7219
13. A.M.C. Emons, & B.M. Mulder, How the deposition of cellulose microfibrils build cell wall architecture, *Trends in Plant Science* **2000,** 35-40
14. S.C. Mueller & R.M. Brown, Jr. Evidence for an intramembranous component associated with a cellulose microfibril-synthesizing complex in higher plants, *J. Cell Biol.* **1980** 84, 315-326
15. A.M.C. Emons, The role of rosettes and terminal globules in cellulose synthesis, In: *Biosynthesis and Biodegradation of Cellulose and Cellulosic Material*, P. Weimer & C.H. Haigler, **1991** pp. 71-98
16. D.P. Delmer, Cellulose biosynthesis, Exiting times for a difficult field of study, *Annu. Rev. Plant Physiol. Plant Mol. Biol.* **1999** 50, 245-276
17. S. Kimura, W. Laosinchai, T. Itoh, X. Cui, C.R. Linder, R.M. Brown, Immunogold labeling of rosette terminal cellulose-synthesizing complexes in the vascular plant Vigna angularis, *Plant Cell* **1999,** 11, 2075-2085
18. R.J. Cyr, Microtubules in plant morphogenensis: role of the cortical arrray, *Annu Rev. Cell Biol.* **1994**, 10, 153-180
19. I.B. Heath, A unified hypothesis for the role of membrane bound enzyme complexes and microtubules in plant cell wall synthesis. *Journal of Theoretical Biology* **1974**, 48, 445-449
20. L.A. Staehelin, & T.H Giddings Jr. Membrane mediated control of cell wall microfibrillar order, In: *Developmental order: Its origin and regulation*, S. Subtelny, & P.B. Green, (eds), **1982** pp. 133-147

21. C. Wymer, & C Lloyd, Dynamic microtubules: implications for cell wall patterns, *Trends in Plant Science* **1996**, 7, 222-228
22. B. Satiat-Jeunemaître, Microtubule and microfibril arrangements during cell wall expansion along the growth gradient of the mung bean hypocotyl. *Ann Sci. Nat., Botanique*, Paris **1991**, 11: 105-115
23. F.H.A. Wilms, A.M.C. Wolters-Arts & J. Derksen, Orientation of cellulose microfibrils in cortical cells of tobacco explants. Effect of microtubule-depolymerizing drugs. *Planta* **1990**, 182, 1-8
24. K. Sugimoto, R.E. Williamson & G.O. Wasteneys, New techniques enable comparative analysis of microtubule orientation, wall texture, and growth rate in intact roots of *Arabidopsis, Plat Physiology* **2000**, 124, 1493-1506
25. T. Hogetsu, Orientation of wall microfibril deposition in root cells of *Pisum sativum* L.var. Alska, *Plant Cell Physiol.* **1986, 27,** 947-951
26. J.A. Traas & J. Derksen, Microtubules and cellulose microfibrils in plant cells: simultaneous demonstration in dry cleave preparations, *Eur. J. Cell Biol.* **1989,** 48, 159-164
27. T.I. Baskin, H.T.H.M. Meekes, B.M. Liang & R.E. Sharp, Regulation of growth anisotropy in well-watered and water-stressed maize roots: II Role of cortical microtubules and cellulose microfibrils. *Plant Physiol.* **1999,** 681-692

Molecular Breeding of Woody Plants
N. Morohoshi and A. Komamine, editors.

OCCURENCE OF HIGH CRYSTALLINE CELLULOSE IN THE MOST PRIMITIVE TUNICATE, APPENDICULARIAN

Satoshi Kimura & Takao Itoh

Wood Research Institute, Kyoto University,
Uji, Kyoto, 611-0011, Japan

ABSTRACT

Using electron diffraction analysis, highly crystalline cellulose Iβ was found in the house (a special structure in which the tunicate lives) of an appendicularian, *Oikopleura rufescens*. The cellulose microfibrils of 20 nm in width were observed in a random array or highly organized with 2 to 10 μm rectangular spacings in the house. The bundled cellulose microfibrils formed in the inlet filters, which are highly ordered meshwork structure. This provides first account for the existence of cellulose in the house of the appendicularian. The evidence shows that the house and tunic are homologous tissue among the tunicates, and that the common ancestor of the tunicates (ascidians, thaliaceans, and appendicularians) had already possessed the ability of the cellulose biosynthesis. This investigation will offer a key to understanding whether the cellulose synthetic ability is universal or not in the tunicates, and how the cellulose biosynthetic ability is transmitted into the tunicates.

KEYWORDS

Appendicularian, Cellulose, electron diffraction, house, tunicate

INTRODUCTION

The tunicates (urochordates) are the only animals known to produce highly crystalline cellulose[1]. The name "Tunicata" is derived from the unique integumentary tissue called the “tunic”, which contains the cellulose microfibrils. To date, cellulose I microfibrils have been found in almost all of the ascidians and thaliaceans [2-3]. Thus, the cellulosic composition of the tunic is considered to be a characteristic common to ascidians and thaliaceans in animals belonging to the subphylum Tunicata.

The appendicularians is another group in the Tunicata. Although molecular phylogeny based on 18S rDNA sequences suggested that appendicularians share a common ancestor with the other groups of tunicates [4-5], they do not possess the tunic as an integumentary tissue. On the other hand, the appendicularians secrete a balloon-like, gelatinous structure called a “house” that acts as a feeding apparatus [6]. It is possible that the house corresponds to a kind of tunic in the appendicularians, but it is not yet clear whether the house contains cellulose. Therefore, it is necessary to clarify the existence of cellulose in the appendicularian house to understand the evolutionary pathway of cellulose biosynthesis in the tunicates.

The present study focused on the existence and characterization of cellulose in the appendicularian house. Our investigation provides a key to understanding whether the ability for cellulose synthesis is universal in the tunicates.

MATERIALS AND METHODS

Sample collection and fixation

Appendicularians with their houses were collected in hand-held glass jars by skin divers at depths not exceeding 1m off Maeda point, Okinawa, Japan. This method enabled us to obtain samples with minimal disturbance or mechanical damage to the fragile animals and their houses. After collection, the specimens (houses and animals) were immediately fixed in 2.5% glutaraldehyde-Millipore filtered seawater.

Light and electron microscopy

For replica preparation and electron diffraction analysis, the houses were treated with KOH and $NaClO_2$ solution to remove non-cellulosic materials as described previously [3]. A JEM-2000EXII transmission electron microscope was used for observation of the replica and selected area electron diffraction at an accelerating voltage of 100 kV. Some houses were observed using a polarization and Nomarski differential interference contrast microscope before or after treatment with KOH and $NaClO_2$.

RESULTS & DISCUSSION

The house of *O. rufescens* is a typical oikopleurid house, which is a spherical, gelatinous structure and possesses two inlet filters (Fig. 1). After treatment with KOH and $NaClO_2$ solutions, insoluble gelatinous materials with an irregular meshwork structure were obtained. The insoluble materials were composed of numerous microfibrils that were randomly arrayed (Fig. 2 background) or bundled (Fig. 2, arrows). The former were derived from a gelatinous region and the latter from the inlet filters. Electron diffraction images of the bundled microfibrils showed them to be comprised of highly crystalline cellulose I (inset of Fig. 2). The sharpness and strength of the 002 spot derived from cellulose Iβ allomorph and the lack of super lattice reflections derived from Iα showed that the cellulose microfibrils were composed of purely crystalline Iβ cellulose (arrowhead in inset of Fig. 2). The diffractograms derived from cellulose Iβ were also obtained from the random oriented microfibrils with circular diffraction patterns (data not shown). The mean diameter of individual cellulose microfibrils was 23 nm (n=50, SD=3). The width of microfibril bundles was variable from 2 to 10 μm (n=20). Our observations indicated that the cellulose synthetic ability is a characteristic common to all tunicates, i.e. ascidians, thaliaceans and appendicularians. The crystalline features and the dimensions of cellulose microfibril in *O. rufescens* were similar to those of ascidians and thaliaceans [2-3]. A similar biosynthetic system such as a linear type of cellulose synthesizing complex

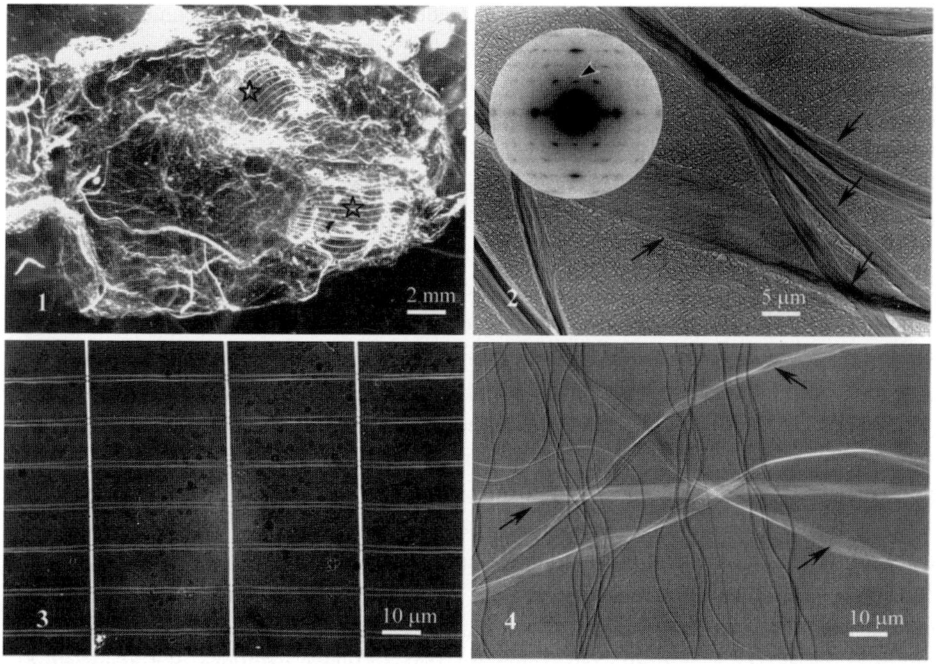

Figure 1.
Dark field macro photograph of the house of *O. rufescens* without chemical treatment. The house of *O. rufescens* is a typical oikopleurid house in both size and morphology. The two asterisks indicate the meshwork of inlet filters.

Figure 2.
Electron micrographs of the house of *O. rufescens* after treatment with KOH and $NaClO_2$. The insoluble residue from the house is composed of numerous microfibrils. The microfibrils were observed with a random arrangement (background of figure) or as bundles of two thickness (arrows). The inset shows the electron diffraction pattern from bundled microfibrils. The diffractograms indicated highly crystalline and almost pure cellulose Iβ allomorph with a 002 spot (arrowhead).

Figures 3 and 4.
Light micrographs of inlet filter in the house of *O. rufescens* before (3) and after (4) treatment with KOH and $NaClO_2$. The inlet filter is a highly ordered meshwork structure and composed of cellulose bundles with two ranges of thickness (3, polarization microscope). The meshwork of the inlet filter collapsed after chemical treatment (4, Nomarski differential interference contrast). Thicker cellulose bundles are visible with a flattened ribbon shape (4, arrows).

(linear TC), which has been found in the epidermal cells of the ascidian [7], may be involved in the formation of cellulose microfibrils in the appendicularian house.

The inlet filter of the house formed as a highly ordered meshwork structure similar to a woven textile consisting of a single warp thread 5 μm in width and two tightly closed woof threads 3 μm in width (Fig. 3). The two woof threads were woven along the warp at intervals of 22.8 μm (n=100, SD=1.7). The meshworks of inlet filters were unfolded after treatment with KOH and $NaClO_2$ (Fig. 4). The warp of the inlet filter had a flattened ribbon-shape 10 μm in width (Fig. 4, arrows) after treatment. The woof threads maintained the same dimensions after treatment. Some proteins or non-crystalline polysaccharides have been suggested to participate in gluing the cellulose bundles to one another.

The 18S rDNA molecular phylogeny indicated that the tunicates are a monophyletic group and that the appendicularians diverged early from the tunicates [4-5]. The present study indicated that the common ancestor of all tunicates already possessed the ability to produce highly crystalline cellulose I. This suggests that other animals close to the tunicates, such as cephalochordates, vertebrates and hemichordates, may also produce cellulose or possess the cellulose synthase genes. Cellulose synthase genes have been found in bacteria, protists and land plants [8]. The genes have some highly conserved regions and variable regions among species [8]. The cloning of cellulose synthase genes of tunicates as well as finding other animals that produce cellulose are required to understand the evolutionary pathway and diversity of cellulose synthetic ability among the cellulosic organisms.

ACKNOWLEDGMENT

This work was supported by a Grant-in-Aid for "Research for the Future" Program (no. JSPS-RFTF 96L00605) from the Ministry of Education, Sciences, Sports and Culture of Japan to T.Itoh

REFERECES

1. P. A. Richmond, 'Occurrence and functions of native cellulose', In: C. Haigler and P. J. Weimer (eds) Biosynthesis and biodegradation of cellulose, Marcel Dekker Inc., New York, Basal, Hong Kong, 1991, pp 5-23.
2. T. Okamoto, J. Sugiyama & T. Itoh, 'Structural diversity of ascidian cellulose', *Wood Res*, 1996, 83: 27-29.
3. E. Hirose, S. Kimura, T. Itoh & J. Nishikawa, 'Tunic of pyrosomas, doliolids and salps (Thaliacea, Urochordata): morphology and cellulosic components', *Biological Bulletin*,1999, 196: 113-120.
4. H. Wada & N. Satoh, 'Details of the evolutionary history from invertebrates to vertebrates, as deduced from the sequences of 18SrDNA', *Proc. Natl. Acad. Sci. USA*, 1994, 91: 1801-1804.
5. H. Wada, 'Evolutionary history of free-swimming and sessile lifestyles in

urochordates as deduced from 18S rDNA molecular phylogeny', *Mol. Biol. Evol.*, 1998, 15: 1189-1194.

6. P. R. Flood & D. Deibel, 'The appendicularian house', In: Bone Q (ed) The biology of pelagic tunicates, Oxford University Press, Oxford, 1998, pp 105-124.
7. S. Kimura & T. Itoh, 'New cellulose synthesizing complexes (=terminal complexes) involved in animal cellulose biosynthesis in the tunicate, *Metandrocarpa uedai*', *Protoplasma*, 1996, 194: 151-163.
8. D. P. Delmer, 'Cellulose biosynthesis: exciting times for a difficult field of study', *Annu. Rev. Plant Physiol. Mol. Biol.*, 1999, 50: 245-276.

Molecular Breeding of Woody Plants
N. Morohoshi and A. Komamine, editors.

THE ROLE OF CORTICAL MICROTUBULES IN WOOD FORMATION

Ryo FUNADA

Laboratory of Wood Biology, Graduate School of Agriculture, Hokkaido University, Sapporo 060-8589, Japan

ABSTRACT

During differentiation of cambial cells into secondary xylem cells in trees (wood formation), newly deposited cellulose microfibrils on the innermost surface of cell walls change their orientation progressively. The orientation of cellulose microfibrils in the secondary wall, referred to as the microfibril angle, is one of the most important characteristics that determine the properties of wood. In particular, the angle in the thickest middle layer (S_2 layer) of secondary wall is closely related to mechanical properties. Therefore, the control of orientation of cellulose microfibrils allows us to change the quality of wood and its products. Recent observations in a wide variety of trees have revealed that there is a close relationship between the orientation of cortical microtubules, which are one component of cytoskeleton, and the orientation of newly deposited cellulose microfibrils in differentiating secondary xylem cells. Such a close relationship has been found in both normal, compression wood and tension wood. These evidences indicate that cortical microtubules play an important role in the control of the orientation of cellulose microfibrils in secondary xylem cells. Therefore, biotechnological control of the orientation of cellulose microfibrils in the cell walls by manipulation of cortical microtubules provides new tools to improve the quality of wood and its products.

KEYWORDS

Cambium, cell wall, cellulose microfibrils, cortical microtubules, cytoskeleton, xylem differentiation, wood formation, wood quality

INTRODUCTION

Wood has been used for millennia as a raw material such as timber for construction, furniture, pulp for paper, chemicals and fuels. Therefore, there is a still great demand for wood as renewable material and energy. In addition, recent environmental interest, global warming, has focused attention on woods as main CO_2-sinks in the biosphere. Woods are expected to be an important role in removing the excess of atmospheric CO_2 generated by the burning of fossil fuels.

Wood quality is of great importance for the forest products industry. However, wood quality varies markedly within a single species. Such a variation is depending on environmental factors, which affect the cambial growth of trees. In addition, wood quality varies among species and within individual trees according to cambial age, stem position and distance from the crown. Tree-to-tree variability within species under identical growth conditions shows that genetic factors also influence wood quality.

Therefore, wood quality might be improved or unified not only by silvicultural treatments but also by breeding programs to select genetically desirable trees (Panshin and de Zeeuw 1980, Zobel and van Buijtenen 1989, Zobel and Jett 1995). Variations in wood quality, in particular in mechanical properties, are largely due to differences in wood structure, such as the shape and size of xylem cells, the thickness of cell wall, and the microfibril angle (the angle between cellulose microfibrils and the main cell axis). Thus, wood structure is one of most important targets to control wood quality.

Wood is produced by the vascular cambium (cambium) of trees. The division of cambial cells produces the secondary phloem on the outside and the secondary xylem on the inside. The amount of secondary xylem produced is usually much more than the amount of secondary phloem (Larson 1994). Thus, matured secondary xylem cells are usually used as wood. As soon as cambial cells lose the ability of division, they start to differentiate. Fusiform cambial cells differentiate to tracheids, vessel elements, wood fibers and axial parenchyma cells, while ray cambial cells differentiate to ray parenchyma cells and ray tracheids. The stages of differentiation of secondary xylem cells can be categorized as follows: cell expansion or elongation, cell wall thickening, cell wall sculpture, lignification, and cell death (Panshin and de Zeeuw 1980, Thomas 1991). During differentiation, secondary xylem cells form cell walls with a highly organized layered structure. After the completion of lignification, cell death (cell autolysis) occurs immediately in most secondary xylem cells that function in support of the tree and in conduction of water, such as tracheids and vessel elements. However, cells that function in storage, such as ray parenchyma cells, remain alive for several years without immediate autolysis of cell organelles (Murakami et al. 1999). Although wood structure has been extensively studied, the process of its formation is not yet fully understood (Catesson 1994, Higuchi 1997). Therefore, we need more detailed cellular and molecular information of wood formation for the improvement of more desirable wood quality by biotechnological manipulation.

Recent observations reveal that the arrangement of components of cytoskeleton is closely related to the process of cambial division and differentiation of secondary xylem cells in gymnosperm (Funada 2000, 2001, Funada et al. 2000, Oribe et al. 2001) and angiosperm trees (Chaffey 1999, 2000, Chaffey and Barlow 2000). Here I summarize briefly the role of components of cytoskeleton in different developmental process of wood formation, in particular the interaction of newly deposited cellulose microfibrils and cortical microtubules in secondary xylem cells.

ARRANGEMENT OF CELLULOSE MICROFIBRILS

During differentiation, cells increase in diameter and length until they become their final size and shape. The driving force for the enlargement of cells is the turgor pressure within cells. When the turgor pressure exceeds the yield point of cell wall, the cell can enlarge. The turgor pressure is exerted equally in all directions within the cell. Therefore, if there is no reinforcement system, cells might be enlarged spherically. However, a framework of cellulose microfibrils, which have very high tensile strength, reinforces the cell wall and resists in response to the turgor pressure. The predominant orientation of cellulose microfibrils in the cell wall is generally perpendicular to the direction of cell enlargement. Therefore, the orientation of cellulose microfibrils determines the direction and extent of cell enlargement, indicating its important role in the control of the final size and shape of cells.

Fusiform cambial derivatives increase in diameter and length. For example, tracheids and vessel elements increase considerably in diameter but no or only slightly increase in length (Kitin et al. 1999). The cell wall during the cell enlargement, referred to as the primary wall, consists of loose aggregates of cellulose microfibrils, allowing relatively easy enlargement of the cells (Harada and Côté 1985). The cellulose microfibrils on the innermost surface of the primary wall of cambial derivatives are not well-ordered (Abe et al. 1995b). The predominant orientation of cellulose microfibrils in tracheids at the early stage of cell enlargement is longitudinal to the cell axis. Longitudinally oriented cellulose microfibrils might restrain the longitudinal enlargement by the turgor pressure. Therefore, the primary wall of cambial derivatives has the structure that allows easy lateral expansion. As the cell expands, the cellulose microfibrils on the innermost surface change their orientation from longitudinal to transverse to the cell axis. The transversely oriented cellulose microfibrils at the final stage of cell enlargement might prevent further lateral expansion.

When expansion of tracheids or wood fibers ceases, well-ordered cellulose microfibrils are deposited on the innermost surfaces of cell wall (Abe et al. 1991, 1992, 1995a,b, 1997, Kataoka et al. 1992, Prodhan et al. 1995a,b, Awano et al. 2000). Cellulose microfibrils are oriented transversely to the cell axis. Such well-ordered cellulose microfibrils are considered to be the secondary wall due to their texture (Harada and Côté 1985). Once the deposition of well-ordered cellulose microfibrils starts, no further lateral expansion of cells is restricted.

Continuous deposition of the secondary wall increases the thickness of the cell wall. During formation of the secondary wall in tracheids or wood fibers, newly deposited cellulose microfibrils change their orientation progressively (Figure 1). Such a successive change in the orientation of cellulose microfibrils develops a three-layered structure of secondary wall, namely the outer layer (S_1), the middle layer (S_2) and the inner layer (S_3) of secondary wall, in tracheids or wood fibers. The cellulose microfibrils change their orientation rotatively from a flat S-helix (S_1 layer) to a steep Z-helix (S_2 layer) during formation of the secondary wall. The successive changes in the orientation of cellulose microfibrils can be considered to reflect the semi-helicoidal structure of the cell wall (Roland et al. 1987). With cessation of rotation, the thick S_2 layer develops because of repeated deposition of cellulose microfibrils with a consistent texture. The microfibril angle in the S_2 layer is usually 5-25°. Then the orientation of the cellulose microfibrils changes from a steep Z-helix to a flat S-helix (S_3 layer) at the final stage of secondary wall formation.

Several growth conditions vary the microfibril angles of the cell wall. For example, the S_3 layer is absent from the compression wood tracheids, which are formed at the lower side of inclined stem in conifers. In addition, the microfibril angle in the S_2 layer of the compression wood tracheids is about 45° to the tracheid axis, which is larger than the S_2 layer of the normal wood tracheids (Timell 1986). The difference in the rotation of cellulose microfibrils during the formation of secondary wall leads to variations in microfibril angles on the innermost surface of cell wall.

ARRANGEMENT OF CORTICAL MICROTUBULES

It has been widely accepted that cortical microtubules, one of components of cytoskeleton, are involved in cell wall formation. Co-alignment of cortical microtubules and cellulose microfibrils has been observed in the cells of a wide variety of plants (Giddings and Staehelin 1991). Cortical microtubules might guide the movement of cellulose synthase complexes in the plasma membrane.

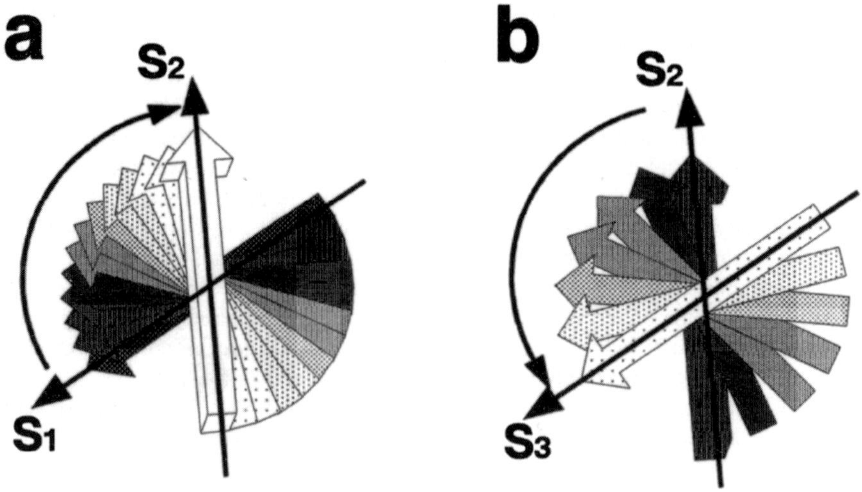

Figure 1. Schematic model of progressive changes in the orientation of newly deposited cellulose microfibrils (arrows), viewed from the lumen side, in differentiating tracheids during the formation of secondary wall. a; The orientation of cellulose microfibrils changes from a flat S-helix to a steep Z-helix (from S_1 layer to S_2 layer). The continuous deposition of cellulose microfibrils with a steep Z-helix produces a thick cell wall layer. b; The orientation of cellulose microfibrils changes from a steep Z-helix to a flat S-helix (from S_2 layer to S_3 layer). The shift from a steep Z-helix to a flat S-helix is more abrupt that from a flat S-helix to a steep Z-helix.

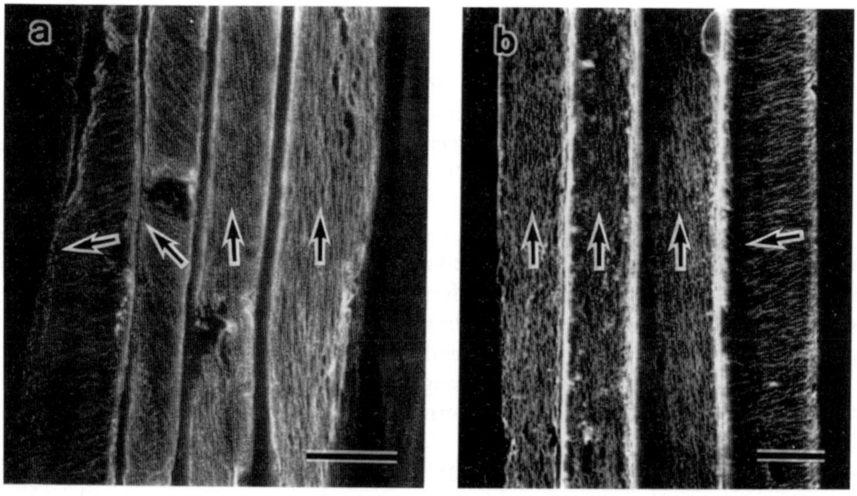

Figure 2. The successive changes in the orientation of cortical microtubules (viewed from the lumen side) in differentiating tracheids during the formation of secondary wall. a; The orientation of cortical microtubules (arrows) changes from a flat helix to a steep Z-helix. b; The orientation of cortical microtubules (arrows) changes from a steep Z-helix to a flat helix at the final stage of secondary wall formation. Scale bar = 25 μm.

The cortical microtubules in fusiform cambial cells are not well ordered (Abe et al. 1995a,b, Chaffey et al. 1997a,b,c, 1998, 1999a,b, Funada et al. 1997). They are occasionally overlapped (Chaffey et al. 1997a,b). The network of randomly oriented cortical microtubules is consistent with the loose aggregates of cellulose microfibrils of the primary wall of cambial cells and cambial derivatives. The predominant orientation of cortical microtubules is longitudinal to the cell axis in fusiform cambial cells and differentiating tracheids at the early stage of cell enlargement (Abe et al. 1995a,b, Funada et al. 1997). In contrast, no obviously preferred angle for cortical microtubules was observed in fusiform cambial cells in hardwoods (Chaffey et al. 1997a, 1998, 1999a). Such a difference may be due to difference between gymnosperm and angiosperm trees. The cortical microtubules in cambial derivatives become gradually more ordered during xylem (Abe et al. 1995b, Chaffey et al. 1997a,b, 1999b, Funada et al. 1997) and phloem differentiation (Chaffey et al. 1997a,b, 1999a). The shift in arrangement of cortical microtubules from random to ordered may be a marker for the commencement of differentiation of cambial cells (Chaffey et al. 1997a). The predominant orientation of cortical microtubules changes progressively from longitudinal to transverse as tracheids laterally expand (Abe et al. 1995a,b, Funada et al. 1997). These changes in the orientation of cortical microtubules resemble the shift of newly deposited cellulose microfibrils in tracheids during the formation of primary wall. The cortical microtubules might be involved in determining the orientation of cellulose microfibrils. It is likely that the ordered and transversely oriented cellulose microfibrils controlled by cortical microtubules impede the further lateral expansion of tracheids.

A close relationship between cortical microtubules and the most recent deposited cellulose microfibrils has been observed in differentiating tracheids and wood fibers during the formation of secondary wall by transmission electron microscopy (Cronshaw 1965, Nobuchi and Fujita 1972, Robrads and Kidwai 1972, Fujita et al. 1974, Barnett 1981, Inomata et al. 1992, Chaffey et al. 1999b) or scanning electron microscopy (Hirakawa 1984, Abe et al. 1994, Prodhan et al. 1995a). Cortical microtubules are abundant throughout the formation of secondary wall (Barnett 1981, Abe et al. 1994, Prodhan et al. 1995a, Chaffey et al. 1997a, 1999b). Confocal laser scanning microscopy has revealed that the orientation of cortical microtubules changes dynamically during the formation of secondary wall of tracheids (Figure 2; Abe et al. 1995a, Furusawa et al. 1998, Funada 2000, Funada et al. 2000). The orientation changes rotatively from a flat helix to a steep Z-helix after cessation of expansion of tracheids. Then, the shift of cortical microtubules is arrested to keep a steep Z-helix, about 5-25° to the cell axis. At the final stage of differentiation, the orientation of cortical microtubules changes from a steep Z-helix to a flat helix. These observations provide evidence that the orientation of cortical microtubules changes in a manner similar to the progressive changes in the orientation of cellulose microfibrils during the formation of secondary wall as shown in Figure 1. Therefore, the orientation of cortical microtubules controls the ordered orientation of cellulose microfibrils in the semi-helicoidal wall of secondary xylem cells. In addition, the rotational motion of cortical microtubules is associated with the pattern of layered structure of cell wall.

A similar close relationship between cortical microtubules and cellulose microfibrils has been found in reaction wood, namely in compression wood tracheids of conifers (Furusawa et al. 1998) and in tension wood fibers of hardwoods (Nobuchi and Fujita 1972, Fujita et al. 1974, Prodhan et al. 1995a), which are formed in response to the directional change in gravity. In compression wood tracheids, cortical microtubules are oriented transversely during the early stage of secondary wall formation and then their orientation changes progressively until an angle becomes about 45° in a Z-helix. This angle is larger than that of the S_2 layer of the normal wood tracheids. The orientation of cortical microtubules in compression wood tracheids is consistent with the characteristic microfibril angle of the S_2 layer, about 45° in a Z-helix, of these tracheids. The same orientation of cortical microtubules is retained until xylem differentiation is complete. This result is consistent with the absence of the S_3 layer, whose cellulose microfibrils are oriented in a flat helix, in compression wood tracheids. Several stimuli, such as gravity and plant hormones, alter the pattern of orientation of cortical microtubules. Such an alternation reflects the characteristic features of cell wall structure.

FUTURE PROSPECTS

Wood structure is closely related to wood quality. In particular, the microfibril angle of S_2 layer, which is the thickest layer in the secondary wall, are of major importance. The microfibril angles of the S_2 layer are negatively correlated with the modulus of elasticity of wood and thus wood with large microfibril angles has low strength. In addition, wood with large microfibril angles of the S_2 layer exhibits great longitudinal shrinkage. Thus, large changes in dimensions due to shrinkage with changes in moisture content result in warping and collapse in wood. Therefore, the control of microfibril angle in the secondary wall, in particular in the S_2 layer, provides a powerful tool for changing the mechanical properties of wood.

Recent evidence presented here provides the involvement of cortical microtubules in the control of the direction of cell enlargement and the structure of secondary wall, in particular the microfibril angles and cell wall thickness, in secondary xylem cells. The cortical microtubules play an important role in the control of the orientation of newly deposited cellulose microfibrils in secondary xylem cells (Chaffey 2000, Funada 2000, 2001, Funada et al. 2000). In addition, localized cortical microtubules are closely associated with the development of localized deposition of cellulose microfibrils such as helical thickenings (Uehara and Hogetsu 1993, Furusawa et al. 1998, Chaffey et al. 1999b), bordered pits (Robards and Humpherson 1967, Uehara and Hogetsu 1993, Abe et al. 1995b, Chaffey et al. 1997b, Funada et al. 1997) and perforation plates (Chaffey et al. 1997c). Therefore, biotechnological control of orientation and localization of cellulose microfibrils in the cell walls by manipulation of cortical microtubules provides new tools to improve the quality of wood and its products. Such approaches might design better trees with more desirable properties. For these approaches, we need more detailed molecular information of directional factors that are responsible for the orientation of cortical microtubules.

ACKNOWLEDGMENTS

The author thanks Dr. H. Abe, Dr. A.K.M.A. Prodhan, Miss. H. Imaizumi, Mr. O. Furusawa, Mr. H. Miura, Mr. M. Shibagaki, Mr. T. Miura, Mr. F. Katoh and Mr. E. Fukatsu for their cooperation during the preparation of this review. The author's research has been supported by Grants-in-Aid for Scientific Research from the Ministry of Education, Culture, Sports, Science and Technology, Japan (nos. 06404013 and 09760152) and the Japan Society for the Promotion of Science (no. JSPS-RFTF 96L00605).

REFERENCES

Abe, H., Ohtani, J., Fukazawa, K.: FE-SEM observation on the microfibrillar orientation in the secondary wall of tracheids, IAWA Bulletin new series, 12, 431-438 (1991).

Abe, H., Ohtani, J., Fukazawa, K.: Microfibrilar orientation of the innermost surface of conifer tracheid walls, IAWA Bulletin new series, 13, 411-417 (1992).

Abe, H., Ohtani, J., Fukazawa, K.: A scanning electron microscopic study of changes in microtubule distributions during secondary wall formation in tracheids, IAWA Journal, 15, 185-189 (1994).

Abe, H., Funada, R., Imaizumi, H., Ohtani, J., Fukazawa, K.: Dynamic changes in the arrangement of cortical microtubules in conifer tracheids during differentiation, Planta, 197, 418-421 (1995a).

Abe, H., Funada, R., Ohtani, J., Fukazawa, K.: Changes in the arrangement of microtubules and microfibrils in differentiating conifer tracheids during the expansion of cells, Annals of Botany, 75, 305-310 (1995b).

Abe, H., Funada, R., Ohtani, J., Fukazawa, K.: Changes in the arrangement of cellulose microfibrils associated with the cessation of cell expansion in tracheids, Trees, 11, 328-332 (1997).

Awano, T., Takabe, K., Fujita, M., Daniel, G.: Deposition of glucuronoxylans on the secondary cell wall of Japanese beech as observed by immune-scanning electron microscopy, Protoplasma, 212, 72-79 (2000).

Barnett, J.R.: Secondary xylem cell development. In: Barnett, J.R. (ed.) Xylem Cell Development, Castle House Publications, Tunbridge Wells, pp. 47-95 (1981).

Catesson, A.M.: Cambial ultrastructure and biochemistry: changes in relation to vascular tissue differentiation and the seasonal cycle, International Journal of Plant Science, 155, 251-261 (1994).

Chaffey, N.: Cambium: old challenges – new opportunities, Trees, 13, 138-151 (1999).

Chaffey, N.J.: Cytoskeleton, cell walls and cambium: new insights into secondary xylem differentiation, In: Savidge, R., Barnett, J., Napier, R. (eds.) Cell and Molecular Biology of Wood Formation. BIOS Scientific Publishers, Oxford, pp. 31-42 (2000).

Chaffey, N.J., Barlow, P.W.: Actin in the secondary vascular system, In: Baluska, F, Barlow, P.W., Staiger, C, Volkmann, D (eds.) Actin: A Dynamic Framework for Multiple Plant Cell Functions, Kluwer Academic Publishers, Dordrecht, pp. 587-600 (2000).

Chaffey, N., Barlow, P., Barnett, J.: Cortical microtubules rearrange during differentiation of vascular cambial derivatives, microfilaments do not, Trees, 11, 333-341 (1997a).

Chaffey, N.J., Barnett, J.R., Barlow, P.W.: Cortical microtubule involvement in bordered pit formation in secondary xylem vessel elements of *Aesculus hippocastanum* L. (Hippocastanaceae): a correlative study using electron microscopy and indirect immunofluorescence microscopy, Protoplasma, 197, 64-75 (1997b).

Chaffey, N.J., Barnett, J.R., Barlow, P.W.: Visualization of the cytoskeleton within the secondary vascular system of hardwood species, Journal of Microscopy, 187, 77-84 (1997c).

Chaffey, N.J., Barlow, P.W., Barnett, J.R.: A seasonal cycle of cell wall structure is accompanied by a cyclical rearrangement of cortical microtubules in fusiform cambial cells within taproots of *Aesculus hippocastanum* L. (Hippocastanaceae), New Phytologist, 139, 623-635 (1998).

Chaffey, N., Barlow, P., Barnett, J.: Structure-function relationships during secondary phloem development in an angiosperm tree, *Aesculus hippocastanum*: microtubules and cell walls, Tree Physiology 20, 777-786 (1999a).

Chaffey, N., Barnett, J., Barlow, P.: A cytoskeletal basis for wood formation in angiosperm trees: the involvement of cortical microtubules, Planta 208, 19-30 (1999b).

Cronshaw, J.: Cytoplasmic fine structure and cell wall development in differentiating xylem elements, In: Côté W.A.Jr. (ed.) Cellular Ultrastructure of Woody Plants, Syracuse University Press, Syracuse, pp. 99-124 (1965).

Fujita, M., Saiki, H., Harada, H.: Electron microscopy of microtubules and cellulose microfibrils in secondary wall formation of poplar tension wood fibers, Mokuzai Gakkaishi, 20, 147-156 (1974).

Funada, R.: Control of wood structure, In: Nick, P. (ed.) Plant Microtubules: Potential for Biotechnology, Springer-Verlag, Heidelberg, Germany, pp. 51-81 (2000).

Funada, R.: Immunolocalisation and visualisation of the cytoskeleton in gymnosperms using confocal laser scanning microscopy (CLSM), In: Chaffey, N (ed.) Wood Formation in Trees: Developmental Cell Biology Techniques, Harwood Academic Publishers, Reading (2001, in press).

Funada, R., Abe, H., Furusawa, O., Imaizumi, H., Fukazawa, K., Ohtani, J. : The orientation and localization of cortical microtubules in differentiating conifer tracheids during cell expansion, Plant and Cell Physiology, 38, 210-212 (1997).

Funada, R., Furusawa, O., Shibagaki, M., Miura, H., Miura, T., Abe, H., Ohtani, J.: The role of cytoskeleton in secondary xylem differentiation in conifers, In: Savidge, R., Barnett, J., Napier, R. (eds.) Cell and Molecular Biology of Wood Formation, Bios Scientific Publishers, Oxford, pp. 255-264 (2000).

Furusawa, O., Funada, R., Murakami, Y., Ohtani, J.: Arrangement of cortical microtubules in compression wood tracheids of *Taxus cuspidata* visualized by confocal laser microscopy, Journal of Wood Science, 44, 230-233 (1998).

Giddings, T.H. Jr., Staehelin, L.A.: Microtubule-mediated control of microfibril deposition: a re-examination of the hypothesis, In: Lloyd, C.W. (ed.) The Cytoskeletal Basis of Plant Growth and Form, Academic Press, London, pp. 85-99 (1991).

Harada, H., Côté, W.A. Jr.: Structure of wood. In: Higuchi, T. (ed.) Biosynthesis and Biodegradation of Wood Components, Academic Press, Orlando, pp. 1-42 (1985).

Higuchi, T.: Biochemistry and Molecular Biology of Wood, Springer-Verlag, Berlin, pp. 1-362 (1997).

Hirakawa, Y.: A SEM observation of microtubules in xylem cells forming secondary walls of trees, Research Bulletin of College Experimental Forests, Hokkaido University, 41, 535-550 (1984).

Inomata, F., Takabe, K., Saiki, H.: Cell wall formation of conifer tracheid as revealed by rapid-freeze and substitution method, Journal of Electron Microscopy, 41, 369-374 (1992).

Kataoka, Y., Saiki, H., Fujita, M.: Arrangement and superimposition of cellulose microfibrils in the secondary walls of coniferous tracheids, Mokuzai Gakkaishi, 38, 327-335 (1992).

Kitin, P., Funada, R., Sano, Y., Beeckman, H., Ohtani, J.: Variations in the lengths of fusiform cambial cells and vessel elements in *Kalopanax pictus*, Annals of Botany, 84, 621-632 (1999).

Murakami, Y., Funada, R., Sano, Y., Ohtani, J.: The differentiation of contact cells and isolation cells in the xylem ray parenchyma of *Populus maximowiczii*, Annals of Botany, 84, 429-435 (1999).

Nobuchi, T., Fujita, M.: Cytological structure of differentiating tension wood fibers of *Populus euroamericana*, Mokuzai Gakkaishi, 18, 137-144 (1972).

Oribe, Y., Funada, R., Shibagaki, M., Kubo, T.: Cambial reactivation in the partially heated stem in an evergreen conifer, *Abies sachalinensis*, Planta, 212, 684-691 (2001).

Panshin, A.J., de Zeeuw, C.: Textbook of Wood Technology, 4th ed. MacGraw-Hill, New York, pp. 1-722 (1980).

Prodhan, A.K.M.A., Funada, R., Ohtani, J., Abe, H., Fukazawa, K.: Orientation of microfibrils and microtubules in developing tension-wood fibres of Japanese ash (*Fraxinus mandshurica* var. *japonica*), Planta, 196, 577-585 (1995a).

Prodhan, A.K.M.A., Ohtani, J., Funada, R., Abe, H., Fukazawa, K.: Ultrastructural investigation of tension wood fibre in *Fraxinus mandshurica* Rupr. var. *japonica* Maxim., Annals of Botany, 75, 311-317 (1995b).

Robards, A.W., Humpherson, P.G.: Microtubules and angiosperm bordered pit formation, Planta, 77, 233-238 (1967).

Robards, A.W., Kidwai, P.A.: Microtubules and microfibrils in xylem fibres during secondary wall formation, Cytobiologie, 6, 1-21 (1972).

Roland, J.C., Reis, D., Vian, B., Satiat-Jeunemaitre, B., Mosiniak, M.: Morphogenesis of plant cell walls at the supermolecular level: internal geometry and versatility of helicoidal expression, Protoplasma, 140, 75-91 (1987).

Thomas, R.J.: Wood: formation and morphology, In: Lewin, M, Goldstein, I.S. (eds.) Wood Structure and Composition, Marcel Dekker, New York, pp. 7-47 (1991).

Timell, T.E.: Compression Wood in Gymnosperms 1, Springer-Verlag, Berlin, pp. 1-706 (1986).

Uehara, K., Hogetsu, T.: Arrangement of cortical microtubules during formation of bordered pit in the tracheids of *Taxus*, Protoplasma, 172, 145-153 (1993).

Zobel, B.J., van Buijtenen, J.P.: Wood Variation: Its Causes and Control, Springer-Verlag, Berlin, pp. 1-363 (1989).

Zobel, B.J., Jett, J.B.: Genetics of Wood Production, Springer-Verlag, Berlin, pp. 1-337 (1995).

Molecular Breeding of Woody Plants
N. Morohoshi and A. Komamine, editors.

XYLAN AND LIGNIN DEPOSITION ON THE SECONDARY WALL OF *FAGUS CRENATA* FIBERS

Tatsuya Awano*, Keiji Takabe & Minoru Fujita

Laboratory of Plant Cell Structure, Division of Forest and Biomaterials Science, Graduate School of Agriculture, Kyoto University, Oiwake-cho, Kitashirakawa, Sakyo-ku, Kyoto 606-8502, JAPAN

ABSTRACT

The secondary wall ultrastructure of *Fagus crenata* fibers was examined using field emission scanning electron microscopy (FESEM). Specimens were treated with sodium chlorite and xylanase to remove lignin and xylan, respectively.

Microfibrils were clearly visible at the innermost surface of the differentiating fiber secondary wall and there were no globular substances observed in the control specimen. After delignification or xylanase-degradation, microfibrils remained almost the same size and had the same appearance as controls. Anti-xylan antiserum immunolabeling, however, indicated that the microfibrils were coated with very thin layer of xylan[1].

Microfibrils were not apparent in the secondary wall of the mature fiber in control specimens. The secondary wall appeared to be a single homogeneous substance. Microfibrils with many globular substances were observed in the delignified specimens and their diameter was larger than that of microfibrils at the surface of the differentiating secondary wall. Following xylanase treatment, the microfibrils had a smooth surface without any globules, indicating that the globular structure is xylan.

On the basis of these results, we propose the following mechanism for secondary wall assembly. Cellulose microfibrils synthesized on the plasma membrane are released into the innermost surface of the secondary wall and coated with a very thin layer of xylan that was previously deposited there. Successive deposition of xylan into the cell wall increases the diameter of the microfibrils. The large amount of xylan deposited on the microfibrils has a globular appearance. Lignin deposition occurs simultaneously with xylan deposition and, finally, microfibrils with globular xylan are masked with lignin, resulting in the homogeneous appearance of the cell wall.

KEYWORDS

Cellulose, delignification, lignin, secondary wall, xylan, xylanase

INTRODUCTION

Xylan is the main hemicellulosic component of hardwoods and accounts for approximately 30% of the woody cell wall. The distribution and function of xylan in the cell wall has been studied extensively since the early 1980's. Vian *et al.* (1983, 1986)[2-3] examined the localization of xylan in the fiber cell walls of linden wood using the xylanase-gold labelling technique. Xylan was preferentially located in the S_1 - S_2 transition zone in which microfibrils have a helicoidal alignment. Vian *et al.* hypothesized that xylan acts as a twisting agent for cellulose microfibrils. The role of xylan during secondary wall formation was reviewed by Reis *et al.* (1994)[4].

Although the role of xylan at the molecular level has been discussed, the mode of deposition of xylan has not been described in detail. We previously examined the immunolocalization of glucuronoxylan in the differentiating xylem of Japanese beech wood and demonstrated that glucuronoxylan was exclusively located in the secondary wall of xylem elements and gradually increased during the course of secondary wall formation[5]. The increase of xylan in the differentiating secondary wall was supported by immuno-scanning electron microscopy[1]. Microfibrils in newly formed secondary walls as well as in mature secondary walls were labelled with anti-xylan antiserum. The intensity of labelling was greater in mature secondary walls than in differentiating cell walls, suggesting that deposition of xylan into the cell wall occurs continuously after cellulose microfibril deposition. The diameter of the microfibrils increased during the course of secondary wall formation. Successive xylan deposition might be responsible for the increase in diameter of the microfibrils.

Microfibrils observed using field emission scanning electron microscopy (FESEM) are not composed of pure cellulose, but were cellulose microfibrils coated with xylan. Xylan, however, is not the only molecule that causes microfibril thickening. Lignin also deposits around microfibrils and contributes to thickening of the microfibrils. When mature cell walls were observed with FESEM, microfibrils are not apparent. In mature cell walls, lignin might cover other cell wall components and contribute to cell wall consolidation. Therefore, lignin removal without loss of other cell wall components is necessary to visualize cellulose microfibrils and hemicellulose using FESEM.

In the present study, mildly-delignified beech sections were observed using FESEM. Delignified sections were treated with xylanase to remove xylan, which is the major hemicellulosic components of beech wood. FESEM observation of these sections indicated the spatial relations of cellulose microfibrils, xylan, and lignin.

MATERIALS AND METHODS

Plant material

Small blocks (5 x 5 x 15 mm) containing differentiating xylem were taken from a living Japanese beech tree (*Fagus crenata*) grown in the Kamigamo Experimental Forest (Kyoto, Japan). Blocks were fixed and stored in 3% glutaraldehyde in 0.1 M phosphate buffer (pH 7.2) at 4 °C . Specimen blocks, stored as described previously, were sectioned in the radial plane at a thickness of 100 μm using a sliding microtome. These sections were used for the following extraction procedure.

Delignification and xylanase degradation

Approximately 1 g of never-dried sections were incubated with 50 ml of 8% $NaClO_2$ in 1.5% acetic acid at 40 °C for 72 h. After incubation, the sections were washed several times with distilled water.

Following sodium chlorite delignification, parts of some sections were incubated with xylanase from *Trichoderma viride* (X-3876, Sigma Chemical Co., St. Louis, MO). Approximately 20 mg of sections were incubated with 12.5 U xylanase in 5 ml of 0.1 M carbonate buffer (pH 4.5) for 1 wk. The enzyme reaction was stopped by thoroughly washing with distilled water.

FESEM observation

Sections were postfixed with 2% osmium tetroxide for 2 h at room temperature. Thereafter they were dehydrated through a graded ethanol series and dried using a critical point dryer (HCP-2, Hitachi) with liquid CO_2 as the drying agent. Dried sections were coated with approximately 1 nm thick platinum/carbon by rotary shadowing at an angle of 60° using a freeze etching apparatus (JFD-9010, JEOL). Sections were examined using FESEM (S-4500, Hitachi) at an accelerating voltage of 1.5 kV and 3 to 5 mm of working distance.

RESULTS & DISCUSSION

Secondary wall of the differentiating fiber

The innermost surfaces of the differentiating secondary walls can be regarded as newly formed cell walls. Microfibrils were clearly visible in the control sections and were highly oriented to the cell axis in the S_2 layer (Fig. 1a). After delignification, microfibrils remained almost the same diameter and had the same appearance as the controls (Fig. 1b), indicating that lignin deposition had not occurred in the innermost surface of the cell wall. Xylanase treatment did not change the diameter or appearance of the microfibrils (Fig. 1c). Microfibrils on the innermost surface of cell walls, however, are immunolabelled with anti-xylan antiserum[1]. Therefore, they were coated with a very thin layer of xylan that could not be resolved by FESEM.

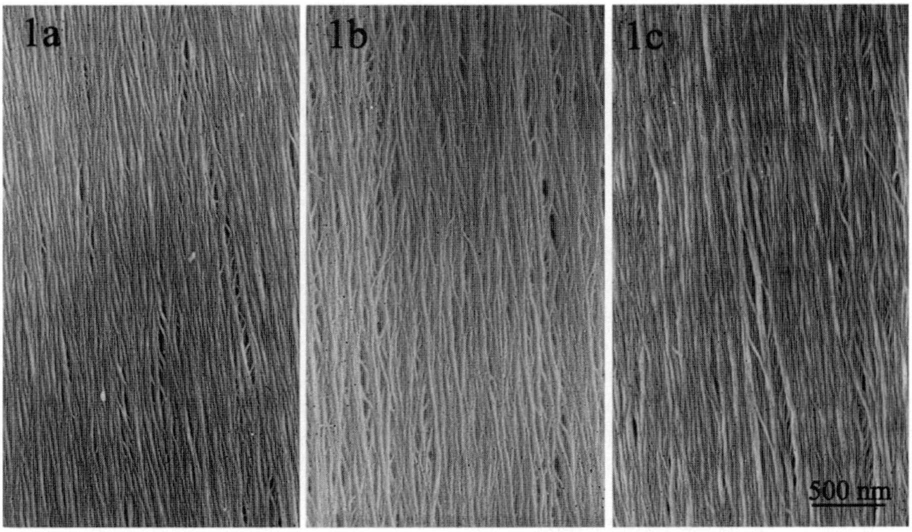

Figure 1. FESEM observation of the innermost surface of the S_2-forming fiber secondary wall. **a** Control section. Microfibrils were clearly visible. **b** Delignified section. Microfibrils had almost the same diameter and appearance as the control sections. **c** Delignified and xylanase-treated section. No structural change was observed.

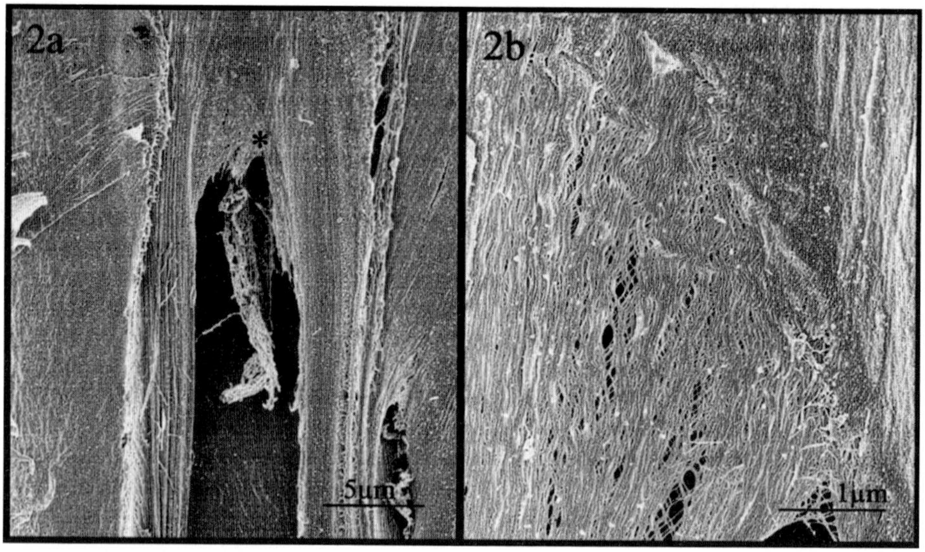

Figure 2. FESEM observation of an obliquely sectioned surface of the S_2-forming fiber secondary wall. Delignified section. **a** Overview of an S_2-forming fiber. Obliquely sectioned region (*) shows various stages of cell wall formation. **b** Detail of an obliquely-sectioned differentiating secondary wall. Microfibrils in the inner part had a smooth surface (left), but those in the outer part had globular substances on their surface (right). The globular substances gradually increased toward the outer part of the secondary wall.

The outer part of the differentiating secondary wall is in the advanced stage of cell wall formation. Although microfibrils were clearly visible near the inner region, they were gradually indiscernible toward the outer part of the secondary wall (data not shown). After delignification of the developing secondary walls, microfibrils with many globular substances were visible in the outer part of the wall, while the microfibrils were observed on the smooth surface in the inner part of the wall (Fig. 2). This finding suggests that lignin deposition occurs only on the outer part of cell walls and does not start in the inner part. The globular substances on the microfibrils disappeared following xylanase treatment, indicating that these substances are xylan. Therefore, xylan not only covers the newly-deposited microfibrils in the inner part of the secondary walls, but also deposits heavily on microfibrils in the outer part of the secondary wall. Observation of obliquely-sectioned secondary walls showed various stages of cell wall formation simultaneously. The gradual increases in the globular substances toward the outer part of secondary wall (Fig. 2) suggests that larger amounts of xylan penetrated into the cell wall and continuously deposited on microfibrils.

Secondary wall of the mature fiber

No fibrillar structure was visible on the innermost surface of the fiber cell wall because a warty layer covered the surface (Fig. 3a). This layer did not disappear following xylanase treatment without delignification (data not shown). Therefore, this layer might be composed of lignin or other phenolics. In delignified sections, microfibrils with many globular substances were visible (Fig. 3b). The size of the globular substance was almost the same

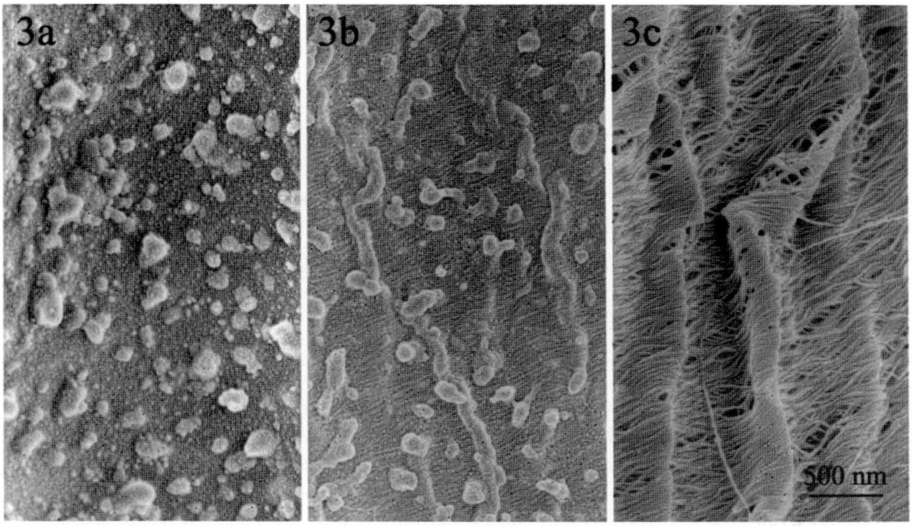

Figure 3. FESEM observation of the innermost surface of a mature fiber. **a** Control section. Microfibrils were not apparent. The surface was covered by a warty layer. **b** Delignified section. Microfibrils with globular substances were visible, and warts remained there. **c** Delignified and xylanase-treated section. Microfibrils were visible and there were no globular substances. Microfibrils were wavy due to extraction of large amount of lignin and xylan. No warts were visible.

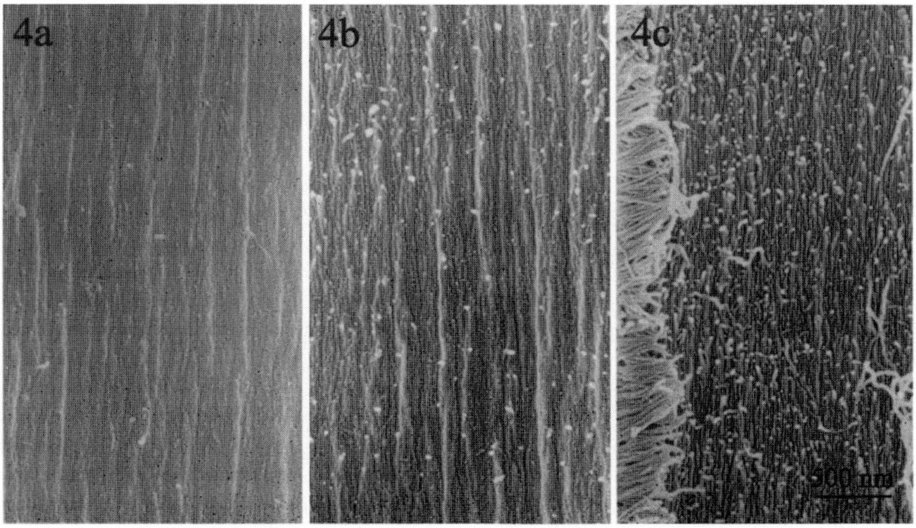

Figure 4. FESEM observation of a sectioned surface of the mature fiber secondary wall. **a** Control section. Microfibrils were not apparent. **b** Delignified section. Microfibrils with globular substances on their surface were visible. **c** Delignified and xylanase-treated sections. Microfibrils were visible and had a smooth surface.

as the width of the microfibrils. Following xylanase treatment, the globular substance and warts disappeared and therefore microfibrils had a smooth surface (Fig. 3c). This indicates that these globular substances on the microfibrils are xylan. In addition, the inner surface had a wavy structure, which could be due to the extraction of a large amount of lignin and xylan.

Microfibrils in the mature S_2 layer were not visible in control sections (Fig. 4a). In delignified sections, microfibrils with many globular substances on their surface were clearly visible (Fig. 4b). Following xylanase treatment, the microfibrils had a smooth surface, indicating that the globular substance is xylan (Fig. 4c).

Mechanism of secondary wall assembly

There are two phases of xylan deposition. First, xylan coats the surface of cellulose microfibrils that have just been released into the innermost surface of the cell wall. The thin layer of xylan could not be resolved by FESEM, but was immunolabelled with anti-xylan antiserum[1]. Second, large amounts of xylan penetrated into the cell wall because there is no room for xylan to coat the cellulose microfibril surface. This type of xylan accumulates onto the microfibrils and has a globular appearance. These globular substances, however, are not be observed in the control sections, which suggests that lignin deposition occurs simultaneously with xylan penetration.

On the basis of the present results, we propose the following mechanism for secondary wall assembly. Cellulose microfibrils synthesized on the plasma membrane are released into the innermost surface of the secondary wall. The cellulose microfibrils are coated with a very thin layer of xylan that was previously deposited there. Large amounts of xylan, however, penetrate the cell wall and accumulate on the microfibrils producing a globular appearance. Lignin deposition occurs simultaneously with xylan penetration and, finally, microfibrils with globular xylan are masked by lignin, resulting in the homogeneous appearance of the cell wall.

ACKNOWLEDGEMENTS

The authors are grateful to Dr. T. Higasa (Research Institute for Food Science, Kyoto University, Japan) for sharing useful information on the FESEM operation.

REFERENCES

1. T. Awano, K. Takabe, M. Fujita & G. Daniel, 'Deposition and localization of glucuronoxylans in the secondary cell wall of Japanese beech as observed using immuno FE-SEM', *Protoplasma*, **2000**, 212, 72-79.
2. B. Vian, J. Brillouet & B. Satiat-Jeunemaitre, 'Ultrastructural visualization of xylans in cell walls of hardwood by means of xylanase-gold complex', *Biol. Cell,* **1983**, 49, 179-182.
3. B. Vian, D. Reis, M. Mosiniak & J. C. Roland, 'The glucuronoxylans and the helicoidal shift in cellulose microfibrils in linden wood: Cytochemistry in muro and on isolated molecules', *Protoplasma,* **1986**, 131, 185-199.
4. D. Reis, B. Vian, & J. P. Roland, 'Cellulose-glucuronoxylans and plant cell wall structure', *Micron,* **1994**, 25, 171-187.
5. T. Awano, K. Takabe & M. Fujita, 'Localization of glucuronoxylans in Japanese beech visualized by immunogold labelling', *Protoplasma*, **1998**, 202, 213-222.

Molecular Breeding of Woody Plants
N. Morohoshi and A. Komamine, editors.

ISOLATION OF MONOCLONAL ANTIBODIES RECOGNIZING XYLEM CELL WALL COMPONENTS BY USING A PHAGE DISPLAY SUBTRACTION METHOD

Naoki Shinohara*[1], Taku Demura[2] & Hiroo Fukuda[1]

[1]*Department of Biological Science, Graduate School of Science, University of Tokyo, 7-3-1, Hongo, Bunkyo-ku, Tokyo 113-0033, Japan*

[2]*Morphogenesis Research Group, Plant Science Center, RIKEN, 2-1, Hirosawa, Wako-shi, Saitama 351-0198, Japan*

ABSTRACT

Monoclonal antibodies (mAbs) recognizing differentiated cell-specific wall components are useful in distinguishing types of cells and dissecting developmental processes. We tried to generate such mAbs using a strategy consisting of (i) isolation of cell walls from synchronously differentiating cells of an *in vitro Zinnia* culture system, which reflects vascular differentiation in plants (ii) construction of phage display library of recombinant antibody against the cell walls (iii) screening via biopannings with subtractive procedures using non-differentiating cell walls. The advantage of this strategy is that purposive mAbs can be easily screened without purifying antigens of interest. Moreover, this approach can be applied to search for novel subpopulations of cells that are difficult to address by conventional methods because of small subpopulations in heterogeneous mixtures.

As a result, we succeeded in isolation of three mAbs, designated CN 8, XD 3, and XD 27. Immunolocalization analyses in *Zinnia* plants revealed that CN 8 epitope localized in walls of immature tracheary elements (TEs) and xylem parenchyma cells, XD 3 epitope localized in walls of TE precursors and immature fiber cells, and XD 27 epitope localized only in walls of immature TEs. In the *Zinnia* culture system, these three mAbs distinguished subpopulations of cells in different developmental stages. These results demonstrate that some wall components change dynamically in association with xylem cell differentiation, as cell-surface antigens of animal cells. These mAbs, therefore, are useful as molecular markers to dissect the vascular developmental stage and also as tools to isolate specific xylem cells using a cell sorter.

KEYWORDS

Cell wall, monoclonal antibody, plant development, phage display, xylem differentiation.

INTRODUCTION

The walls of plant cells are supermolecular structures, comprising polysaccharides, proteins, and phenolic compounds[1,2]. Recent studies have suggested that structure and composition of the cell wall change dynamically during plant development, which makes it likely that these changes play essential roles in development[3]. Thus antibodies raised against some wall components can be useful as markers to distinguish types of cells and to dissect developmental processes[4]. Indeed, monoclonal antibodies (mAbs) raised against some wall components have revealed dynamic changes in the wall components coupled with developmental events[5-9]. However, it is still hard to approach from phenomena to the related molecules, i.e. to isolate antibodies against ill-defined wall components that mark specific developmental stages of cells. To isolate such mAbs surely, we devised a novel strategy consisting of the following procedures: (*i*) to use a cell culture system that allows us to define the specific stages of cell differentiation and to prepare the homogeneous walls from differentiating cells, (*ii*) to make an antibody library to cell walls using phage display technology, and (*iii*) to screen differentiating cell-specific mAbs using a subtractive procedure.

Single mesophyll cells isolated from *Zinnia elegans* transdifferentiate into tracheary elements (TEs) when cultured in a medium containing auxin and cytokinin[10]. This *in vitro* differentiation reflects vascular differentiation in plant[11,12]. Because it occurs synchronously and at high frequency, we can isolate a relatively large amount of cells that are at the same stage of differentiation. We tried to isolate cell wall fractions from differentiating cells that had not yet exhibited morphological features unique to TEs and then to prepare mAbs against the cell wall at the early stage of differentiation.

To isolate the mAbs, we adopted an antibody phage display method[13] coupled with a subtractive procedure. The reason is that an amount of developmental stage-specific epitopes in the cell walls may be limited while that of non-specific epitopes may be much more, and therefore we needed a method allowing large scale generation of mAbs and an efficient screening. Indeed, the phage display subtraction method has been used successfully for the isolation of blood cell-specific[14,15] and tumor-specific[16] mAbs. Using this procedure, we successfully isolated differentiation-specific mAbs that marked different developmental stages of xylogenesis.

MATERIALS & METHODS

Most of the experimental procedures are described in reference 17. Thus here we mention the essential parts of the experiment in brief.

Cell wall preparation

To prepare cell wall for immunization, cells were cultured for 42 h in the TE-inductive medium. The cells were sonicated keeping under 4°C and then washed with HEPES buffer containing 1% (w/v) TritonX-100. After sufficient washing, the cell wall was prepared as white precipitant. (The cell wall is designated "differentiating cell wall." The wall from cells cultured for 42 h in non-inductive medium is also designated "undifferentiating cell wall.")

Generation of mAbs

Poly $(A)^+$ RNA extracted from spleens of mice immunized with the differentiating cell wall was used for construction of phage antibody libraries. The construction was performed using the Recombinant Phage Antibody System (Amersham Pharmacia).

Screening with subtraction

To remove non-specific bounder, phage antibody suspension was incubated with an excess of the undifferentiating cell wall. Then the suspension was incubated with the differentiating cell wall to select out differentiation-specific phage antibodies. The screened differentiation-specific monoclonal phage antibodies were used for production of the recombinant mAbs (scFvs).

ELISA against whole cells

Fixed cells were sequentially treated with 1 μg/ml of each mAb and a peroxidase-labeled secondary antibody after inactivation of cellular intrinsic peroxidase activity.

Immunohistochemical analyses

Cultured *Zinnia* cells were fixed. The shoot apexes of *Zinnia* plants were embedded in paraffin, sectioned, and deparaffinized. The fixed cells and deparaffinized sections were reacted with mAbs and then an alkaline-phosphatese conjugated antibody. Detection of alkaline-phosphatase activity was performed using ELF-97 (Molecular Probes).

RESULTS & DISCCUSION

Changes of three epitopes in the Zinnia mesophyll cell system

Using a phage display subtraction method, we succeeded in isolating mAbs that recognize wall components of xylem cells. ELISA against whole cells revealed that epitopes of all the three mAbs increased specifically in cells cultured in the TE-inductive medium in accordance with or prior to secondary wall thickenings, but not in those in a non-inductive medium. Of the three epitopes, the XD 3 epitope increased earliest during the *in vitro* TE-differentiation (Fig 1.).

Distinction of stages in TE-differentiation

Immunolocalization analyses with *Zinnia* plants revealed that CN 8 epitope localized in walls of immature tracheary elements (TEs) and xylem parenchyma cells, that XD 3 epitope localized in walls of TE precursors and immature fiber cells, and that XD 27 epitope localized only in walls of immature TEs. Interestingly, these three epitopes accumulate transiently in TEs. In other words, they localized in TE precursors or immature TEs but no longer in mature TEs (Table 1.). In the *Zinnia* cultured cell, these three mAbs distinguished subpopulations of cells in different developmental stages. In *Zinnia* plant, they recognized different xylem cell types in plants (Table 1.). These results demonstrate that some wall components vary in association with xylem cell differentiation.

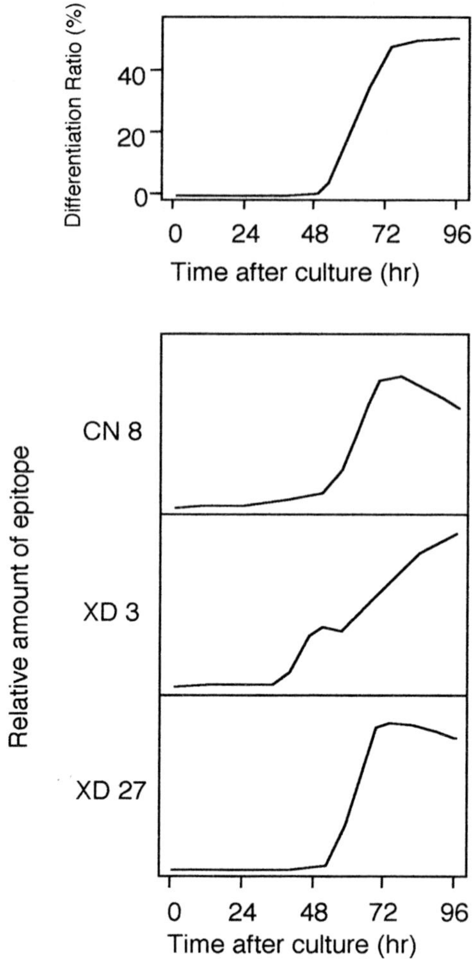

Fig 1. Time course of changes in epitopes in cultured *Zinnia* cells.

mAb	plants	cultured cells
CN 8	immature TE xylem parenchyma	immature TE xylem parenchyma cell (?)
XD 3	TE precursor immature fiber collenchyma epidermis	TE precursor xylem parenchyma cell (?)
XD 27	immature TE	immature TE

Table 1. Epitope expression in plants and cultured cells of *Zinnia*

CONCLUSIONS

Our results indicated that the phage display subtraction method enabled to approach from phenomena to the related molecules. Immunolocalization analyses of the isolated three mAbs, designated CN 8, XD 3, and XD 27, showed that these mAbs distinguished cells in different developmental stages of xylogenesis both *in situ* and *in vitro*. Thus these mAbs are useful as molecular markers to dissect the vascular developmental stage of xylem differentiation.

ACKNOWLEDGEMENT

We are grateful to Dr. Tatsushi Mogi for his valuable suggestion on protein expression analysis. We also thank Dr. Jin Nakashima for his help on microscopy. This work was supported in part by Japan Society for the Promotion of Science Research Fellowships for Young Scientists.

REFERENCES

1. E. J. Varner & L. S. Lin, 'Plant cell wall architecture', *Cell*, **1989**, 56, 231–239.
2. N. C. Carpita & D. M. Gibeaut, 'Structural models of primary cell walls in flowering plants: consistency of molecular structure with the physical properties of the walls during growth', *Plant J.*, **1993**, 3, 1–30.
3. R. I. Pennell, 'Cell walls: structures and signals', *Curr. Opin. Plant Biol.*, **1998**, 1, 504–510.
4. J. P. Knox, 'The use of antibodies to study the architecture and developmental regulation of plant cell walls', *Int. Rev. Cytol.*, **1997**, 171, 79–120.
5. J. P. Knox, P. J. Linstead, P. C. Cooper & K. Roberts, 'Developmentally regulated epitopes of cell surface arabinogalactan proteins and thier relation to root tissue pattern formation', *Plant J.*, **1991**, 1, 317–326.

6. M. Smallwood, A. Beven, N. Donovan, S. J. Neill, J. Peart, K. Roberts & J. P. Knox, 'Localization of cell wall proteins in realtion to the developental anatomy of carrot root apex', *Plant J.*, **1994**, 5, 237–246.
7. M. Kreuger & G. J. van Holst, 'Arabinogalactan-protein epitope in somatic embryogenesis of *Daucus carota* L', *Planta*, **1995**, 197, 135–141.
8. M. N. V. Williams, G. Freshour, A. G. Darvill, P. Albersheim & M. G. Hahn, 'An antibody Fab selected from a recombinant phage display library detects deesterified pectic polysaccharide rhamnogalacturonan II in plant cells', *Plant Cell*, **1996**, 8, 673–685.
9. P. F. McCabe, T. A. Valentine, L. S. Forsberg & R. I. Pennell, 'Soluble signals from cells identified at the cell wall establish a developmental pathway in carrot', *Plant Cell*, **1997**, 9, 2225–2241.
10. H. Fukuda & A. Komamine, 'Establishment of experimental system for the study of tracheary element differentiation from single cells isolated from the mesophyll of *Zinnia elegans*', *Plant Physiol.*, **1980**, 65, 57–60.
11. H. Fukuda, 'Tracheary element differentiation', *Plant Cell*, **1997**, 9, 1147–1156.
12. K. Roberts & M. C. MaCann, 'Xylogenesis: the birth of a corpse', *Curr. Opin. Plant Biol.*, **2000**, 3, 517-522.
13. H. R. Hoogenboom, 'Designing and optimizing library selection strategies for generating high-affinity antibodies', *Trends Biotechnol.*, **1997**, 15, 62–70.
14. J. de Kruif, L. Terstappen, E. Boel & T. Logtenberg, 'Rapid selection of cell subpopulation-specific human monoclonal antibodies from a synthetic phage antibody library', *Proc. Natl. Acad.Sci. USA*, **1995**, 92, 3938–3942.
15. D. L. Siegel, T. Y. Chang, S. L. Russell & V. Y. Bunya, 'Isolation of cell surface-specific human monoclonal antibodies using phage display and magnetically-activated cell sorting: applications in immunohematology', *J. Immunol. Methods*, **1997**, 206, 73–85.
16. X. Cai & A. Garen, 'Anti-melanoma antibodies from melanoma patients immunized with genetically modified autologous tumor cells: selection of specific antibodies from single-chain Fv fusion phage libraries', *Proc. Natl. Acad. Sci. USA* , **1995**, 92, 6537–6541.
17. N. Shinohara, T. Demura & H. Fukuda, 'Isolation of a vascular cell wall-specific monoclonal antibody recognizing a cell polarity by using a phage display subtraction method', *Proc. Natl. Acad. Sci. USA* , **2000**, 97, 2585-2590.

Molecular Breeding of Woody Plants
N. Morohoshi and A. Komamine, editors.

ON THE MECHANISM TO REGULATE THE RATIO OF SYRINGYL TO GUAIACYL MOIETIES IN LIGNIN

Kazuhiko Fukushima

Graduate School of Bioagricultural Sciences, Nagoya University, Nagoya 464-8601 Japan

ABSTRACT

Now, lignin biosynthetic pathway is believed to form the metabolic grids. It is very important to elucidate the actual pathway that regulates the ratio of syringyl to guaiacyl moieties in lignin. Here, possible regulating systems proposed on the basis of the enzymatic, genetic and feeding studies are introduced and discussed. In addition, our recent results are also introduced.

KEYWORDS

Lignin, biosynthesis, guaiacyl, syringyl, *p*-hydroxyphenyl, ferulic acid, sinapic acid, coniferyl alcohol, sinapyl alcohol, xylem

KNOWLEDGE ON THE REGULATION OF SYRINGYL TO GUAIACYL RATIO IN LIGNIN

Conventional pathway that ragulates the syringyl to guaiacyl ratio

The biosynthesis of lignin precursors, monolignols are considered as being started from phenylalanine (3). In monocotyledons, tyrosine (4) is also the starting material for the biosynthesis of lignin precursor (Neish, 1961). These amino acids are derived from the shikimate (1)/chorismate (2) pathway as shown in Figure 1. (Lewis, 1999, Whetten and Sederoff, 1995). Phenylalanine (3) is deaminated by phenylalanine ammonia-lyase (PAL, EC 4.3.1.5) to yield cinnamate (5), this is the first step in phenylpropanoid metabolism and a key enzyme in the biosynthesis of phenylpropanoid compounds. Cinnamate (5) is hydroxylated by 4-hydroxylase (C4H) to form *p*-coumarate (6). *p*-Coumarate (6) can be synthesized to thioester, *p*-coumaroyl-CoA (11) by 4-coumarate-CoA ligase (4CL, EC 6.2.1.12). This compound is the common precursor for synthesis of flavonoids, stilbens and other phenylpropanoids as well as lignin.

p-Coumarate (6) is hydroxylated at 3 position to yield caffeate (7) by cinnamate 3 hydroxyrase (C3H) and the hydroxy groupe can be methylated by an *O*-methyltransferase (COMT, EC 2.1.1.6) to produce ferulate (8). Ferulate can be hydroxylated by ferulate 5-hydroxylase (F5H) to form 5-hydroxyferulate (9) and the hydroxy group at 5 position in 5-hydroxyferulate (9) can be methylated by COMT to produce sinapate (10). The hydroxy cinnamic acids (6-10) can be synthesized to corresponding thioesters (11-15). Reduction of the CoA thioesters to the corresponding aldehydes is catalyzed by cinnamoyl-CoA reductase (CCR, EC 1.2.1.44). The aldehydes are further reduced by the reaction of cinnamyl alcohol dehydrogenase (CAD, EC 1.1.1.195) to give rise to monolignols (21, 23 and 25) which are the direct

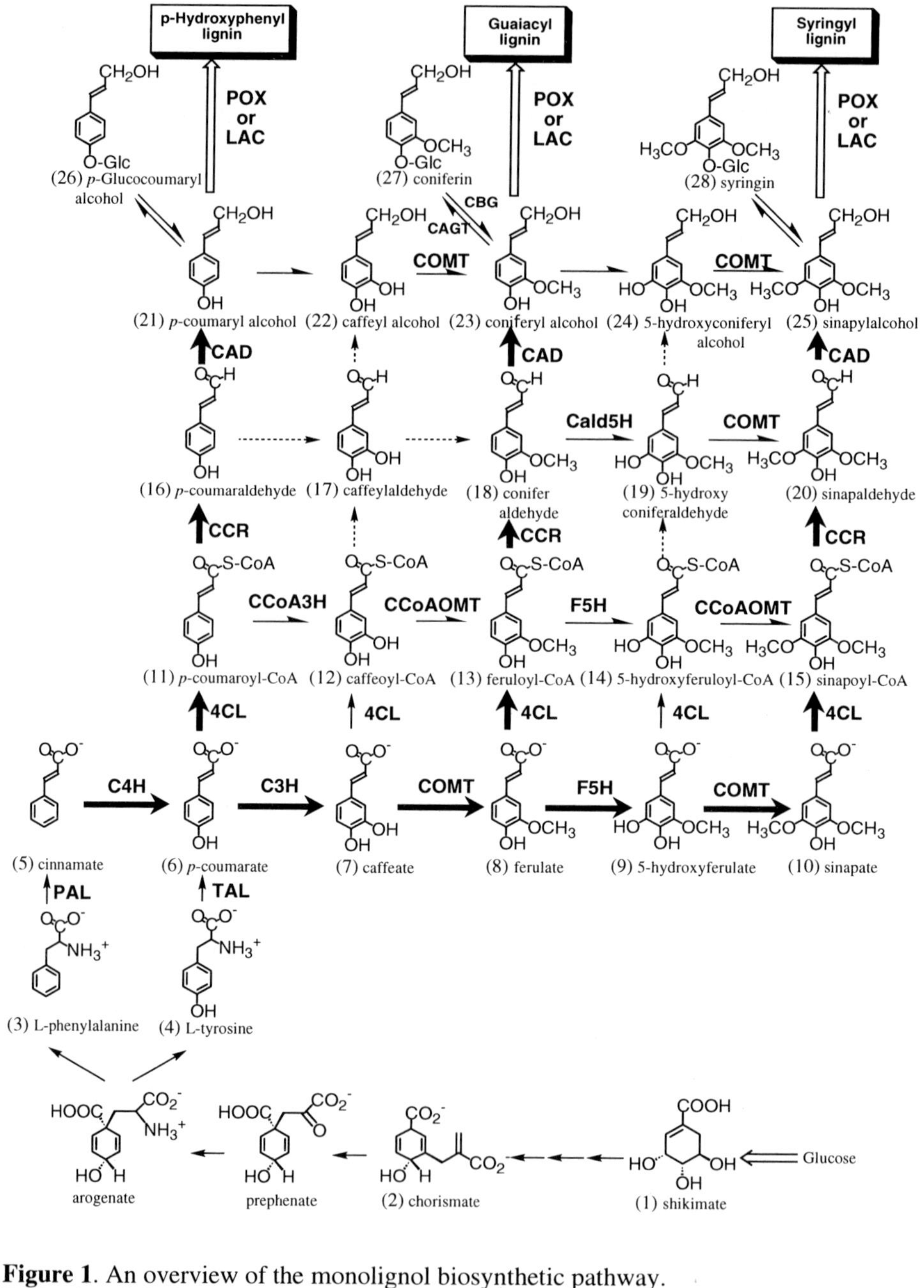

Figure 1. An overview of the monolignol biosynthetic pathway.
PAL, phenylalanine ammonia-lyase; TAL, tyrosine ammonia-lyase; C4H, cinnamate 4-hydroxylase; C3H, cinnamate 3 hydroxyrase; COMT, caffeate *O*-methyltransferase; F5H, ferulate 5-hydroxylase; 4CL, 4-coumarate-CoA ligase; CCoA3H, caffeoyl-CoA 3-hydroxydase; CCoAOMT, caffeoyl-CoA *O*-metyltransferase; CCR, cinnamoyl-CoA reductase; Cald5H, ; CAD, cinnamyl alcohol dehydrogenase; CAGT, monolignol UDP-glucose-glucosyltransferase; CBG, coniferin specific beta-glucosidase; POX, peroxidase; LAC, laccase. (Bold arrows: conventional pathway)

precursors of lignin. In brown midrib mutant (bm3) of maize, 5-hydroxyconiferyl alcohol (24) is also involved in dehydrogenative polymerization (Suzuki *et al.*, 1995).

Findings of alternative pathways that regulate the syringyl to guaiacyl ratio

In the last decade our understanding of lignin biosynthesis has rapidly progressed. The genetic manipulation of genes encoding enzymes of the conventional lignin pathway has generated unexpected results which have led the scientific community to re-evaluate lignin biosynthesis (Sederoff *et al.* 1999, Grima-Pettenati *et al.* 1999).

Grand *et al.* (1983) had isolated the isoenzymes of 4-coumarate CoA ligase (4CL) specific to sinapate in lignifying tissues of hybrid poplar, this was a strong proof of the conventional pathway (Higuchi 1997, 1990) that syringyl moieties are formed at cinnamate pathway and resulting sinapic acid (10) is converted to sinapyl alcohol (25). However, it was found that most angiospermous plants cannot synthesize sinapoyl-CoA (15) directly via sinapic acid (10) by 4-coumarate CoA ligase (Allina *et al.* 1998, Hu *et al.* 1998, Kajita *et al.* 1996; Lee *et al.* 1997, Meyermansa et al., 2001).

At present, four routes that regulate the ratios of syringyl to guaiacyl moieties are proposed at the stages of cinnamate (Higuchi 1997, 1990), cinnamoyl CoA (Ye 1997, Ye *et al.* 1994, 1995, Zong *et al.* 1998), cinnamylaldehyde (Osakabe *et al.* 1999, Li *et al.* 2000, Humphreys *et al.* 1999) and cinnamyl alcohol (Chen *et al.* 1999ab, Matsui *et al.* 2000). *O*-Methylation may occur by COMT or caffeoyl-CoA *O*-metyltransferase (CCoAOMT, EC 2.1.1.104) in each stage.

Enzymes involved in the regulation of syringyl to guaiacyl ratio

Today, the center of interest in lignin biosynthesis is how guaiacyl intermediates are hydroxylated and methylated to the syringyl monolignol in angiosperms.

In the *Arabidopsis* mutants defective of ferulate 5-hydroxylase, the amount of syringyl units was below the detection limits (nitrobenzene oxidation) (Chapple *et al.* 1992; Lee *et al.* 1997), indicating that the pathways leading to the synthesis of sinapyl alcohol had been blocked. However, this enzyme is effectively catalyzed coniferyl aldehyde and coniferyl alcohol rather than ferulate (Humphreys et al. 1999). This 5-hydroxylase like this might play an important role in the conversion of coniferyl alcohol to sinapyl alcohol, though there are no evidence whether the enzymes involved in cinnamyl alcohol pathway are the same ones involved in other pathways. Recently, it was found *O*-methyltransferase (COMT1) from tabacco could *O*-methylate the 5-hydroxyconiferyl alcohol (Maury et al. 1999). More-extensive studies are required to clarify the mechanism of regulation of the composition of lignin.

Meng and campbell (1998) suggested from the viewpoint of plant evolution that CCoAOMT in aspen developing secondary xylem is likely to be responsible for biosynthesis of lignin precursors in the guaiacyl pathway and may represent a more primitive enzyme form leftover from very early land plant, and COMT is more likely to be responsible for syringyl lignin precursor biosynthesis in the evolutionarily advanced plants such as angiosperms, since it can catalyze methylation of 5-hydroxyferuloyl-CoA quite effectively.

Selective labeling of syringyl and guaicyl lignins

Selective radiolabeling of lignin structural units in the differentiating xylem of magnolia, lilac, pine, ginkgo could be generally achieved by feeding of the monolignol glucosides labeled with ^3H or ^{14}C (Fukushima 1997, Fukushima and Terashima 1990, 1991a, Terashima et al. 1986, 1988, Terashima and Fukushima 1988 1989). In these studies, *p*-glucocoumaryl alcohol (26), coniferin (27) and syringin (28) were the effective precursors for *p*-hydroxyphenyl (H), guaiacyl (G) and syringyl (S) moieties respectively, though specificy of the labeling to corresponding units from the precursor was not complete. The labeled lignin *in situ* could be also visualized by microautoradiography, that is, H lignin is deposited in the compound middle lamella at the early stage, S lignin is deposited in the secondary wall at the later stage of cell wall differentiation and G lignin is deposited in both layers through lignification. The lignification process in compression wood in which *p*-hydroxyphenyl units are characteristic was also visualized in *Pinus thunbergii* (Fukushima and Terashima, 1991b). It is concluded that the supply of three kind of monolignols occurs under definite biochemical regulation to give a macromolecule which is specific in its morphologfical location.

Supply and transport of lignin precursor

Autoradiographic study with tritiated phenyl alanine showed that the Golgi-body and smooth-endoplasmic reticulumm are involved in the biosynthesis and/or transport of monolignols (Takabe *et al.* 1985). This suggested the route of transportation of lignin precursors in the cells, however, there was an alternative possibility that this route is involving in other biosynthesis, because phenylalanine is also the precursor of proteins. Localizations of enzymes involved in lignification were also investigated by immunocytochemistry. By using the antibodies against the enzymes involved in lignification of mesophyll cells of *Zinnia elegans* (Nakashima *et al.* 1997), differentiating xylem of *Eucalyptus globulus* and *Poplus euramericana* (Takabe *et al.* 1999), it was found that the enzymes, PAL, OMT, 4CL and CAD, are synthesized at polysomes and then are dispersed into the cytosol of lignifying cells. On the contrary, peroxidase involved in dehydrogenative polymerization of monolignols is synthesized in rough endoplasmic reticulum and transported to lignifying cell walls via the Golgi apparatus (Takabe *et al.* 1999).

Monolignol glucosides have been isolated from all gymnosperms examined, but found in a limited kind of angiosperms (Freudenberg and Harkin 1963, Terazawa *et al.* 1984). The role of monolignol glucosides such as coniferin (28) and syringin (29) existing in the stems of trees has not been explained well to date. It has been confirmed by microautoradiographic studies that radio-labeled monolignol glucosides are incorporated immediately into cell wall lignin in the differentiating xylem where lignification was proceeding (Fukushima 1997, Fukushima and Terashima 1990, 1991a, Terashima *et al.* 1986, 1988, Terashima and Fukushima 1988 1989). It was reported that coniferin specific β-glucosidase was localized in the lignifying tracheids of *Pinus contorta* (Dharmawardhana *et al.* 1995). From these results, monolignil glucosides are considered to be involving in the transport and storage form of lignin precursors (Whetten and Sederoff 1995). On the other hand, it was reported by Savidge (1989) that coniferin (27) is a biochemical indicator of tracheid differentiation. Terazawa and Miyake (1984) suggested that syringin (28) in the angiospermous trees may not be correlate with the lignification because of the different seasonal changes from those of coniferin (27) in conifers. The distributions of monolignol glucosides in the stems of *Magnolia kobus* and *Pinus thunbergii* in different season were examined in detail, and it is concluded that coniferin (27) in the stems is considered a lignin precursor, but accumulated syringin cannot be regarded as a lignin precursor because of the high content in winter season and localization in non-woody tissues (Fukushima *et al.*1996, 1997).

Recently, Förster *et al.* (1999) proposed new hypothesis that UDP-glucose: coniferyl alcohol glucosyltransferase (CAGT) can also catalyze coniferyl aldehyde (18) to form its glucoside, and the glucoside may play an alternative strage form of lignin precursor. Coniferyl aldehyde (18) is the important intermediate of monolignol biosynthesis because the pathway is branched off here into coniferyl alcohol (guaiacyl lignin) and sinapyl alcohol via cinnamyl aldehyde pathway (Osakabe *et al.* 1999, Li *et al.* 2000, Humphreys *et al.* 1999).

Coniferin-β-glucosidase (CBG) catalyzes the hydrolysis of monolignol glucosides to release the cinnamyl alcohols for oxidative polymerization to lignin. This enzyme localizes in lignifying xylem of *Pinus controla* (Dharmawardhana *et al.* 1995). CBG is considered to be a secretory protein targeted to the ER (Dharmawardhana *et al.* 1999).

Alteration of the syringyl to guaiacyl ratio by gene manipulation

Transgenic *Populus tremuloides* in which expression of a lignin biosynthetic pathway gene *Pt4CL1* encoding 4CL has been downregulated by antisense inhibition was produced. The trees with suppressed Pt4CL1 expression exhibited up to a 45% reduction of lignin, but this was compensated for by a 15% increase in cellulose (Hu *et al.* 1999). In 4CL antisence *Arabidopsis*, nitrobenzene oxidation of the cell wall of stems revealed a significant decrease in guaiacyl units in 4CL-suppressed plants, however, levels of syringyl lignin units were unchanged in even the severely 4CL-suppressed plants (Lee *et al.* 1997). These results suggested that an uncharacterised metabolic route to sinapyl alcohol (25), which is independent of 4CL, might exist in *Arabidopsis*. In the brownish tissue of the transgenic *Nicotiana tabacum* in which the activity of 4CL is very low, the levels of cinnamyl aldehyde residues and syringyl units in lignin and three hydroxycinnamic acids, *p*-coumaric (6), ferulic (8), and sinapic acid (10) which were bound to cell walls were apparently increased as a result of down-regulation of the expression of the gene for 4CL (Kajita *et al.* 1996, 1997).

Chemical analysis of woody tissue of stems in the transgenic tabacco plants, which represented the expression of an antisence bi-OMT indicated a reduced content of syringyl units in lignin (Dwivedi *et al.* 1994). The reduction in CCoAOMT also led to a dramatic alteration in lignin composition, especially guaiacyl lignin was preferentially reduced, which resulted in an increase in the S/G (syringyl/guaiacyl) ratio (Zhong *et al.* 1998). In the transformants (pale rose colored xylem) of *Populus trichocarpa* x *P. deltoides*, the COMT activity was reduced by 95%, and the syringyl/guaiacyl ratio was reduced by sixfold (due to a decrease of S and an increase of G) and the 5-hydroxyguaiacyl residue was detected in the thioacidolysis products (Van Doorsselaere *et al.*1995). Mutations at the COMT gene give a brown midrib3 (bm3) phenotype of maize, and the gene genetically recognized as bm3 is the same as the one coding for COMT (Vignols *et al.* 1995).

New biotechnological strategies including the genetic manipulation of more than one gene simultaneously and the use of xylem-specific promoters must be employed (Grima-Pettenati *et al.* 1999).

PROPOSAL BY OUR GROUPE

Regulation at the cinnamyl alcohol level

Feeding experiments with deuterium labeled ciniferyl alcohol provide conclusive evidence for a novel alternative pathway for regulation of the syringyl to guaiacyl ratios in lignin of *Magnolia* and *Eucalyptus* (Chen *et al.*, 1999a b). Pentadeutero [γ-D,D, OCD_3]-coniferyl alcohol (23) was supplied to shoots of trees and the incorporation of the labeled precursor into lignin was traced by gas chromatography-mass spectrometry. In addition to the direct incorporation of the labeled precursor into guaiacyl units, we detected a significant amount of pentadeuterium-labeled syringyl units with two gamma-deuterium atoms.

This pathway from coniferyl alcohol (23) to sinapyl alcohol (25) might be composed of two step reactions (Matsui *et al.* 2000), that is, the hydroxylation and *O*-methylation as well as cinnamate pathway. Since coniferin (27), both a storage and a transport forms of coniferyl alcohol, accumulates at the early stage of the differentiation of xylem prior to lignification in some angiosperms, such as *Magnolia* (Fukushima *et al.* 1997), the conversion at the cinnamyl alcohol pathway might play an important role in the synthesis of sinapyl alcohol (25).

Evidence of the conventional pathway of the monolignol biosynthesis

For many angiosperms, a 4-coumarate:CoA ligase (4CL; EC 6.2.1.12) isozyme that exhibits activity with a sinapic acid (10) substrate has not been detected *in vitro* (Allina *et al.* 1998, Hu *et al.* 1998, Kajita *et al.* 1996; Lee *et al.* 1997, Meyermansa *et al.*, 2000). Therefore, the possible role of sinapic acid as a precursor of syringyl lignin has been questioned. To further examine this issue, tetradeutero ferulic acid [β-D, $-OCD_3$] and heptadeutero sinapic acid [β-D, $(-OCD_3)_2$] were synthesized and fed to shoots of oleander (*Nerium indicum* Mill.) and *Robinia* (*Robinia pseudoacacia* L.) trees, and the incorporation of these labeled precursors into lignin was traced. Lignin in the newly formed xylem of the treated stems was analyzed by derivatization followed by reductive cleavage. The two labeled methoxyl groups of sinapic acid (10) were useful in monitoring the conversion of sinapic acid into syringyl lignin. When heptadeutero sinapic acid was fed, syringyl units containing seven deuterium labels were detected. These results suggest that sinapic acid is converted to sinapoyl-CoA (15) during lignin biosynthesis in oleander and *Robinia* trees *in vivo* (Yamauchi *et al.* another chapter in this issue).

Formation of syringyl muclei without F5H and OMTs

In addition to the enzymatic formation of syringyl units, the transfer or rearrangement of methoxyl group on guaiacyl ring during the polymerization step may also perticipate in the formation of syringyl units. When coniferyl alcohol (23) was dehydrogenatively polymerized very slowly with a horseradish peroxidase according to the procedure previously described (Nakatsubo and Higuchi, 1975, Wayman and Obiaga, 1974), small amounts of syringyl (S) and *p*-hydroxyphenyl (H) units were formed in the product (**endwise DHP**) in addition to guaiacyl (G) units. The products of G, S and H types from both thioacidolysis (Lapierre *et al.*, 1985) and nitrobenzen oxidation were identified and analyzed quantitatively by GC-MS. In both degradation procedures, the relative compositions (mol%) of S units were 10% of total products from **endwise DHP** and trace amounts of H units were also detected. Another type of DHP were also synthesized by bulk polymerization of coniferyl alcohol (23), however, only G units were detected (Saito *et al.*, 2001).

Recently it has been well investigated that coniferyl alcohol (23) is converted to syringyl moieties with enzymes such as 5-hydroxylase or *O*-methyltransferase (Humphreys *et al.*, 1999; Matsui *et al.*, 2000). However, our results suggest that the modification of aromatic ring without these enzymes may occur even in the lignification of plant cell wall, although the contribution of this conversion would be very small in vivo.

Further studies are required to clarify the mechanism and find the appropriate condition for the formation of S and H units during the polymerization of coniferyl alcohol (23).

CONCLUSIONS

At this moment, we can not discard any of pathways proposed on the monolignol pathways. Each pathway has the possibility to regulate the composition of lignin structural units.

I emphasize here that there are the problems both in the feeding experiments *in vivo* and in the enzymatic studies *in vitro*. For feeding experiments *in vivo*, we must not always accept the results of the feeding with exogenous precursors as endogenous ones. For enzymatic studies *in vitro*, we must consider if the isozyme examined is really the target enzyme or not, namely the isozyme operates in intact site of lignification or not.

REFERENCES

Allina, S. M., Pri-Hadash, A., Theilmann, D. A., Ellis, B. E., Douglas, C. J. (1998) 4-Coumarate:coenzyme A ligase in hybrid poplar properties of native emzymes, cDNA cloning, and analysis of recombinant enzymes. Plant Physiol., 116, 743-754

Chapple, C.C.S., Vogt, T., Ellis, B,E., Somerville, C.R. (1992) An *Arabidopsis* mutant defective in the general phenylpropanoid pathway. *Plant Cell,* **4**, 1413-1424

Chen, F., Yasuda, S., Fukushima, K. (1999a) Evidence for a novel biosynthetic pathway that regulates the ratio of syringyl to guaiacyl residues in lignin in the differentiating xylem of *Magnolia kobus* DC., *Planta*, **207**, 597-603

Chen, F., Yasuda, S., Fukushima, K. (1999b) The structural conversion of lignin subunit at cinnamyl alcohol stage in *Eucalyptus camaldulensis*, *J. Wood Sci.*, **45**, 487-491

Dharmawardhana, D. P., Ellis, B. E., Carlson, J. E. (1995) A _-gulucosidase from lodgepole pine xylem specific for the lignin precursor coniferin, *Plant Physiology*, **107**, 331-339

Dharmawardhana, D. P., Ellis, B. E., Carlson, J. E. (1999) cDNA cloning and heterologous expression of coniferin beta-glucosidase. *Plant Molecular Biology*. **40**, 365-372.

Dwivedi, U. N., Campbell, W. H., Yu, J., Datla, R. S. S., Bugos, R. C., Chiang, V. L., Podila, G. K . (1994) Modification of lignin biosynthesis in transgenic Nicotiana through expression of an antisense O-methyltransferase gene from Populus. *Plant Molecular Biology*, **26**, 61-71.

Freudenberg, K. Harkin, G. M. (1963) The glucosides of cambial sap of spuruce. *Phytochemistry*, **2**, 189-193

Fukushima, K., Biolabelling of polyphenols in woody plants I, *POLYGHENOLS ACTUALITES*, **16**, 11-15 (1997)

Fukushima, K., Taguchi, S., Matsui, N., Yasuda. S. (1996) Heterogeneous Distribution of monolignol glucosides in the stems of Magnolia kobus. *Mokuzai Gakkaishi*, **42**, 1029-1031

Fukushima, K., Taguchi, S., Matsui, N., Yasuda. S. (1997) Distribution and seasonal changes of monolignol glucosides in *Pinus thunbergii*. *Mokuzai Gakkaishi*, **43**, 254-259

Fukushima, K., Terashima, N. (1990) Heterogeneity in firmation of lignin. XIII. Formation of p-hydroxyphenyl lignin in various hardwoods visualized by microautoradiography. *J. Wood Chem. Technol.*, **10**, 413-433

Fukushima, K., Terashima, N. (1991a) Heterogeneity in formation of lignin. XIV. Formation and structure of lignin in differentiating xylem of Ginkgo biloba. *Holzforschung*, **45**, 87-94

Fukushima, K., Terashima, N. (1991b) Heterogeneity in formation of lignin. XV. Formation and structure of lignin in compression wood of Pinus Thunbergii studied by microautoradiography. *Wood Sci. Technol.*, **25**, 371-381

Förster, H., Pommer, U., Savidge, R. A. (1999) Metabolic activity of uridine 5'-diphosphoglucose: cinnamyl alcohol glucosyltransferase as an intrinsic indicator of cambial growth in conifers, In Plant Polyphenol 2, Eds. Gross, G. G., Hemingway, R. W., Yoshioda, T. (Kluwer academic / Plenum publishers, New York) pp.371-392

Grand, C., Boudet, A., Boudet A. M. (1983) Isoenzymes of hedroxycinnamate : CoA ligase from poplar stems properties and tissue distribution. *Planta*, **158**, 225-229

Grima-Pettenati, ,J., Goffner, D. (1999) Lignin genetic engineering revisited. *Plant Science*, **145**, 51-65.

Higuchi, T. (1990) Lignin biochemistry: Biosynthesis and biodegradation. *Wood Sci Technol.*, **24**, 23-63

Higuchi, T. (1997) Biochemistry and molecular biology of wood (Springer-Verlag, Berlin-Heidelberg-New York) pp.93-181

Hu, W-J., Harding, S. A., Lung, J., Popko, Jacqueline L., Ralph, J., Stokke, D. D., Tsai, C-J. Chiang, V. L. (1999) Repression of lignin biosynthesis promotes cellulose accumulation and growth in transgenic trees. Nature Biotechnology. **17**. 808-812.

Hu, W-J., Kawaoka, A., Tsai, C-J., Lung, J., Osakabe, K., Ebinuma, H., Chiang, V. L. (1998) Compartmentalized expression of two structurally and functionally distinct 4-coumarate:CoA ligase genes in aspen (Populus tremuloides). *Proc. Natl. Acad. Sci USA*. **95**, 5407-5412.

Humphreys, J. M., Hemm, M. R., Chapple, C. (1999) New routes for lignin biosynthesis defined by biochemical characterization of recombinant ferulate 5-hydroxylase, a multifunctional cytochrome P450-dependent monooxygenase. *Proc. Natl. Acad. Sci. USA*, **96**, 10045-10050.

Kajita, S., Hishiyama, S., Tomimura, Y., Katayama, Y., Omori, S. (1997) Structural characterization of modified lignin in transgenic tobacco plants in which the activity of 4-coumarate:coenzyme A ligase is depressed. *Plant Physiology*, **114**, 871-879.

Kajita, S., Katayama, Y., Omori, S. (1996) Alterations in the biosynthesis of lignin in transgenic plants with chimeric genes for 4-coumarate:Coenzyme A ligase. *Plant & Cell Physiology*. **37**, 957-965.

Lapierre,C. Monties, B. Rolando,C. (1985) Thioacidolysis of lignin : comparison with acidolysis, *J. Wood Chem.Technol.,* **5**, 277-292.

Lee, D., Meyer, K., Chapple, C., Douglas, C. J. (1997) Antisense suppression of 4-coumarate:coenzyme A ligase activity in arabidopsis leads to altered lignin subunit composition. *Plant Cell*, **9**, 1985-1998.

Lewis, N., Davin, L. B., Sarkanen, S. (1999) The nature and function of lignins, In "Comprehensive Natural Products Chemistry", Barton, Sir D. H. R. and Nakanishi, K. eds-in-chief, Vol. 3, Carbohydrates and their derivatives includeing tannins, Cellulose and Related Lignins, pp.617-745

Li, L., Popko, J. L., Umezawa, T., Chiang, V. L. (2000) 5-Hydroxyconiferyl aldehyde modulates enzymatic methylation for syringyl monolignol formation, a new view of monolignol biosynthesis in angiosperms, *J. Biol. Chem.*, **275**, 6537-6545

Matsui, N., Chen, F., Yasuda, S., Fukushima, K. (2000) Conversion of guaiacyl to syringyl moieties on the cinnamyl alcohol pathway during the biosynthesis of lignin in angiosperms, *Planta*, **210**, 831-835

Maury, S., Geoffroy, P., Legrand, M., (1999) Tobacco O-methyltransferases involved in phenylpropanoid metabolism. The different caffeoyl-coenzyme A/5-hydroxyferuloyl-coenzyme A 3/5-O-methyltransferase and caffeic acid/5-hydroxyferulic acid 3/5-O-methyltransferase classes have distinct substrate specificities and expression patterns. *Plant Physiology*, **121**, 215-223.

Meng, H., Campbell, W. H. (1998) Substrate profiles and expression of caffeoyl coenzyme A and caffeic acid 0-methyltransferases in secondary xylem of aspen during seasonal development. *Plant Molecular Biology*. **38,** 513-520.

Meyermansa, H., Morreelab, K., Lapierrec, C., Pollet, B., De Bruynd. A., Bussone, R., Herdewijne, P., Devreesefg, B., Van Beeumenf, J., Maritahi, L., Ralph, J., Chena, C., Burggraevea, B., Van Montagu , A, Messensa, E., Boerjan, W. (2000) Modifications in Lignin and Accumulation of Phenolic Glucosides in Poplar Xylem upon Down-regulation of Caffeoyl-Coenzyme A O-Methyltransferase, an Enzyme Involved in Lignin Biosynthesis J. Biol. Chem., 275, 36899-36909

Nakashima, J., Awano, T., Takabe, K., Fujita, M., Saiki H. (1997) Immunocytochemical localization of phenylalanine ammonia-lyase and cinnamyl alcohol dehydrogenase in differentiating tracheary elements derived from Zinnia mesophyll cells. *Plant Cell Physiol.*, **38**, 113-123

Nakatsubo, F., Higuchi, T..(1975) Holzforschung. 29, 64-68.

Osakabe, K., Tsao, C. C., Li, L., Popko J. L., Umezawa, T., Carraway D. T. Smeltzer, R. H., Joshi, C. P., Chiang, V. L. (1999) Coniferyl aldehyde 5-hydroxylation and methylation direct syringyl lignin biosynthesis in angiosperms. Proc. Natl. Acad. Sci. USA, **96**, 8955-8960.

Saito, K., Yasuda, S., Ona, T. and Fukushima, K. (2001) Quite new mechanism for the generation of syringyl units in lignin. Proc. 11th ISWPC, oral session.

Savidge, R. A. (1989) Coniferin, a biochemical indicator of commitment to tracheid differentiation in conifers. *Can. J. Bot.* **67**, 2663-2668

Sederoff, R.R. MacKay, J.J. Ralph, J. Hatfield, R.D. (1999) Unexpected variations in lignin, *Curr. Opinion Plant Sci.*, **2** , 145-152.

Suzuki, S., Lam, T. B. T., Iiyama, K. (1995) 5-Hydroxyguaiacyl nuclei as aromatic constituents of native lignin. *Phytochemistry*, **46**, 695-700.

Takabe, K., Ito, M., Takeuchi, M., Fujita, M., Katayama, Y. (1999) Distribution of enzymes involved in lignification of cell wall. Proc. 44th Lignin Symposium, Gifu, Japan. pp.5-8

Takabe, K., Fujita, M., Harada, H., Saiki, H. (1985) Autoradiographic investigationof lignification in the cell walls of cryptomeria (*Cryptomeria japonica* D. Don). *Mokuzai Gakksishi*, **31**, 613-619

Terashima, N., Fukushima, K. (1986) Heterogeneity in formation of lignin. VII. An autoradiographic study on the formation of guaiacyl and syringyl lignin in poplar. *J.Wood Chem. Technol.*, **6**, 495-504

Terashima, N., Fukushima, K. (1988) Heterogeneity in formation of lignin. XI. An autoradiographic study of the heterogeneous formation and structure of pine lignin. *Wood Sci. Technol.*, **22**, 259-270

Terashima, N., Fukushima, K. (1989) Biogenesis and structure of macromolecular lignin in the cell wall of tree xylem as studied by microautoradiography. In "Plant Cell Wall Polymers, Biogenesis and Biodegradation" N. G. Lewis and M.G. Paice eds., ACS Symposium Series 399, American Chemical Society, Washington D.C., pp.160-168

Terashima, N., Fukushima, K., Sano, Y., Takabe, K. (1986) Heterogeneity in formation of lignin. VIII. An autoradiographic study on the formation of guaiacyl and syringyl lignin in *Magnolia kobus* DC. *Holzforschung*, **40 Suppl**. 101-105

Terashima, N., Fukushima, K., Sano, Y., Takabe, K. (1988) Heterogeneity in formation of lignin. X. Visualization of lignification process in differentiating xylem of pine by microautoradiography. *Holzforschung*, **42**, 347-350

Terazawa, M., Miyake, M. (1984) Phenolic compounds in living tissue of woods. II. Seasonal variations of phenolic glycosides in the cambial sap of woods. *Mokuzai Gakkaishi*, **30**, 329-334

Terazawa, M., Okuyama, H., Miyake, M. (1984) Phenolic compounds in living tissue of woods. I. Phenolic _-glycosides of 4-hydroxycinnamyl alcohol derivatives in the cambial sap of woods. *Mokuzai Gakkaishi*, **30**, 322-328

Van Doorsselaere, J., Baucher, M., Chognot, E., Chabbert, B., Tollier, M-T., Petit-Conil, M., Leple, J-C., Pilate, G., Cornu D., Monties, B., Van Montagu, M., Inze, D., Boerjan, W., Jouanin, L. (1995) A novel lignin in poplar trees with a reduced caffeic acid/5-hydroxyferulic acid O-methyltransferase activity. *Plant Journal*. **8**, 855-864.

Vignols, F., Rigau, J., Torres M. A., Capellades, M., Puigdomenech, P. (1995) The brown midrib3 (bm3) Mutation in Maize Occurs in the Gene Encoding Caffeic Acid O-Methyltransferase. *Plant Cell*, 7, 407-416.

Wayman, M., Obiaga, T. I. (1974) Can. J. Chem. 52, 2102-2110

Whetten, R., Sederoff, R. (1995) Lignin biosynthesis. *Plant Cell*, **7**, 1001-1003

Ye, Z-H. (1997) Association of caffeoyl coenzyme A 3-O-methyltransferase expression with lignifying tissues in several dicot plants. *Plant Physiology*, **115**, 1341-1350.

Ye, Z-H., Kneusel, R. E., Matern, U., Varner, J. E. (1994) An alternative methylation pathway in lignin biosynthesis in Zinnia. *Plant Cell*. **6**, 1427-1439.

Ye, Z-H., Varner, J. E. (1995) Differential expression of two O-methyltransferases in lignin biosynthesis in Zinnia elegans. *Plant Physiology*, **108**, 459-467.

Zhong, R., Morrison, W. H. III., Negrel, J., Ye, Z-H. (1998) Dual methylation pathways in lignin biosynthesis. *Plant Cell*, **10**, 2033-2045.

Molecular Breeding of Woody Plants
N. Morohoshi and A. Komamine, editors.

THE BEHAVIOR OF EXOGENOUS SINAPIC ACID IN THE DIFFERENTIATING XYLEM OF ANGIOSPERM

Kazuchika Yamauchi, Seiichi Yasuda and Kazuhiko Fukushima

Graduate school of Bioagriaultural Sciences, Nagoya University
Furo-cho Chikusa-ku Nagoya 464-8601, Japan

ABSTRACT

For many angiosperms, a 4-coumarate-CoA ligase (4CL; EC 6.2.1.12) isozyme that exhibits activity with a sinapic acid substrate has not been detected *in vitro*. Therefore, the possible role of sinapic acid as a precursor of syringyl lignin has been questioned. To further examine this issue, heptadeutero sinapic acid [β-D, $(\text{-OCD}_3)_2$] were synthesized and fed to shoots of oleander (*Nerium indicum* Mill.) and Robinia (*Robinia pseudoacacia* L.) trees, and the incorporation of these labeled precursors into lignin was traced. Lignin in the newly formed xylem of the treated stems was analyzed by derivatization followed by reductive cleavage (DFRC). The two labeled methoxyl groups of sinapic acid were useful in monitoring the conversion of sinapic acid into syringyl lignin. When heptadeutero sinapic acid was fed, syringyl units containing seven deuterium labels were detected. These results suggest that sinapic acid is converted to sinapoyl-CoA during lignin biosynthesis in oleander and *Robinia* trees *in vivo*.

KEYWORDS

Robinia, oleander, sinapic acid, lignin biosynthesis, tracer experiment

INTRODUCTION

According to the traditional lignin biosynthetic pathway which was established by Brown, Neish and Higuchi [1-3], ferulic acid and sinapic acid have been proposed as the precursors of guaiacylpropane and syringylpropane lignin monomers, respectively. However, in recent studies, 4CL activity with a sinapic acid substrate was not detectable for several plants [4,5]. If sinapic acid cannot be a precursor of syringyl lignin monomers, another pathway must be responsible for syringyl monolignol biosynthesis. Our previous studies have shown that the conversion of guaiacylpropane units to syringylpropane units may occur at the monolignol stage [6,7]. And some route in which sinapic acid is not used for lignin biosynthesis has been proposed [8,9].

To clarify these questions, we fed heptadeuterium-labeled sinapic acid [β-D, $(\text{-OCD}_3)_2$] (HSA) to oleander and Robinia trees, and traced the incorporation of the labeled precursor into lignin. Lignins were degraded by derivatization followed by reductive cleavage (DFRC), and the degradation products were analyzed by GC-MS. The result of the present study provided evidence of a route from sinapic acid to sinapyl alcohol via sinapoyl-CoA in oleander and Robinia.

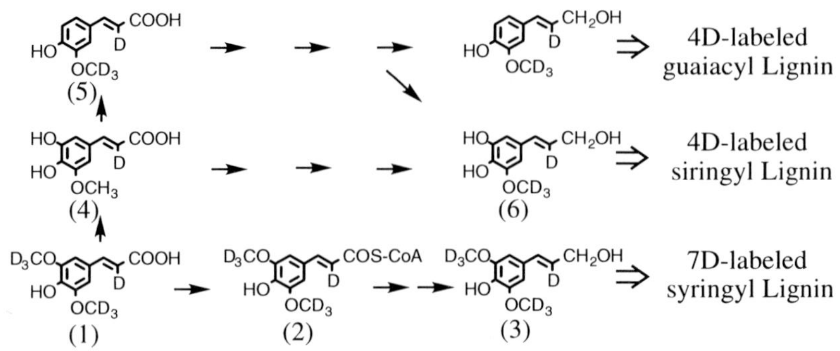

Figure 1. Possible pathways for the incorporation of D-labeled precursor (1) into lignin.

RATIOALE

As shown in Figure 1, HSA (1) is incorporated into lignin after it is administered to living trees. If the methoxyl deuterium atoms of HAS are not released during incorporation into lignin, syringyl units containing seven deuteriums (7D) should be detected by MS. This observation would support the existence of a route from sinapic acid to sinapyl alcohol via sinapoyl-CoA (1→2→3 in Figure 1), and it would also support the traditionally accepted pathway, in which sinapic acid is the precursor of syringyl lignin, as put forth by Higuchi [3]. If the conversion to syringylpropane can occur via routes (1→4→6) or (1→4→5→6), one deuterium-labeled methoxyl group should be released and syringyl units containing four deuteriums (4D) should be detected. This result would suggest that syringyl lignin can be generated from sinapic acid without synthesis of sinapoyl-CoA. On the other hand, if sinapic acid is not a precursor of syringyl lignin, then no labeled syringyl lignin should be detected at all.

MATERIALS & METHODS

Synthesis of labeled precursors

Hexadeutero syringaldehyde [(-OCD_3)$_2$] was synthesized from 2,6-dibromophenol as described previously [10]. HSA [β-D, (-OCD_3)$_2$] was synthesized by reacting hexadeutero syringaldehyde [(-OCD_3)$_2$] with malonic acid-d_4 (99%; ISOTEC Inc.). The methoxyl groups of both precursors were completely labeled with deuterium. About 60 - 70% of the β-hydrogens were substituted with deuterium at the side chain.

Plant material and administration of precursors.

The upper parts of 2-year-old shoots of oleander (*Nerium indicum* Mill.) and Robinia (*Robinia pseudoacacia* L.) trees that were growing on the campus of Nagoya University were removed in July 1999. A small depression was made at the top of the remaining stem, and the depression was filled with 50 (Robinia) or 25 ml (oleander) of a 4 mM aqueous solution of the precursor in a 4:6 (v/v) mixture of 66.7 mM KH_2PO_4 and 66.7 mM Na_2HPO_4 (pH 7.1). After 4 weeks, each shoot was harvested as described previously [6].

Analysis of lignin

The DFRC method of Lu and Ralph [11] was used to degrade the lignin, as described previously [6].

RESULTS & DISCUSION

The mass spectra of guaiacylpropane and syringylpropane monomers derived by DFRC from the differentiating xylem of Robinia after feeding of non-labeled ferulic acid are shown in Figures 2. Each of these monomers occurred in two isomeric forms (*cis* and *trans*). The *trans* isomers, *trans*-4-acetoxyconiferyl acetate (G_t) and *trans*-4-acetoxysinapyl acetate (S_t), were used in estimating incorporation of D from the precursors. The mass spectra of the G_t and S_t monomers gave base peaks (M) at m/z 222 and 252, respectively. Partial mass spectra of Robinia DFRC products resulting from administration of HSA are shown in Figures 3. No difference in D-enrichment was observed for the *cis* and *trans* isomers. When HSA was fed, no D-enriched peaks were present in both the G_t and S_t, monomers, but 4D (m+4, m/z 256) and 7D (m+7, m/z 259) labels were detected in the S_t monomers.

The observation of 7D-labeled syringylpropane units suggests that sinapic acid can be converted into syringyl lignin without loss of methoxyl groups [(1→2→3); Figure 1]. It seems likely that sinapic acid was converted into sinapyl alcohol via sinapoyl-CoA. This result supports the pathway proposed by Higuchi [3]. The observation of 4D-

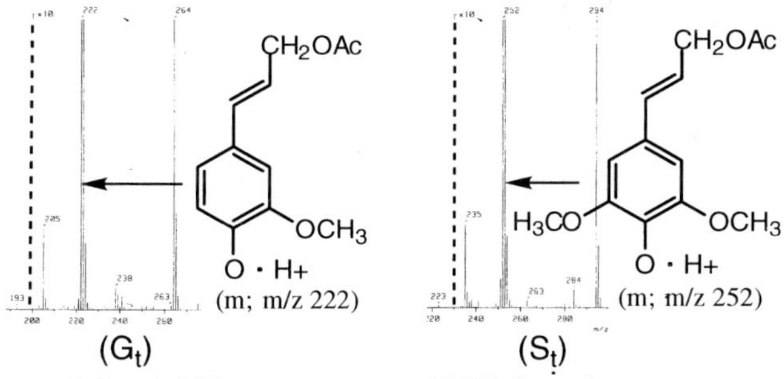

Figure 2. Partial EI mass spectra of DFRC products.
G_t and S_t are from Robinia fed with non-labeled ferulic acid

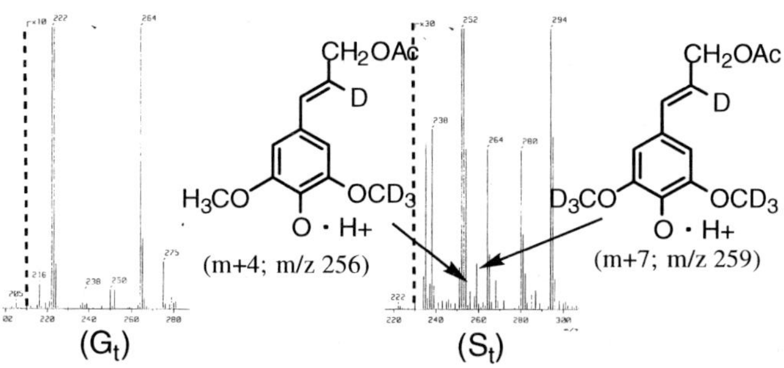

Figure 3. Partial EI mass spectra of DFRC products.
G_t and S_t are from Robinia fed with D-labeled sinapic acid (HSA).

labeled S_t suggests that sinapic acid is converted into syringyl lignin, releasing only one methoxyl group. Sinapic acid may be incorporated into syringyl lignin after conversion into 5-hydroxyferulic acid [(1→4→6); Figure 1] or ferulic acid [(1→4→5→6); Figure 1]. If the sinapic acid were demethylated and dehydroxylated to form ferulic acid, it could be incorporated into 4D-labeled guaiacylpropane lignin. However, 4D-labeled Gt could not be detected. These observations suggest that sinapic acid can be demethylated to form 5-hydroxyferulic acid, but is not dehydroxylated to form ferulic acid. Results for oleander were similar to those found for Robinia.

These results strongly indicated the existence of the pathway from sinapic acid to sinapyl alcohol via sinapoyl-CoA in Robinia and oleander trees.

REFFERENCES

1. S. A. Brown and A. C. Neish, Studies of lignin biosynthesis using isotopic carbon IV. Formation from some aromatic monomers, *Can. J. Biochem. Physiol.*,**1955**, 33, 948-962
2. T. Higuchi and S. A. Brown, Studies of lignin biosynthesis using isotopic carbon XII. The biosynthesis and metabolism of sinapic acid, *Can. J. Biochem. Physiol.*,**1963**, 41, 613-620
3. T. Higuchi, Lignin biochemistry: Biosynthesis and biodegradation, *Wood Science and Technology,* **1990**, 24, 23-63
4. H. Kutsuki, M. Shimada and T. Higuchi, Distribution and Roles of *p*-Hydroxycinnamate: CoA Ligase in Lignin Biosynthesis, *Phytochemistry,* **1982**, 21(2), 267-271
5. S. M. Allina, A. Pri-Hadash, D. A. Theilmann, B. E. Ellis and C. J. Douglas, 4-Coumarate: Coenzyme A Ligase in Hybrid Poplar, *Plant physiol.*, **1998**, 116, 743-754
6. F. Chen, S. Yasuda and K. Fukushima, Evidence for a novel biosynthetic pathway that regulates the ratio of syringyl to guiacyl residues in lignin in the differentiating xylem of *Magnolia kobus* DC., *Planta*, **1999**, 207, 597-603
7. N. Matsui, F. Chen, S. Yasuda and K. Fukushima, Conversion of guaiacyl to syringyl moieties on the cinnamyl alcohol pathway during the biosynthsis of lignin in angiosperms, *Planta*, **2000**, 210, 831-835
8. L. Li, J. L. Popko, T. Umezawa and V. L. Chiang, 5-Hydroxyconiferyl Aldehyde Modulates Enzymatic Methylation for Syringyl Monolibnol Formation, a New View of Monolignol Biosynthesis in Angiosperms, *J. Biol. Chem.*, **2000**, 275(9), 6537-6545
9. J. H. Meyermans, M. R. Hemm and C. Chapple, Modifications in Lignin and Accumulation of Phenolic Glucosides in Poplar Xylem upon Down-regulation of Caffeoyl-Coenzyme A *O*-Metheyltransferase, an Enzyme Involved in Lignin Biosynthesis, *J. Biol. Chem.*, **2000**, 275(47), 36899-36909
10. K. Yamauchi, S. Yasuda and K. Fukushima, Synthesis of both methoxyl labeled sinapic acid with six deuteriums, convenient precursor to elucidate the metabolism of phenyl propanoids that have syringyl nuclei., in preparation.
11. F. Lu and J. Ralph, Derivatization Followed by Reductive Cleavage (DFRC Method), a New Method for Lignin Analysis: Protocol for Analysis of DFRC Monomers, *J. Agric. Food Chem.,* **1997**, 45, 2590-2592

Molecular Breeding of Woody Plants
N. Morohoshi and A. Komamine, editors.

FUNCTIONAL ANALYSIS OF *PHENYLALANINE AMMONIA-LYASE* GENE PROMOTER OF POPULAR

Mikiko Oyanagi & Yoshihiro Ozeki*

Department of Biotechnology, Faculty of Technology, Tokyo University of Agriculture and Technology, Koganei 184-8588, Japan.

ABSTRACT

Phenlyalanine ammonia-lyase (PAL) catalyzes the first step in the phenylpropanoid pathway and plays an important role in the biosynthesis of phenylpropanoid and flavonoid compounds. *Cis*-elements of the promoter region of *g2bPAL* gene, a *PAL* gene obtained from popular, *Populus kitakamiensis*, were analyzed by transient expression in carrot protoplasts and transgenic tobacco. Four conserved *cis*-elements, M-, G-, L- and P'-boxes in the PAL promoters of several plant species, were found in the proximal region of the *g2bPAL* promoter. Deletion of G-box in the promoter resulted in the loss of *g2bPAL* promoter activity in carrot protoplasts. In transgenic tobacco plants harboring *g2bPAL*-GUS, deletion of P'-box resulted in strong GUS expression in the endodermis of the stems and in the root. The findings indicate that the P'-box may play an important role in the regulation of *g2bPAL* expression to synthesize lignin during xylem development.

KEYWORDS

Phenylalanine ammonia-lyase, phenylpropanoid pathway, popular, *Populus*, xylem

INTRODUCTION

The phenylpropanoid pathway produces protectants against environmental stresses such as pathogen infection, mechanical injury and ultraviolet (UV) irradiation. It also produces lignin during xylem differentiation and anthocyanin during petal development. The protectants are produced in response to environment cues, but lignin and anthocyanin are produced at the particular stage of programmed differentiation.

The expression mechanism of *phenylalanine ammonia-lyase* (*PAL*) gene, which encodes the gene involved in the first step of the phenylpropanoid pathway, has been well studied recently in various plant species, because it functions at the branching point from the primary metabolic pathway to the phenylpropanoid pathway. However, most of the research has been focused on *cis*-elements of PAL promoters induced by environmental stresses and *trans*-regulatory factors which act on them [1, 2, 3, 4, 5]. Only a few researchers have studied the expression mechanisms of *PAL* promoters during differentiation [6, 7].

During the growth of higher plants, cell walls are mechanically strengthened by accumulation of many kinds of substances on the cortex. The most remarkable example is the accumulation of lignin, called lignification at the cell walls of vessels, tracheids and fiber. Often parenchymatous and pith cells are also lignified. After lignification, the protoplasm dies, but translocation of water and solutes in water are not prevented.

Lignification strengthens tissues, prevents decomposition of cellulose, and increases chemical resistance.

In a previous study, Osakabe *et al.* cloned four *PAL* genes *(g1PAL, g2aPAL, g2bPAL, g4PAL)* in popular, *Populus kitakamiensis*, and characterized their structure [8]. The transcript level of *g2bPAL* was higher than that of the other *PAL* genes. In this report, *cis*-elements of the promoter region of the *g2bPAL* gene were analyzed by transient expression in carrot protoplasts and by using transgenic tobacco.

MATERIALS & METHODS

Preparation of deletion mutant of *g2bPAL* promoter and construction of promoter-GUS and -luc fusions

The *g2bPAL* promoter region was amplified by PCR using primers, 5'-CAT-GAT-ATC-CAG-ACA-CAG-TCG-3' [corresponding to –617 to –597 of *g2bPAL* (accession no D43802), +1 is located at the first letter of ATG of the first methionine, underlines showing EcoRV site of *g2bPAL* promoter region] and 5'-ggc-tct-aga-GTA-AAG-CTA-TGA-TCC-TCT-ATG-3' [upper case letters corresponding to –6 to –26 of *g2bPAL*, and lower case letters corresponding to additional XbaI site for cloning work], to clone λ DNA containing *g2bPAL* gene [8] as a template, and an amplified fragment was digested with XbaI and EcoRV and cloned into pBluescript SK. M-, G-, L- and P'-box deleted promoters were prepared by PCR using primers, M1 and M2, G1 and G2, L1 and L2, P1 and P2 (Fig. 1), respectively, to clone the *g2bPAL* promoter region in pBluescript SK as a template, followed by self-ligation and introduction into *E. coli*. After confirmation of nucleotide sequences by sequencer, DNA fragments of the promoters were prepared with HindIII and XbaI digestion, and pBI-H1 [9] and p35Sluc/19 [10] were digested with HindIII and XbaI followed by agarose gel electrophoresis to remove the 35S promoter fragment from vectors. DNA fragments of the *g2bPAL* promoters prepared as above and vectors were ligated and introduced into *E. coli*. All the plasmids were prepared and restriction endonuclease maps were determined to confirm the accuracy of the construction. g2b-luc is a native *g2bPAL* promoter, and –M-luc, –G-luc, –L-luc and –P'-luc are M-, G-, L- and P'-box deleted promoters, respectively, fused to the *luciferase* gene for transient expression experiments. g2b/pAB, –M/pAB, –G/pAB, –L/pAB and –P'/pAB are the promoters fused to the GUS gene used to prepare stable tobacco transformants.

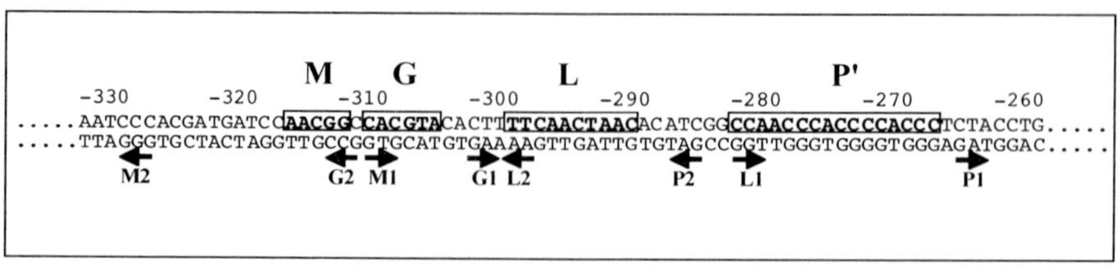

Figure. 1. Putative *cis*-elements found in the proximal region of *g2bPAL* promoter. Arrows indicate the 5' ends of PCR-primer sequences of M1 and M2, G1 and G2, L1 and L2, P1 and P2 used to prepare deletion mutants of the promoter.

Transient expression assay and preparation of stable tobacco transformants and GUS assay

Preparation of carrot protoplasts, electroporation and firefly luciferase activity assay were performed as previously described [10].

g2b/pAB, –M/pAB, –G/pAB, –P/pAB and –L/pAB were introduced into *Agrobacterium tumefaciens* EHA101 by electroporation. Transformed tobacco plants (*Nicotiana tabacum* cv SR1) were produced by the leaf disk method using transformed *A. tumefaciens*.

The genomic DNAs were extracted from randomly selected independent plants from transformed tobacco plants to confirm the introduced gene by Southern analysis. The leaf, stem and root (0.3 g each) were taken from each transformed tobacco plants, in which the gene was correctly introduced, and the proteins were extracted with the extraction buffer in GUS Light kit (Tropix, Inc.). The protein concentration was determined using Bio-Rad's protein assay kit and then using GUS-Light solution and the luminometer Lumat LB905 (Berthold Japan). The GUS activity was measured for 5 seconds, and expressed as luminescence value per protein weight.

For GUS staining of the tissues of transgenic tobacco, the plants were randomly selected from several transgenic plants, and slices of the stem and root were prepared from each selected tobacco. Then they were placed in GUS staining solution (38.3 mM X-gluc : GUS staining buffer = 13 : 487) at 37 ℃ for several hours, decolorized with 70% ethanol for several hours, and the GUS expression pattern was observed by microscopy

RESULTS & DISCUSSION

Sequence analysis of *g2bPAL* promoter

We analyzed the sequence of *g2bPAL* promoter that is expressed most strongly in popular. Four consensus sequence motifs, AACNG motif (M-box) , the DNA binding consensus sequence of *myb* protein, a gene-regulatory protein in animals [11] , G-box (CACGTG) homolog [12] , L-box ($^{T}/_{C}C^{T}/_{C}{}^{C}/_{T}$ACCTACC) and P'-box [similar to P-box ($^{C}/_{T}T^{T}/_{C}{}^{C}/_{T}{}^{A}/_{C}{}^{A}/_{C}C^{A}/_{C}A^{A}/_{C}C^{C}/_{A}{}^{C}/_{A}C$) [11]] were detected. The last two have been reported in regions other than the promoter region of a number of known *PAL* genes [13] . In the deletion mutants lacking the P- and L-boxes, the transcription activity of the carrot *PAL* promoter was not increased by UV irradiation [5] .

Transient assay of deletion mutant of *g2bPAL* promoter in carrot protoplasts

We prepared individual box-deleted mutants of the *g2bPAL* promoter gene, which was fused to the *luciferase* gene as a reporter. The transient expression of luciferase was examined after electroporation into carrot protoplasts (Fig. 2). Deletion of the G-box lowered the reporter gene activity compared to the native promoter. However, this does not confirm that the box plays a role in the regulation of gene expression in plants, because we used "undifferentiated" carrot protoplasts and analyzed only the transient expression of the promoter. In the present study, in order to analyze the role of the boxes in stable expression, we prepared individual box-deleted mutants of the *g2bPAL* promoter gene fused to the β-glucuronidase (GUS) gene as a reporter. After introducing them to tobacco plants using *A. tumefaciens*, we analyzed the reporter gene activity and specificity of the expression in each mutant.

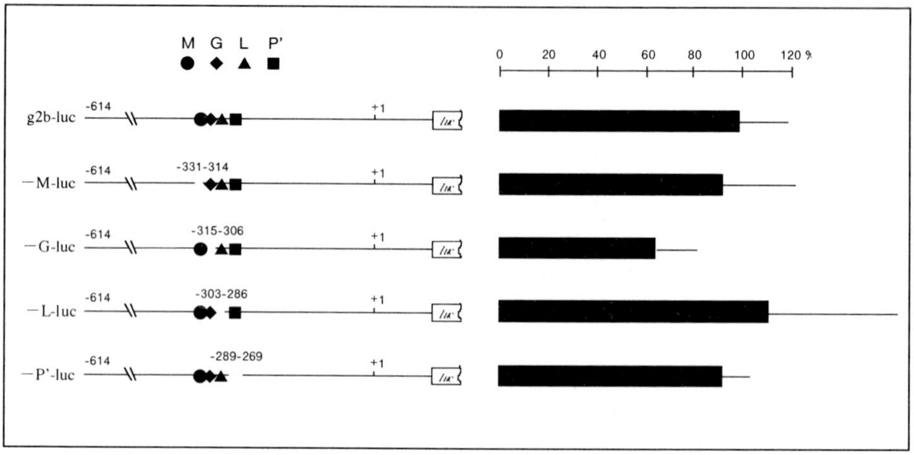

Figure. 2. Relative amounts of Luc activity/GUS activity of deletion mutants of popular *g2bPAL* promoter in carrot protoplasts.

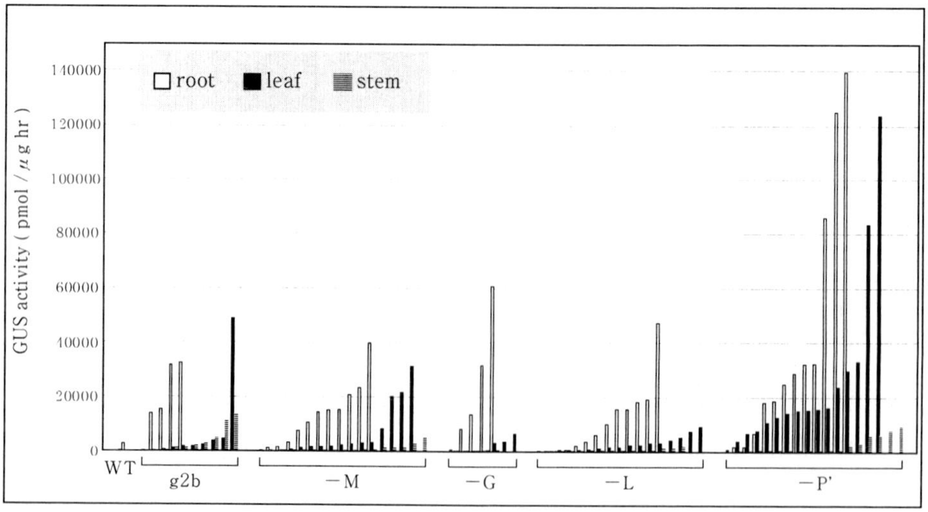

Figure. 3. GUS activity of transgenic tobacco transformed deletion mutants of *g2bPAL* promoter. Each column shows GUS activity of root, leaf and stem of individual transgenic tobacco plantlets harboring the control native *g2bPAL* promoter (g2b) and the mutant promoters lacking the M, G, L and P' boxes, –M, –L, –G and –P', respectively.

Analysis of GUS expression in transgenic tobacco plants

The shoot regeneration efficiency of tobacco leaf discs from the G-box-deleted mutant (–G) and that of the control native *g2bPAL* promoter (g2b) was about one half to one third of the mutants lacking the M, L and P' boxes, –M, –L and –P', respectively, and the transgenic tobacco –G and g2b grew very slowly (data not shown). It is uncertain whether the tobacco cells in the leaf discs transformed with the constructs harboring the M, L and P'-boxes grows well or not. However, the fact that deletion of only one box, M, L or P'-box, did not suppress the growth indicates complicated interaction among these boxes.

After identifying the deletion mutant of *g2bPAL* promoter in their genomes by Southern hybridization (data not shown), we measured the GUS activity in leaves, stems and roots of each transgenic tobacco plant (Fig. 3). The transgenic tobacco plant –P' showed higher GUS activity than the other transgenic plants.

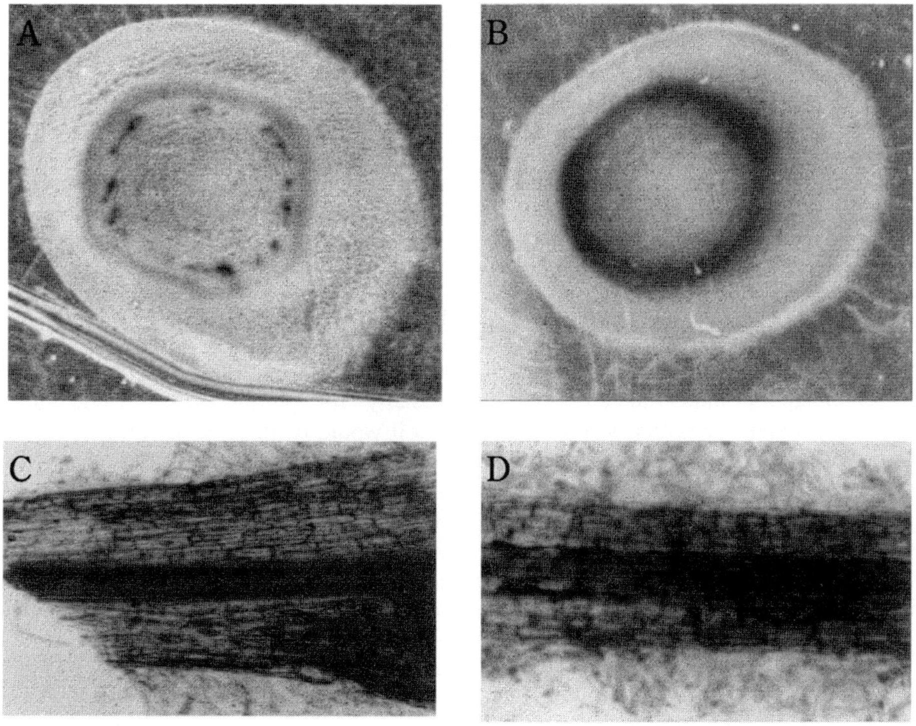

Figure. 4. Histochemical localization of GUS activity in the tobacco plants transformed with deletion mutants of the *g2bPAL* promoter. A, cross-section through a stem of a plant transgenic for –M-box promoter ; B, cross-section through a stem of a plant transgenic for –P'-box promoter ; C, a root of a plant transgenic for *g2bPAL* promoter ; D, a root of a plant transgenic for –P'-box promoter.

Slices of stems and roots from each transgenic tobacco were placed in a GUS staining solution (Fig. 4). GUS reporter gene was expressed at only the vascular bundle of stems in the control native *g2bPAL* promoter, –G, –L, and –M (Fig. 4 A), but at the endodermis of the transgenic plant –P' and the region surrounding it (Fig. 4 B). By contrast, the expression of GUS reporter gene was weak in the roots of transgenic plants except –P', but was very strong in the whole cells of roots of –P' (Fig. 4 D). This indicates that the P'-box contributed to the change in the region of GUS gene expression and to the loss of the specific expression in xylem.

Although deletion of P'-box in popular *g2bPAL* promoter caused loss of the specific expression in xylem, it increased the GUS gene activity (Fig. 3). These results suggested that the P'-box in popular *g2bPAL* promoter may play an important role in the accurate expression of *g2bPAL* in cells of xylem and the severe repression in cells of the other tissues except for xylem.

g2bPAL promoter was not activated by light in the transgenic tobacco plants

Lignin synthesis is induced by light [14, 15], suggesting that *PAL* genes may also be induced by light. Based on this assumption, we analyzed the light inducibility of popular *g2bPAL* promoter by light using transgenic tobacco plants with deletion mutants of the *g2bPAL* promoter. However, the activity of GUS gene expression was not induced by light in any of the transgenic plants, although GUS expression was reduced by light in some of the transgenic tobacco plants obtained here (data not shown).

CONCLUSIONS

Recently, many studies have been carried about on the genes that encode enzymes related to the phenylpropanoid pathway and lignin biosynthesis pathway, and the introduction of these genes into plants. By artificially changing the activity of the gene expression and resulting suppression of lignin formation, such transgenic plants would be suitable for industrial application, especially for production of wood pulp. However, lignin is essential to life of plant, and it would be impossible for the plant to grow without it. For example, the transgenic plants harboring the 35S promoter with anti-sense *PAL* cDNA tended to have scorched leaves, to grow very slowly and to have an unusual shape (Ozeki *et al.*, unpublished data). Thus, it is difficult to produce transgenic plants containing a low amount of lignin by using the 35S promoter, which is strongly expressed in all plant tissues. In order to regulate the accumulation of lignin in plants, it is necessary to regulate the expression of the lignin biosynthesis pathway more strictly and exactly.

Here, we demonstrated the P'-box in popular *g2bPAL* promoter to be a *cis*-element related to the gene activity and/or specificity of expression. Further analyses focused on *cis*-elements related to *PAL* gene expression may reveal the regulatory mechanisms of the level and tissue-specificity of expression. The information will enable us to design a suitable *PAL* promoter, which suppresses the production of excessive amounts of lignin, but permits synthesis of minimum amounts of lignin in the tissues and organs need for normal plant growth.

ACKNOWLEDGEMENTS

We are grateful to Professor Noriyuki Morohoshi (Tokyo University of Agriculture and Technology) for the gift of λ clone containing the *g2bPAL* gene. We are indebted to Dr. Shinya Kawai (Tokyo University of Agriculture and Technology) for his helpful discussion. We acknowledge financial support from the "Research for the Future" Program of the Japan Society for the Promotion of Science.

REFERENCES

1. H. Kato, M. Wada, K. Muraya, K. Malik, T. Shiraishi, Y. Ichinose & T. Yamada, 'Characterization of nuclear factors for elicitor-mediated activation of the promoter of the pea phenylalanine ammonia-lyase gene 1', *Plant Physiol.*, **1995**, 108, 129-139.
2. E. Logemann, M. Parniske & K. Hahlbrock, 'Modes of expression and common structural features of the complete phenylalanine ammonia-lyase gene family in parsley', *Proc. Natl. Acad. Sci. USA*, **1995**, 92, 5905-5909.
3. R. Lois, A. Dietrich, K. Hahlbrock & W. Schulz, 'A phenylalanine ammonia-lyase gene from parsley : structure, regulation and identification of elicitor and light responsive *cis*-acting elements', *EMBO J.*, **1989**, 8, 1641-1648.
4. A. Seguin, G. Laible, A. Leyva, R. A. Dixon & C. J. Lamb, 'Characterization of a gene encoding a DNA-binding protein that interacts in vitro with vascular specific *cis* elements of the phenylalanine ammonia-lyase promoter', *Plant Mol. Biol.*, **1997**, 35, 281-291.
5. J. Takeda, Y. Ozeki & K. Yoshida, 'Action spectrum for induction of promoter activity of phenylalanine ammonia-lyase gene by UV in carrot suspension cells', *Photochem. Photobiol.*, **1997**, 66, 464-470.
6. A. Leyva, X. Liang, J. A. Pintor-Toro, R. A. Dixon & C. J. Lamb, '*Cis*-element combinations determine phenylalanine ammonia-lyase gene tissue-specific expression patterns', *Plant Cell*, **1992**, 4, 263-271.
7. X. W. Liang, M. Dron, J. Schmid, R. A. Dixon & C. J. Lamb, 'Developmental and environmental regulation of a phenylalanine ammonia-lyase-beta-glucuronidase gene fusion in transgenic tobacco plants', *Proc. Natl. Acad. Sci. USA*, **1989**, 86, 9284-9288.
8. Y. Osakabe, K. Osakabe, S. Kawai, Y. Katayama & N. Morohoshi, 'Characterization of the structure and determination of mRNA levels of the phenylalanine ammonia-lyase gene family from *Populus kitakamiensis*', *Plant Mol. Biol.*, **1995**, 28, 1133-1141.
9. T. Kimura, S. Takeda, J. Kyozuka, T. Asahi, K. Shimamoto & K. Nakamura, 'The presequence of a precursor to the δ -subunit of sweet potato mitochondrial F_1ATPase is not sufficient for the transport of β -glucuronidase (GUS) into mitochondria of tobacco, rice and yeast cells', *Plant Cell Physiol.*, **1993**, 34, 345-355.
10. Y. Ozeki, E. Davies & J. Takeda, 'Structure and expression of chalcone synthase gene in carrot suspension cultured cells regulated by 2,4-D', *Plant Cell Physiol.*, **1993**, 34, 1029-1037.
11. H. Biedenkapp, U. Borgmeyer, A. E. Sippel & K. -H. Klempnauer, 'Vial *myb* oncogene encodes a sequence-specific DNA-binding activity', *Nature*, **1988**, 335, 835-837.
12. F. Katagiri & N. H. Chua, 'Plant transcription factors: present knowledge and future challenges', *Trends Genet.*, **1992**, 8, 22-27.
13. Y. Ozeki, Y. Ito, N. Sasaki, M. Oyanagi, H. Akimoto, Y. Chikagawa & J. Takeda,

'Phenylalanine ammonia-lyase genes involved in anthocyanin synthesis and the regulation of its expression in suspension cultured carrot cells', *J. Plant Res.*, **2000**, 113, 319-326.

14. S. C. Grace & B. A. Logan, 'Energy dissipation and radical scavenging by the plant phenylpropanoid pathway', *Philos. Trans. R. Soc. Lond. B. Biol. Sci.*, **2000**, 355, 1499-1510.
15. U. Ryser, M. Schorderet, G. F. Zhao, D. Studer, K. Ruel, G. Hauf & B. Keller, 'Structural cell-wall proteins in protoxylem development: evidence for a repair process mediated by a glycine-rich protein', *Plant J.*, **1997**, 12, 97-111.

Molecular Breeding of Woody Plants
N. Morohoshi and A. Komamine, editors.

XYLEM PEROXIDASES: PURIFICATION AND ALTERED EXPRESSION

Jørgen H. Christensen, Marc Van Montagu, Guy Bauw & Wout Boerjan*

Vakgroep Moleculaire Genetica, Departement Plantengenetica, Vlaams Interuniversitair Instituut voor Biotechnologie (VIB), Universiteit Gent, K.L. Ledeganckstraat 35, B-9000 Gent, Belgium.

ABSTRACT

The peroxidase-dependent oxidation of the lignin monomer analogue syringaldazine (SYR) shows an appealing co-localization with lignification in most species analyzed. We have isolated two SYR-oxidizing peroxidases from the xylem of poplar (*P. trichocarpa* 'Trichobel'). We have shown that these were the only isoenzymes able to catalyze this reaction in the xylem, and that the corresponding activity correlates with actively lignifying cells within the poplar xylem. Furthermore, we have isolated a cDNA that codes for one of these enzymes (PXP 3-4) and demonstrated that the mRNA is expressed in the bark and the xylem of the stem and in the xylem of the roots. The cDNA encodes a peroxidase that is expressed as a preprotein with signalpeptides at both termini. The peroxidase cDNA *PXP 3-4* was expressed in *P. tremula* × *P. alba* in sense orientation, using constructs encoding either the native enzyme or a version without the C-terminal propeptide. Also plants expressing this cDNA in antisense orientation were generated. Overexpressing lines with more than 800-fold higher peroxidase activity were identified. In the bark of these plants, half of the extracted proteins corresponded to PXP 3-4, which is more than 0.5 mg peroxidase per g of bark. All overexpressing lines were phenotypically normal. No alteration was observed in lignin amount, condensation or monolignol composition and the metabolic profiles and the redox state of these plants were unaltered.

KEYWORDS

cDNAs, expression, lignin, phenolics, poplar xylem peroxidases, purification, redox state, SYR oxidation, transgene expression.

INTRODUCTION

Classical peroxidases (class III; EC 1.11.1.7, donor:hydrogen peroxide oxidoreductase) are ubiquitous secretory heme-containing oxidative plant enzymes[1], that are encoded by multigene families. The Arabidopsis genome is expected to code for approximately 80 different peroxidase genes[2]. The enzymatic reaction mechanism for peroxidation is very well described and involves three consecutive enzymatic redox stages, resulting in the consumption of one equivalent of H_2O_2 and the oxidation of two equivalents of reducing substrate (AH) ($H_2O_2 + 2\ AH \rightarrow 2\ H_2O + 2A\cdot$). In this reaction, peroxidases accept a broad spectrum of reducing substrates[3]. Also oxidase reactions have been observed for many peroxidases.

In contrast to the detailed molecular description of peroxidases[4], the biological function of most peroxidase isoenzymes remains unknown or has not been conclusively determined. Based on *in vitro* evidence, a plethora of putative functions has been suggested. However, their non-specific way of action, their abundance, their close sequence and immunological relations and the high reactivity of the formed radicals have impeded the assignment of biological functions to specific isoenzymes.

The most studied function for peroxidases is their putative role in lignin polymerization. Lignin is a polymeric constituent of the plant cell wall that serves supportive and impermeability functions in the vascular tissue and plays a role in defense against pathogens[5]. Peroxidases have been shown to catalyze the polymerization of lignin monomers (*p*-coumaryl alcohol, coniferyl alcohol, and sinapyl alcohol) *in vitro* to form a lignin-like polymer[6,7]. Tobacco plants expressing a peroxidase cDNA in antisense orientation showed a reduced deposition of lignin after wounding[8]. By expressing the same peroxidase in sense orientation, lignin was ectopically formed in the pith, a tissue that is not lignified in control plants[9]. These results demonstrate that peroxidases can form lignin-like polymers *in vivo* and are involved in this process upon wounding. However, their putative contribution to the developmentally regulated lignification process has not been proven yet.

We have undertaken a project to identify lignification-specific peroxidase isoenzymes. Previously, we have reported on the isolation and characterization of all the identified anionic isoenzymes from the xylem of poplar[10], and thereby identified the isoenzymes able to oxidize the lignin monomer analogue syringaldazine (SYR). This enzymatic activity was previously correlated with lignification in poplar[11].

Here, we firstly report on the isolation and characterization of these enzymes and their corresponding cDNAs and, secondly, on the further analysis of the SYR-oxidizing peroxidases PXP 3-4 by transgene technology.

RESULT & DISCUSSION

Isolation and characterization of proteins and cDNAs

As the first stage in identifying lignin-specific peroxidase isoenzymes, the anionic peroxidases from the xylem of poplar (*Populus trichocarpa* 'Trichobel') were purified to homogeneity and characterized. A fairly standard procedure, exploiting methods often used to purify plant peroxidases (ionexchange, hydrophobic interaction, and lectin binding), allowed us to co-purify all the detected anionic peroxidases from the xylem of poplar stems. These peroxidases were subsequently separated by ion exchange chromatography into six pure peroxidase fractions corresponding to five different poplar xylem peroxidases (PXP 1, PXP 2, PXP 3-4, PXP 5, and PXP 6)[10] (Fig. 1). Two of the five isolated peroxidases (PXP 3-4 and PXP 5) could oxidize the lignin monomer analogue SYR, an activity previously correlated with lignification in poplar[11]. Because these isoenzymes were specifically or preferentially expressed in xylem, PXP 3-4 and PXP 5 were suggested to be involved in lignin polymerization and selected for further studies.

With the peptide information obtained from PXP 3-4 and PXP 5, we were able to isolate four different full-length peroxidase cDNAs from a *Populus trichocarpa* 'Trichobel' xylem cDNA library (*PXP 1, PXP 3-4, PXP 11,* and *PXP 22*). Two cDNAs, designated *PXP 1* and *PXP 3-4*, encoded all peptide sequences obtained from the purified isoenzymes PXP 1 and PXP 3-4, respectively, from which PXP 3-4 was one of the SYR-oxidizing enzymes. By RFLP-PCR, the expression of *PXP 3-4* was shown in the bark and xylem of the stem and in the xylem of the root.

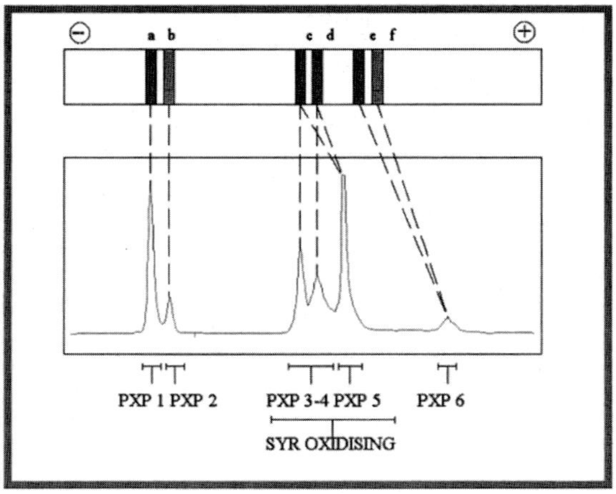

Figure 1. Correlation of peroxidase isoenzymes, gel activities and SYR oxidation. The upper panel represents a lane on an activity gel loaded with a xylem protein extract. The lower panel represents a chromatogram from the Mono Q separation of the co-purified peroxidases. The dashed lines indicate the relation between gel activities and the Mono-Q separated peroxidase isoenzymes. PXP 1 to PXP 6 represent the gene products determined from the molecular characterization. The SYR-oxidizing isoenzymes are indicated.

All the isolated cDNAs encoded an N-terminal ER-targeting signal peptide and an additional C-terminal propeptide. The latter is believed to be a vacuolar-targeting signal[1]. This putative localization seemed, at the most, to be transient because the corresponding *in situ* activity was shown in lignifying and lignified xylem cells, in which the mature xylem cells are dead and devoid of cellular organelles. This led to the hypothesis that these peroxidases were expressed to the vacuoles, but would be liberated to the cell wall upon tonoplast rupture during programmed cell death.

Alteration of *PXP 3-4* expression

To further investigate the function of PXP 3-4 in lignin formation and the significance of the C-terminal vacuolar-targeting signal, we generated transgenic plants that expressed *PXP 3-4* to high levels, with and without this peptide as well as plants with reduced expression of this peroxidase.

Transgenic lines expressing *PXP 3-4* in sense orientation with and without the C-terminal propeptide contained active peroxidases at high levels in all tissues examined. The level of peroxidase accumulation was dependant on tissue age with the highest activity encountered at the base of the stem (Fig. 2). This might indicate that PXP 3-4 is a highly stable enzyme that accumulates during the lifetime of the plant. In the bark, at the base of one-year-old plants, peroxidase concentrations up to 0.6 μg/mg tissue were found, whereas in xylem extracts from the base of the stem, peroxidase activities of more than 800-fold the wild-type total peroxidase activity levels were measured.

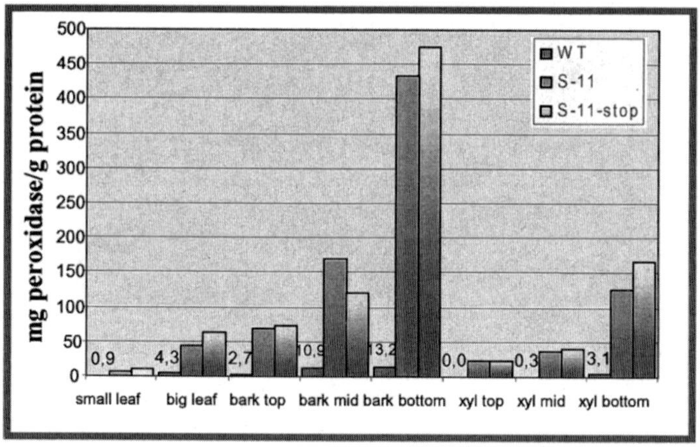

Figure 2. Peroxidase amounts in different tissues of wild type (WT), poplar that overexpress PXP 3-4 with (S-11) and without C-terminal peptide (S-11-stop), shown in relation to extracted protein content. Dev, developing; mat, mature; phl, phloem; xyl, xylem.

PXP 3-4 accumulation in the different tissues was similar in plants expressing *PXP 3-4* with or without the C-terminal pro-peptide (Fig. 2). The transgenic lines showed in general no visible phenotypes. The analysis of plants with lower level of *PXP 3-4* expression is still to preliminary and will not be discussed.

Although no morphological phenotypes could be ascribed to the overexpression of either version of *PXP 3-4*, altered levels of potential peroxidase substrates were anticipated. For this reason, the overexpressing lines were analyzed for the accumulation of phenols by metabolic profiling. Sixty-three major products (presumably mainly phenols) from the xylem of overexpressing plants were separated by HPLC, quantified and compared to wild-type plants. However, none of these extractable compounds showed significant differences from wild-types, suggesting that the overproduced protein did not perform additional reactions. Ascorbic acid (AA) has been shown to inhibit peroxidase reactions *in vitro* by reducing oxidized substrates (here lignin precursors, LP) back to their initial oxidation state, generating dehydroascorbic acid (DHAA) ($LP + AA \rightarrow LP\cdot + AA \rightarrow LP + DHAA$). The analysis of bark samples revealed no measurable difference between overexpressing and wild-type plants. Taking AA/DHAA as an indicator for the general level of peroxidation, these results indicate that no additional peroxidation occurs in plants with elevated peroxidase levels. Finally, Klason lignin analysis showed that the amount of lignin was unaltered, and by thioacidolysis the S/G ratio and the degree of condensation (S+G) was as well shown to be similar to that of wild-type. All together, the obtained results indicate that H_2O_2 is the limiting factor for peroxidation – and not the level of peroxidase.

A pure peroxidase preparation could be obtained with a simple precipitation protocol consisting of 2-3 acetone precipitation steps (Fig. 3), making these overproducing plants a convenient material for the isolation of large quantities of PXP 3-4.

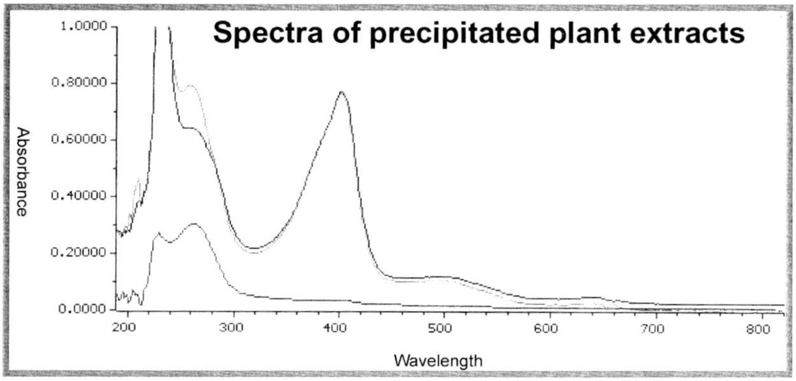

Figure 3. Spectra of acetone-precipitated bark proteins. The additional absorbance is derived from the heme group of the peroxidase. Bark proteins were precipitated three times with acetone and dissolved in the original volume and spectra were taken (200 - 800 nm). The lower spectrum is from WT plants and the two upper spectra are from plants that expresses PXP 3-4 with and without C-terminal peptide, respectively.

CONCLUSIONS AND PERSPECTIVES

One of the main arguments for peroxidases being responsible for lignin polymerization is the appealing strict correlation between the enzymatic *in situ* peroxidation of SYR and lignifying tissues[12]. We have, for the first time, been able to isolate the molecular background for this enzymatic *in situ* activity from a lignifying tissue.

Transgenic plants that accumulate high levels of this putative lignification peroxidase are indistinguishable from wild-type plants (except for peroxidase levels). Ectopic expression of peroxidases was expected to confer a highly pleiotropic phenotype, due to the unspecific nature of peroxidases. In contrast, expressing *PXP 3-4* either to the cell wall or putatively to the vacuole, yielded phenotypically normal plants, as are the redox state, the levels of the major phenolic compounds, and the lignin content of the xylem. These results indicate that PXP 3-4 does not show peroxidase oxidase activity against important cellular compounds and, consequently, is expected to exerts a function only when H_2O_2 is formed. These results are very promising for applications, because high peroxidase levels have been shown to confer higher resistance to insect damage[13]. However, these plants showed unwanted side effects due to the peroxidase oxidase reaction.

All data obtained so far have been derived from transgenic plant material grown under normal "non-stressed" conditions. Further research will be directed towards the analysis of plants that have been challenged with a number of biotic and abiotic stresses known to induce H_2O_2 production. In the stressed plants, altered phenotypes can be expected either as a consequence of fast H_2O_2 scavenging (influence on stress signal transduction?) or due to the oxidation/polymerization of certain compounds that are not oxidized in wild-type plants.

ACKNOWLEDGMENTS

This research was supported by funds from the Danish Agricultural and Veterinary Research Council, the AIR2-CT93-1661 and COPOL programs from the European Commission, and Novo Nordisk (Denmark).

REFERENCES

1. K. G. Welinder, Plant peroxidases: structure - function relationships, In: *Plant Peroxidases 1980 - 1990, Topics and Detailed Literature on Molecular, Biochemical and Physiological Aspects.*, C. Penel, T. Gaspar & H. Greppin (eds.), Université de Genève, Genève, Switzerland, **1992,** pp 1-24.
2. M. Tognolli, S. Overney, C. Penel, H. Greppin & P. Simon, A genetic and enzymatic survey of *Arabidopsis thaliana* peroxidases, *Plant Peroxidase Newsletters,* **2000,** 14, 3-12.
3. T. Gaspar, C. Penel, T. Thorpe & H. Greppin, *Peroxidases 1970 - 1980,* Université de Genève, Genève, Switzerland, **1972.**
4. H. B. Dunford, *Heme peroxidases*, Wiley-VSH, New York, **1999.**
5. M. Baucher, B. Monties, M. Van Montagu & W. Boerjan, Biosynthesis and genetic engineering of lignin, *Crit. Rev. Plant Sci.,* **1998,** 17, 125-197.
6. R. Sterjiades, J. F. D. Dean, G. Gamble, D. S. Himmelsbach & K. E. L. Eriksson, Extracellular laccases and peroxidases from sycamore maple (*Acer pseudoplatanus*) cell-suspension cultures. Reactions with monolignols and lignin model compounds, *Planta,* **1993**, 190, 75-87.
7. Y. Tsutsumi, K. Matsui & K. Sakai, Substrate-specific peroxidases in woody angiosperms and gymnosperms participate in regulating the dehydrogenative polymerization of syringyl and guaiacyl type lignins, *Holzforschung,* **1998,** 52, 275-281.
8. L. M. Lagrimini, V. Gingas, F. Finger, S. Rothstein & T. T. Y. Liu, Characterization of antisense transformed plants deficient in the tobacco anionic peroxidase, *Plant Physiol.*, **1997,** 114, 1187-1196.
9. L. M. Lagrimini, Wound-induced deposition of polyphenols in transgenic plants overexpressing peroxidase, *Plant Physiol.*, **1991,** 96, 577-583.
10. J. H. Christensen, G. Bauw, K. G. Welinder, M. Van Montagu & W. Boerjan, Purification and characterization of peroxidases correlated with lignification in poplar xylem, *Plant Physiol.,* **1998,** 118, 125-135.
11. R. Goldberg, A. M. Catesson, & Y. Czaninski, Some properties of syringaldazine oxidase, a peroxidase specifically involved in the lignification processes., *Z. Pflanzenphysiol.,* **1983,** 110, 267-279.
12. J. M. Harkin & J. R. Obst, Lignification in trees: indication of exclusive peroxidase participation., *Science*, **1973,** 180, 296-297.
13. P. F. Dowd, L. M. Lagrimini & D. A. Herms, Differential leaf resistance to insects of transgenic sweetgum (*Liquidambar styraciflua*) expressing tobacco anionic peroxidase, *Cell. Mol. Life Sci.*, **1998,** 54, 712-720.

Molecular Breeding of Woody Plants
N. Morohoshi and A. Komamine, editors.

IMMUNOLOCALIZATION OF ENZYMES INVOLVED IN LIGNIFICATION

Keiji Takabe[1*], Miyuki Takeuchi[1], Takahiko Sato[1], Masaki Ito[2], and Minoru Fujita[1]

[1] *Graduate School of Agriculture, Kyoto University, Sakyo, Kyoto 606-8502, JAPAN*

[2] *Yatsushiro Mill, Nippon Paper Industries, Yatsushiro, Kumamoto 866-0823, JAPAN*

ABSTRACT

Localization of the enzymes involved in lignification is essential to better understand the dynamic aspect of lignin biosynthesis in the cell, as well as to better understand the regulation of lignification in woody plants. Lignification has three steps: 1) biosynthesis of the monolignols, 2) transport and secretion of monolignols, and 3) dehydrogenative polymerization of the monolignols. Though most of the enzymes involved in lignification have been identified, their localization in the cell and cell wall remains equivocal. To obtain more detailed information, we investigated the localization of enzymes involved in lignification within the cell and the cell wall in poplar trees using immunocytochemistry.

Immunolabeling of phenylalanine ammonia-lyase, caffeate *O*-methyltransferase, 4-coumalate:CoA ligase, and cinnnamyl alcohol dehydrogenase was localized in the differentiating xylem. These enzymes were particularly abundant during secondary wall formation. Immunolabeling was observed on the polysomes and in the cytosol of the cells during secondary wall formation, indicating that these enzymes are synthesized in the polysomes and released in the cytosol. The synthesis of monolignols might occur in the cytosol.

Immunolabeling of peroxidase was also localized in the differentiating xylem, particularly during secondary wall formation. The labeling, however, was observed in the rough-endoplasmic reticulum, the Golgi apparatus, and the plasma membrane, indicating that peroxidase is synthesized in the r-ER, transported to the Golgi apparatus, and localized at the plasma membrane by fusion of the Golgi vesicles to the membrane.

The most interesting feature of the present results is that the enzymes involved in monolignol synthesis are separated from the enzyme involved in polymerization of monolignols in the cell by the membrane. Monolignols passing through the plasma membrane are polymerized dehydrogenatively in the cell wall in the presence of peroxidase and hydrogen peroxide.

KEYWORDS

Antibody, differentiating xylem, immunocytochemistry, lignification, monolignol, poplar

INTRODUCTION

Lignin biosynthesis is considered to involve three stages: biosynthesis of monolignols in the cell, transport and secretion of monolignols toward the cell wall, and dehydrogenative polymerization of monolignols within the cell wall [1,2]. Feeding experiments using radiolabeled lignin precursors identified most of the enzymes involved in the synthesis of monolignols. Based on these results, Neish and Higuchi reviewed the biosynthetic pathway of monolignols [3,4,5]. Recent advances in molecular genetic technology and refined biochemistry in higher plants have been instrumental in demonstrating an alternative biosynthetic pathway of monolignols, such as methylation at the cinnamoyl-CoA level by caffeoyl-coenzyme A *O*-methyltransferase (CCoAOMT) [6,7,8], and have suggested the conversion of guaiacyl to syringyl moieties in the cinnamyl alcohol pathway [9]. Higuchi and Umezawa reviewed the metabolic grid in the biosynthesis of monolignols [10,11].

Although there are many papers related to the identification and characterization of the enzymes involved in the biosynthesis of monolignols, localization of the enzymes in the cells remains uncertain. Osakabe et al. observed the immunolocalization of phenylalanine ammonia-lyase (PAL) in poplar and demonstrated that PAL labeling is localized in the xylem and the phloem fibers [12]. Samaj et al. demonstrated that cinnamyl alcohol dehydrogenase 2 (CAD2) is localized in the cambium and in fully or partially lignified cells [13]. Electron microscopic observation indicated that CAD2 is localized in the endoplasmic reticulum and the Golgi-derived vesicles in cambial, ray, and young xylem cells. Nakashima et al. reported the immunolocalization of PAL and CAD in *Zinnia* tracheary elements, and demonstrated that both enzymes are localized in the cytoplasmic matrix, Golgi-derived vesicles, and secondary wall thickenings [14]. They suggested that there are at least two types of each enzyme. One type is distributed in cytosol, whereas the other is secreted to the secondary wall thickenings by exocytosis of Golgi vesicles. Because the localization of CAD is still controversial, additional investigation is necessary to determine its subcellular localization. Recently, Takeuchi et al. investigated the localization of caffeate *O*-methyltransferase (COMT) in poplar and reported that immunolabeling was observed in the cytosol during secondary wall formation [15]. Although many enzymes are involved in the biosynthesis of monolignols, the localization in the cells of only a few enzymes has been investigated. Further studies are required to determine the localization of whole enzymes in the cell to understand the precise metabolic pathway of monolignols in the cell. Therefore, we carefully prepared antibodies raised against the enzymes involved in monolignol synthesis and observed the immunolocalization of the enzymes in poplar.

The second stage of lignin biosynthesis also remains uncertain. Autoradiographic investigations revealed that radioactivity derived from tritiated lignin precursors were located in the Golgi apparatus and lignifying cell walls [16,17,18,19]. These findings suggest that the Golgi apparatus is involved in transportation of monolignols to the cell wall. Most of the radioactive lignin precursors incorporated into the cell, however, might be extracted during the exchanges of organic solvents for embedding the specimens into resin. Therefore, further study is necessary to reveal the mechanisms for transportation of monolignols from the cytoplasm to the developing cell walls.

The final stage of lignin biosynthesis is dehydrogenative polymerization of monolignols. Peroxidases (POX) are thought to be involved in this reaction. Cytochemical investigations using diaminobenzidine as a substrate indicated that POXs are localized in the cell walls [20, 21]. Diaminobenzidine reacts with all POX isoforms. Therefore, an improved method is required to detect the POX specifically involved in dehydrogenative polymerization of the monolignols. In the last 15 years, many studies have focused on the identification and characterization of POX involved in lignification [22, 23, 24, 25, 26, 27]. The anionic POX, prxA3a and prxA4a, were isolated from poplar, and prxA3a was shown to be expressed in the younger part of the stem including the shoot tip, though prxA4a was expressed only in the shoot tip. The expression of prxA3a corresponded to lignification [28]. Therefore, we produced an antibody against prxA3a to reveal its localization during cell wall formation in poplar.

MATERIALS & METHODS

Plant material

Differentiating xylem of poplar (*Populus kitakamiensis*) was cut into small pieces with a razor blade and plunged immediately into liquid propane cooled with liquid nitrogen. The specimens were then immersed into 0.5% glutaraldehyde in acetone cooled at –80 °C for 2 days. Thereafter they were washed with acetone three times, placed in ethanol, and embedded in LR white resin (London Resin Co. Ltd., Basingatoke, UK).

Preparation of antibodies against the enzymes involved in lignification

Whole amino acid sequences of PAL, 4-coumalate:CoA ligase (4CL), CAD, and POX were provided by Dr. Katayama, Tokyo University of Agriculture and Technology, Tokyo, Japan. Three amino acid sequences were selected for epitopes from each whole sequence using the Epitope Advisor software (FQS, Fukuoka, Japan). These epitopes were each composed of 10 amino acids (Table 1). The peptide of each epitope was synthesized and then conjugated with keyhole lympet hemocyanin.

Table 1 Epitopes selected from whole amino acid sequence of PAL, COMT, 4CL, CAD and PRX

Enzyme	Amino acid sequence for epitope		
PAL	TSHRRTKQGG	DPLQKPKQDR	GEKVKSPGEE
COMT	Recombinant COMT protein		
4CL	NLSKHSSKPC	ATSRTIDKEG	KAPSGKILRKNLKEKLPGI
CAD	VYTDGKPTQG	VISSSDKKRE	
POX	NTDTIESEKE	SGSGNPDPTL	NCRRVNDNST

The antibodies were raised in mice (8 weeks of age, BALB/c, female). Preimmune serum was taken from each mouse before the first antigen injection. Each peptide- keyhole limpet hemocyanin was dissolved in phosphate-buffered saline (PBS), mixed with the same volume of Freund's complete adjuvant, and intraperitoneally injected into the mice. This injection was repeated several times at 2 weeks intervals in the same way except that Freund's incomplete adjuvant was used. Three days after the final injection, antiserum was taken from the immunized mice. The antiserum was purified on an affinity column of the corresponding peptide-sepharose 4B.

Escherichia coli (E. coli) cells transformed to express COMT-histidine fusion protein were provided by Dr. Katayama, Tokyo University of Agriculture and Technology, Tokyo, Japan. COMT-histidine fusion protein was collected and then the antibody against COMT was obtained according to the method of Takeuchi et al. [15].

Immunoassay

The specificity of the purified antibodies was examined using Western blot analysis. Enzyme extracts from poplar differentiating xylem were prepared according to Takeuchi et al. [15]. The enzyme extracts were separated using sodium dodecyl sulfate-polyacrylamide gel electrophoresis (SDS-PAGE) or native PAGE performed with 12% polyacrylamide. Separated proteins were transferred to polyvinylidine difluoride membranes, which were then incubated with the purified antibody, followed by incubation with alkaline phosphatase-conjugated goat antiserum to mouse IgG (Cappel). Thereafter the membrane was treated with p-nitroblue tetrazolium and 5-bromo-4-chloro-3-indolylphosphate for detection of antiserum.

Immunocytochemistry

Sections (1 μm) were cut from the differentiating xylem of poplar embedded in LR white resin, and mounted on a glass slide. The sections were immunolabeled according to Takeuchi et al. [15]. Fluorescein-labeled secondary antibody was purchased from Amersham, diluted 1:500, and used for detection of the primary antibody. The sections were observed under a confocal laser scanning microscope (Fluoview FV300, Olympus, Japan).

Ultra-thin sections were cut from the poplar differentiating xylem, mounted on nickel grids and immunolabeled with each purified antibody according to Takeuchi et al. [15]. The sections were stained with 2% uranyl acetate for 10 min , then with Reynolds' lead citrate for 1 min , and observed under a transmission electron microscope (JEOL 2000EX, JEOL, Japan) at 100 kV.

RESULTS & DISCUSSION

Characterization of antibodies against the enzymes involved in lignification

Western blot analysis revealed that the antibody raised against the peptides synthesized according to the amino acid sequence of PAL recognized a 75-kDa protein. PAL has a subunit molecular weight of approximately 77000 [27]. This antibody inhibited PAL activity in the extracts from the differentiating xylem of poplar. These results suggested that the antibody raised against the peptides synthesized according to the amino acid sequence of PAL recognized only PAL subunits in poplar.

E. coli cells transformed to express COMT-histidine fusion protein produced a 14-kDa protein composed of a part of COMT and histidine. The antibody raised against a part of COMT detected the 14-kDa protein among the whole proteins extracted from *E. coli* using Western blot analysis. It was also immunoreactive to a 40-kDa protein among the whole proteins extracted from the differentiating xylem of poplar. COMT has a molecular weight around 40 kDa [29,30,31,32]. Therefore, the antibody raised against a part of COMT might recognize the COMT protein in the poplar extracts.

The antibody raised against the peptides according to the amino acid sequence of CAD recognized 40-kDa and 43-kDa proteins. Goffner et al. reported two distinct isoforms of CAD, CAD 1 and CAD 2, from *Eucalyptus gunnii* Hook. CAD2 has a high affinity to coniferyl, sinapyl and *p*-coumaryl alcohols, and corresponding aldehydes. CAD 2 is a dimer composed of two subunits of 42 to 43 kDa proteins [33]. Hibino et al. also demonstrated that CAD of *Aralia cordata* has two closely spaced bands of approximately 39000 *Mr.* using SDS-PAGE, and indicated that CAD is composed of two heterogeneous subunits of very similar ***Mr*** [34]. Therefore, the antibody against CAD might have detected the two subunits of CAD.

The POX antibody recognized a 50-kDa protein on nitrocellulose blots following SDS-PAGE. The antibody also recognized the first migrating protein on nitrocellulose blots following native PAGE. The first migrating position had POX activity, demonstrated using a peroxidase staining kit. Furthermore, the antibody inhibited POX activity in extracts from the differentiating xylem of poplar. Therefore, the antibody against POX specifically recognized a first migrating POX among the isoforms of the enzyme.

Immunolocalization of PAL, COMT, 4CL, CAD, and POX

Under light microscopic observation, there was PAL immunolabeling in the differentiating xylem of poplar, particularly in the ray cells. Under electron microscopic observation, PAL immunolabeling was observed in the cytosol of ray cells in the differentiating xylem. It was also detected in the cytosol of fiber and vessel cells during secondary wall formation (Figure 1).

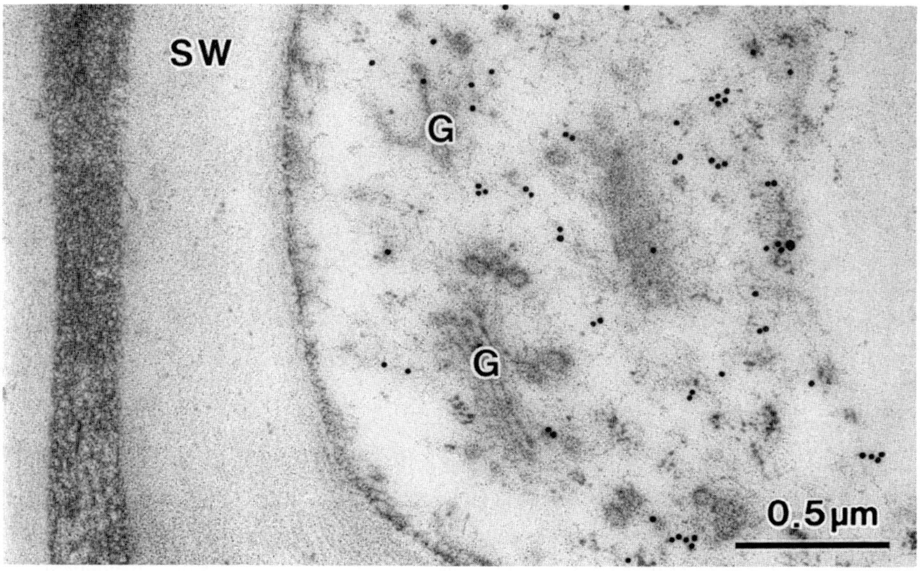

Figure 1. PAL immunolabeling of differentiating fiber. The labeling is mainly distributed in the cytosol. SW: secondary wall, G: Golgi apparatus.

COMT immunolabeling was also observed in the differentiating xylem, particularly during secondary wall formation. Under electron microscopic observation, COMT labeling was not observed in the Golgi apparatus, rough endoplasmic reticulum, or developing cell walls, but located in the cytosol and polysomes. Immunolabeling of 4CL was similar to that of COMT.

CAD immunolabeling was observed in the differentiating xylem of poplar, both during primary and secondary wall formation. Under electron microscopic observation, the labeling was observed in the cytosol and polysomes. It was rarely observed in the Golgi apparatus, rough endoplasmic reticulum, or developing cell wall.

Under light microscopic observation, POX immunolabeling was localized in the differentiating xylem, particularly during secondary wall formation. The labeling was restricted to the boundary between the cell and the newly formed cell wall. Under electron microscopic observation, the labeling was clearly observed on the plasma membrane and between the plasma membrane and newly formed cell wall (Figure 2). Rough endoplasmic reticulum and the Golgi apparatus were weakly labeled. Hydropathy plot analysis of POX strongly suggested a transmembrane helix near the N-terminus, which was inside of the membrane.

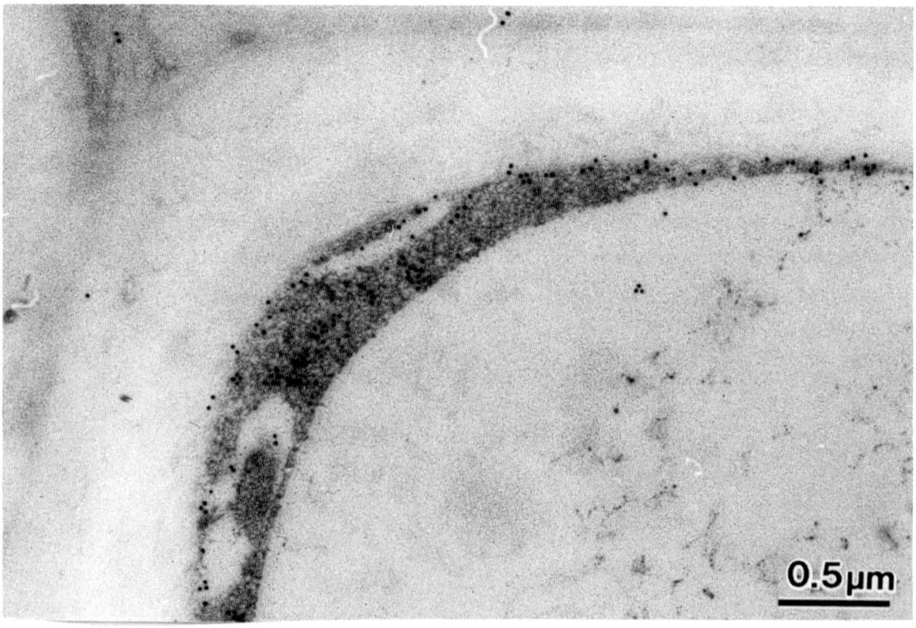

Figure 2. POX immunolabeling of differentiating fiber. Labeling is concentrated on the plasma membrane.

Lignification of the cell wall is considered to start at the cell corner middle lamella when the S_1 layer is deposited, and to proceed toward the compound middle lamella [35,36,37,38,39,40]. Lignification of the secondary wall starts at the outer portion of the S_1 layer and proceeds toward the lumen, lagging behind the thickening of the cell wall. Most of the enzymes involved in lignification localize in the cells during secondary wall formation, indicating that the enzymes localize in the lignifying cells.

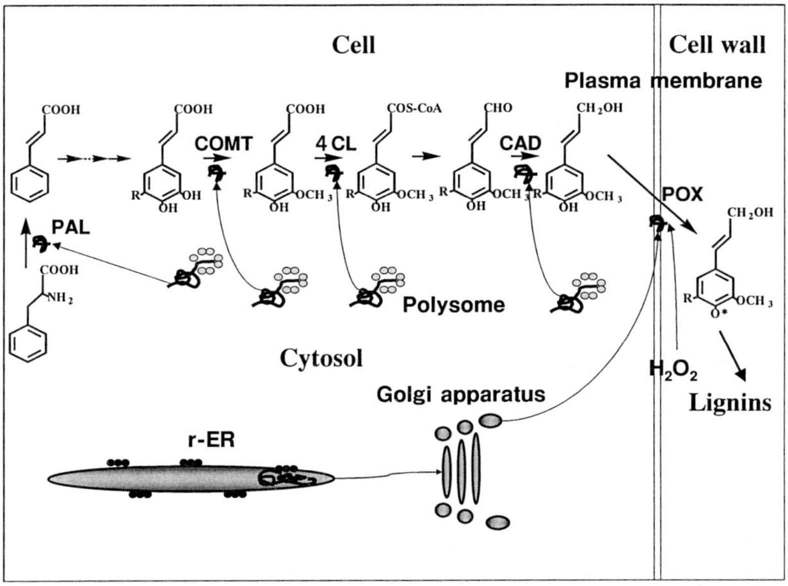

Figure 3. Tentative model of lignin biosynthesis.

Electron microscopic observation indicated that the enzymes involved in the biosynthesis of monolignols are synthesized at the polysomes and released into the cytosol during secondary wall formation. This suggests that the synthesis of monolignols occurs in the cytosol. Metabolic flux of monolignol synthesis might be regulated not only by the presence of enzymes in the cell, but also by the substrate specificity of each enzyme to the metabolic intermediates from phenylalanine to monolignols.

In contrast to the enzymes involved in synthesis of monolignols, POX is synthesized in the rough endoplasmic reticulum and transported to the Golgi apparatus where glycosylation of POX might occur. Thereafter, the processed POX is transported to the plasma membrane by fusion of the Golgi-vesicles to the plasma membrane. POX immunolabeling is usually observed below the membrane of the rough endoplasmic reticulum, indicating that the synthesis and processing of POX might localize to the rough endoplasmic reticulum membranes and the Golgi apparatus. Therefore, POX is separated from the monolignols synthesized in the cytosol by the membrane of the rough endoplasmic reticulum and the Golgi apparatus in the cell. When POX is transported to the plasma membrane, it locates at the outer surface of the plasma membrane. This is supported by hydropathy plot analysis of POX. Monolignols might pass through the plasma membrane via mechanisms that are still unknown, and localize at the boundary between the plasma membrane and newly formed cell wall. Then, POX exposed to the newly formed cell wall might oxidize the monolignols in the presence of hydrogen peroxide to form monolignol radicals. The radicals are then transported to the outer portion of cell wall and polymerized. Figure 3 shows a schematic representation of lignin biosynthesis based on our results.

CONCLUSIONS

Lignification is controlled by the cell. The enzymes involved in the biosynthesis of monolignols are synthesized at polysomes and released into the cytosol during secondary wall formation, indicating that synthesis of monolignols occurs in the cytosol. In contrast, the enzyme involved in dehydrogenative polymerization of monolignols, POX, is synthesized in the rough endoplasmic reticulum, transported to the Golgi apparatus, and finally localized in the plasma membrane. Monolignols are transported to the boundary between the plasma membrane and newly formed cell wall, oxidized in the presence of POX and hydrogen peroxide, and then polymerized within the cell wall.

ACKNOWLEDGEMENTS

This work was supported in part by the program Research for the Future, Japanese Society for the Promotion of Science (JSPS-RFTF96100604). The authors wish to thank Professor Yoshihiro Katayama, Tokyo University of Agriculture and Technology, Tokyo, Japan, for his gift of *E. coli* cells, and for his information regarding the amino acid sequences of PAL, 4CL, CAD, and POX.

REFERENCES

1. A. M. Boudet, C. lapierre & J. Grima-Pettenati, Biochemistry and molecular biology of lignification, New Phytol. **1995**, 129, 203-236.
2. R. Whetten & R. Sederoff, Lignin biosynthesis, Plant Cell, **1995**, 7, 1001-1013.
3. A. C. Neish, Monomeric intermediates in the biosynthesis of lignin, in *Constitution and Biosynthesis of Lignin*, K. Freudenberg & A. C. Neish (eds), Springer-Verlag, New York, **1968**, pp3-43.
4. T. Higuchi, Biosynthesis of lignin, in: *Biosynthesis and Biodegradation of Wood Components*, T. Higuchi (ed), Academic Press, New York, **1985**, pp141-160.
5. T. Higuchi, Lignin biochemistry: Biosynthesis and biodegradation, Wood Sci. Technol., **1990**, 24, 23-63.
6. Z.-H. Ye, R. E. Kneusel, U. Matern, & J. E. Varner, An alternative methylation pathway in lignin biosynthesis in *Zinnia*, The Plant Cell, **1994**, 6, 1427-1439.
7. Z.-H. Ye & J. E. Varner, Differential expression of two *O*-methyltransferases in lignin biosynthesis in *Zinnia elegans*, Plant Physiol., **1995**, 108, 459-467.
8. Z.-H. Ye, Association of caffeoyl coenzyme A 3-*O*-methyltransferase expression with lignifying tissues in several dicot plants, Plant Physiol., **1997**, 115, 1341-1350.
9. N. Matsui, F. Chen, S. Yasuda & K. Fukushima, Conversion of guaiacyl to syringyl moieties on the cinnamyl alcohol pathway during the biosynthesis of lignin in angiosperms, Planta, **2000**, 210, 831-835.
10. T. Higuchi, Pathways for monolignol biosynthesis via metabolic grids: Coniferyl aldehyde 5-hydroxylase, a key enzyme in syringyl lignin biosynthesis, Proceedings of the 44th Lignin Symposium, **1999**, pp177-182.

11. T. Umezawa, Biochemistry and molecular biology of cinnamate pathway, Mokuzai Kenkyu, 2000, 36, 17-33.
12. Y. Osakabe, K. Nanto, H. Kitamura, S. Kawai, Y. Kondo, T. Fujii, K. Takabe, Y. Katayama & N. Morohoshi, Immunocytochemical localization of phenylalanine ammonia-lyase in tissues of *Populus kitakamiensis*, Planta, **1996**, 200, 13-19.
13. J. Samaj, S. Hawkins, V. Lauvergeat, J. Grima-Pettenati & A. Boudet, Immunolocalization of cinnnamyl alcohol dehydrogenase 2 (CAD 2) indicates a good correlation with cell-specific activity of CAD 2 promoter in transgenic poplar shoots, **1998**, 204, 437-443.
14. J. Nakashima, T. Awano, K. Takabe & M. Fujita, Immunocytochemical localization of phenylalanine ammonia-lyase and cinnamyl alcohol dehydrogenase in differentiating tracheary elements derived from *Zinnia* mesophyll cells, Plant Cell Physiol., **1997**, 38, 113-123.
15. M. Takeuchi, K. Takabe & M. Fujita, Immunolacalization of *O*-methyltransferase and peroxidase in differentiating xylem of poplar, Holzforschung, **2001**, in press.
16. J. D. Pickett-Heaps, Xylem wall deposition Radioautographic investigations using lignin precursors, Protoplasma, **1968**, 65, 181-205.
17. F. B. P. Wooding, Radioautographic and chemical studies of incorporation into sycamore vascular tissue walls, J. Cell Sci., **1968**, 3, 71-80.
18. M. Fujita & H. Harada, Autoradiographic investigations of cell wall development. II Tritiated phenylalanine and feruric acid assimilation in relation to lignification, Mokuzai Gakkaishi, **1979**, 25, 89-94.
19. K. Takabe, M. Fujita, H. Harada & H. Saiki, Autoradiographic investigation of lignification in the cell walls of cryptomeria, Mokuzai Gakkaishi, **1985**, 31, 613-619.
20. P. K. Hepler, R. M. Rice & W. A. Terranova, Cytochemical localization of peroxidase activity in wound vessel members of *Coleus*, Can. J. Bot., **1972**, 50, 977-983.
21. J. L. Hall & R. Sexton, Cytochemical localization of peroxidase activity in root cells, Planta, **1972**, 108, 103-120.
22. A. Imberty, R. Goldberg & A.-M. Catesson, Isolation and characterization of *Populus* isoperoxidases involved in the last step of lignin formation, Planta, **1985**, 164, 221-226.
23. L. M. Langrimini, W. Burkhart, M. Moyer & S. Rothstein, Molecular cloning of complementary DNA encoding the lignin-forming peroxidase from tobacco: Molecular analysis and tissue-specific expression, Proc. Natl. Acad. Sci. USA, **1987**, 84, 7542-7546.
24. K. Osakabe, H. Koyama, S. Kawai, Y. Katayama & N. Morohoshi, Molecular cloning and the nucleotide sequences of two novel cDNAs that encode anionic peroxidases of Populus kitakamiensis, Plant Science, **1994**, 103, 167-175.
25. S. Kawai, Y. Matsumoto, S. Kajita, K. Yamada, Y. Katayama & N. Morohoshi, Nucleotide sequence for the genomic DNA encoding an anionic peroxidase gene from a hybrid poplar, *Populus kitakamiensis*, Biosci. Biotech. Biochem., **1993**, 57, 131-133.
26. Y. Sato, M. Sugiyama, A. Komamine & H. Fukuda, Separation and characterization of the isoenzymes of wall-bound peroxidase from cultured Zinnia cells during tracheary element differentiation, Planta, **1995**, 196, 141-147.

27. J. H. Christensen, G. Bauw, K. G. Welinder, M. V. Montagu & W. Boerjan, Purification and characterization of peroxidases correlated with lignification in poplar xylem, Plant Physiol., 1998, 118, 125-135.
28. K. Osakabe, H. Koyama, S. Kawai, Y. Katayama & N. Morohoshi, Molecular cloning of two tandemly arranged peroxidase genes from *Populus kitakamiensis* and their differential regulation in the stem, Plant Molecular Biology, **1995,** 28, 677-689.
29. R. C. Bugos, V. L. Chiang & W. H. Champbell, cDNA cloning, sequence analysis and seasonal expression of lignin bispecific caffeic acid/5-hydroxyferuric acid *O*-methytransferase of aspen, Plant Mol. Biol., **1991**, 17, 1203-1215.
30. R. C. Bugos, V. L. Chiang & W. H. Champbell, Characterization of bispecific caffeic acid/5-hydroxyferuric acid *O*-methytransferase of aspen, Phytochem., **1992**, 31, 1495-1498.
31. J. V. Doorsselaere, B. Dumas, M. Baucher, B. Fritig, M. Legrand, M. V. Montagu & D. Inze, One step purification and characterization of a lignin-specific *O*-methyltransferase from poplar, Gene, **1993**, 133, 213-217.
32. T. Hayakawa, K. nanto, S. Kawai, Y. katayama & N. Morohoshi, Molecular cloning and tissue specific expression of two genes that encode caffeic acid *O*-methyltransferase from *Populus kitakamiensis*, Plant Sci., **1996**, 113, 157-165.
33. D. Goffner, I. Joffroy, J. Grima-Pettenati, C. Halpin, M. E. Knight, W. Schuch & A. M. Boudet, Purification and characterization of isoforms of cinnamyl alcohol dehydrogenase from *Eucalyptus* xylem, Planta, **1992**, 188, 48-53.
34. T. Hibino, D. Shibata, T. Umezawa & T. Higuchi, Purification and partial sequences of *Aralia cordata* cinnamyl alcohol dehydrogenase, *Phytochemistry*, **1993**, 32, 565-567.
35. A. B. Wardrop, The phase of lignification in the differentiation of wood fibers, TAPPI, **1957**, 40, 225-243.
36. N. P. Kutscha & J. M. Schwarzmann, The lignification sequence in normal wood of balsam fir (*Abies balsamea*), Holzforschung, **1975**, 29, 79-84.
37. H. Imagawa, k. Fukazawa & S. Ishida, Study on the lignification in tracheids of Japanese larch, *Larix leptolepis* Gord., Research Bulletins of the College Experimental Forests, Hokkaido University, Japan, **1975**, 33, 127-138.
38. K. Takabe, M. Fujita, H. Harada & H. Saiki, Lignification process of Japanese black pine (*Pinus Thunbergii* Parl.) tracheids, Mokuzai Gakkaishi, **1981**, 27, 813-820.
39. S. Saka & R. J. Thomas, A study of lignification in loblolly pine tracheid by the SEM-EDXA technique, Wood Sci. Technol., **1982**, 16, 167-179.
40. K. Takabe, M. Fujita, H. Harada & H. Saiki, Lignification process in cryptomeria (*Cryptomeria japonica* D. Don) trachied: Electron microscopic obserbation of lignin skeleton of differentiating xylem, Research Bulletins of the College Experimental Forests, Hokkaido University, Japan, **1986**, 43, 783-788.

Molecular Breeding of Woody Plants
N. Morohoshi and A. Komamine, editors.

LIGNIN BIOSYNTHESIS IN POPLAR: GENETIC ENGINEERING AND EFFECTS ON KRAFT PULPING

Wout Boerjan[1]*, Hugo Meyermans[1], Cuiying Chen[1], Marie Baucher[1], Jan Van Doorsselaere[1], Kris Morreel[1], Eric Messens[1], Catherine Lapierre[2], Brigitte Pollet[2], Lise Jouanin[3], Jean-Charles Leplé[4], John Ralph[5], Jane Marita[5], Emma Guiney[6], Wolfgang Schuch[7], Michel Petit-Conil[8] & Gilles Pilate[4]

[1] *Vakgroep Moleculaire Genetica & Departement Plantengenetica, Vlaams Interuniversitair Instituut voor Biotechnologie (VIB), Universiteit Gent, K.L. Ledeganckstraat 35, B-9000 Gent, Belgium*

[2] *Laboratoire de Chimie Biologique, Institut National Agronomique, Paris-Grignon, F-78850 Thiverval-Grignon, France*

[3] *Laboratoire de Biologie Cellulaire, Institut National de la Recherche Agronomique, F-78026 Versailles Cedex, France*

[4] *Unité d'Amélioration, Génétique et Physiologie Forestières, Institut National de la Recherche Agronomique , F-45166 Olivet, France*

[5] *U.S. Dairy Forage Research Center, Agricultural Research Service, U.S. Department of Agriculture, Madison, WI 53706, USA*

[6] *Crop Genetics Research, Syngenta, Jealott's Hill International Research Centre, Bracknell, Berkshire, RG42 6EY, United Kingdom*

[7] *Syngenta Wheat Improvement Centre, Norwich Research Park, Colney, Norwich NR4 7UH, United Kingdom*

[8] *Centre Technique du Papier, F-38004 Grenoble Cedex 09, France*

ABSTRACT

Transgenic poplar, downregulated in cinnamyl alcohol dehydrogenase (CAD) or caffeic acid-*O*-methyltransferase (COMT) expression have been grown in field trials. Wood of these trees has been evaluated for Kraft pulping. The results show that lignin is more easily extracted from wood of the CAD-downregulated trees, whereas wood from COMT-downregulated trees is less suitable for Kraft pulping. Detailed NMR analyses of lignin from COMT-downregulated poplars reveal the presence of benzodioxane structures, which are derived from coupling of 5-hydroxyconiferyl alcohol with the lignin polymer, showing that monolignols other than *p*-coumaryl, coniferyl and sinapyl alcohol can be incorporated into lignin. Analysis of poplar downregulated for CCoAOMT shows that CCoAOMT is involved in the synthesis of both syringyl and guaiacyl units, and that sinapic is probably not an important precursor for syringyl lignin synthesis.

KEYWORDS

CAD, CCoAOMT, COMT, Kraft pulping, lignin, poplar

INTRODUCTION

Lignin is a complex and heterogeneous polymer that is predominantly present in the walls of secondary thickened cells [1]. It is mainly derived from three cinnamyl alcohol units, the monolignols *p*-coumaryl alcohol, coniferyl alcohol and sinapyl alcohol (Fig. 1). When incorporated into lignin, these form the *p*-hydroxyphenyl (H), guaiacyl (G) and syringyl (S) units, respectively. The biosynthesis pathway of the monolignols and their polymerization are not completely understood, because of the fragmented knowledge of the kinetic properties of the different enzymes of the monolignol biosynthesis pathway, their cell-specific and conditional expression patterns, the concentration of pathway intermediates, the regulatory mechanisms that act at the enzyme activity and gene expression levels, and possible metabolic channelling.

For the production of high-quality papers, lignin needs to be chemically removed from cellulose. Because this process is energy consuming and polluting, it may be beneficial to modify lignin content or its composition in trees by genetic engineering to facilitate its extraction from the pulp. One way to proceed towards this goal is to inhibit the production of enzymes that play important roles in the biosynthesis pathway of the monolignols in transgenic plants and to analyze wood of these plants for lignin amount and composition and for its suitability for chemical pulping.

Here we present data on Kraft pulping of wood from transgenic field-grown poplars downregulated for cinnamyl alcohol dehydrogenase (CAD) or caffeic acid-*O*-methyltransferase (COMT). In addition, NMR data on the structure of lignin from COMT-downregulated poplars are presented, as well as a summary of the results obtained by analyzing the role of caffeoyl-coenzyme A-*O*-methyltransferase (CCoAOMT) in lignin biosynthesis.

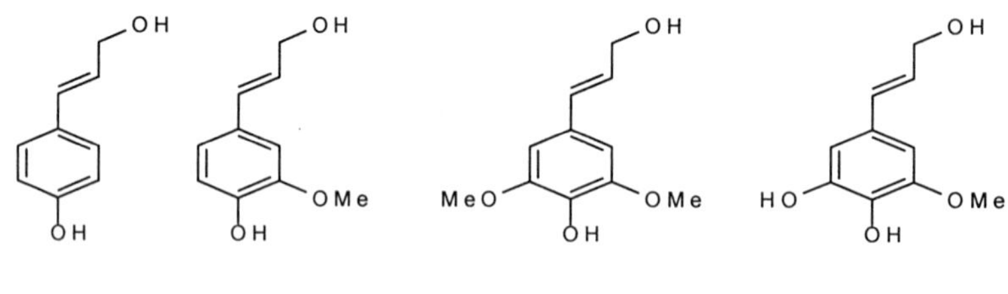

Figure 1. The three main monolignols *p*-coumaryl alcohol (left), coniferyl alcohol (second) and sinapyl alcohol (third) and the unusual 5-hydroxyconiferyl alcohol (right), which is incorporated in lignin of COMT-downregulated plants.

RESULTS

Kraft pulping of wood from transgenic plants with modified lignin synthesis

Previously, we have generated transgenic poplars downregulated for CAD and COMT [2-4]. Transgenic poplars with a reduced CAD activity had a slightly reduced in lignin content, but no major differences in S/G or S+G were detected by thioacidolysis. The lignin contained more free phenolic groups on S and G units, which may contribute to the solubility of lignin in the pulping liquor [4]. The lignin of the wood of these greenhouse-grown transgenic plants appeared to be more easily extracted by simulated Kraft pulping experiments [2, 4]. Lignin of transgenic poplars downregulated for COMT was more condensed (lower S+G), had a reduced S/G ratio and had fewer free phenolic groups on G units. A novel 5-hydroxyguaiacyl unit (5-OHG), derived from 5-hydroxyconiferyl alcohol (Fig. 1), was released by thioacidolysis [3, 4]. Wood from these greenhouse-grown COMT-downregulated poplars was less suitable for Kraft pulping [4].

The Kraft pulping analyses of these plants had been performed on greenhouse-grown plant material. Field trials with two COMT-downregulated lines, two CAD-downregulated lines and the wild-type were subsequently generated to evaluate whether (i) the beneficial effect on Kraft pulping had remained stable after 4-years of growth in the field and (ii) whether the genetic alteration had any effect on the overall growth and development of the transgenic plants or on pest and disease resistance. Fig. 2 shows results of Kraft pulping experiments on wood of the transgenic poplars. When the chemical charge (% active alkali) of the pulp liquor increases, the kappa number, which reflects the residual lignin present in the pulp, decreases. For the transgenic poplars ASB2B and ASB10B [3], having a reduced COMT activity, the kappa numbers were higher for a given chemical charge, indicating that more chemicals were needed to extract a given amount of lignin from the pulp. For the transgenic poplars with reduced CAD activity (ASCADT21 and ASCADT52), a lower kappa number was achieved for a given chemical charge, indicating that fewer chemicals were needed to remove a given amount of lignin from the pulp [5].

Fig. 3 shows that the degree of polymerization (DP) of the cellulose fibres (cellulose DP) remained the same for all transgenic lines for a given chemical charge. The selectivity curve (Fig. 4) indicates that when the wood is pulped to a given kappa number, the CAD-downregulated poplars have a slightly higher cellulose DP than the wild-type controls, whereas the antisense COMT poplars have a lower cellulose DP.

Fig. 5 shows that the screened yield of the antisense CAD poplars remained high at low chemical charges, whereas for the COMT transgenic poplars and the wild-type control, the screened yield decreases with lower chemical charges, indicating that less uncooked particles are present in the pulp of the antisense CAD poplars at low chemical charges.

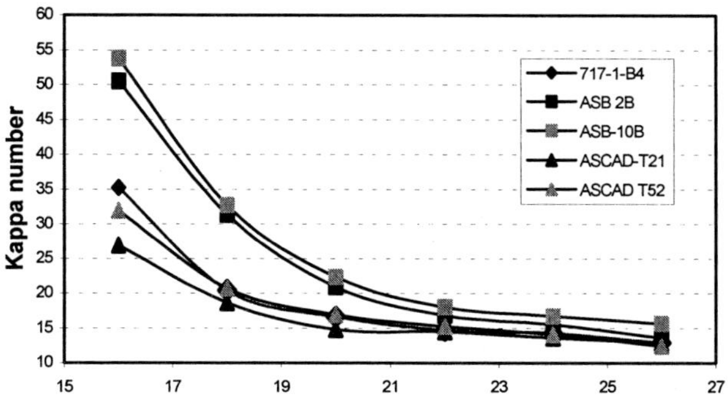

Figure 2. Evolution of kappa number with active alkali charge for Kraft pulping of field-grown poplars [5].

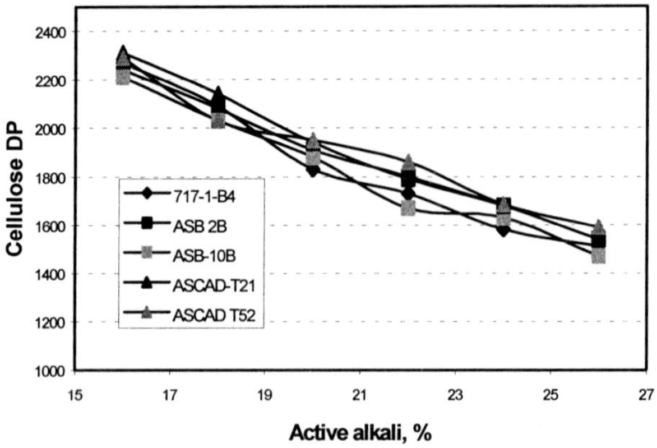

Figure 3. Evolution of cellulose degree of polymerization with active alkali charge for Kraft pulping of field-grown poplars [5].

No differences were observed in growth of the transgenic plants over the 4-year period in the field, nor were any negative effects on disease or pest resistance observed (data not shown). In conclusion, transgenic poplars downregulated for CAD are beneficial for Kraft pulping, whereas wood of COMT-downregulated lines is more difficult to delignify.

5-hydroxyconiferyl alcohol is incorporated in lignin of transgenic poplars with reduced COMT activity

We have previously shown that upon thioacidolysis of lignin isolated from antisense COMT poplars, a novel unit, 5-OHG, is released in addition to the regular G and S units. Fig. 6 shows that NMR of COMT-downregulated poplar lignin reveals the presence of benzodioxane structures that are not present in lignin from wild-type control poplars. The presence of benzodioxane structures strongly suggests that 5-hydroxyconiferyl alcohol incorporates as a monomer into the lignin, demonstrating the metabolic plasticity of plants to allow polymerization of products of incomplete synthesis along with the traditional monolignols [6].

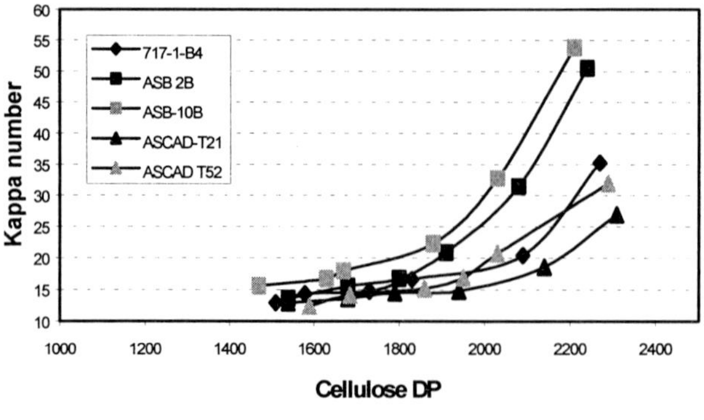

Figure 4. Selectivity curve: kappa number versus cellulose DP for Kraft pulping of field-grown poplars [5].

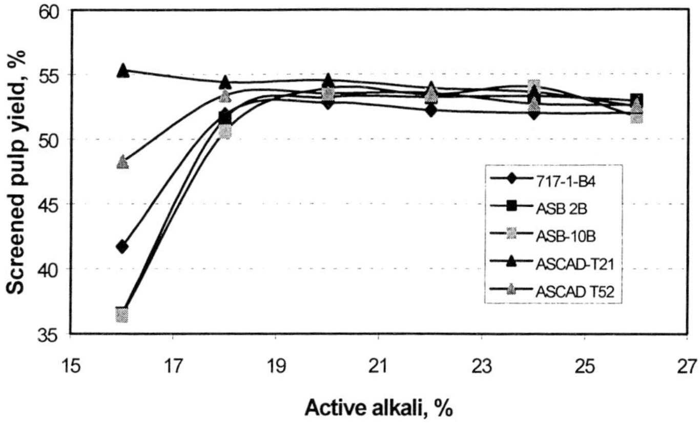

Figure 5. Evolution of screened yield with kappa number for Kraft pulping of field-grown poplars [5].

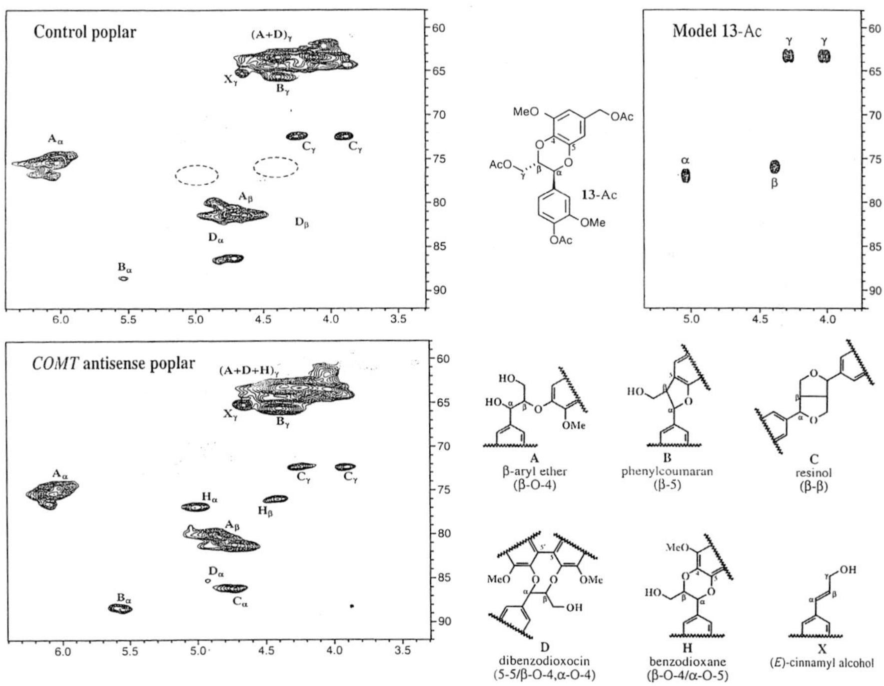

Figure 6. Partial spectra from gradient HMQC NMR experiments showing new peaks for benzodioxane units H in acetylated lignin from COMT-downregulated poplar. Similar peaks were revealed with a benzodioxane model 13-Ac [6].

Tissue-specific and conditional expression of *CCoAOMT* in poplar

It has long been thought that COMT is the only methylating enzyme in monolignol biosynthesis and that it methylates caffeic acid and 5-hydroxyferulic acid for the synthesis of monolignol precursors. Recently, enzymatic assays have shown that COMT preferentially methylates 5-hydroxyconiferaldehyde, 5-hydroxyconiferyl alcohol or caffeoyl aldehyde [7-9]. Furthermore, it has been shown that another methylating enzyme, CCoAOMT, methylates caffeoyl-CoA and 5-hydroxyferuloyl-CoA to feruloyl-CoA and sinapoyl-CoA *in vitro*, respectively, and therefore may also play an important role in lignin synthesis [10, 11]. To investigate whether CCoAOMT is involved in monolignol biosynthesis, we first analyzed whether its expression coincides with lignification. In a next phase, transgenic poplars were made with a reduced expression of *CCoAOMT*. We have shown that in poplar, CCoAOMT is encoded by two genes [12, 13]. We have fused the two *CCoAOMT* promoters to the ß-glucuronidase-coding sequence and transformed these constructs into poplar [14]. In addition, we have immunolocalized CCoAOMT in poplar stems [14]. *CCoAOMT* promoter activity correlated well with the presence of CCoAOMT protein. Both genes were similarly expressed in xylem and differentially in phloem. In xylem, expression was preferentially observed in vessels and contact rays (Fig. 7). Furthermore, the expression of both genes was induced concomitantly with *de novo* lignin deposition after wounding, fungal infection, and bending (data not shown). Importantly, upon bending and leaning of the stem, the cell-specific expression pattern was lost and both genes were expressed in all cell types of the xylem (Fig. 7). These expression characteristics may explain at least in part the heterogeneity in lignin composition that is observed between cell types and upon different environmental conditions [14].

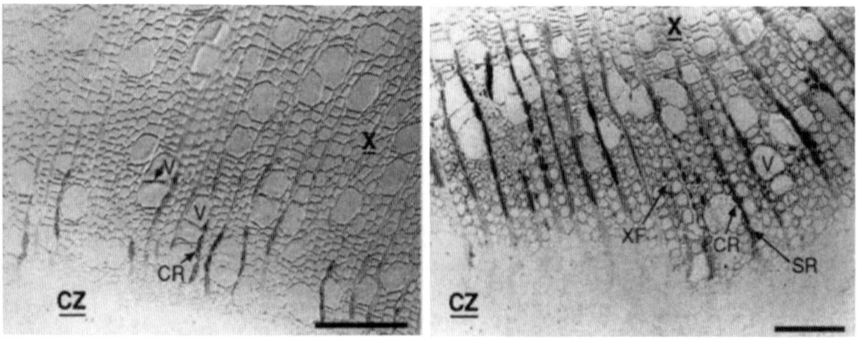

Figure 7. Immunolocalization of CCoAOMT in wild-type poplar wood. Left: non-bent stem; right: bent stem. CR, contact ray; CZ, cambial zone; SR, storage ray; V, vessel; X, xylem; XF, xylem fiber [14].

Downregulation of CCoAOMT in poplar

To clarify the *in vivo* role of CCoAOMT in lignin biosynthesis, transgenic poplars with 10% residual CCoAOMT protein levels in the stem xylem were generated [15]. Upon analysis of the xylem, the affected transgenic lines had a 12% reduction in Klason lignin content, an 11% increase in the S/G ratio in the non-condensed lignin fraction, an increase in lignin-attached *p*-hydroxybenzoic acid, but otherwise a lignin composition similar to that of wild-type. Stem xylem of the CCoAOMT-downregulated lines had a pink-red coloration, which coincided with an enhanced fluorescence of mature vessel cell walls (data not shown). The localization of the fluorescence in the vessel cell wall coincides with the cell-specific expression pattern of CCoAOMT (see above). The reduced production of CCoAOMT caused an accumulation of O^4-β-D-glucopyranosyl-sinapic acid, O^3-β-D-glucopyranosyl-caffeic acid and O^4-β-D-glucopyranosyl-vanillic acid as detected by comparative HPLC profiles of methanol-soluble phenolics extracted from differentiating xylem, followed by structure identification by NMR and mass spectroscopy. The glucosides of caffeic acid and sinapic acid also accumulated when feeding wild-type poplars with caffeic acid and sinapic acid, respectively, suggesting that the accumulation of the glucosides is generated by excess of the free acids. This result shows that endogenously produced sinapic acid is not committed to S unit biosynthesis, supporting the recently obtained *in vitro* enzymatic data that suggest that the path from caffeic acid to sinapic does not play a major role in lignin biosynthesis [7].

ACKNOWLEDGMENTS

This work was carried out in the framework of the European Union Research Programs AGRE-0021-C, FAIR-CT95-0424.

REFERENCES

1. M. Baucher, B. Monties, M. Van Montagu & W. Boerjan, Biosynthesis and genetic engineering of lignin. *Crit. Rev. Plant Sci.*, **1998**, 17, 125-197.
2. M. Baucher, B. Chabbert, G. Pilate, J. Van Doorsselaere, M.-T. Tollier, M. Petit-Conil, D. Cornu, B. Monties, M. Van Montagu, D. Inzé, L. Jouanin & W. Boerjan, Red xylem and higher lignin extractability by down-regulating a cinnamyl alcohol dehydrogenase in poplar (*Populus tremula x P. alba*). *Plant Physiol.*, **1996**, 112, 1479-1490.
3. J. Van Doorsselaere, M. Baucher, E. Chognot, B. Chabbert, M.-T. Tollier, M. Petit-Conil, J.-C. Leplé, G. Pilate, D. Cornu, B. Monties, M. Van Montagu, D. Inzé, W. Boerjan & L. Jouanin, A novel lignin in poplar trees with a reduced caffeic acid/5-hydroxyferulic acid *O*-methyltransferase activity. *Plant J.*, **1995**, 8, 855-864.
4. C. Lapierre, B. Pollet, M. Petit-Conil, G. Toval, J. Romero, G. Pilate, J.-C. Leplé, W. Boerjan, V. Ferret, V. De Nadai & L. Jouanin, Structural alterations of lignins in transgenic poplars with depressed cinnamyl alcohol dehydrogenase or caffeic acid *O*-methyltransferase activity have opposite impact on the efficiency of industrial Kraft pulping. *Plant Physiol.*, **1999**, 119, 153-163.
5. M. Petit-Conil, C. Chirat, C. Lapierre, B. Pollet, I. Mila, L. Jouanin, E. Guiney, W. Schuch, W. Boerjan, J. C. Leplé & G. Pilate, Designing lignin structure for modifying pulping performances: application to poplar tree. In *Proceedings of the TAPPI Pulping Conference*, Technical Association for Paper and Pulp Industry (ed.). Atlanta, TAPPI Press, **2000**, in press.

6. J. Ralph, C. Lapierre, F. Lu, J. M. Marita, G. Pilate, J. Van Doorsselaere, W. Boerjan & L. Jouanin, NMR evidence for benzodioxane structures resulting from incorporation of 5-hydroxyconiferyl alcohol into lignins of *O*-methyltransferase-deficient plants. *J. Agric. Food Chem.*, **2001**, 49, 86-91.
7. L. Li, J. L. Popko, T. Umezawa & V. L. Chiang, 5-Hydroxyconiferyl aldehyde modulates enzymatic methylation for syringyl monolignol formation, a new view of monolignol biosynthesis in angiosperms. *J. Biol. Chem.*, **2000**, 275, 6537-6545.
8. J. M. Humphreys, M. R. Hemm & C. Chapple, New routes for lignin biosynthesis defined by biochemical characterization of recombinant ferulate 5-hydroxylase, a multifunctional cytochrome P450-dependent monooxygenase. *Proc. Natl. Acad. Sci. USA*, **1999**, 96, 10045-10050.
9. K. Parvathi, F. Chen, D. Guo, J. W. Blount & R. A. Dixon, Substrate preferences of *O*-methyltransferases in alfalfa suggest new pathways for 3-*O*-methylation of monolignols. *Plant J.*, **2001**, 25, 193-202.
10. Z.-H. Ye, R. E. Kneusel, U. Matern & J. E. Varner, An alternative methylation pathway in lignin biosynthesis in *Zinnia*. *Plant Cell*, **1994**, 6, 1427-1439.
11. R. Zhong, W. H. Morrison III, J. Negrel & Z.-H. Ye, Dual methylation pathways in lignin biosynthesis. *Plant Cell*, **1998**, 10, 2033-2046.
12. C. Chen, H. Meyermans, J. Van Doorsselaere, M. Van Montagu & W. Boerjan, A gene encoding caffeoyl coenzyme A 3-*O*-methyltransferase (CCoAOMT) from *Populus trichocarpa* (Accession No. AJ223621) (PGR98-104). *Plant Physiol.*, **1998**, 117, 719.
13. C. Chen, W. Ardiles-Diaz, M. Van Montagu & W. Boerjan, A poplar gene for caffeoyl-coenzyme A 3-*O*-methyltransferase (Accession No. AJ223620) (PGR99-085). *Plant Physiol.*, **1999**, 120, 635.
14. C. Chen, H. Meyermans, B. Burggraeve, R. M. De Rycke, K. Inoue, V. De Vleesschauwer, M. Steenackers, M. C. Van Montagu, G. J. Engler & W. A. Boerjan, Cell-specific and conditional expression of caffeoyl-CoA *O*-methyltransferase in poplar. *Plant Physiol.*, **2000**, 123, 853-867.
15. H. Meyermans, K. Morreel, C. Lapierre, B. Pollet, A. De Bruyn, R. Busson, P. Herdewijn, B. Devreese, J. Van Beeumen, J. M. Marita, J. Ralph, C. Chen, B. Burggraeve, M. Van Montagu, E. Messens & W. Boerjan, Changes in lignin and *in vivo* accumulation of phenolic glucosides in poplar xylem upon downregulation of caffeoyl-CoA-*O*-methyltransferase, an enzyme involved in lignin biosynthesis. *J. Biol. Chem.*, **2000**, 275, 36899-36909.

Molecular Breeding of Woody Plants
N. Morohoshi and A. Komamine, editors.

ANALYSIS OF TRANSGENIC POPLAR IN WHICH THE EXPRESSION OF PEROXIDASE GENE IS SUPPRESSED

Li Yahong, Yukiko Tsuji, Nobuyuki Nishikubo, Shinya Kajita and Noriyuki Morohoshi

Graduate School of Bio-Application and System Engineering. Tokyo University of Agriculture & Technology, 2-24-16, Nakamachi, Koganei-shi, Tokyo 184-8588, Japan

ABSTRACT

Aims of our research are to produce a tree with a lower lignin content by controlling lignin biosynthesis genes using antisense RNA and to clear the relationship between lignification and the expression of peroxidase gene.

The isolation and sequencing of some peroxidase genes from hybrid aspen (*Populus kitakamiensis*) and also the isolation of promoter region of these genes were achieved. From the results of northern hybridization and expression of these genes, it was concluded that the peroxidase gene involved in lignification was *prxA3a*. A new vector, containing the original peroxidase gene (*prxA3a*) promoter and the antisense *prxA3a* gene involved in lignification was constructed.

Transgenic poplars formed had lower total peroxidase activity compared with that of the control. After peroxidase isozyme analysis by the isoelectric focusing, it was clear that the enzyme activity of a peroxidase band (pI 3.8) was suppressed in the transgenic poplars. The lignin content of the transgenic plants decreased 3-26% compared with that of the control, when the potassium permanganate oxidation method was used to determine the lignin content. From the microscopic observation with cross-sections after staining by color reactions, it was cleared that the peroxidase activity and the staining of mäule and phloroglucinol color reactions were disappeared in the xylem region near cambial zone in transgenic poplar samples.

The fact shows that the transformants in which the expression of the peroxidase gene was suppressed by antisense RNA had peroxidase activity decreased and lignin formation suppressed in early stage of secondary xylem formation, resulting that transgenic poplars with a lowered lignin content were formed.

KEYWORDS

Acidic peroxidase, *Agrobacterium tumefaciens* LBA4404, antisense RNA, hybrid poplar (*Populus kitakamiensis*), lignification, permanganate oxidation, Ti-plasmid.

INTRODUCTION

The exhaustion of energy and resources, and aggravation of the environmental situation on the earth are serious problems and will continue in the near future. One way to solve these problems is to achieve an increased biomass yield and to develop useful techniques for biomass utilization, because the biomass must take the place of energy and resources from fossil fuel at present. As trees are a major biomass, it is very important to improve trees with genetic engineering techniques and to establish some useful conversion systems for biomass in near future.

The lignin biosynthetic pathway occupies a huge secondary metabolism system in trees. Inhibition of lignin biosynthesis to form a useful tree with a lower lignin content and being able to use polysaccharide materials effectively was a focus of this research. *Populus kitakamiensis* was used as a plant material, because had fast growth, strong cold resistance, low lignin content, and the ease of tissue culture and of the infection by *Agrobacterum* sp. A target gene for the lignin biosynthetic pathway was the peroxidase gene involved in the last step in lignification [1,2]. As there are many peroxidase isozymes in plants, we must first decide the peroxidase isozyme involved in lignification. In this point Mäder and Füssl [3] and Lagrimini et al. [4] suggested that anionic peroxidases were involved in lignin biosynthesis. Therefore, we first tried to isolate peroxidase genes and analyze the expression of anionic peroxidase isozyme genes in the hybrid aspen.

In order to form an improved plant by genetic engineering generally; we need to establish three technical pre-requisites, which are as follows;

1) To establish the developmental and regenerative techniques from callus initiation to a mature plant; 2) to establish a stable transformation system for foreign gene integration into the plant tissue; and 3) to isolate and analyze target genes and their promoters. We have already been established these three requisites and, therefore, will discuss here whether the peroxidase activity and lignin content in the transgenic poplars can be suppressed by the antisense RNA method or not, where the expression of peroxidase genes is suppressed in the growing tissue of transgenic poplars and what characteristics of lignin chemical structure the transformants have.

MATERIALS & METHODS

Plant materials, regeneration and growth conditions

Hybrid aspen (*Populus kitakamiensis*) plantlets were grown from shoot meristem by use of the method developed by Ebinuma et al. [5]. Eighteen-hr fluorescent light period was used. The callus of the hybrid aspen grew on the Murashige-Skoog (MS) medium containing 2,4-dichlorophenoxyacetic acid (2,4-D: 0.5 mg/l). Formation of adventitious buds from the callus was carried out in MS medium containing benzyladenine (BA: 0.1 mg/l) and zeatin (1.0 mg/l). Rooting of shoots was achieved in MS medium containing 1-naphthaleneacetic acid (NAA: 1.5 mg/l).

Formation of transgenic plants

Peroxidase genes, *prxA1, prxA2a, prxA2b, prxA3a, prxA4a* and *HPOX14* were isolated from the hybrid poplar and *prxA3a* was selected as the peroxidase gene involved in lignin biosynthesis based on Northern hybridization and expression tests of the peroxidase promoter genes in the hybrid poplar [6]. The vector used for the transduction was constructed with the pBI121. Transformation of plant tissue was achieved by use of the binary vector system of Ti-plasmid method [7]. The transformants (7 lines called POXs 27, 29, 43, 44, 47, 49 and 53) were grown to calli, to shoots *in vitro* (about 6 and10cm high), and to young plants *ex vitro* (about 60-100cm high), as the control was.

Chemical and biochemical analyses of transformants

Fresh plant materials of shoots were weighed and frozen with liquid nitrogen, and then ground with a mortar and pestle. Samples were pre-extracted with a Tris buffer solution (100mM Tris-HCl, pH7.0). The content of proteins was determined by use of PROTEIN ASSAY KID (BIO-RAD).

Peroxidase activity was determined with guaiacol (13 mM) as a substrate. Isoelectric focusing patterns were produced on the gel plates stained with guaiacol and H_2O_2 (5 mM).

Extractives of the samples were obtained by extracting the protein-free wood meals with an ethanol-benzene (1: 2) solution.

Lignin content was determined by the acetyl bromide method [8, 9] and a modified permanganate oxidation degradation method, which is performed according to Miksche et al. [12] and Morohoshi et al. [10, 11] in order to give higher yield of degradation products.

Thioacidolysis of the samples was carried out according to Lapierre [13].

For visible and fluorescent microscopic observation, 10μm thick cross-sections were prepared with a sliding microtome following extracting with a benzene-ethanol (2:1) solution. They were then used for the observation with a visible light and fluorescent microscope for both fiber and vessel secondary walls after staining by mäule and phloroglucinol-HCl color reactions [14, 15]. To determine the presence of peroxidases, the cross sections of 1.5 μm thick in transgenic poplars were observed by light microscopy after staining with 3, 3', 5, 5'-tetramethylbenzidine (TMB).

RESULTS & DISCUSSION

Regeneration and selection of transgenic poplars

An antisense vector with the original promoter of *prxA3a* was constructed and integrated into a wild poplar. After transformation, the seedlings of transgenic poplars were grown by the similar procedure, as the wild was [16]. The seven lines of transformants had similar growth speed in comparison with the wild. In order to verify whether 7 transformants include a part of the vector as a foreign gene or not, the transgenic samples were subjected to southern hybridization analysis and amplification test by PCR with the inserted NPT II gene used as a selection maker. It was shown that all lines were transformed (data not shown).

Suppression of peroxidase activity in transgenic poplars

Peroxidase activity in the stem of the transformants decreased about 0-44% compared with the control and in the leaf did about 0-35%. Of 6 lines, POX 47 is the most suppressed line in the stem and leaf (Table 1). These results suggested that the *prxA3a* gene expressed stronger in the stem than in the leaf, indicating that this gene specifically expresses in the xylem and phloem constructing vascular bundle. This explanation is supported by the result of expression test of the *prxA3a* gene [17]. It is possible that POX 47 is the best transformant in which its lignin biosynthesis was suppressed *in vivo* by the antisense RNA method.

From the analysis of isoelectric focusing, some transformants clearly decreased the staining of a pI 3.8 band being a peroxidase isozyme in comparison with the control.

POXs 47 and 49 especially decreased strongly. This result also shows that POX 47 is one of transgenic poplar in which the expression of the peroxidase gene is suppressed strongly.

Lignin content of transgenic poplars

To evaluate the lignin content of transgenic poplars, they were subjected to acetyl

Table 1. Total peroxidase activity of transgenic and control poplars
[Extractive-free wood meals prepared from 10cm high transformant and control seedlings *in vitro* were subjected to the measurement of the peroxidase activity. The numbers (%) in this table indicate relative peroxidase activity to that of the control.]

Samples	Relative peroxidase activity (%)	
	In stem	In leaf
Control	100	100
POX 27	69.1	89.1
POX 29	63.6	87.3
POX 44	99.6	98.9
POX 47	56.4	65.5
POX 49	74.5	74.5
POX 53	75.5	97.3

Table 2. Lignin content determined by acetyl bromide and permanganate oxidation methods.
[Lignin content (%) of transgenic poplars (7 lines) was calculated as relative values to 100% of control sample. These values are an average number in the replicates of three times.]

Methods / Samples	Acetyl bromide (%)	Permanganate oxidation (%)
Control	100	100
POX 27	87	-
POX 29	79	-
POX 43	100	85
POX 44	93	97
POX 47	79	74
POX 49	82	78
POX 53	95	74

bromide and potassium permanganate oxidation analyses. From the result of acetyl bromide, the lignin content of the transformants decreased 0-21% in comparison with that of the control. POXs 29 and 47 had more decreasing values than others (Table 2) did. The potassium permanganate oxidation analysis was carried out with the transgenic poplars growing 6cm and 60-100cm high. The lignin content by the permanganate oxidation method is calculated from the total yield of monomer and dimer degradation products (Monomers are the mono- or di-methyl esters of anisic acid, veratric acid, 4-methyl-o-phthalic acid, trimethyl gallic acid, 4-methoxy isophthalic acid, isohemipinic acid, methahemipinic acid and methoxy trimesic acid. Dimers are that of biphenyl ether compounds and dihydro diveratric acid.). The analytical data of 60cm high samples are shown in Table 2. In the 6cm high samples *in vitro* (POXs 43, 44, 49 and 53), the lignin contents decreased about 20-60% compared with control and were variable in repeated analyses, but not variable in 60-100cm high samples. The result from permanganate oxidation in Table 2 showed that the lignin content of the transgenic poplars decreased about 3-26% in comparison with that of the control. Of them, POXs 47 and 53 had lower lignin content than that of others. Conclusively, POX 47 is the best transgenic poplar in which the expression of peroxidase gene is suppressed by the antisense RNA method, because it had the lowest lignin content in both 6 and 60cm high samples.

Chemical property of the transgenic poplars

In order to analyze the lignin chemical structure of transformants in detail, the samples were subjected to potassium permanganate oxidation degradation and thioacidolysis. The result of thioacidolysis showed that the transgenic poplars had higher value than that of the control, indicating that the transgenic poplars have the lignin structure including more ß-O-4 linkage structures among monomer units (Table

Table 3. Characteristics of lignin chemical structure in transgenic poplars. [* UCO is given by the total yield of anisic acid, veratric acid and trimethyl gallic acid. ** CO is given by the total yield of 4-methyl-o-phthalic acid, 4-methoxy isophthalic acid, isohemipinic acid, methahemipinic acid, methoxy trimesic acid, biphenyl ether and dihydro diveratric acid. All values of the transformants are calculated as a relative value to the control.]

Samples	Degradation products yield by thioacidolysis (%)	The amount of uncondensed unit (UCO)*	The amount of condensed unit (CO)**	UCO/CO
Control	100	100	100	1.00
POX43	121	88	79	1.11
POX44	188	104	74	1.42
POX47	190	77	69	1.11
POX49	203	91	77	1.18
POX53	186	72	76	0. 95

Table 4. Comparison of the ratio of syringyl to guaiacyl units (S/V) in transgenic and control poplar lignins.
[S/V value is calculated from the yields of veratric acid and trimethyl gallic acid in permanganate oxidation degradation.]

Samples	Syringy/Guaiacyl (S/V)	The relative values to (S/V) in control (%)
Control	0.161	100
POX43	0.230	142
POX44	0.187	116
POX47	0.298	184
POX49	0.304	189
POX53	0.230	143

3). From the analysis of permanganate oxidation the values of uncondensed (UCO) and condensed (CO) lignin units were determined from the amounts of degradation products and then the value of UCO/CO was calculated (Table 3). It is an important property that the lignin chemical structure of transgenic poplar has more uncondensed lignin and ß-O-4 linkage units than that of the control has. Because it shows that the lignin of transgenic poplars is not bulky-like polymer, but rather endwise-like polymer compared with the control. Therefore, the transgenic poplars must probably have a special lignin possessing flexible three-dimensional structure and the ease for pulping. The result supporting this explanation concerning the pulping test has been reported by our group [18].

The ratio of syringyl to guaiacyl units in transgenic lignins was determined from the degradation products of permanganate oxidation in 6cm high seedlings of the transgenic poplars (Table 4). The S/V values of 6cm high seedlings *in vitro* were very low compared with those of mature samples, indicating that the lignin of young seedlings contains smaller syringyl units than that of mature one does. This phenomenon is attributable to the unmaturity of young seedlings, because it is believed that guaiacyl units are incorporated faster than syringyl units into lignin formation in the early stage of lignin biosynthesis. In the comparison between 60cm high transgenic and control samples, it was also clear that the lignin of transgenic poplars is syringyl-rich. This fact support the result obtained from thioacidolysis and permanganate oxidation concerning the value of UCO/CO.

Histochemical analysis of the transformed tissues

The cross sections, which were prepared from the part of the stem indicated by the bar shown in Fig. 1, were observed with light and fluorescent microscopes without staining of color reactions. Shape of cells in primary and secondary xylems was not different between wild and transformant (Fig. 1). In the observation of cross sections after staining of mäule and phloroglucinol color reactions, we could find uncolored region in the secondary xylem that is near cambial zone (shown by arrows in the transformant picture of Fig. 2). The cells in the uncolored region must be unlignified and no staining by both color reactions is probably attributable to the absence of peroxidase or of a monomer, coniferyl alcohol, as a precursor of dehydrogenation polymerization in lignin biosynthesis. In the other hand, we could detect the peroxidase activity (colored zone) in the secondary xylem which is near cambial zone in the wild sample, when the cross sections were observed by the light microscopy after staining with 3, 3', 5, 5'-tetramethylbenzidine (TMB). The colored zone is the part indicated by an arrow in the picture of wild in Fig.3. But we could not find the colored zone in the transformant sample (POX29). This uncolored zone is indicated by an arrow in the picture B of the transformant in Fig. 3. These regions colored and uncolored by staining with TMB correspond the region uncolored by mäule and phloroglucinol color reactions as shown in Fig. 2. These facts show that the lignification in the transformant is suppressed in the early stage of secondary xylem formation, and the decrease of lignin content in the transformants is attributable to the absence of peroxidase in the secondary xylem tissue.

CONCLUSIONS

In order to produce a tree with lowed lignin content, peroxidase genes involved in lignin biosynthesis were isolated from the hybrid poplar. Five peroxidase genes were isolated and the DNA sequences and their promoters were analyzed. As the promoter of

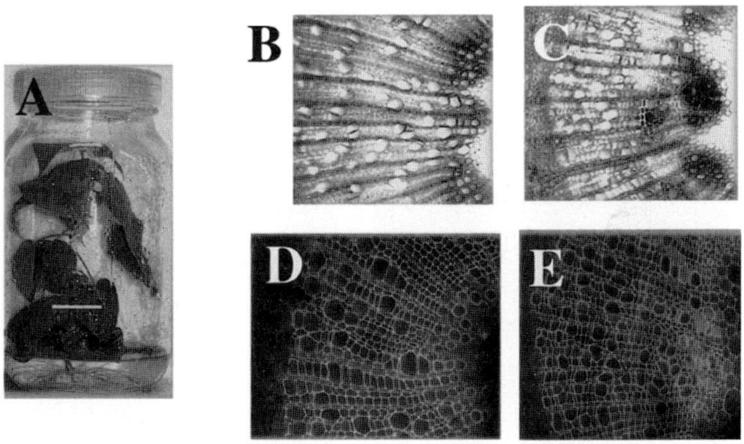

Figure 1. Observation of the cross-sections in wild and transformant samples by light microscopy.
[A: Seedlings of a transformant. A bar indicates the cutting place of 10cm high seedlings to prepare the samples of microscopic observation. B: Wild sample observed with light microscope. C: Transformant (POX27) observed with light microscope. D: Wild observed with fluorescent microscope. E: POX27 observed with fluorescent microscope.]

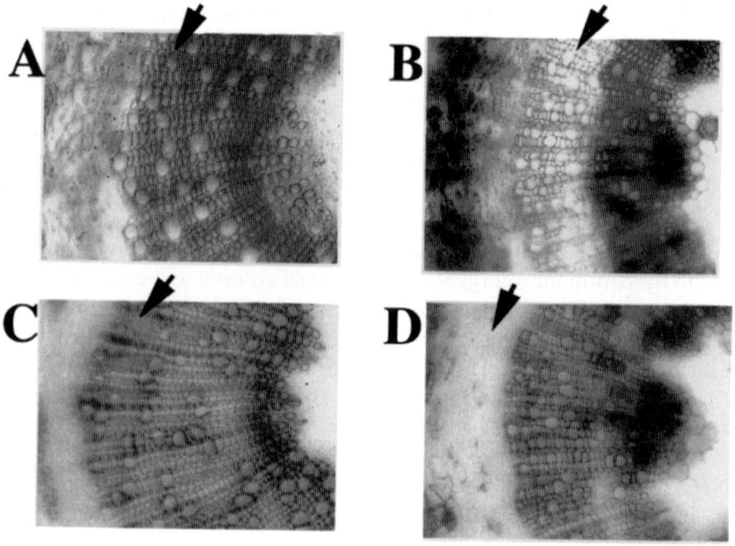

Figure 2. Observation of cross-sections by light microscopy after staining by mäule and phloroglucinol-HCl color reactions.
[The cross sections for microscopic observation are the similar ones using in Fig 1. A: Wild sample after staining by mäule reaction. B: Transformant, POX27 after mäule reaction. C. Wild after phloroglucinol-HCl reaction. D: Transformant, POX27 after phloroglucinol-HCl reaction.]

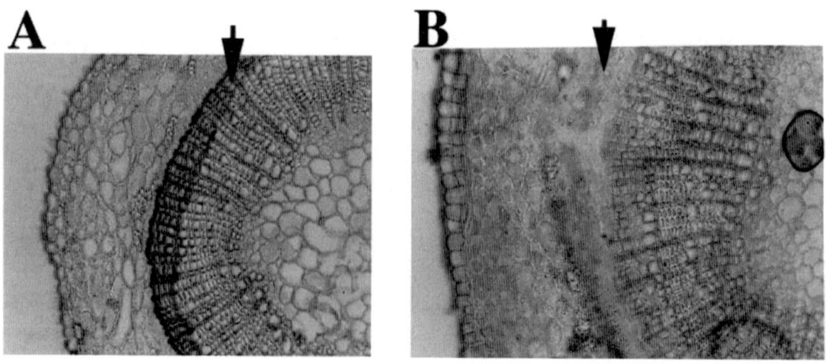

Figure 3. Histochemical localization of peroxidase activity in the cross-sections of wild and transformant (POX29) samples.
[The cross sections are the similar ones using in Figs. 1 and 2. A: Wild after staining with 3,3', 5, 5'-tetramethylbenzidine (TMB). B: Transformant, POX29 after staining with TMB.]

prxA3a was expressed in the stem specifically, it was selected as being involved in lignin biosynthesis. By using the antisense RNA method, 7 lines of the transformed poplar in which introduction of the vector were verified, were formed. These transformants had lower peroxidase activities (decreasing 0-44 % in the stem), and lower lignin content (decreasing 3-26%) than that of the control. From these results, it was cleared that POXs 47 and 53 had the lowest lignin content. Especially, POX47 was the best transgenic poplar in which the expression of peroxidase gene is suppressed by the antisense RNA method.

Furthermore, the transformants were subjected to light microscopic observation after color reactions, it was ascertained that the peroxidase activity of the ransformants is suppressed in the secondary xylem part near cambial zone and the lignification is inhibited in that region.

These results obtained will give us hope that we may succeed in forming lower lignin content trees by means of genetic engineering techniques. In our future research, we need to analyze the mature transgenic poplars in detail to get final evaluation of transformants in which the expression of peroxidase gene is suppressed.

ACKNOWLEDGEMENTS

The authors acknowledge financial support from the Japan Society for the Promotion of Science (JSPS-RFTF96L00605).

REFERENCES

1. K. Freudenberg, 'Biosynthesis and constitution of lignin', *Nature*, **1959**, 183, 1152-1155.
2. T. Higuchi & Y. Ito, 'Dehydrogenation products of coniferyl alcohol formed by the action of mushroom phenoloxidase, rhus-laccase and radish peroxidase', *J. Biochem*, **1985**, 45, 575-579.
3. M. Mäder & R. Füssl, 'Role of peroxidase in lignification of tobacco cells. II. Regulation by phenolic compounds', *Plant Physiol.*, **1982**, 70, 1132-1134
4. L. M. Lagrimini & S. Rothstein, 'Tissue specificity of tobacco peroxidase isoenzymes and their induction by wounding and tobacco mosaic virus infection', *Plant Physiol.*, **1987**, 84, 438-442.
5. H. Ebinuma, K. Sugita, E. Matunaga et al., 'Selection of marker-free transgenic plant using the isopentenyl transferase gene', *J. Proc. Natl. Acad. Sci. USA*, **1997**, 94, 2117-2121.
6. K Osakabe, H. Koyama, S. Kawai et al., 'Molecular cloning and the nucleotide sequences of two novel cDNAs that encode anionic peroxidases of *Populus kitakamiensis*', *J. Plant Science*, **1994**, 103, 167-175.
7. S. Kajita, K. Osakabe, Y. Katayama et al., 'Agrobacterium-mediated transformation of poplar using a disarmed binary vector and the over expression of a specific member of a family of poplar peroxidase gene in transgenic poplar cell', *J. Plant Science*, **1998**, 103, 231-239.
8. D. B. Johnson, W. E. Moore & L. C. Zank, 'The spectrophotometric determination of lignin small wood samples', *Tappi*, **1961**, 44, 793-798.
9. K. Iiyama & A. F. A. Wallis, 'An improved acetyl bromide procedure for determining lignin in woods and wood pulps', *Wood Sci. Technol.*, **1988**, 22, 271-

280.
10. N. Morohoshi & W. G. Glasser, 'The structure of lignins in pulps. Part 4 –Comparative evaluation of five lignin depolymerization techniques', *Wood Sci. Technol.*, **1979**, 13, 165-178l.
11. N. Morohoshi & W. G. Glasser, 'The structure of lignins in pulps. Part 5 -Gas and gel permeation chromatography of permanganate oxidation products', *Wood Sci. Technol.*, **1979**, 13, 249-264.
12. M. Erickson, S. Larsson & G. E. Miksche, 'Gaschromatografishe Analyse von Lignin oxydations produkten.VII. Ein verbessertes Verfahren zur Charakterisierung von Ligninen durch Methylierung und oxydativen Abbau', *Acta Chemica Scandinavica*, **1973**, 27, 127-140.
13. C. Lapierre, B. Montiies & C. Rolando, 'Thioacidolysis of poplar lignin-identification of monomeric syringyl products and characterization of guaiacyl-syringyl lignin fractions' *Holzforschurrg*, **1986**, 40, 113-118
14. K. Takabe, S. Miyauchj, R. Tsunoda & K. Fukazawa, 'Distribution of guaiacyl and syringyl lignin in Japanese beech (*Fagus crenata*): Variation within an annualring', *IAWA Bulletin n. s.*, **1992**, 13, 105-112.
15. L. M. Stivastava, 'Histochemical studies on lignin', *Tappi*, **1966**, 49, 173-183
16. N. Morohoshi, 'Formation and Characterization of Transformed Woody Plants Inhibiting Lignin Biosynthesis' in 'Challenge of Plant and Agricultural Sciences to the Crises of Biosphere on the Earth in the 21st Century', K. Watanabe & A. Komamine ed., EUREKAH-COM/LANDES BIOSCIENCE, **2000**, pp. 275-280.
17. K. Osakabe, H. Koyama, S.Kawai et al., 'Molecular cloning of two tandemly arranged peroxidase genes from Populus kitakamiensis and their differential regulation in the stem', *J. Plant Molecular Biology*, **1995**, 28, 677-689.
18. T. Tamura, N. Morohoshi & S. Yasuda, 'Suitability of peroxidase-suppressed transgenic hybrid aspen (*Populus Sieboldii* x *P. gradidentata*) for pulping', *Holzforschung*, **2001**, 55, in press.

Molecular Breeding of Woody Plants
N. Morohoshi and A. Komamine, editors.

TRANSCRIPTIONAL REGULATION OF LIGNIN BIOSYNTHESIS BY TOBACCO LIM PROTEIN IN TRANSGENIC WOODY PLANT

Akiyoshi Kawaoka*, Kazuya Nanto, Koichi Sugita, Saori Endo, Keiko Yamada-Watanabe, Etsuko Matsunaga & Hiroyasu Ebinuma

Pulp and Paper Research Laboratory, Nippon Paper Industries, 5-21-1 Oji, Kita-ku, Tokyo 114-0002, Japan

ABSTRACT

Lignin is a complex phenolic plant polymer that is essential for mechanical support, defense, and water transport in higher plants. The AC-rich motif, Pal-box is an important *cis*-acting element for gene expression involved in phenylpropanoid biosynthesis. A cDNA clone (Ntlim1) encoding a Pal-box binding protein was isolated by Southwestern screening. The deduced amino acid sequence is highly similar to the members of the LIM protein family that contain zinc finger motif. Moreover, the Ntlim1 had a specific DNA binding ability and transiently activated the transcription of β-glu-curonidase reporter gene driven by the Pal-box sequence in tobacco protoplasts. The results of transient expression assays with tobacco cultured cells showed that fusion proteins between GFP and Ntlim1 can enter nuclei. The transgenic tobacco with antisense Ntlim1 showed low level of transcripts from some key phenylpropanoid pathway genes such as, phenylalanine ammonia-lyase (PAL), hydroxycinnamate CoA ligase (4CL) and cinnamyl alcohol dehydrogenase (CAD). A 27% reduction of lignin content was observed in the transgenic tobacco with antisense Ntlim1. We succeeded the production of transgenic woody plants (*Eucalyptus camaldulensis)* with low lignin content by introduction of the antisense Ntlim1.

KEYWORDS

Transcription factor, lignification, Pal-box, transgenic Eucalyptus

INTRODUCTION

Lignin is a complex phenolic polymer that reinforces the walls of certain cells in the vascular tissues of higher plants. Lignin plays an important role in mechanical support, water transport and pathogen resistance. In woody plants, high levels of lignin are synthesized in wood and account for 15-36% of dry weight of wood materials. In the pulp and paper industry, lignin must be removed by harsh chemical treatments, which is a costly process both to the mill and the environment. Genetic modification of the lignin content and composition of trees is thus receiving considerable attention in current forest biological studies and wood formation, although the effects of lignin content on fitness are still unclear [14]. To understand regulation of lignin biosynthesis, we focused on a transcription factor that controls gene expression encoding enzymes in this pathway. The horseradish peroxidase (HRP) *prxC2* gene was probably involved in lignification [5,6]. Functional analysis of *prxC2* promoter suggested that Pal-box like motif, at position between -178 and -136 bp from the translation start site is a *cis*-acting element for its expression [4]. We have also detected nuclear proteins, which bound to the Pal-box sequence. The Pal-box motif, CCA (C/A) (A/T) A (A/C) C (C/T) CC, is thought to be a reserved sequence involved in gene expression of phenylpropanoid biosynthesis, such as PAL, 4CL, CHS and CAD [9].

Eucalypts plants with the fast growth rates and short rotations, are most important plantation hardwoods in the world. *E. camaldulensis* has been successfully grown in Australia. Genetic engineering could assist the genetic improvement of woody plants for producing the high-quality pulp with lignin modification.

We describe here the isolation of a cDNA encoding a tobacco protein Ntlim1, which specifically bound to the Pal-box sequence [7]. The Ntlim1 protein acts as a transcription factor involved in lignin biosynthesis. Suppression of this gene by antisense method affected expression levels of PAL, 4CL and CAD, and showed 27% and 20% reduction of lignin content in harbaceous plant tobacco and woody plant *E. camaldulensis*, respectively.

MATERIALS & METHODS

Plant material

Nicotiana tabacum cv Petit Havana SR-1 was used, and were grown in a greenhouse at 25°C under a 16 hours light and 8 hours dark photo-period. Tobacco cultured cell BY-2 was used for the transient assay.

Transformation of tobacco

Chimeric genes of antisense and sense *Ntlim1* cDNA under the control of the 35S promoter were ligated into the binary vector pBI121 (Clontech). Tobacco was transformed with *Agrobacterium tumefaciens* LBA-4404 using the leaf disk transformation method.

Transformation of Eucalyptus

The seeds of *E. camaldulensis* were obtained from Koei Syubyo Boeki Co. ltd. (Tokyo, Japan). Seeds were sterilized by immersion in 2% (v/v) sodium hypochlorite following 70% (v/v) ethanol treatment and finally rinsing three times with sterile water. The sterilized seeds were sown on half strength macronutrients of Murashige and Skoog (MS) medium and the hypocotyl explants of germinated seedling were used as material for transformation. As a basal medium, half-strength macronutrients of MS medium containing 10 mM NH_4^+, 30 mM NO_3^- and 2% sucurose was used. For shoot regeneration, basal medium supplemented 2 mg/L trans-zeatin and 0.3 mg/L naphthaleneacetic acid (NAA) was used. We prepared rooting medium that the basal medium contained 50 μg/L indolebutyric acid (IBA). Regenerated Eucalyptus plants were grown in a greenhouse.

Lignin analysis

Lignin determination was performed on the dried insoluble cell wall residues (CWR) of samples soxhlet extracted with toluene/ethanol, ethanol and water. Klason lignin was measured by the method of Effland (1977) [3].

RESULTS & DISCUSSION

Isolation of a cDNA encoding Pal-box binding protein

A λgt11 expression library constructed with mRNA from tobacco stems was screened by using the probe P. One positive clone, #101, was obtained from the screening of one million recombinant phages. The 1 kb cDNA insert of clone #101 was amplified by PCR. The cDNA insert of the #101 contained an open reading frame that encodes a protein of 200 amino acid residues (Figure 1). From a GenBank database search, the deduced amino acid sequence of #101 produced high-score segment pairs with LIM protein, such as SF3 of late pollen genes in the sunflower [1]. We designated the protein encoded by this cDNA, Ntlim1. The LIM motif defines one class of zinc-binding domain and was originally recognized in, and named after, the protein products of the *lin-11*, *isl-1* and *mec-3* genes [13]. These all genes have a homeodomain. The gene products of *lin-11* and *mec-3* transcriptionally regulate genes involved in cell fate determination and differentiation in *Caenorhabditis elegans*, and the *isl-1* gene encodes a rat insulin I gene enhancer-binding protein.

The Ntlim1 protein localization in tobacco cells was investigated using chimeric genes between the Ntlim1 protein and Green Fluorescent Protein (GFP). The GFP-LIM and LIM-GFP were translationally ligated to the GFP gene at the N- or C-terminal of Ntlim1, respectively. These genes were introduced into tobacco cultured BY-2 cells by particle bombardment. At 48 hr after introduction, fluorescence microscopy studies showed both types of fusion proteins at high levels exhibiting bright green mainly in nuclei, whereas no green fluorescence was detected in the untransformed BY-2 cells. We also introduced GFP alone into BY-2 cells and strong green fluorescence was observed in the cytoplasm of these cells. Ntlim1 appears to be a nuclear protein [8].

To determine the number of genes in the tobacco genome that are related to the gene for Ntlim1, Southern hybridization was carried out. Three hybridizing bands were obtained with genomic DNA digested with either *Eco*RV or *Dra*I. This result suggests that there are few copies of *Ntlim1* in the tobacco genome [7].

Total RNA was extracted from tobacco roots, stems and leaves. Ten μg of the total RNA was subjected to electrophoresis and blotted, and the accumulation of *Ntlim1* mRNA in each organ was determined by Northern hybridization. The *Ntlim1* mRNA was abundant in stems and also detected in leaves and roots [7].

```
MAFAGTTQKCMACDKTVYLVDKLTADNRIYHKACFRCHHC  40
KGTVKLGNYNSFEGVLYCRPHFDQLFKQTGSLDKSFEGTP  80
KIVKPQKPIDSEKPQVAKVTSMFGGTREKCFGCKKTVYPT 120
EKVSANGTPYHKSCFQCSHGGCVISPSNYTAHEGRLYCKH 160
HHIQLIKEKGNLSKLEGDHEMNSTTTTEVTAESYTADQVD 200
```

Figure 1. Amino acid sequence of Ntlim1.
The LIM domains are indicated by underlines. The acidic domain of C-terminal is boxed.

Ntlim1 is a transcription factor that binds a specific DNA sequence

There is still enough evidence to proof that LIM domain is able to bind to DNA. To determine whether the Ntlim1 protein indeed binds to the Pal-box like sequence, we prepared protein extracts from *E. coli* carrying the expression. The fusion protein, GST-Ntlim1 was purified through Glutathione-Sepharose 4B resin. After digestion with thrombin protease, a band of the purified Ntlim1 protein was observed by SDS-PAGE. The DNA binding ability of the purified Ntlim1 was tested by gel-retardation assay using digoxigen-labeled probe P. The retarded band showing DNA binding activity of Ntlim1, which disappeared in the presence of an excess amount of non-labeled probe P, was observed. This result indicates the LIM domain has DNA-binding ability.

There are three notable domains of Ntlim1, LIM1 domain (amino acids 10-61), LIM2 domain (amino acids 110-161) and the acidic domain (amino acids 164-200). Although the isoelectric point (pI) of the entire Ntlim1 protein is 8.39, the pI of the acidic domain at the C-terminal region is 3.92 by computer analysis of Genetyx (Software Development Co., Ltd, Tokyo, Japan). To determine whether Ntlim1 functions as a transcription activator in vivo, we constructed an effector plasmid under the control of cauliflower mosaic virus (CaMV) 35S promoter, and chimeric GUS gene as a reporter plasmid that a three-copy sequence of the Pal-box like motif (-CCACTTGAGTAC-) was ligated to the *Eco*RV site at -90 bp of the CaMV 35S promoter. The correctness of orientation of the inserted oligonucleotides was checked by sequencing. Co-introduction of reporter with the full length *Ntlim1* effector construct increased GUS activity approximately two to four-fold compared to the negative control. These results indicate that Ntlim1 protein has a function as a transcription activator [7]. The most likely candidate for this domain is the C-terminal acidic polypeptide, although this is only 37 amino acids. No homology in the C-terminal region of Ntlim1 to known transcription activation domain or distinctive feature in its amino acid sequence has been observed. Random acidic amino acid peptides from *E.coli* have been reported to substitute for the acidic activator domain of Gal4 in yeast [10].

Ntlim1 affects gene expression of phenylpropanoid biosynthesis and lignin content in transgenic tobacco

To investigate the molecular effects of Ntlim1 on lignin biosynthesis, a 1.0-kb fragment of Ntlim1 was inserted in the antisense orientation between the CaMV 35S promoter and the 3' terminator sequence of the nopaline synthase gene. This construct was introduced into tobacco plants by Agrobacterium-mediated transformation. Twenty independent primary transgenic plants (T1) were produced and all lines contained the respective T-DNA determined by PCR analysis. Subsequently, seven antisense lines were screened by measuring the *Ntlim1* transcript levels in the stems, and were grown to maturity in a greenhouse. Their self-pollinated seeds (T2) were collected and used for further investigation. Ten T2 kanamycin-resistant transgenic plants were grown in a greenhouse. All of them showed a normal growth rate and flowered at similar times. No remarkable abnormal growth or flower color changes were found in the transgenic lines, except that several lines showed a slightly increased stem height. We isolated cDNA clones encoding some of the structural genes, *PAL*, *4CL* and *CAD*, involved in phenylpropanoid biosynthesis from tobacco by PCR. The expression levels of these structural genes, the transgene, and the endogenous *Ntlim1* were estimated by Northern blotting using random-primed probes. Total RNAs were extracted from stems in the kanamycin-resistant T2-transgenic antisense plants. Signals of the antisense Ntlim1 gene were not detected in either plant carrying the antisense Ntlim1. The antisense Ntlim1 transcript is probably unstable in tobacco cells. In antisense plants, endogenous *Ntlim1* (1.2 kb) expression was suppressed and the mRNA levels of *PAL, 4CL* and *CAD* were also lower than those in the wild-type plant [7]. These results imply that Ntlim1 is able to regulate the expression of some genes involved in the phenylpropanoid biosynthesis pathway which have Pal-box in their 5' upstream region.

The analysis of the levels and monomeric composition of lignin in the transgenic plants was performed with the cell wall residues (CWR) of xylem tissues in stems. Lignin determinations were carried out on T2 kanamycin-resistant, different transgenic lines after 10 weeks of growth in greenhouse. The wild-type and transgenic plants with sense and antisense *Ntlim1* were grown to maturity in the greenhouse. All of them showed similar height at a similar growth rate and flowered at similar times. No remarkable abnormal growth was observed in transgenic line except slightly increased height in A9. The lignin content of line S1 and A2 were slightly reduced, but no differences could be detected between the sense line S2 and control plant WT. In contrast, 27% of reduction in A9 was observed compared to control (Table 1).

To determine the monomeric composition of lignin in the CWR, we examined the CWR from each plant by alkaline nitrobenzene oxidation analysis. Vanillin and syringaldehyde are generated from non-condensed fraction of lignin by this procedure [2]. The relative abundance of guaiacyl and syringyl units in lignin can be summerized as S/V ratio (the ratio of the amounts of syringaldehyde and vanillin). No remarkable change of S/V ratio was detected in each transgenic line except A9 that showed lower value. In particular, the amounts of total aldehydes were decreased in S1 and A9. These data indicates that lignin in the transgenic plants with decreased expression level of Ntlim1 contained more condensed units that could not be degradated by oxidation procedures.

Table 1. Lignin content in transgenic plant.

Line	lignin content[a]			
WT	21.2	±	0.6	(100)
S1	19.3	±	0.1	(91)
S2	22.3	±	0.5	(105)
A2	19.1	±	0.1	(90)
A9	15.5	±	0.3	(73)

[a] Lignin content is expressed as a percentage (w/w) of cell wall residues (CWR) by Klason lignin.

Low lignin content in trnasgenic Eucalyptus with the antisense Ntlim1

The antisense Ntlim1 constructs were introduced into Eucalyptus plants by the Agrobacterium mediated transformation. Forty-five independent primary transgenic plants were produced and all lines contained the respective T-DNA by PCR analysis. Subsequently, fifteen lines were screened by measuring the 4CL activity levels in the stems and were grown in the greenhouse. No remarkable abnormal growth was found in the transgenic lines. The lignin content of the cell wall residue (CWR) of stem xylem tissues was measured by the gravimetric Klason procedure. Lignin determinations were carried out on fifteen transgenic lines after 4 months of growth in the greenhouse. About 20% reductions were observed in some lines that showed lower 4CL-activity levels.

The expression levels of the transgene, endogenous *lim* gene of *E. camaldulensis*, and these structural genes were investigated by Northern blotting using heterologous probes. Total RNAs were extracted from stems in the transgenic plants. From the Northern analysis, *E. camaldulensis* has a *Ntlim1* homologous gene at the size of about 2.0 kb. In the Eucalyptus plants carrying the antisense *Ntlim1* gene, the signals of antisense Ntlim1 were not detected and the endogenous lim gene of expression was relatevely suppressed and the mRNA levels of *PAL, 4CL* and *CAD* were also at low levels. These results imply that *Ntlim1* is able to simultaneously regulate the expression of many genes involved in phenylpropanoid biosynthesis.

CONCLUSIONS

We have isolated Pal-box binding protein, Ntlim1 from tobacco. Final point regards the possible regulation of genes in phenylpropanoid and lignin biosynthesis by the cloned Ntlim1. The Ntlim1 expressed constitutively and mostly in stem. Perhaps Ntlim1 binds Pal-box motif at any time in lignifying tissues, such as xylem. In the case of

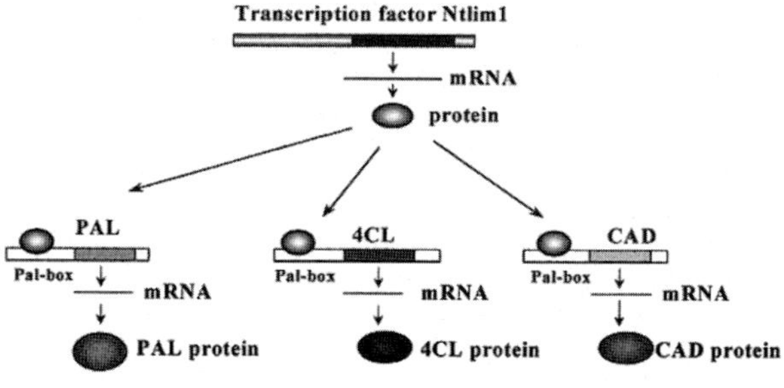

Figure 2. Regulation of gene expression by the transcription factor, Ntlim1. Transcription factor Ntlim1 can associate with gene expression involved in lignin biosynthesis by binding to Pal-box

suffering circumstantial stimuli, for instance UV light, elicitor or wounding, various transcription factor(s) binds to another *cis*-acting element such as G-box or ABRE (abscisic acid responsive element) etc., then enzyme would be activated by these machinery [11,12]. The Ntlim1 may act as a transcription factor to maintain basal enzyme activities of target enzymes (Figure 2).

ACKNOWLEDGEMENTS

We thank Kazuya Yoshida and Pulla Kaothien, Nara Institute of Science and Technology for their helpful discussion. We also thank Yukifumi Uesono, University of Tokyo for his helpful advice. We are grateful to Seiichiro Hasezawa and Toshiyuki Nagata, University of Tokyo for their providing tobacco cultured cell BY-2. We are also grateful to Masako Obata, Nippon Paper Industries for their technical advice.

REFERENCES

1. R. Baltz, C. Domon, D. T. N. Pillay, & A. Steinmetz, 'Characterization of a pollen-specific cDNA from sunflower encoding a zinc finger protein' *Plant J.* **1992,** 2, 713-721.
2. V. L. Chiang & M. Funaoka, 'The formation and quantity of diphenylmethane type structures in residual lignin during kraft delignification of Douglas-fir' *Holzforschung,* **1998,** 42, 385-391.
3. M. J. Effland ' Modified procedure to determine acid insoluble lignin in wood and pulp' . *Tappi,* **1977,** 60, 143-144.
4. P. Kaothien, Y. Shimokawatoko, A. Kawaoka, K. Yoshida & A. Shinmyo, 'A.*cis*-element containing PAL-box functions in the expression of the wound-inducible peroxidase gene of horseradish' *Plant Cell Reports,* **2000,** 19, 558-562.
5. A. Kawaoka, S. Sato, K. Nakahara, N. Matsushima, N. Okada, M. Sekine, A. Shinmyo & Takano, M. 'Expression and promoter activity of genes for isozymes of horseradish peroxidase' *Plant Cell Physiol.,* **1992,** 33, 1143-1150.
6. A. Kawaoka, T. Kawamoto, M, Sekine, K. Yoshida, M. Takano & A. Shinmyo, 'A cis-acting element and trans-acting factor involved in the wound-induced expression of a horseradish peroxidase gene' *Plant J.,* **1994.** 6, 87-97.
7. A. Kawaoka, P. Kaothien, K. Yoshida, S. Endo, K. Yamada & H. Ebinuma, Functional analysis of tobacco LIM protein Ntlim1 involved in lignin biosynthesis. *Plant J.,* **2000,** 22, 289-301.
8. A. Kawaoka & H. Ebinuma, 'Transcriptional control of lignin biosynthesis by tobacco LIM protein' *Phytochemistry,* **2001**, in press.
9. R. Lois, A. Dietrich, K. Hahlblock & W. Schulz, 'A phenylalanine ammonia-lyase gene from parsley: structure, regulation and identification of elicitor and light responsive cis-acting elements' *EMBO J.,* **1989,** 1641-1648.
10. J. Ma & M. Plasshne, 'Deletion analysis of GAL4 defines two transcriptional activating segments' *Cell,* **1987,** 48, 847-853.
11. A. E. Menkens, U. Schindler & A. R. Cashmore, 'The G-box: A ubiquitous regulatory DNA element in plants bound by GBF family of bZIP proteins' *Trends Biochem. Sci.,* **1995,** 13, 506-510.
12. K. Shinozaki & K. Yamaguchi-Shinozaki, 'Gene expression and signal transduction in water-stress response' *Plant Physiol.,* **1997,** 115, 327-334.
13. M. Taira, J-L. Evrard, A. Steinmetz & I. B. Dawid, 'Classification of LIM proteins' *Trend in Genetics,* **1995,** *11,* 431-432.
14. R. W. Whetten, J.J. MacKay & R. R. Sederoff, 'Recent advances in understanding lignin biosynthesis' *Annu. Rev. Plant Physiol. Plant Mol. Biol.,* **1997,** 49, 585-609.

Molecular Breeding of Woody Plants
N. Morohoshi and A. Komamine, editors.

GENETIC ENGINEERING OF *PINUS RADIATA* AND *PICEA ABIES*, PRODUCTION OF TRANSGENIC PLANTS AND GENE EXPRESSION STUDIES

Christian Walter[1], Sharon Bishop-Hurley[2], Julia Charity[1], Jens Find[3], Lynette Grace[1], Kai Höfig[1], Lyn Holland[1], Ralf Möller[1], Judy Moody[1], Armin Wagner[1] & Adrian Walden[4]

[1]*New Zealand Forest Research Institute Ltd, Rotorua, New Zealand*

[2] *108 Waters Hall, Plant Pathology, University of Columbia, Missouri, MO65211, USA*

[3]*Tissue Culture Laboratory, Botanic Garden of Copenhagen, 0. Farimagsgade 2B, 1353 Kbh.K, Denmark*

[4] *Vialactia Biosciences, Level 4 Clinical Building University of Auckland School of Medicine, 85 Parl Road, Grafton, Auckland, New Zealand.*

ABSTRACT

Plantation forestry, based on successful breeding of superior tree genotypes, is becoming more widely used by international forestry companies, since it offers the possibility to grow and manage forests of high economic value and superior quality. However, a number of highly desirable traits are not readily available in the breeding population and may be introduced using desirable genes from other organisms. Forest molecular biology, and in particular tree genetic engineering is now at a stage where the technologies are readily available to transfer specific traits of commercial and scientific interest into forest trees.

Our efforts are aimed at the genetic engineering of plantation grown conifer trees such as *Pinus radiata*, *Pinus taeda* and *Picea abies*. Stable transformation technologies have been developed for embryogenic tissue using either *Agrobacterium tumefaciens* or a Biolistic® particle delivery system. Many genes from other organisms combined with promoters of various origins, were transferred into conifer tissue and transgenic plants recovered. Examples are genes for resistance against herbicides, genes involved in reproductive development, and genes involved in lignin formation. Analysis of transgenic tissue and plants has confirmed successful transfer of genes and their expression. For example, a herbicide resistance gene was introduced into *Pinus radiata* and *Picea abies*, and transgenic plants were regenerated and exposed to an operational concentration of the respective herbicide. Spray tests demonstrated the expression of the herbicide resistance gene and the newly acquired resistance of transgenic conifers to the herbicide.

Another example is the expression of the endogenous *cad* gene in sense or antisense orientation in transgenic *Pinus radiata* tissue and plants. Biochemical analyses indicate that the influence of the inserted gene construct on the expression of the endogenous *cad* gene can be dependent on the developmental stage of the plant.

A further challenge is provided by the growing numbers of potentially useful genes and promoter / gene combinations offering potential for forest trees. Early screening technologies that allow the quick and economic screening of candidate genes need to be developed to enable researchers to make the right choice of a gene for a specific purpose. To this end we are developing tissue culture protocols to produce secondary wall-forming cells and tracheary elements *in vitro*. These cells may be genetically transformed and may provide an excellent screening and gene expression analysis tool by avoiding the long periods of time otherwise necessary to regenerate and analyse transgenic plants.

INTRODUCTION

Genetic engineering has contributed to significant improvements in agricultural crops, and plants with engineered resistance against herbicides or insects are used in commercial plantations worldwide [1,2]. This relatively new technology has the potential to improve quality and yield of agricultural products, and newly developed products for human consumption hold the promise to significantly contribute to human health and welfare [3]. The use of agrochemicals can be reduced, leading to a more environmentally acceptable agriculture that is truly sustainable.

The development of molecular biology platforms including genetic engineering has somewhat lagged behind in forestry, mainly due to additional challenges related to the long rotation time of these plants, long breeding times and difficulties with tissue culture and genetic transformation protocols [4].

Conventional breeding has been the predominant technique to improve genetic gain in plantation forestry, and many techniques have successfully been applied to improve gain and various growth and performance characteristics [5,6]. Conventional tree

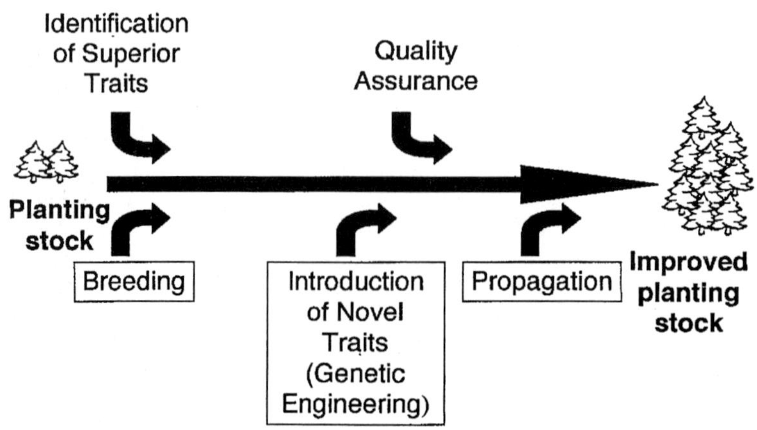

Figure 1: The production of superior tree germplasm. Major techniques used to provide improved planting stock

improvement programs aiming at the production of superior germplasm (Figure 1) have traditionally made use of the identification of superior traits. Also, various breeding techniques and methods of propagation (including both micro and macro propagation) to provide superior planting stock for commercial plantations are used. More recent developments, in particular in the area of molecular biology, have added techniques for quality assurance such as marker-aided selection (MAS) and genetic fingerprinting [7]. Also, over the past 10 years significant progress in developing genetic engineering protocols has been made and they are now available for most major forest tree species of commercial importance. These can provide techniques to transfer traits that are not readily available in the existing breeding population [8,9].

In this paper we review genetic engineering technologies developed at the New Zealand Forest Research Institute and present results from the analysis of transgenic tissue and plants. Further, we discuss results obtained from conifer promoter analysis in a heterologous plant species, and present strategies for more efficient and faster functional analysis of candidate genes.

GENETIC TRANSFORMATION OF EMBRYOGENIC TISSUE USING BIOLISTICS® OR *AGROBACTERIUM TUMEFACIENS*

Embryogenic tissue on a maintenance medium [10] is used for transformation with Biolistic® [11] techniques, or via *Agrobacterium tumefaciens* [4 and Charity, in preparation]. Transformation takes place with embryogenic tissue at an early stage in development and involves a plasmid vector carrying the *npt*II gene for resistance against aminoglycoside antibiotics such as kanamycin and geneticin. Geneticin was found to be the better selective agent for the selection of transgenic conifer tissue. The exact concentration ranges from 5 to 35mg / l, depending on the conifer species and the genotype. Kanamycin was also tested as a selective agent, however it allowed the emergence of escapes ie tissue that grew on selective media but was not transformed. Geneticin selection never allowed such growth. The antibiotic hygromycin was also successfully used to select transgenic radiata pine tissue after transformation with a vector containing the *aph*4 gene [12]. Tissue resistant to the antibiotic proliferated and became visible usually 4-6 weeks after transformation. At this stage it was transferred to fresh media for further proliferation, analysis and regeneration. For information on protocols, see [12,13,14 and Charity in preparation]. In summary, a range of conifer species were successfully transformed by Biolistic® or *Agrobacterium* related techniques (Table 1).

Analysis of transgenic material

Transgenic tissue and plants were analysed using a variety of techniques to confirm transgenic nature and function of the introduced gene product. This includes histochemical and fluorometric uidA analysis, Southern and Northern hybridisation, RT-PCR, nptII and *Bacillus thuringiensis* toxin (*cry*IAc) ELISA, and spray testing with herbicides for functional analysis.

Initial molecular analysis with putative transgenic lines was carried out using PCR analysis. DNA isolated from tissue pieces growing on selective media were probed with

Species	Transformation technique	Transgenic tissue selected	Molecular analysis	Regeneration of transgenic plants
Pinus radiata	Biolistic®	+	+	+
Picea abies	Biolistic®	+	+	+
Pinus taeda	Biolistic®	+	+	-
Abies nordmanniana	Biolistic®	+	+	+
Pinus radiata	*Agrobacterium*	+	+	+

Table 1: Genetic transformation of various conifer species, analysis and regeneration of plants.

primers initiating the amplification of the *npt*II or *uid*A gene and products were analysed by gel electrophoresis. For example, tissue selected on geneticin after *Agrobacterium* transformation was tested for the presence of *npt*II DNA and confirmed to contain the transgene of interest (Figure 2).

Controls including primers for the *vir*D gene from *A. tumefaciens* did not amplify a corresponding DNA fragment indicating the absence of *A. tumefaciens* in transgenic tissue.

Further analysis of transgenic selected tissue involved histochemical staining for *uid*A expression and blue colouration was found in most selected lines at varying intensity (data not shown).

Southern hybridisation analysis of transgenic *P. radiata* plants derived from a Biolistic® transformation experiment revealed medium to high copy numbers of the integrated *npt*II gene (Figure 3).

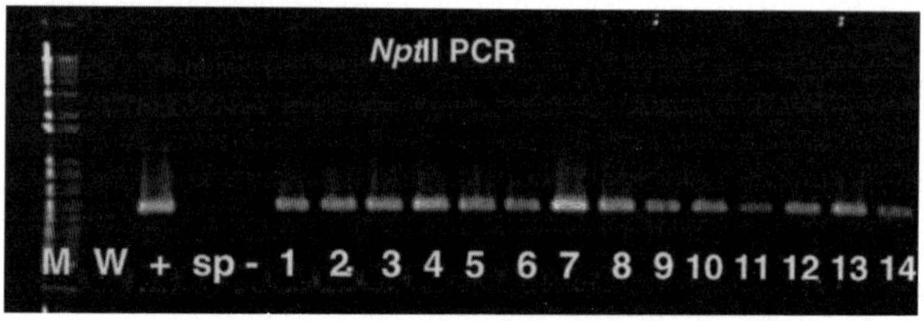

Figure 2: PCR analysis of embryogenic tissue of *Pinus radiata* transformed with *Agrobacterium tumefaciens*. Abbreviations: M: Molecular marker; W: Water-control; +: Positive (plasmid) control; sp: Space (no DNA); -: Untransformed tissue control; 1-14: Tissue from individually transformed lines.

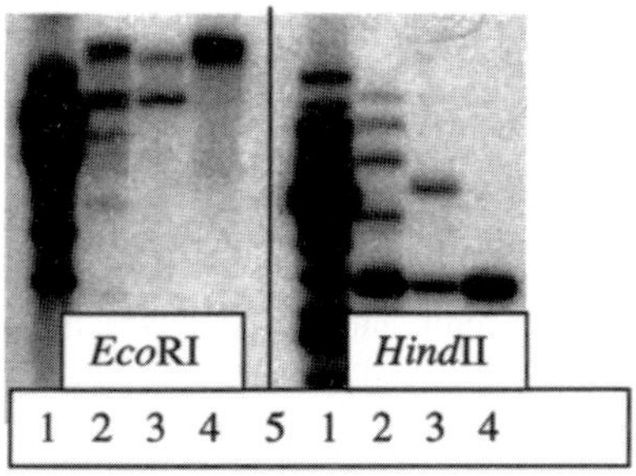

Figure 3: Southern hybridisation of transgenic *P. radiata* plants. Genomic DNA was digested with either the enzyme *Eco*RI or *Hind*III. 1-4: Transgenic lines, 5: non-transformed control.

This is typical for transclones that are produced using Biolistic® techniques, in contrast to *Agrobacterium* transformation where single or low copy numbers of the transgene can usually be expected. High copy numbers can have negative effects on gene expression stability and correct long-term gene expression. This is particularly important with trees that have generation times of 30 years and more.
Transgenic conifer tissue has also been assayed using ELISA (Enzyme Linked ImmunoSorbent Assays) to confirm the expression of *npt*II and the presence of the neomycin-phosphotransferase enzyme in transgenic tissue and plants [14]. The assay is very specific and sensitive and was positive for most transgenic tissue lines and also transgenic plants in the field and the greenhouse (Figure 4).

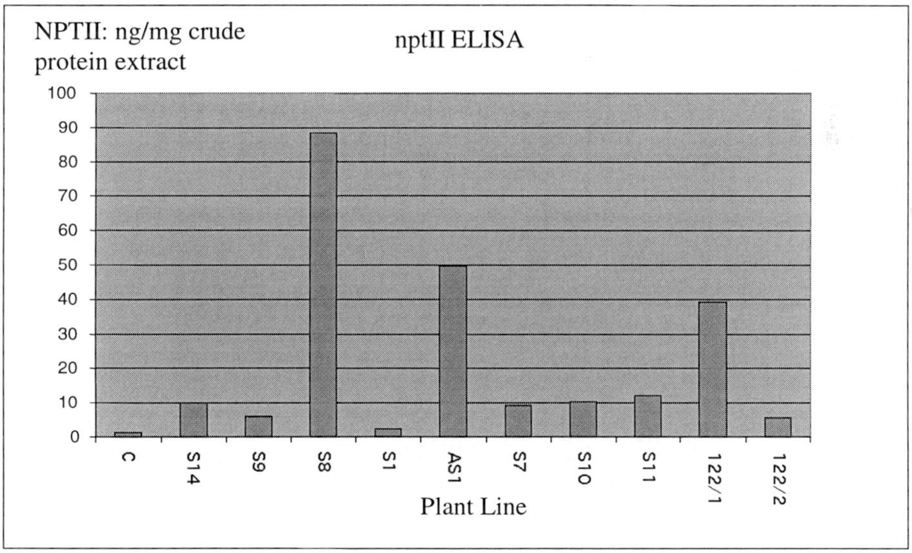

Figure 4: *npt*II ELISA with transgenic *Picea abies* plants grown in the greenhouse. Abbreviations: C: Non-transformed control; 1 and 2: Positive control plants (assayed positive in earlier experiments); a-h: transformed *Picea abies* plants.

In this assay, most transgenic lines, with exception of line S1 showed levels of the NPTII protein significantly higher than the control. Line S1 may have lost expression of the *npt*II gene, or the expression is very low compared to the other lines. The results also show that foreign gene expression is highly variable within the different lines tested. This may reflect a position effect of the introduced gene, copy number effects or silencing of transgenes after integration.

The *npt*II ELISA is now routinely used to confirm transgenic nature of selected lines early in the process. A similar technique was used to confirm the expression of the *Bacillus thuringiensis* toxin gene (*cry*IAc) in *P. radiata* embryogenic tissue (Figure 5). The 11 lines tested were co-transformed with a vector containing the selection gene *npt*II and a vector containing the Bt gene. Nine of the 11 lines confirmed transgenic by *npt*II-PCR, showed high levels of Bt toxin present in the cells. A further line, confirmed transgenic by *npt*II and *cry*IAc –PCR, did not express the *cry*IAc gene as evidenced by ELISA. Another line was negative in *cry*IAc PCR but positive in *npt*II PCR. The results indicate that the co-transformation frequency for the two unlinked genes was 91%.

FUNCTIONAL ANALYSIS: EXPRESSION OF A HERBICIDE RESISTANCE GENE IN TRANSGENIC CONIFERS:

P. radiata and *Picea abies* embryogenic tissue was transformed with a construct containing the *npt*II selection gene, and a second construct containing the *bar* gene for resistance against the herbicide phosphinothricin [15]. The gene product of the *bar* gene, a phosphinothricin-acetyl-transferase modifies phosphinothricin by the addition of an acetyl group. Acetylated phosphinothricin is no longer an active herbicide and is rapidly degraded by plants and microorganisms [16, 17].

Transgenic *P. radiata* and *Picea abies* plants were regenerated from 35 independent transformation events, and propagated in a greenhouse. Samples were spray tested with operational concentrations of the herbicide to confirm resistance. Six

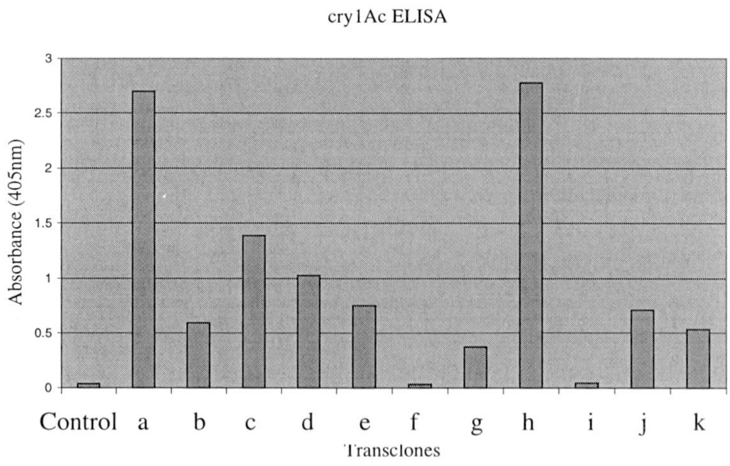

Figure 5: ELISA with *P. radiata* embryogenic tissue transformed with the *cry*1A gene. Abbreviations: control: Non-transformed control tissue; a-k: Transformed embryogenic lines.

weeks after spraying, all non-transformed control plants were dead, whereas transgenic plants continued to grow unaffected by the herbicide. This is the first demonstration of a commercially useful gene expressed functionally in a conifer. The experiments also demonstrated that even very low expression levels of *bar* (as determined by Northern blotting experiments) were sufficient to render transgenic plants resistant against the herbicide [15].

FUNCTIONAL ANALYSIS: EXPRESSION OF ENDOGENOUS *PINUS RADIATA* GENES AND THE FUNCTION OF THEIR PROMOTERS IN A HETEROLOGOUS SYSTEM:

Using a differential screening approach, a series of different cDNAs specifically expressed in male cone tissues of *P. radiata* were isolated. Northern hybridisation experiments confirmed the expression of the respective genes in male cone tissue, and absence of expression in other tissue types. Analysis of DNA sequences and homologies between the sequences, groups one of the cDNAs isolated (PrCHS1) with plant chalcone synthase genes, and another sequence (PrLTP2) with a clade of lipid transfer protein genes. *In situ* expression analysis confirmed the expression of PrCHS1 mainly in the tapetum of developing male reproductive structures, whereas PrLTP2 was confirmed to express mainly in developing microspores [18].

Based on sequence information and using a Clontech Genome Walker kit, the genomic upstream sequences of PrCHS1 and PrLTP2 were amplified and cloned upstream of a *uid*A reporter gene (Höfig, in preparation). *Arabidopsis thaliana* plants were transformed with these constructs using a vacuum infiltration technique [19]. Transgenic seed were selected by germination on antibiotic containing media and resulting plants were selfed to produce heterozygous transgenic populations. Plants from the T3 generation were grown to maturity and their *uid*A expression patterns analysed. Microscopical studies included phase-contrast and a very sensitive ultra-darkfield technology. The results indicate that the PrCHS1 promoter directs expression towards the *Arabidopsis* tapetum and to microspores at an early stage in development (pollen mother cells). The activity of the PrLTP2 promoter was also observed in these tissues, however the onset of expression was later, at the tetrad stage of microspore development.

This study demonstrates that sequences upstream of differentially expressed conifer genes are evolutionary conserved with respect to their function in two such distant species as *A. thaliana* and *P. radiata*. Further experiments will be based on these sequences to specifically direct expression of genes to the tapetum cells. As an example, a cytotoxin gene fused to this promoter has the potential to abolish tapetum function and lead to male sterility.

FUNCTIONAL ANALYSIS OF THE CAD GENE IN TRANSGENIC RADIATA PINE

Lignin biosynthesis is essential in terrestrial plants, contributing significantly to the properties of wood forming tissue. In pulp and paper making however, lignin needs to be removed. This is an expensive process and it generates environmentally unfriendly waste products. Therefore, various approaches were made to reduce lignin content or change its composition to improve the process and efficiency of pulp and paper production.

We are using both *sense* and *antisense* suppression to understand and manipulate lignin biosynthesis related gene expression.

Little data is available on the effectiveness of different suppression mechanisms based on *sense* or *antisense* constructs in gymnosperms. We compared the effect of both *sense* and *antisense* constructs in suppressing the endogenous expression of *cad*, a lignin related gene encoding for Cinnamyl Alcohol Dehydrogenase, in *P. radiata*. More than 100 transgenic *P. radiata* lines containing a *cad* sense or *antisense* construct were generated. CAD assays were performed to quantify the activity of intracellular CAD in transgenic and non-transgenic control tissue at different stages in development. Our results indicated that the *sense* construct used in this study was more efficient in suppressing the endogenous CAD activity compared to the *antisense* construct. The resulting CAD activity in suppressed lines was also monitored over time [Wagner in prep]. Figure 6 summarises the development of the CAD activity in line *cad sense1* in different tissue types over a period of approximately 18 months.

This line presented an over-expression phenotype at the embryogenic tissue stage, severe suppression at the seedling stage (approximately 6 weeks old plantlets) and an easing suppression level after 18 month of development. As a general rule, it was found that the suppression level was strongest in tissues with high endogenous CAD activity.

Results demonstrate that suppression can be dependent on developmental changes in the plant. The reduced level of suppression in older material may be a function of the promoter activity, or an indication for silencing of sense-overexpression constructs. Long term studies with transgenic trees may be necessary to fully understand this phenomenon.

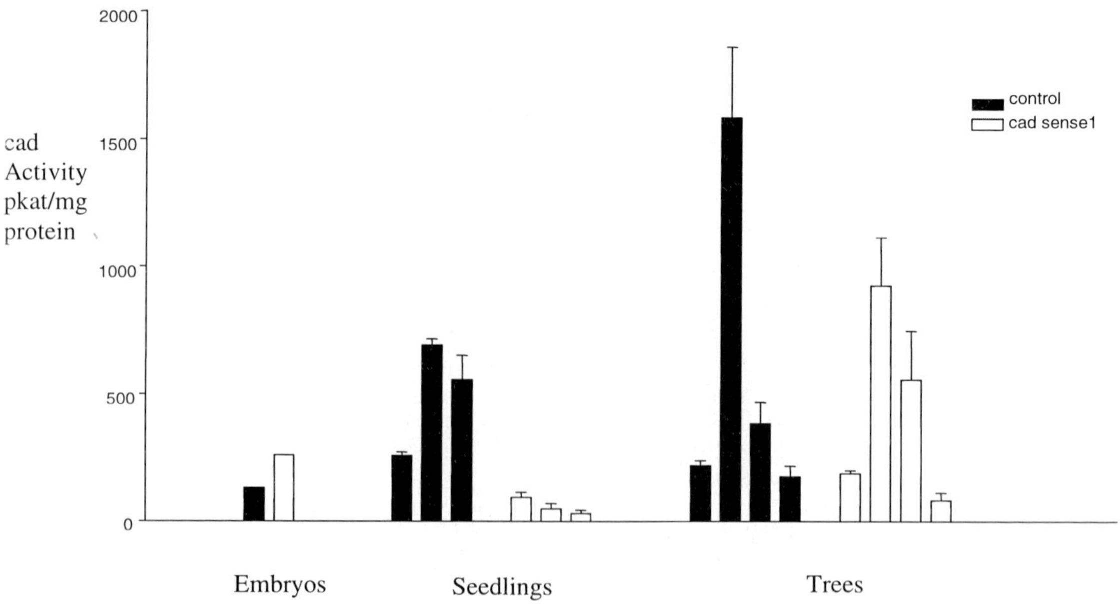

Figure 6: Development of CAD activity in transgenic line *cad sense1* and wildtype control G95-9 over time

SECONDARY CELL WALL FORMING TISSUE CULTURE CELLS AS AN EARLY ASSAY SYSTEM FOR WOOD DEVELOPMENT RELATED GENES:

The analysis of foreign gene expression in transgenic trees such as shown above, is time consuming and the long generation times of conifers often make efficient screening of candidate genes impossible. The problem is evident when the large number of potentially useful genes isolated from a variety of organisms, is compared with the numbers of genes and promoter / gene constructs that are analysed in current genetic engineering programs.

A possible solution is an approach where the function of genes is tested in tissue culture systems rather than developing and mature trees. Many genes could be tested simultaneously, and the technology should also be much more efficient in terms of time and space needed. Further, field-testing can be avoided and this should reduce cost for compliance with local and national regulations.

We have developed an approach where radiata pine cells are grown in tissue culture, and induced to form secondary cell walls. (Figure 7).

Current research concentrates on the development of a genetic engineering technology for cells developing secondary walls. These should be useful as an assay system for genes and promoters related to secondary cell wall formation. Protocols for the chemical analysis of very small amounts of cells were also developed. This involves pyrolysis followed by GC/MS, to evaluate lignin content of tissue culture samples (Figure 8).

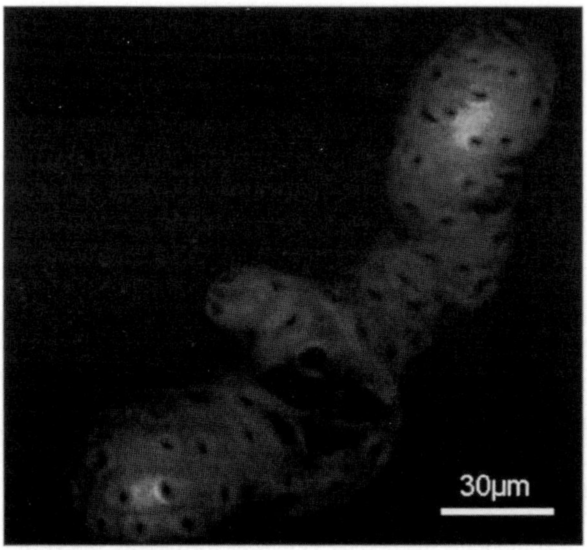

Figure 7: Confocal image of a tracheary element induced from *P. radiata* tissue cultured cells

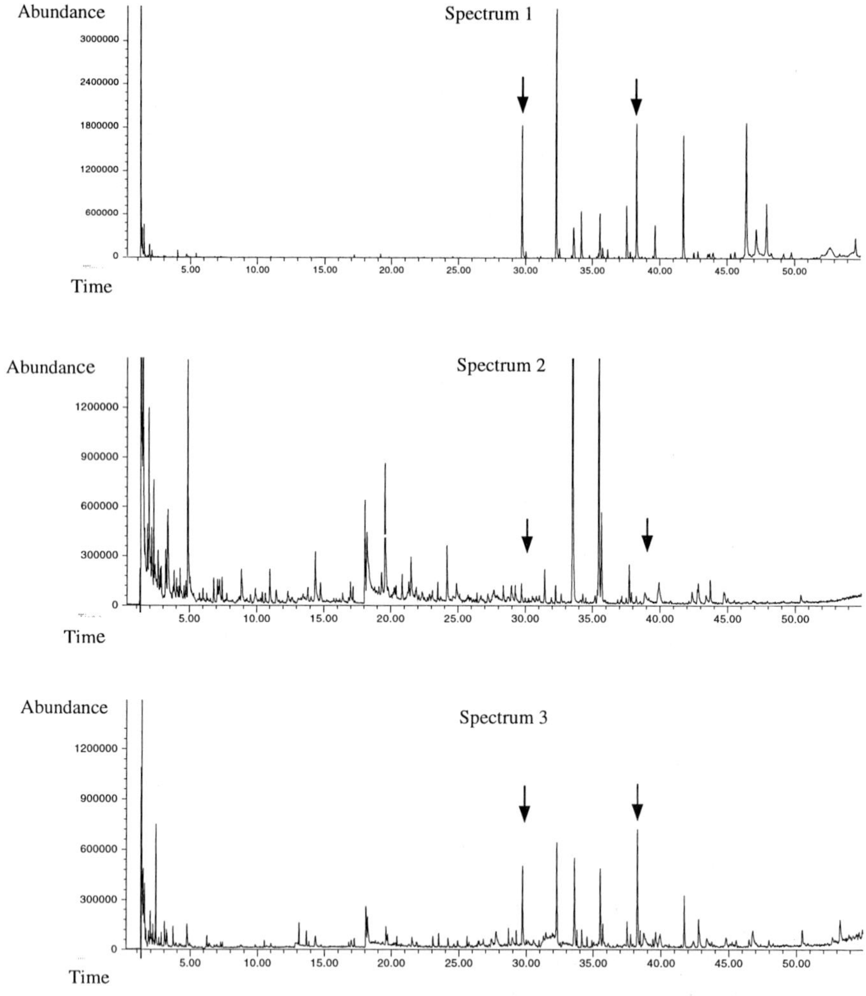

Figure 8: Gas chromatograms of milled wood *lignin P. radiata* (spectrum 1), parenchymatous radiata pine callus cells (spectrum 2) and secondary cell wall forming radiata pine callus cells (spectrum 3) after pyrolysis.

Spectrum 1 shows a typical chromatogram of milled wood lignin of *P. radiata* after pyrolysis. Spectrum 2 was obtained after pyrolysis of parenchymatous radiata pine callus cells grown on maintenance medium. The peaks, typical for lignin were significantly reduced in spectrum 2, indicating a low lignin content. After transfer of the cells to a medium inducing secondary cell wall formation, the corresponding spectrum shows typical lignin peaks (spectrum 3). The peaks found at the retention times of 29.75 and 38.28 are characteristic for guaiacol and 4-vinyl guaiacol respectively, which are breakdown products of softwood lignin.

CONCLUSIONS:

Protocols for genetic engineering of conifers have been developed and a range of genes have been successfully transferred into embryogenic tissue. Transgenic plants have been regenerated and molecular and functional analysis have confirmed the expression and function of introduced genes. Future challenges include the development of techniques to ensure correct expression of transgenes over the life span of engineered trees, and functional analysis of many candidate genes using quick and efficient functional assays.

REFERENCES

1. C. James, A. F. Krattiger, 'Global review of the field-testing and commercialisation of transgenic plants: 1986 to 1995, the first decade of crop biotechnology', International Service for the Acquisition of Agri-Biotech Applications (ISAAA) Brief No. 1. ISAAA, Ithaca, N. Y., 1996.
2. H.W. Kendall, R. Beachy, T. Eisner, F. Gould, R. Herdt, P. H. Raven, J. S. Schell, M. S. Swaminathan, 'Bioengineering of crops: report of the World Bank. Environmentally and Socially Sustainable Development Monographs', 1997. Series No. 23. World Bank, Washington, D. C.
3. M. L. Guerinot, 'the Green Revolution strikes Gold', Science 2000, 287, 7984-89.
4. C. Walter, S. D. Carson, M. I. Menzies, T. Richardson, M. Carson, 'Review: Application of biotechnology to forestry - molecular biology of conifers', World Journal of Microbiology and Biotechnology 1998, 14, 321-330.
5. M. J. Carson, R. D. Burdon, S. D. Carson, A. Firth, C. J. A. Shelbourne, T. G. Vincent, 'Realising genetic gains in production forests', in: Proceedings IUFRO working parties on Douglas fir, Lodgepole pine, *Sitka* and *Abies* spp. Breeding Genetic Resources. Session: Genetic gains in production forests. Olympia, Washington, 1989
6. C. J. A. Shelbourne, M. J. Carson, M. D. Wilcox, 'New techniques in the genetic improvement of Radiata pine. *Commonwealth Forest Review*, 1989, *68*: 3.
7. P. H. Wilcox, H. V. Amerson, G. Kuhlman, G. H. Liu, D. M. O'Malley, R. R. Sederoff, 'Detection of a major gene from resistance to fusiform rust disease in loblolly pine by genome mapping', Proceedings of the National Academy of Science, 1996, USA **93**, 3859-3864.
8. C. Walter, L. J. Grace, in: Molecular Biology of Woody Plants, Vol 2, S. M. Jain and S. C. Minocha (eds), 1999, Kluwer, pp 79-104.
9. T. Tzfira, A. Zuker, A. Altman, 'Forest-tree biotechnology; genetic transformation and its application to future forests', TIBTECH 1998, 16, 439-446.
10. D. R. Smith, 'Growth medium US patent number: 5, 565, 355', 1996.
11. T.M. Klein, E. D. Wolf, R. Wu, J. C. Sandford, 'High-velocity microprojectiles for delivering nucleic acids into living cells', Nature 1987, 327, 70-73.
12. A.Wagner, J. Moody, L. J. Grace, C. Walter, 'Stable transformation of *Pinus radiata* based on selection with Hygromycin B', New Zealand Journal of Forestry Science 1997, 27(3), 280-288.
13. C. Walter, L. J. Grace, A. Wagner, A. R. Walden, D. W. R. White, S. S. Donaldson, H. H. Hinton, R. C. Gardner, D. R. Smith, 'Stable transformation and regeneration of transgenic plants of *Pinus radiata* D. Don', Plant Cell Reports 1998, 17, 460-468.

14. C. Walter, L.J. Grace, S.S. Donaldson, J. Moody, J. E. Gemmell, S. van der Maas, H. Kvaalen, A. Loenneborg, 'An efficient biolistic transformation protocol for *Picea abies* (L) Karst embryogenic tissue and regeneration of transgenic plants', Canadian Journal of Forest Research 1998, 29, (10), 1539-1546.
15. S. L. Bishop-Hurley, R. J. Zabkievicz, L. J. Grace, R. C. Gardner, C. Walter, 'Conifer genetic engineering: transgenic *Pinus radiata* (D Don) and *Picea abies* (Karst) plants are resistant to the herbicide Buster', Plant Cell Reports, 2001, in print.
16. M. De Block, J. Botterman, M. Vandewiele, J. Dockx, C. Thoen, Gosselé, N. Rao Movva, C. Thompson, M. van Montagu, J. Leemans, 'Engineering herbicide resistance in plants by expression of a detoxifying enzyme', The EMBO Journal 1987, 6, 2513-2518.
17. W. Gotz, E. Dorn, E. Ebert, K. H. Leist, H. Kocher, 'HOE 39866, a new non-selective herbicide: Chemical and toxicological properties; Mode of action and metabolism', Proceedings of the Ninth conference of the Asian Pacific Weed Science Society, 1983.
18. A. R. Walden, C. Walter, R. C. Gardner, 'Genes expressed in *Pinus radiata* male cones include homologous to anther specific and pathogenesis response genes', Plant Physiology, 1999, 121, 1103-116.
19. N. Bechthold, G. Pelletier, 'In planta *Agrobacterium*-mediated transformation of adult *Arabidopsis thaliana* plants by vacuum infiltration', Methods Mol Biol 1998, 82, 259-266.

Molecular Breeding of Woody Plants
N. Morohoshi and A. Komamine, editors.

ANALYSIS OF WOOD DEVELOPMENT WITH A GENOMIC APPROACH: EUCALYPTUS ESTS AND TAC GENOMIC LIBRARY

Shigeru Sato, Keiko Horikiri, Kyoko Miyashita, Naoko Ishige, Takayuki Asada & Takashi Hibino

Forestry Research Institute Oji Paper Co. Ltd.
24-9 Nobobo-cho, Kameyama, Mie, 519-0212, Japan

ABSTRACT

A forest tree is one of the valuable biomass resources in industry. Understanding the physiology of wood development is important for utilization of the biomass. To get comprehensive understanding the development, we are collecting the genome-scale information in forest trees. ESTs are being obtained from differentiating xylem and leaves of several Eucalyptus species that are used as model trees. Purposes of this project are 1; to identify a large number of ESTs, 2; to analyze their expression profiles using cDNA arrays, 3; to compare Eucalyptus ESTs with other plant ESTs. On the basis of the genome-scale information, we want to focus on gene expression peculiar to forest trees. We are also constructing TAC (Transformation-competent Artificial Chromosome) genomic libraries in several Eucalyptus species. The current status of this project will be discussed.

KEYWARDS

Eucalyptus, genome, woody plant, EST, TAC, xylem

INTRODUCTION

Oji Paper is advancing the overseas forest plantation business for securing the forest resource and to protect the environment. We are making an effort to achieve 200,000 ha of overseas forest plantations by 2010. We will thus be able to acquire 70 percent of our paper and paperboard resources from collected post-consumer paper, while the remaining 30 percent will be from sawmill residue and planted trees from our own forests overseas. Oji's Forestry Research Institute supports the technical aspects of the forest plantation project in which we are engaged so that we can secure ample forest resources and preserve the environment. For example, whenever we plant trees overseas, we first conduct a soil survey, select a tree species that is appropriate for the land and climate, and conduct basic research on raising seedlings, breeding, and plantation.

To breed genetically improved *Eucalyptus* trees, we developed regeneration and transformation system in several *Eucalyptus* species (Fig. 1). It had been said that tissue culture was difficult in the *Eucalyptus* species. To dissolve the problem, we produced shoot primordia by using a new culture method [2]. Shoot primordium is like a meristem cluster and can regenerate to plantlet easily. In addition, we developed the precocious

Fig. 1 ***Eucalyptus*** **transformation system**

branch method that can introduce genes into *Eucalyptus* by *Agrobacterium*-mediated transformation. By these systems, we are creating the genetically improved *Eucalyptus* that can be produced low-cost and high-quality pulp.

A forest tree is one of the valuable biomass resources in industry. Understanding the physiology of wood development is important for utilization of the biomass. ESTs are one of the efficient approaches to obtain information about gene expression in wood development. Recently, Several woody plant ESTs has been reported and revealed a lot of unique unknown genes of them [3,4,7]. Functional analysis of the genes will contribute the understanding the wood development.

The *Eucalyptus* species are widely distributed in the Temperate Zone in the Southern Hemisphere and used as a raw material of pulp in Asian-Oceanic region. They grow more rapidly than pine and poplar, which are used as the material in North America and Europe.

The information of gene expression in *Eucalyptus* would become a useful tool in breeding not only *Eucalyptus* species but also other woody plants. However, a little is known about the information in *Eucalyptus* species.

To get comprehensive understanding the wood development, we are collecting the genome-scale information in *Eucalyptus*. ESTs are being obtained from differentiating xylem and leaves of several *Eucalyptus* species. Purposes of this project are 1; to identify a large number of ESTs, 2; to analyze their expression profiles using cDNA arrays, 3; to compare *Eucalyptus* ESTs with other plant ESTs. On the basis of the genome-scale information, we will focus on gene expression peculiar to forest trees. We are also constructing TAC genomic libraries in several *Eucalyptus* species.

A scheme of our molecular breeding is shown in Fig. 2. We are constructing *Eucalyptus* EST database and TAC genome libraries in this project. Based on the EST database, we will attempt to discover genes for cell wall synthesis and cell elongation. We are also searching loci for wood quality regulation by using molecular markers. The genes and the loci will be transferred into *Eucalyptus* trees.

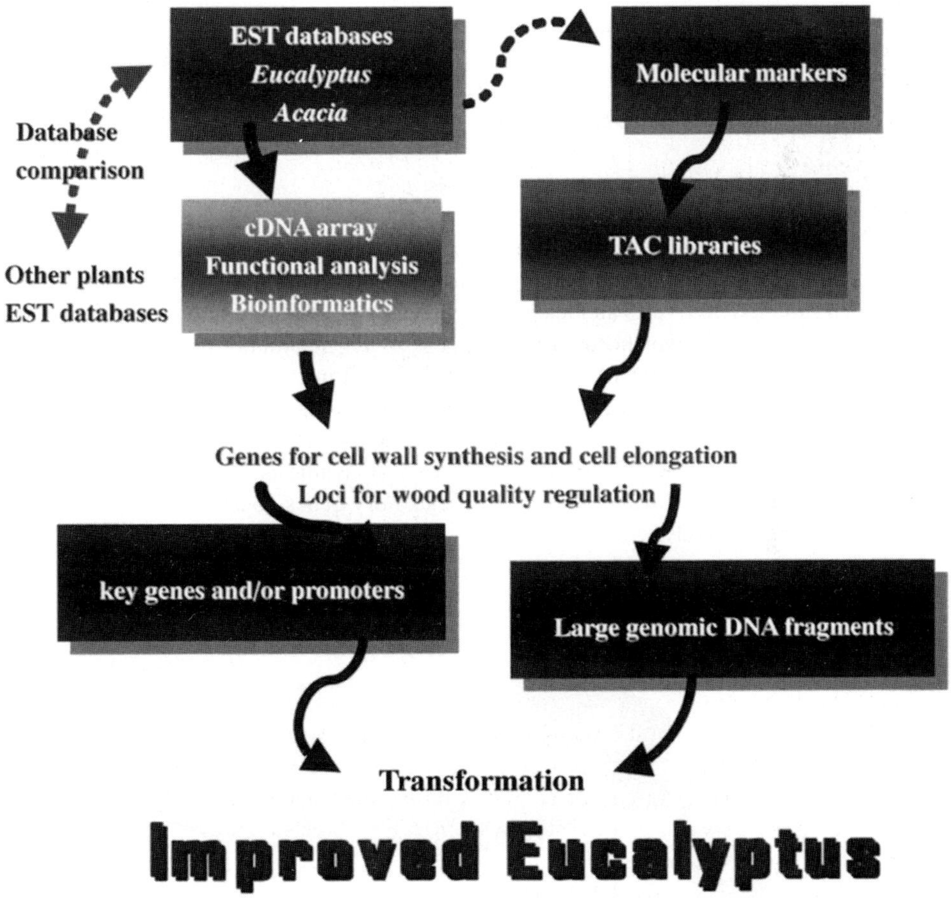

Fig. 2 Scheme of molecular breeding

MATERIALS & METHODS

Plant materials

About a three years old Eucalyptus camaldulensis (OJCA6; cutting clone), that has been grown at our Kameyama experimental field, was used as a material for DNA and RNA extraction.

DNA & RNA extraction

DNA and RNA extraction was performed with the Hibino et al's methods [1].

Construction of libraries

Several cDNA libraries were constructed using the SMART kit (CLONETECH). Large DNA fragments were prepared by the Peterson et al's methods [8].

Sequencing and analysis

Plasmid DNAs from cDNA clones were purified with the Mini-prep kit (Qiagen). Sequencing was performed with ABI 3700 sequencer and Li-cor 4200 sequencer.

RESULTS & DISCUSSION

cDNA sequencing and micro array analysis

We have started sequencing *Eucalyptus* cDNA clones from last October. Our first goal is collection of 50,000 cDNA sequence data in order to obtain a large number of ESTs derived from xylem tissues and leaves, respectively.

To date, we have sequenced 1152 templates of *E. camaldulensis* full-length cDNA library. Sequencing of cDNA clones were performed in the direction from 5-prime to 3-prime. BlastX search revealed that 68.6% of ESTs have similarity to genes whose functions could be assigned. Functions of the remaining 31.4% of ESTs were unknown.

Among the known ESTs, genes related to cell wall synthesis and cell elongation made up 6.5% of the total. Stress inducible and regulatory genes constituted 2.1% and 1.8% of the total respectively.

Unknown ESTs were divided into following 3 classes (Fig.3).

We also started analyzing expression profiles using cDNA arrays.

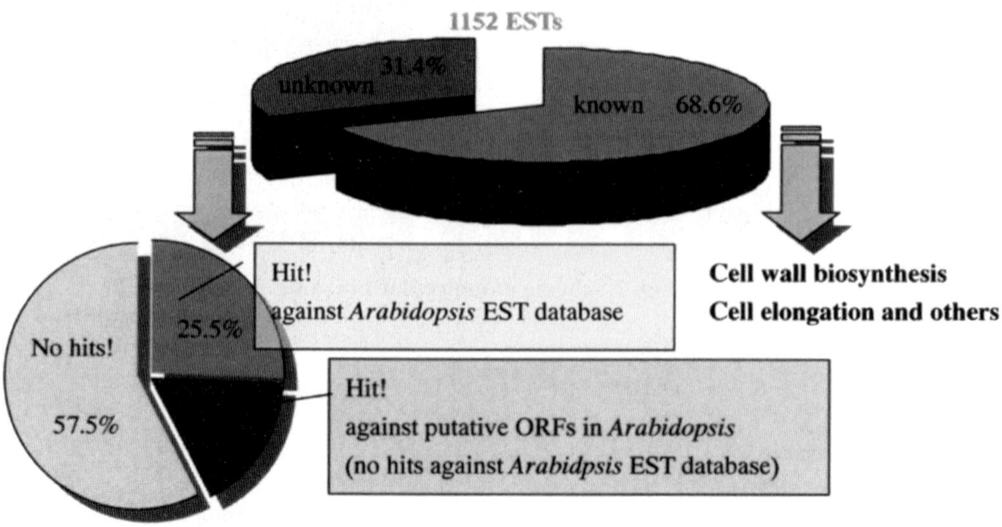

Fig. 3 Classification of the 1,152 ESTs from the xylem tissues

Construction of *Eucalyptus* TAC library

A TAC vector, which developed by Liu et al. [5], is suitable for stable maintenance of large genomic DNA fragments in both *Escherichia coli* and *Agrobacterium tumefacience*. The genome mapping and sequencing project of *A. thaliana* at Kazusa DNA Research Institute (Chiba, Japan) has been using the TAC library for preparing contigs and sequencing parts of chromosome III and V. The vector has the cis sequences required for *Agrobacterium*-mediated gene transfer into plants. Liu et al. showed that large genomic insert (*ca.* 80kb) of *Arabidopsis* TAC clones can be transferred into plants. The TAC system is useful for not only positional cloning of genes but also introducing quantitative trait loci (QTLs) covered with TAC contigs. Moreover, reduction of position effects can be expected in transformants because of the long insert.

To apply this system to our molecular breeding, we are constructing TAC libraries of several *Eucalyptus* species. Fig. 4 shows a scheme of TAC library construction. High molecular weight genomic DNA was extracted from young leaves of *E. camaldulensis* (OJCA6; cutting clone). The DNA was digested partially with *Hin*dIII and size-fractionated in the 50 to 150kb size ranges. The partially digested and size selected DNA fragments were ligated with *Hin*dIII-digested TAC and then used for transformation of *E. coli* DH10B by electroporation. Transformants carrying inserts were selected on LB agar plates containing 25 mg/ml kanamycin and 5% sucrose. We are arraying TAC libraries into 384 well micro-titer plates, now.

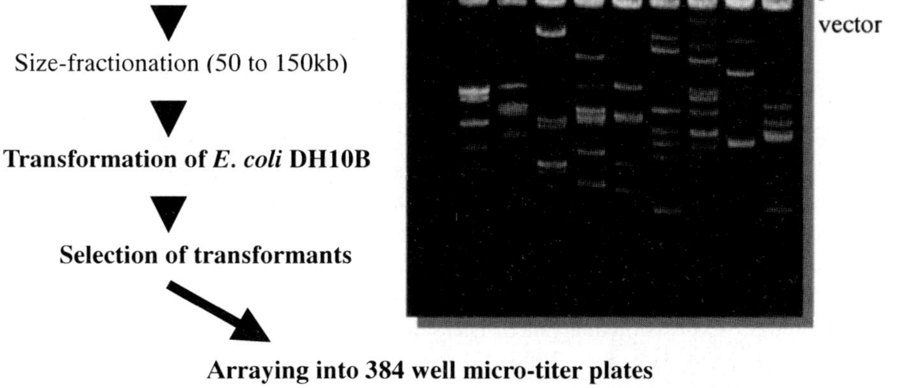

Fig. 4 Construction of *Eucalyptus* TAC genomic library

CONCLUSIONS

To get comprehensive understanding the wood development, we are collecting the genome-scale information in *Eucalyptus*. ESTs are being obtained from differentiating xylem and leaves of several *Eucalyptus* species.

We have sequenced 1146 templates of *E. camaldulensis* full-length cDNA library. Sequencing of cDNA clones were performed in the direction from 5-prime to 3-prime. BlastX search revealed that 68.6% of ESTs have similarity to genes whose functions could be assigned. Functions of the remaining 31.4% of ESTs were unknown.

We are arraying *E. camaldulensis* genomic (TAC) libraries into 384 well microtiter plates, now.

We will construct Eucalyptus EST database by this year.

We will make new and useful transgenic Eucalyptus trees for both industrial materials and one of solutions against current worldwide environmental problem in this century.

ACKNOWLEDGEMENTS

The authors acknowledge technical supports from Dr. Norihiro Mitsukawa (Toyota Central R&D Labs. Inc.)

REFERENCES

1. T. Hibino, K. Ito, M. Shibata, Extraction of nucleic acid from woody plant, JP patent, 08-080191, March, **1996**
2. K. Ito, K. Doi, Y. Tatemichi & M. Shibata, 'Plant regeneration of eucalypts from rotating nodule cultures' *Plant Cell Rep.*, **1996**, 16; 42-45
3. I. Allona, M. Quinn, E. Shoop, K. Swope, S. T. Cyr, J. Carlis, J. Riedl, E. Retzel, M. M. Campbell, R. Sederoff & R. W. Whetten, 'Analysis of xylem formation in pine by cDNA sequencing' *Proc. Natl. Acad. Sci. USA*, **1998**, 95; 9693–9698
4. F. Sterky, S. Regan, J. Karlsson, M. Hertzberg, A. Rohde, A. Holmberg, B. Amini, R. Bhalerao, M. Larsson, R. Villarroel, M. V. Montagu, G. Sandberg, O. Olsson, T. T. Teeri, W. Boerjan, P. Gustafsson, M. Uhlen, B. Sundberg & J. Lundeberg, 'Gene discovery in the wood-forming tissues of poplar: Analysis of 5,692 expressed sequence tags', *Proc. Natl. Acad. Sci. USA*, **1998**, 95; 13330–13335.
5. Y.-G. Liu, Y. Shirano, H. Fukaki, Y. Yanai, M. Tasaka, S. Tabata & D. Shibata, 'Complementation of plant mutants with large genomic DNA fragments by a transformation-competent artificial chromosome vector accelerates positional cloning', *Proc. Natl. Acad. Sci. USA,* **1999**, 96; 6535–6540
6. The Arabidopsis Genome Initiative, 'Analysis of the genome sequence of the flowering plant *Arabidopsis thaliana', Nature*, **2000**, 14; 796-815
7. T. Ujino-Ihara, K. Yoshimura, Y. Ugawa, H. Yoshimaru, K. Nagasaka & Y. Tsumura, 'Expression analysis of ESTs derived from the inner bark of *Cryptomeria japonica*' *Plant Mol. Biol,* **2000**, 43; 451-457.
8. D. G. Peterson, J. P. Tomkins, D. A. Frisch, R. A. Wing & A. H. Paterson, 'Construction of Plant Bacterial Artificial Chromosome (BAC) Libraries: An Illustrated Guide', *J. Agr. Genomics*, **2000**, 5 (http://www.ncgr.org/research/jag/papers00/paper300/indexpage300.html)

Molecular Breeding of Woody Plants
N. Morohoshi and A. Komamine, editors.

MODIFYING *POPULUS* ENVIRONMENTAL RESPONSES: IMPACTS ON WOOD QUANTITY AND QUALITY

Richard B. Hall[1]*, E.R. Hart[1,2], and Ilona Peszlen[1]

[1]*Department of Forestry, Iowa State University, 253 Bessey Hall, Ames, IA 50011-1021, USA*

[2]*Department of Entomology, Iowa State University, Ames, IA 50011-3222, USA*

ABSTRACT

Natural selection for environmental fitness has led to a wide variety of wood properties between and within trees. In the redesign of trees to better serve industry needs, we believe that a focus on modifying genetic control of environmental responses can simultaneously improve total wood yield, tailor wood properties for different uses, and reduce the within tree variation in wood properties. To do that, we need much more understanding of environmental signals, the physiological processes they elicit, and the resulting changes in wood formation. We are studying two types of environmental interactions that we know have large impacts on wood yield, competition effects and responses to pest attack, to determine the corresponding impacts on wood properties. In wild trees, both of theses environmental factors cause a shift in allocation of photosynthate away from stem wood formation to stem height growth and/or replacement of shaded or damaged leaves. Concurrent changes in hormone levels within the tree may impact wood quality traits such as specific gravity and fiber length. We have started evaluating model systems to look at the magnitude of these effects. We discuss results from a spacing trial and an insect impact study in this paper.

KEYWORDS

Biomass, specific gravity, insect impacts, shade-avoidance response, spacing studies

INTRODUCTION

The reason that trees produce wood is not to support our factories; woody stems evolved to support leaf canopies in the sunlight above their competition. Wood production responds to a complex system of environmental impacts on trees and natural selection has led to the best overall wood structure to support particular species of trees in their ecological strategies. Since environmental factors change over time, including changes in physical support requirements as trees grow bigger, wood structure changes throughout the stem of a tree (Fig. 1) [1, 2]. This variability suits the needs of wild trees well. This same variability is often a complication for industrial use, especially when it changes in unpredictable ways between feed stocks derived from different trees of the same species. In domesticating trees to serve our raw material needs, we should consider the whole wood production process as a system that can be re-bioengineered to produce more of what we want at lower unit costs. Understanding and reducing wood variability within and between trees should be one of the strategies. Advances in conventional breeding and genetic engineering have opened up a multitude of options for improving trees for growth rate, response to the environment, and wood quality [3].

However, we need to know more about our wood production system in the tree before we can make all the right changes. This paper will focus on biomass production and specific gravity as they are influenced in *Populus* clones by two environmental factors, defoliation by insects and the perception of competition from neighboring trees. Previous studies of specific gravity variation in poplar trees indicate they produce slightly higher specific gravity near the pith, then produce a zone of lower specific gravity in the remainder of the juvenile core before again producing higher specific gravity in the mature wood [2,4]. However, there can be more than 10% variability across even a short span of growth rings (Fig. 1).

The cottonwood leaf beetle, *Chrysomela scripta* F., is a major defoliator of young plantation *Populus* [5,6]. In Iowa, the insect typically produces three generations per year, but late, mild winters can lead to additional outbreaks at the end of the growing season. Some adult and early instar feeding occurs on recently formed, succulent leaves. Heavy infestations of late instars can spread to stem and branch terminals causing heavy damage. In most cases, trees have enough energy reserves and residual older leaves to refoliate and continue growing. When terminal shoots are heavily damaged, the subsequent flush of distal buds gives rise to a deformed, highly branched crown. Repeated heavy outbreaks during the first few growing seasons appear to have a negative compounding effect on tree growth [7].

Plants, especially shade intolerant ones like poplars, sense the presence of competing neighbors through the phytochrome pigments which "measure" the relative amounts of red and far-red light (R/FR) being received. Reflected light from adjacent plants reduces this R/FR ratio below 1.0. This change signals growth potential to shift into height to "avoid" being shaded by neighbors [8-10]. This confers a selective advantage on trees with a strong response to the competition signal because they are more likely to overtop their neighbors. However, experience with herbaceous crops indicates that

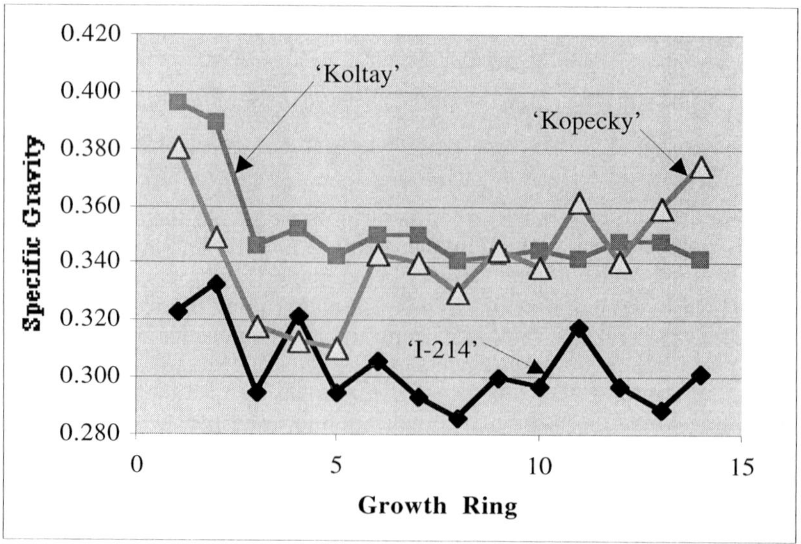

Figure 1. An example of specific gravity variation within trees and between clones in *Populus*. Adapted from previous work by Peszlen [2].

selection for less individual plant response to competition is a key to stand yield improvement [11]. Short-term experiments with poplars confirm that dramatic shifts in the allocation of photosynthetic energy occur when trees respond to the R/FR signal [12]. Good progress is being made in understanding this phenomenon at the molecular level in poplars [13, 14]. The research of Olsen and co-workers demonstrates that simple changes in the phytochrome system such as the over-expression of the phytochrome A gene can reduce the shade-avoidance response, but they are also likely to eliminate winter hardiness. We need much more information before a practical improvement strategy through genetic transformation can be devised. To begin getting that information, we have been trying to dissect what happens in spacing trials as competition develops [15].

MATERIALS & METHODS

Cottonwood leaf beetle impact study

Containerized rooted cuttings were planted in April 1998 for four clones: 'Eugenei' (*P. deltoides* Bartr. X *P. nigra* L.), NM2 (*P. nigra* X *P. maximowiczii* A. Henry), and two pure *P. deltoides* clones 91X04-03 and D105. The NM2 clone typically suffers only a low level of cottonwood leaf beetle damage compared to the other three clones. The planting is on a bottomland site near the Iowa State University Institute for Physical Research and Technology. A split-plot design was used with two treatments: spraying as necessary to control defoliation and control (natural levels of infestation and defoliation). Clones were randomly assigned to positions within each block of the split plots, a clone plot consisted of a 4 x 4 planting of trees at 2 m spacing within a row and 2.5 m between rows. All blocks were surrounded by one row of buffer trees ('I45/51', a *P. deltoides* x *P. nigra* hybrid). Sprayed plots were treated whenever leaf beetle damage began to show in the plantings. Depending on the timing and severity of the beetle outbreak, Novodor® (*Bacillus thuringiensis* (Bt) subspecies *tenebrionis*, Abbott Laboratories, North Chicago, IL), Raven® (Bt subsp. *kurstaki*, Ecogen Inc., Langehorne, PA), or Sevin® XLR PLUS (Aventis CropScience, Research Triangle Park, NC), were applied [16]. However, during the 1999-growing season, exceptionally heavy and desynchronized outbreaks of beetle feeding caused heavy damage in the sprayed plots during the last two weeks of July. There was a partial recovery in the first two weeks of August and then more severe defoliation that forced early growth cessation in all the clones except NM2.

After the third growing season (2000), one tree was sampled on the north and one tree on the south side of each clone plot in four sprayed and four natural infestation control plots. Diameter at breast height (1.3 m) was determined and used to calculate the stem cross-sectional area (basal area). Large diameter (12 mm) increment borings were taken as close to breast height as possible without sampling near the juncture of the stem and a large branch. Core samples were returned to the lab in plastic bags. The bark was removed and the core was cut just outside the pith. Specific gravity was determined by the water displacement method for measuring sample volume [17]. Basal area and specific gravity data were analyzed as a nested design model, with replications nested within the two treatments.

Nelder competition trial

Our first Nelder wheel spacing trials were established in 1994 [15, 18]. Two *P. deltoides* X *P. nigra* clones, Eugenei and I45/51, were planted at the Iowa State University Iden Farm at Ames, IA. In 1995, another Nelder wheel spacing trial was established for the *P. alba* X *P. grandidentata* hybrid clone 'Crandon' on an upland site near the Iowa State University Institute for Physical Research and Technology. The layout for all of these trials was the same, with 17 wheel spokes, 8 trees per spoke and essentially square spacing that increased outward along a spoke from 0.25 m to 3.2 m. Detailed measurements of height and diameter growth were taken on the Iden Farm plots at the end of the second field season as crown competition became intense at the closer spacings. In subsequent years, only qualitative observations of growth were taken on all the plots, until a harvest was conducted on three randomly chosen spokes in the Crandon trial at the end of the sixth growing season. Total fresh weight of stems and live branches was determined for each of the 24 trees. Representative samples of stem and branch material were oven-dried to develop a fresh weight/dry weight conversion factor. Diameters and cross sections of the stems were taken at three positions along the stem: 1.2 m, the base of the last growing season's height increment, and at a point intermediate to the first two that was as close as possible to half way while avoiding stem segments where major branching occurred. Stem taper was calculated as the change in stem diameter per unit length of stem between sample points. The cross section pieces were debarked and specific gravity was determined by the water displacement method for measuring sample volume [17]. Analysis of variance on all data used the statistical model for completely randomized designs even though the Nelder design does violate some elements of randomization.

RESULTS & DISCUSSION

Cottonwood leaf beetle impact study

There were statistically significant differences between clones (P = 0.0001) in basal area production with clone 91X04-03 having the most average growth, followed by NM2 in second place, and Eugenei and D105 in third place and not significantly different from one another (Table 1). On average, the trees that were sprayed to reduce insect feeding produced 27 cm^2 greater basal area growth over three years than did the unprotected trees (statistically significant at P = 0.0001). Growth losses due to the insect feeding ranged from a high of 54.1 % for the fastest growing clone, 91X04-03, to 39.3 % for clone NM2 which the beetles are less prone to attack. Indeed, the basal area growth superiority of clone 91X04-03 compared to NM2 is eliminated in the unprotected plots. So, clearly, the cottonwood leaf beetle can seriously reduce the stem growth of trees if the trees are not protected by sprays or some host-plant genetic resistance.

The impact on wood specific gravity is somewhat more complicated. The statistical analysis is significant for a clone effect (P = 0.0001). It is interesting to note that the fastest growing clone in terms of basal area production had the lowest specific gravity and this negative correlation between clonal growth rate and specific gravity extended across all four clones (r = -0.96, P = 0.01) (Table 1). Further work is needed to determine how pervasive this negative association is between external morphology and specific gravity and how it might be broken by conventional or biotechnological improvement efforts.

Table 1. Basal area production and specific gravity in four clones as influenced by cottonwood leaf beetle feeding in sprayed and non-protected plots. Means ± standard deviations.

		Clone			
		91X04-03	NM2	D105	Eugenei
Basal Area (cm2)	Sprayed	86.2 ± 7.6	65.2 ± 5.7	40.4 ± 6.8	39.5 ± 3.3
	Not Protected	39.5 ± 13.4	39.6 ± 8.5	22.2 ± 3.6	22.6 ± 5.4
	Loss to insects	54.1%	39.3%	45.2%	42.9%
Specific Gravity	Sprayed	0.304 ± 0.006	0.319 ± 0.011	0.353 ± 0.010	0.362 ± 0.010
	Not Protected	0.293 ± 0.011	0.320 ± 0.018	0.345 ± 0.013	0.356 ± 0.016
	Decrease without insect protection	3.87%	-0.48%	2.24%	1.66%

The average difference in specific gravity between sprayed and unprotected trees of the same clone was only significant if we relax our criteria to P = 0.07. But a biologically significant trend appears to be present in the data. The clone (91X04-03) with the largest basal area growth reduction had the largest drop in specific gravity in connection with defoliation by beetles. Clone NM2 showed the least impact from the beetles in both basal area growth and specific gravity. The remaining two clones fit this pattern and the overall correlation between percent loss in basal area and percent loss in specific gravity is 0.95, which is statistically significant at the 1% probability level. It is plausible that the level of insect defoliation and consequent reduction in photosynthate availability is reducing allocations to cell wall thickness. It is also possible that increased auxin flow from shoots that are resprouting after an insect outbreak is increasing the rate of new xylem cell production faster than new allocations of carbohydrates to cell wall development. In either case, these results suggest that cultural or genetic pest management strategies could increase total stem volume production and the specific gravity of the wood produced. It is also possible that these relationships cause some of the variability in wood properties across a growth ring as outbreaks of leaf feeding rise and fall. More detailed studies of the anatomy of wood produced at specific times will be needed. Furthermore, comparison tests, similar to what we report here, should be conducted with the genetically transformed "Bt-poplars" as they become available [19, 20]. This would determine if the protection they offer against the cottonwood leaf beetle also has an effect on wood properties.

Nelder competition trial

Our first results were obtained during the second field season as crown competition became intense at the closer spacings in the Nelder design at our Iden Farm (Fig. 2). For both clones, there was an increase in height growth at spacings below 2 m. Both clones reached their maximum two-year height at spacings of 0.5 to 0.7m. Clone 5326 increased its height growth by 39%. Over this same change in spacing intervals, the

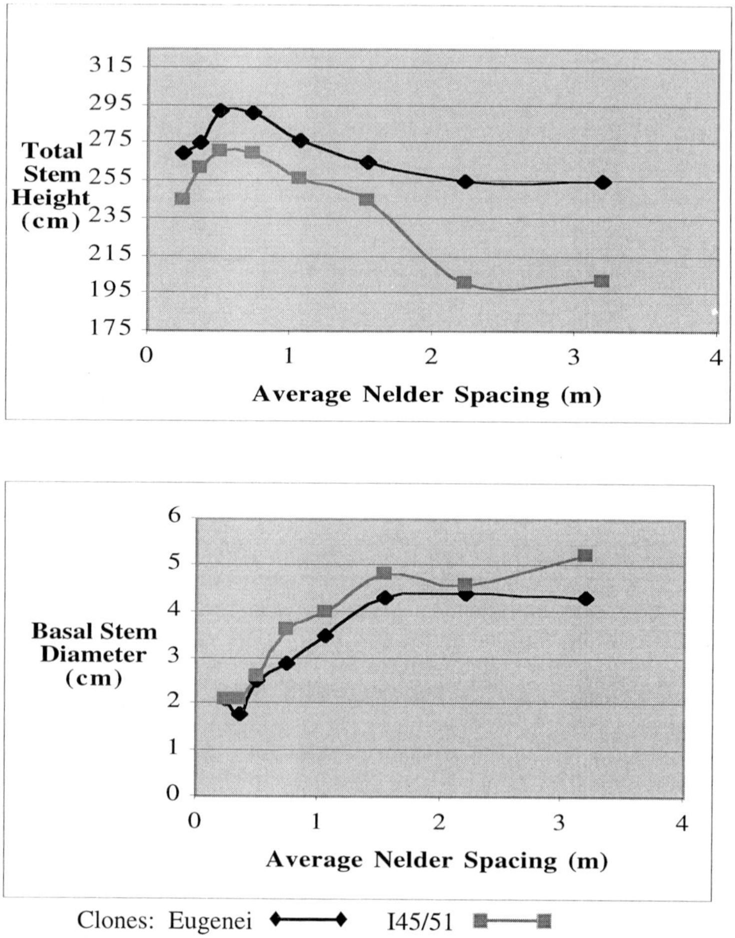

Figure 2. Changes in average stem height and diameter for two poplar clones as a function of spacing in a Nelder (1962) design [18].

R/FR signal is changing from ≥ 1.0 to about 0.8 in the shoot apex zone. Actual shading and reduction of photosynthetically active wavelengths started occurring in the lower crowns at about 1.5 m. At closer spacings, the actual shading reduced whole tree growth capabilities below the height growth stimulating effect of the shade avoidance response; height growth fell off precipitously at the two narrowest spacings. The growth trends for stem diameter are the reverse of height growth over most of the range of spacings, however, both clones continued a fairly consistent, high-rate of diameter growth until spacings fell below 1.5 m. Then diameter growth fell to 49% of maximum in the Eugenei clone and 42% in clone I45/51. The slight rise in diameter growth at the closest spacing for Eugenei was due to shade-induced mortality eliminating the weaker stems, providing more growing space to the stronger stems. These data suggest that as poplars just begin sensing the competition signal, there is a rise in the height growth rate, but no initial loss in diameter growth. As the competition signal becomes stronger, and certainly after actual shading of the lower canopy begins, the loss of diameter growth more than offsets increases in height growth and the total production of harvestable biomass per stem declines. These data also suggest that different poplar clones have different levels of response to competition.

The pattern in height and diameter growth seen at the end of the second growing season continued to shift outward in the Nelder spacings as the trees grew bigger. After six growing seasons in our Nelder trial with the Crandon clone, the shift in stem allocation patterns induced by the shade avoidance mechanisms was taking place over the wider three spacings shown at the top of Fig. 3. At all narrower spacings, actual shading had reduced the live crown to such an extent that overall stem growth was substantially diminished. Therefore, we only studied specific gravity and our other traits of interest in the trees from those last three spacings. Only three replicate spokes of this study have been harvested so far; more replicates should be analyzed before final conclusions are reached. However, some interesting hypotheses are emerging that warrant such further study.

The shade-avoidance phenomenon is definitely expressed in poplars. As the trees in the Nelder spacing trials grew larger, the spacing at which the tallest trees were found moved outward on the spokes to the wider spacings (compare Figs. 2 and 3). At six years of age for the Crandon clone, the optimal spacing for height, stem biomass, and total above ground biomass per tree is at 2.22 m. A similar increase in stem dry weight was observed at some of the closer spacings in a greenhouse Nelder study with potted Douglas-fir seedlings [21]. Whether this represents increased photosynthetic capabilities, or shifts in allocation between stems and roots, remains to be determined.

One might conclude from these data that the shade-avoidance response is increasing total wood yields by making the trees grow taller faster, thereby building an optimum canopy sooner. However, in a shade intolerant species like the poplars, faster height growth means faster senescence of the lower crown [12]. And, of course, maximizing the growth of individual trees is not the objective in production plantations; total harvestable yield of useable raw material is the objective. Extrapolating the data for stem biomass shown in Fig. 3 to a single-spacing, plantation basis, the type of trees produced at the 1.54 m spacing would yield 61.6 t/ha at 6 years-of-age, while the bigger trees grown at the 2.22 m spacing would produce only 47.3 t/ha. Furthermore, without a strong height growth response to competition at narrower spacings, the optimum spacing for biomass production would not increase so rapidly as rotations grew longer. With narrower spacings of trees producing biomass at optimum rates, total productivity could be increased dramatically, just as it has been for modern agriculture crops as they have been bred to respond less to competition in high-density plantings [11].

What would happen to wood quality characteristics and their variation within the tree? Much more study is needed. However, our data in Fig. 3 indicate that the lowest level of specific gravity variation within trees was found at the widest spacing where the trees are expressing less of the shade-avoidance response. Increases in both auxin and gibberellin production and movement down the stem have been shown to be involved in the stem elongation that is characteristic of the shade-avoidance response [13, 22]. It is possible that it is this extra hormone effect, produced by rapidly expanding crowns, coupled with shifts in carbohydrate sinks, that is partially responsible for the poorer quality and higher variability of juvenile wood. We are developing a bioassay to follow these phenomena under more controlled conditions [23].

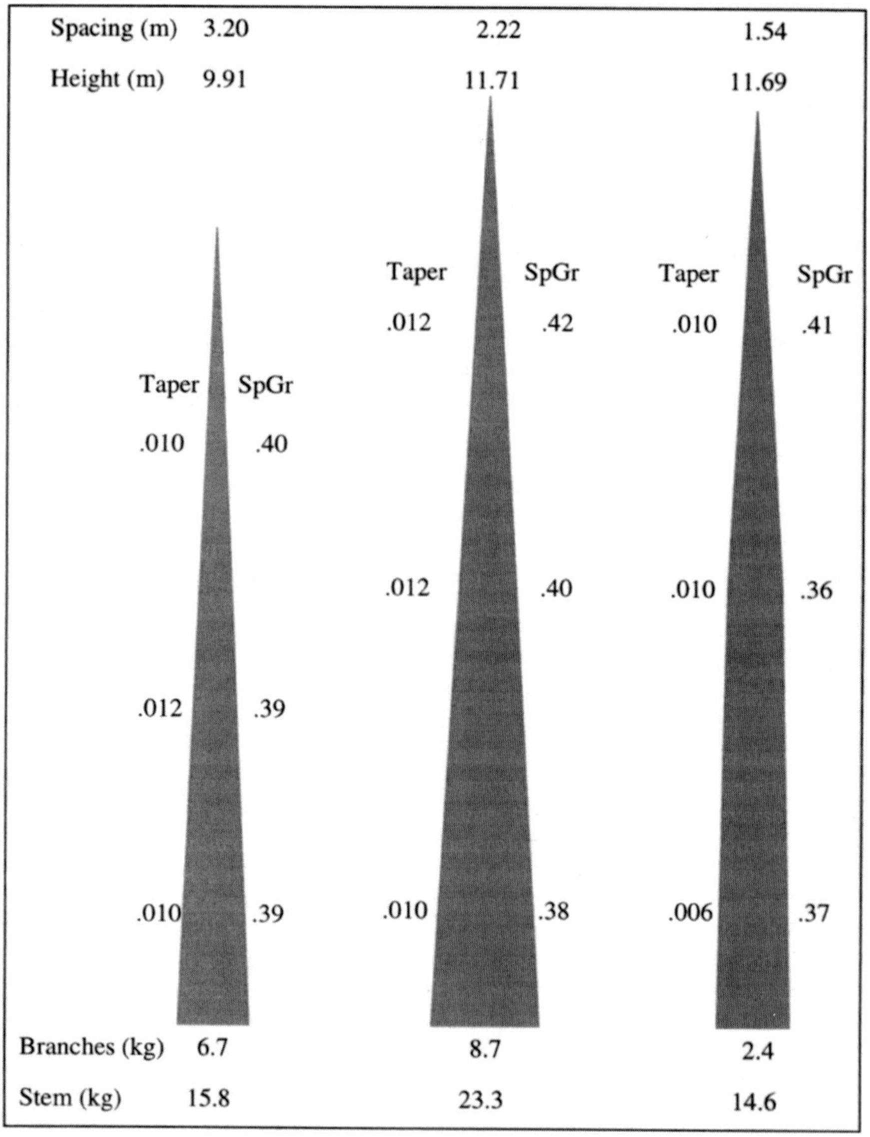

Figure 3. Average values for six-year-old Crandon trees grown at three spacings in a Nelder design. Tree diameters not drawn to scale.

CONCLUSIONS

We have much to learn about how environmental variables interact with tree genetics and physiology to influence the wood quality and variability of our industrial plantations. The two examples reported here need further investigation and we need to begin linking all of the potential interactions into a model of how our wood factory, the tree, really works. As we do that, we should be able to make better progress in increasing wood yields and making wood quality more uniform.

ACKNOWLEDGEMENTS

Our appreciation is extended to our student research aids, Joel Jepsen, Aaron Kejld, Scott Dowling, and Jared Anderson, who maintained the field plots and collected much of the data. Special recognition goes to our former postdoctoral assistant, Joel McMillin, and M.S. alumnus, David Coyle, for designing and managing our insect impact study. Current Ph.D. student Ron Zalesny performed the statistical analyses reported here. Major funding support for this project was provided by the Bioenergy Feedstock Development Program of the U.S. Department of Energy under contract DE-AC05-00OR22725 with University of Tennessee-Battelle LLC. This is Journal Paper No. J-**19288** of the Iowa Agriculture and Home Economics Experiment Station, Ames, IA, project No. 3443, with additional funding support by Hatch, McIntire-Stennis, and State of Iowa Funds.

REFERENCES

1. I. Peszlen, 'Influence of age on selected anatomical properties of *Populus* clones', *IAWA Journal*, **1994**, 15 (3), 311-321.
2. I. Peszlen, 'Variation in specific gravity and mechanical properties of poplar clones', *Drevársky Vyskum – Wood Research*, **1998**, 43 (2), 1-17.
3. R. J. Dinus, P. Payne, M. M. Sewell, V. L. Chiang & G. A. Tuskan, 'Genetic modification of short rotation poplar wood: Properties for ethanol fuel and fiber production', *Critical Rev. Plant Sci.*, **2001**, 20 (1), 51-69.
4. A. D. Yanchuk, B. P. Dancik & M. M. Micko, 'Variation and heritability of wood density and fibre length of trembling aspen in Alberta, Canada', *Silvae Genetica*, **1984**, 33, 11-16.
5. J. D.Solomon, 'Impact of insects on growth and development of young cottonwood plantations', *Res. Pap. SO-217*, **1985**, U.S. Department of Agriculture, Forest Service, Southern Forest Experiment Station, p. 6.
6. B. R. Bingaman & E. R. Hart, 'Feeding and oviposition preferences of adult cottonwood leaf beetles (Coleoptera: Chrysomelidae) among *Populus* clones and leaf age classes', *Environ. Entomol.*, **1992**, 21, 508-517.
7. R. R. Reichenbacker, R. C. Schultz, & E. R. Hart, 'Impact of artificial defoliation on *Populus* growth, biomass production, and total non-structural carbohydrate concentration', *Environ. Entomol.*, **1996**, 25, 632-642.
8. C. L. Ballaré, R. A. Sanchez, A. L. Scopel & C. M. Ghersa, 'Early detection of neighbor plants by phytochrome perception of spectral changes in reflected sunlight', *Plant Cell Environ.*, **1987**, 10, 551-557.
9. H. Smith, 'Light quality, photoperception and plant strategy', *Ann. Rev. Pl. Physiol.*, **1992**, 33, 481-518.
10. A. von Arnim, X.-W. Deng, 'Light control of seedling development', *Ann. Rev. Plant Physiol. Plant Mol. Biol.*, **1996**, 47, 215-243.
11. D. N. Duvick, 'Maize breeding: Past, present, and future', Paper presented at *XX Congresso Nacional de Milho e Sorgo*. Goiania, Goias, Brasil, **1994**

12. I. R. Gilbert, G. P. Seavers, P. G. Jarvis & H. Smith, 'Photomorphogenesis and canopy dynamics: Phytochrome-mediated proximity perception accounts for the growth dynamics of canopies of *Populus trichocarpa x deltoides* 'Beaupré'', *Plant Cell Environ.*, **1995**, 18, 475-497.
13. J. E. Olsen, O. Juntilla, J. Nilsen, M. Eriksson, I. Matinussen, O. Olsson, G. Sandberg & T. Moritz, 'Ectopic expression of oat phytochrome A in hybrid aspen changes critical daylength for growth and prevents cold acclimatization', *Plant J.*, **1997**, 12, 1339-1350.
14. B. E. Frewen, T.H.H. Chen, G. Howe, J. Davis, A. Rohde, W. Boerjan & H.D. Bradshaw, Jr., 'QTL and candidate gene mapping of bud set and bud flush in *Populus*' *Genetics*, **2000**, 154, 837-845.
15. R. B. Hall, 'Use of the crown competition factor concept to select clones and spacings for short-rotation woody crops', *Tree Physiology*, **1994**, 14, 899-909.
16. D. R. Coyle, J. D. McMillin, S. C. Krause, and E. R. Hart, 'Laboratory and field evaluations of two *Bacillus thuringiensis* formulations, Novodor and Raven, for control of cottonwood leaf beetle (Coleoptera: Chrysomelidae)', *J. Econ. Entomol.*, **2000**, 93, 713-720.
17. J. G. Haygreen & J. Bowyer, In: *Forest Products and Wood Science*, Iowa State University Press, Ames, **1996**, pp. 191-212.
18. J. A. Nelder, 'New kinds of systematic designs for spacing experiments' *Biometrics*, **1962**,18, 283-307.
19. R. Meilan, C. Ma, S. Cheng, J. A. Eaton, L. K. Miller, R. P. Crockett, S. P. DiFazio, & S. H. Strauss, 'High levels of RoundupÆ and leaf-beetle resistance in genetically engineered hybrid cottonwoods', Pages 29-38 In: *Hybrid Poplars in the Pacific Northwest: Culture, Commerce and Capability*, K.A. Blatner, J.D. Johnson, & D.M. Baumgartner, (eds.), Washington State University Cooperative Extension Bulletin MISC0272, Pullman, WA, USA, **2000**, pp. 29-38.
20. H. Rao, M. Huang, M. Wang, N. Wu, & Y. Fan, 'Insect resistance transgenic poplar', In: *Molecular Breeding of Woody Plants - Proceedings of the International Wood Biotechnology Symposium*, Morohoshi N and Komamine A (eds.), Elsevier **2002** (this volume).
21. G. A. Ritchie, 'Evidence for red:far red signaling and photomorphogenic growth response in Douglas-fir (*Pseudotsuga menziesii*) seedlings', *Tree Physiology*, **1997**, 17, 161-168.
22. F. D. Beall, E. C. Yeung & R. P. Pharis, 'Far-red light stimulates internode elongation, cell division, cell elongation, and gibberellin levels in bean', *Can. J. Bot.*, **1996**, 74, 743-752.
23. H. Bae, 'Response of plant growth and development to different light conditions in three model plant systems', *Ph.D. Dissertation*, Iowa State University, **2001**, pp. 37-59.

Molecular Breeding of Woody Plants
N. Morohoshi and A. Komamine, editors.

TWO INSECT-RESISTANT GENES WERE TRANSFERRED INTO POPLAR HYBRID AND TRANSGENIC POPLAR SHEW INSECT-RESISTANCE

Hongyu Rao[1], Ningfeng Wu[2], Minren Huang[1], Yunliu Fan[2]& Mingxiu Wang[1]

[1] *The Laboratory of Forest Genetics & Gene Engineering, Nanjing Forestry University, Nanjing 210037, China*

[2] *Institute of Biotechnology Research, Chinese Academy of Agricultural Sciences, Beijing 100081, China*

ABSTRACT

Modified Bt CryIA gene and cowpea trypsin inhibitor (CpTI) gene were co-integrated into the genome of clone NL-80106, *P. deltoides* Marsh. × *P. simonii* Carr. by co-infection with *Agrobacterium tumefaciens* LBA4404. PCR and PCR-Southern blotting analysis showed that both Bt CryIA and CpTI genes were integrated into the transgenic poplar. Insect bioassay indicated that two-gene transgenic poplar were significantly resistant to the larvae of *Lymantria dispar*, whereas the untransformed control plants were sensitive to them. It caused a decreased leaf consumption by larvae, a lower larval weight and a higher larval motility . ELISA analysis of the transgenic clones proved that Bt and CpTI genes were expressed . Bt and CpTI transgenic poplar plants can be used prospectively as the material for selecting highly and long-term insect-resistant poplar clones in the future.

KEY WORDS

P. deltoides × *P. simonii*, *Bacillus thuringensis* (Bt) toxin gene, cowpea trypsin inhibitor (CpTI) gene, insect-resistance

INTRODUCTION

Poplar is an important woody species of artificial forests, with the quantities of fast growth rate, short rotation cycle, and easy vegetative propagation. It is the primary source of wood in many countries and in timber and paper industries. However, diseases and insect pests occur easily in poplar artificial plantations because of large areas and simple diversity. Conventional breeding of poplar takes long time and can not work effective on the genetic improvement of insect-resistance due to the lack of gene resources on poplar genetic background. The method of plant genetic engineering, with the ability to introduce new desirable traits in plant, could contribute to the expansion of the gene pool of poplar species by transferring the useful genes such as those for resistance to disease, insect pest and herbicides into poplar. Insect-resistant transgenic poplar plants were produced by McCown BH et al (1991), Tian Y-C et al (1993) by introducing Cry1 A gene from *Bacillus thuringiensis* (Bt) into poplar. Confalonieri et al (1999) recently reported that the expressing of a soybean Kunitz proteinase inhibitor gene (*KTi3*) in transgenic *Populus nigra* L. did not show any detectable resistance to the tested insects, although transgenic rice plants harboring an introduced potato proteinase inhibitor II gene are insect resistant (Duan X et al., 1996).

The development of resistance to Bt toxins in laboratory and field populations of pests were reported (McGaughey et al., 1992; Tabashnik, 1994). The use of more than a single resistance gene has been presented as one of the resistance management strategies proposed for transgenic plant (Roush, 1997). The technology for introducing and expressing multiple transgenes in crops is available (Hua X et al., 1993; Chen et al., 1998). Hua et al (1993)and Zhao et al (1995) proved respectively that the transgenic tobacco co-expressing Bt and CpTI genes showed stronger resistant to the tested insect compared to the one only expressing Bt gene, which means that the two difference resistance mechanisms are compatible.

We introduced CpTI gene and partially modified *Cry1 A* gene into poplar hybrid clone NL-80106 (*Populus deltoides* L.×*Populus simonii* Carr.) by *Agrobacterium tumefacien*-mediated transformational system. Transgenic poplar plants co-expressing two genes showed strong insect-resistance to the larva of *Lymantria dispar* by fed with the leaves of *in vitro* transgenic plants.

MATERIALS & METHODS

Plant materials

In vitro-grown plantlets of poplar hybrid clone NL-80106 (*Populus deltoides* L.× *Populus simonii* Carr.) are maintained in MS medium at our lab.

Agrobacterium strains and plasmids

Agrobacterium tumefaciens LBA4404 was gifted by Institute of Biotechnology Research, Chinese Academy of Agricultural Sciences. Binary plasmid pFWZ10, harboring partially modified *Cry1 A* gene, and pFWY1, harboring CpTI gene, were cloned and constructed by the laboratory of Prof. Fan Yunliu, Institute of Biotechnology Research, Chinese Academy of Agricultural Sciences. (Fig. 1). The two plasmids were respectively transferred into *A. tumefaciens* LBA4404 by the freeze-melt method to construct the strains of LBA4404 pFWZ10 and LBA4404 pFZY1 for the use of poplar transformation.

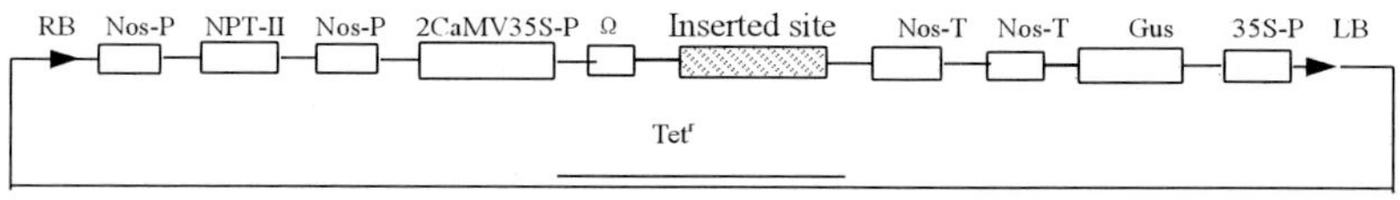

Fig. 1 The scheme of the binary plasmids pFWZ10 and pFZY1
Bt toxin gene and CpTI gene were respectively inserted to the site of 'Inserted site' to construct the binary plasmids of pFWZ10 and pFZY1

Transformation by *Agrobacterium tumefaciens*

Leaf discs (5×10 mm) obtained from *in vitro* plantlets of NL-80106 (*P.deltoides* L. ×*P. simonii* Carr.) were pre-cultured, co-cultivated and transferred to selective regeneration medium using the conditions of Rao HY et al (2000). Two strains of LBA4404 pFWZ10 and LBA4404 CpTI were respectively cultured and re-suspended by MS liquid medium at the same density of OD_{600}=0.3~0.4, then mixed completely and used in poplar transformation.

Molecular analysis

Extraction of DNA from poplar leaves, PCR analysis, PCR-Southern blotting, extraction of leaf soluble protein and ELISA analysis were performed as described by Wu et al (2000).

Insect bioassay analysis

Artificially raised *Lymantria dispar* larva from the Institute of Forestry Protection, Chinese Academy of Forest Sciences were hatched at the same time and uniform in the first-instar for use of insect bioassay.

Detached leaves from *in vitro* control hybrid poplar Nl-80106 or transgenic plantlets, placed in 10 cm plates, were used to feed the first-instar larva of *Lymantria dispar* by 5 larva per plate and two plates each clone.

The plates were incubated at room temperature (25~30℃), and fresh leaves were exchanged to each plate every two days. The data of died and survived larva and the larva developmental stage were observed every two days.

RESULTS

Transformation and plant regeneration

The rooting of NL-80106 was sensitive to km levels, 10 mg/L km in the rooting medium could inhibit its rooting thoroughly (Rao HY et al., 2000). In order to get intact transformational plantlets, km-free rooting medium were used to get the rooted km-resistant buds after they were sub-cultured on the selection medium with km 100 mg/L for 60 days. 154 putative transgenic plant lines were obtained from 867 co-cultivated leaf explants with a transformation frequency of 17.8% (Fig 2).

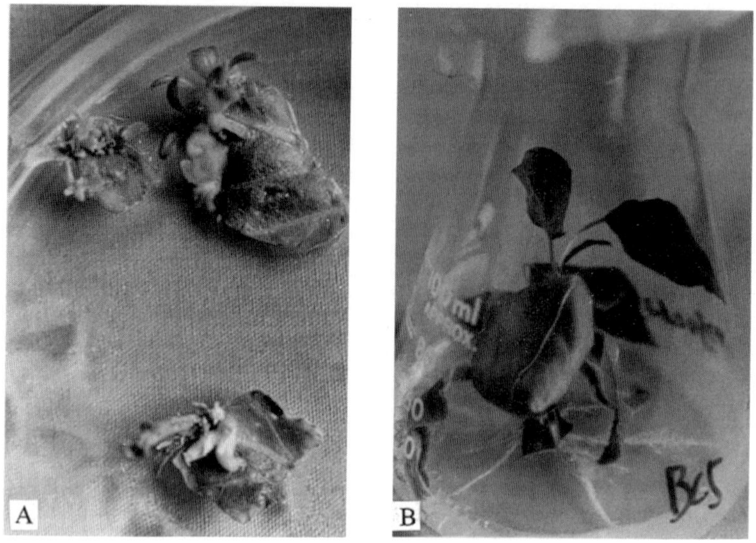

Fig 2 Transformation of poplar infected with LBA4404 carrying *Bt* and *CpTI*
A. Km-resistance buds got from selection medium.
B. Rooting of km-resistance shoots.

Molecular analysis of transgenic plants

DNAs from the different regeneration poplar plants were tested by PCR, using gene specific double primers, which were designed in order to bind an internal part of the *Cry1 A* gene or CpTI gene, thus selectively amplifying 146 bp and 300 bp fragments. 9 of these lines were transformed both with LBA4404 pFWZ10 and LBA4404 CpTI, some respectively transformed, out of 121 plants tested (Fig.3).

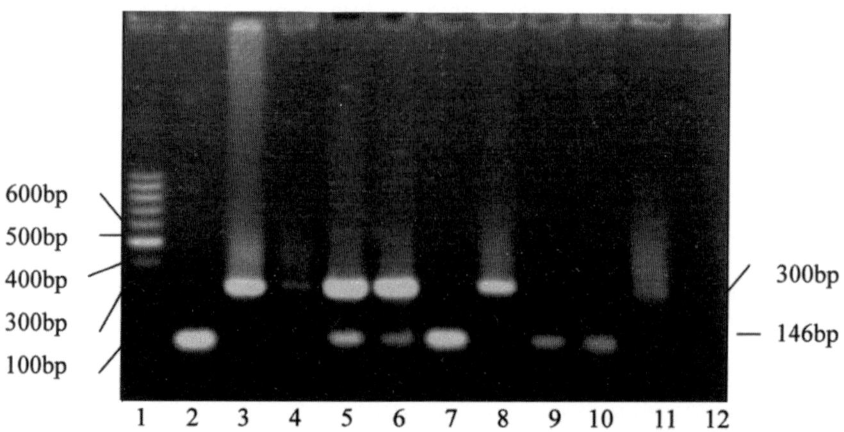

Fig 3. PCR results of km-resistance poplars
1.100bp ladder; 2. pFZY01(CpTI); 3. pFWZ10(Bt); 3-11.km-resistance poplar lines; 12. Untransformed poplar, NL-80106.

PCR products were subsequently blotted and hybridized respectively using 1.8 kbp fragment of *Cry1 A* gene and 430 bp of CpTI gene, as probes (Fig. 4). All 9 lines with both PCR products showing hybridized signals were proved to be the transgenic plants by two genes.

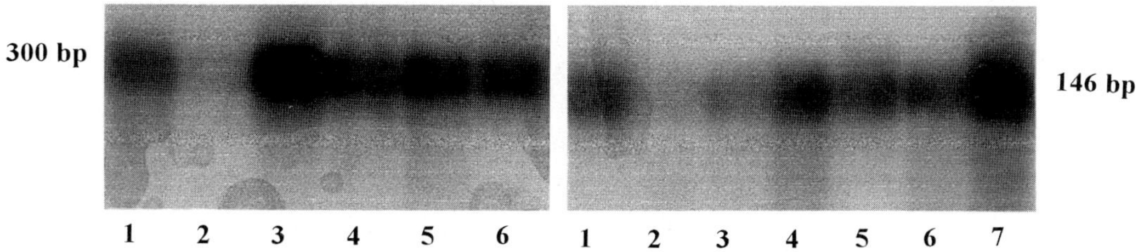

Fig 4. PCR-Southern blotting
A. 1.pFWZ10(Bt); 2.Untransformed poplar, NL-80106; 3-6. km-resistance poplar lines.
B. 1. pFZY01(CpTI); 2. Untransformed poplar; 3-7. km-resistance poplar lines.

Expression of Cry1 A and CpTI proteins

The soluble protein extraction from 3 lines of double gene transformants and 2 lines of untransformed control poplar plants and Cry1 A protein extracted and purified from *Bacillus thuringiensis* were tested by immune dot blotting with specific antibody of Cry1 A or CpTI protein. 3 lines giving a positive reaction of the two antibodies were proved to have the expression of Cry1 A and CpTI protein.

Sandwich ELISA was used to detect the quantity of Cry1 A protein in the soluble protein of the leaves. Cry1 A protein levels in the soluble protein extraction of leaves based on a Cry1 A standard curve made at the same time are shown on Table 1. The expression level of lines BC5, BC27, BC55 respectively reached of the total soluble protein. This expression level is higher than the reports by Perlak et al (1991), Van der Salm et al (1994) and Tian Y et al (2000).

Table 1 Determination of Bt toxin from transgenic poplars by ELISA

Plant line.	Bt protein/total soluble protein of transgenic poplars (%)
BC5	0.048
BC27	0.037
BC55	0.090
Control	0
Blank	0

Insect Bioassays

Preliminary insect bioassay on the leaves of *in vitro*-grown plantlets BC5, BC27 and BC55 carried on the artificially raised *Lymantria dispar* larvae (Fig.5). The results on the 12th day of testing are reported in **Table 2**. Larval mortality was mainly observed in the first week of treatment. According to the preliminary results, these three lines of two-gene transgenic poplar showed obviously stronger resistance to *L. dispar* larva with the larval mortality of 76.7%, 67.1%,89.3% compared to the controls of 5.0% and 9.6%. Survived larva growth were inhibited with the lower larval weight and delayed larval development. Further insect bioassay on *L. dispar* and *Clostera anchoreta* (Fabricius), which heavily destroy poplar in China will be carried out when the transgenic poplar lines are moved to grow in the green house to get enough leaves for testing.

a

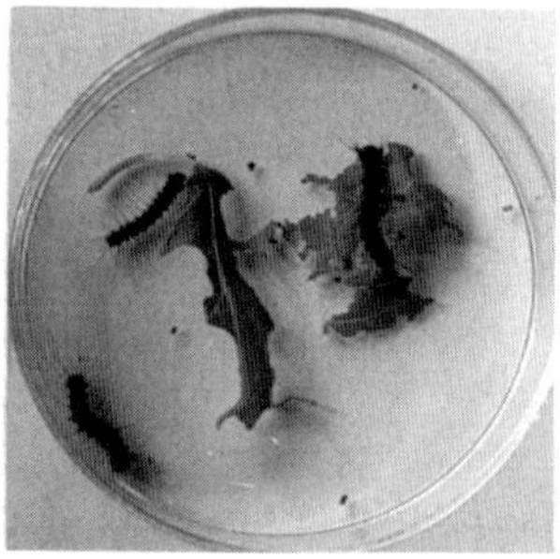

b

Fig 5. *Lymantria dispar* larvae fed with leaves of transgenic poplar line BC5 (a) and with the leaves of untransformed poplar NL-80106 (b).

Table 2 The bioassay carried on *Lymantria dispar* larvae fed with leaves of three different transgenic poplars line for 12 days

Plant line	Larval mortality (%)	Survived larva weight (mg)	Developmental time from first-instar to second –instar (day)
BC5	76.7	3.9	15.7
BC27	67.1	7.9	9.8
BC55	89.3	3.0	10.8
Control 1	5.0	18.1	7.7
Control 2	9.6	15.4	9.5

DISCUSSION

The development of resistance to Bt toxins in laboratory and field populations of pests were reported (McGaughey et al., 1992; Tabashnik, 1994; De Maagd et al., 1999). Poplar has much longer life cycle compared to most of crops. Insect pests of poplar with short life cycle and large population would be easier to induce the resistance to Bt toxins under continual Bt-toxin pressure caused by Bt transgenic poplar. It has been proved to be able to delay insect resistance to Bt toxins by introducing two or more than two insect resistance genes with different insect resistance mechanisms into plants (Zhao et al, 1998). Bt toxin gene and plant proteinase inhibitor gene were integrated to different plants, including cotton, tobacco and poplar and transgenic plants of these two kind of genes showed stronger resistance to insects. Transgenic hybrid poplar transformed by Bt toxin gene and CpTI gene obtained in this paper will be carried out on further study of the effectiveness and durability of resistance to insect pests.

ACKNOWLEDGEMENTS

Thanks are given to Dr Zhao Jun, Chen Rumei and Zhang Chunyi from the Institute of Biology, CAAS; Chen Ying, Wang Guangping and Pan Huixin from Nanjing Forestry University, China for their helps to some parts of this project.

REFERENCES

1. McCown B H, McCabe D E, et al. Stable transformation of *Populus* and incorporation of pest resistance by electric discharge particle acceleration. Plant Cell Rep. **1991**,9:590-595
2. Tian Y-C, Li T-Y ,et al. Insect tolerance of transgenic *Populus nigra* plant transformed with Bacillus thuringiensis toxin gene, Chin J Biotech, **1993**,9:291-298 (in Chinese)
3. Confalonieri M, Allegro G, et al. Regeneration of Populus nigra transgenic plants expressing a Kunitz proteinase inhibitor(Kti3) gene. Mol Breed, **1998,**4:137-145
4. Van der Salm T, Bosch D, et al. Insect resistance of transgenic plants that express modified *Bacillus thuringiensis* CryIAb and CryAc genes: a resistant management strategy. Plant Mol Bio, **1994**,26:51-59
5. Chen L, marmey P,Taylor N J, et al. Expression and inheritance of multiple transgenes in rice plants. Nature Biotechnology. **1998**,16:1060-1064
6. Zhao JZ, Zhao K; et al. Interactions between *Helicover armigera* and transgenic Bt cotton in North China. Chinese Agriculture Sciences. **1999**,9:1-7(in Chinese)
7. Kota M, Daniell H, et al. Overexpression of the *Bacillus thuringiensis* (Bt) Cry2Aa2 protein in chloroplasts confers resistance to plants against susceptible and Bt resistant insects. Proc Nat Aca Sci USA. **1999**,1840-1845
8. Williams H, McGuaghey, et al. Bt resistance management: A plan for reconciling the needs of the many stakeholders in Bt-based products. Nature Biotechnology. **1998**,16:144-146
9. Duan X, Li X,et al. Transgenic rice plants harboring an introduced potato proteinase inhibitor

II gene are insect resistant. Nature Biotechnology. **1996**,14:494-498

10. Richard T. Roush and Anthony M S. Assessing the odds: The emergence of resistance to Bt transgenic plants. Nature Biotechnology. **1997**, 15:816-817
11. Zhao J, Shi,X et al. Insecticidal Activity of Transgenic Tobacco Co-expressing Bt and CptI Genes on *Helicoverpa armigera* and its Role in Delaying the Development of Pest Resistance. Rice Biotechnology Quarterly. **1998**,34:9-10
12. Hua XJ, Chen XB,et al. Transformation of Tobacco by co-transferred with Protein Inhibitor II gene and *δ-endotoxin* gene. Chinese Science Bulletin, **1993,**38(8):747-751(in Chinese)
13. Wu N, Sun Q, et al. Insect-resistant Transgenic Poplar Expressing AaIT Gene. Chin J Biotech. **2000,**16 (2):129-133
14. De Maagd RA, Bosch D and Stiekema W. *Bacillus thuringiensis* toxin-mediated insect resistance in plants. Trend in Plant Science, **1999**,4 (1):9-13
15. Rao H, Chen Y,et al. Genetic transformation of poplar NL-80106 transferred by Bt gene and its insect-resistance. Journal of Plant resources and Environment. **2000,**9 (2):1-5
16. Tian YH, Zhang JB. et al. Studies of Transgenic Hybrid Poplar 741 Carrying Two Insect-resistant Genes. Chinese Acta Botanica Sinica, **2000**,42 (3):263-268

Molecular Breeding of Woody Plants
N. Morohoshi and A. Komamine, editors.

MODIFICATION OF FLOWERING IN TRANSGENIC TREES

Richard Meilan*, Amy M. Brunner, Jeffrey S. Skinner, and Steven H. Strauss

*Forest Science Department, Oregon State University, Corvallis, Oregon, U.S.A, 97331-5752; *meilanr@FSL.orst.edu*

ABSTRACT

The manipulation of flowering can provide many benefits. First, it permits development of a strategy to genetically engineer reproductive sterility. This should help alleviate ecological concerns over the use of transgenic plantations. Additionally, sterility can reduce genetic pollution from plantations, promote vegetative growth, and eliminate nuisance tissues. Flowering control should also allow for shorter breeding cycles.

While each strategy for engineering sterility has advantages, it is unclear which method will work best with trees. Thus, we are testing a variety of techniques, such as tissue-specific ablation; dominant negative mutations; and post-transcriptional gene silencing, including RNA interference. Using the first approach, we have had success with the promoter from *PTD*, a poplar homolog of the *Arabidopsis* gene *APETALA3*. This promoter has directed expression of reporter and cytotoxin genes in floral tissues of *Arabidopsis*, tobacco, and poplar.

Despite indications that one or more of these strategies can be successfully applied to trees, we have not yet demonstrated that any single one fulfills the basic requirements for commercial use as a long-term containment measure. We are conducting research to determine whether sterility can be complete, stable over several rounds of propagation and growing seasons, successfully identified in juvenile trees, and cause no detrimental effects on growth.

Recently we have begun working with genes that affect the onset of flowering. Using real-time PCR to quantify expression of these genes in various tissues collected from female and male trees during different seasons and along an age gradient, we have detected expression changes that correlate with maturation. We are also beginning to experiment with various genomics approaches for identifying other genes that may be useful in engineering flowering control.

KEYWORDS

Ablation, dominant negative mutations, flowering control, gene silencing, *Populus*, RNA interference, sterility

INTRODUCTION

Being able to control flowering in trees is desirable for several reasons [1-3]. First, it will enable the development of trees that are incapable of producing sexual propagules. This would limit gene flow into the wild, helping to mitigate ecological concerns over establishment of transgenic plantations. Second, it is likely to prevent the growth reduction associated with the onset of maturation [4-6]. Third, it could eliminate the production of pollen and other nuisance reproductive structures. Finally, flowering control could lead to more rapid progress through conventional breeding as a result of shorter breeding cycles.

One of the more common ways to genetically engineer reproductive sterility is to ablate cells by expressing a deleterious gene in a tissue-specific fashion [2]. Floral tissue-specific promoters are fused to one of a variety of cytotoxin genes that lead to rapid and early death of the tissues within which the gene product is expressed [3]. The most popular form of sterility employs an RNAse gene that, although isolated from a bacterium, encodes an enzyme that is common in plants and animals [7].

A second way to genetically engineer flowering control is through the use of dominant negative mutations (DNMs). DNMs suppress the function of a gene at the protein level by overexpression of a mutant version of a protein [8]. Inhibition is thought to occur by a variety of means, including formation of an inactive heterodimer, sequestration of protein cofactors, sequestration of metabolites, or stable binding to a DNA regulatory motif. The usefulness of this approach for floral control was demonstrated in *Arabidopsis* with DNM versions of the *AGAMOUS* (*AG*) gene [9]. Expression of a truncated AG protein in which the C-terminal region was deleted resulted in flowers phenotypically similar to those observed in *ag* mutants, suggesting that the truncated version of AG was inhibiting endogenous AG function.

A third technique to control flowering involves post-transcriptional gene silencing (PTGS). Recent studies in a variety of eukaryotic organisms have shown that double-stranded RNA is a potent inducer of PTGS. This approach to induced silencing has been termed RNA interference (RNAi) [10-11]. Recent work in plants using inverted-repeat transgenes showed that RNAi could provide a reliable means for engineering stable suppression of gene activity in plants [12-13].

All of these approaches rely on the use of genes that control floral development, either through the use of floral-specific promoters or coding sequences with high homology to native genes that are targetted for suppression/silencing. In addition, flowering-time genes provide a means of advancing or retarding the onset of reproductive growth. Thus, a major effort in our laboratory is directed at isolating and characterizing genes that regulate flowering time. This paper describes the recent progress we have made in various aspects of flowering control in trees.

CELL ABLATION

In our early attempts at utilizing cell ablation with poplar, we relied on heterologous promoters, which had shown floral-specific expression in tobacco and *Arabidopsis* [14-16], to drive the expression of two cytotoxin genes, *DTA* [17] and barnase [18]. When introduced into transgenic poplars, these fusions resulted in decreased vegetative growth (Fig. 1), suggesting leaky expression in non-target tissues.

Recently, we have begun experimenting with the promoter from a poplar gene, *PTD* (the *Populus trichocarpa* homolog of *DEFICIENS*), that appears to be floral-specific in its expression pattern [19]. We have shown that *PTD* promoter directs expression of the GUS gene early in the development of floral organs in *Arabidopsis* and poplar (Fig. 2A & B). The latter was co-transformed with the *LEAFY* (*LFY*) gene from *Arabidopsis* under the control of the 35S promoter, which has been shown to induce early flowering in poplar [20]. A PTD::*DTA* fusion resulted in either ablated or perturbed development of petals and stamens in *Arabidopsis* (Fig. 2C), and the ablation of petals, stamens, and carpels in transgenic tobacco (Fig. 3A). PTD::*DTA* also appears to prevent flowers

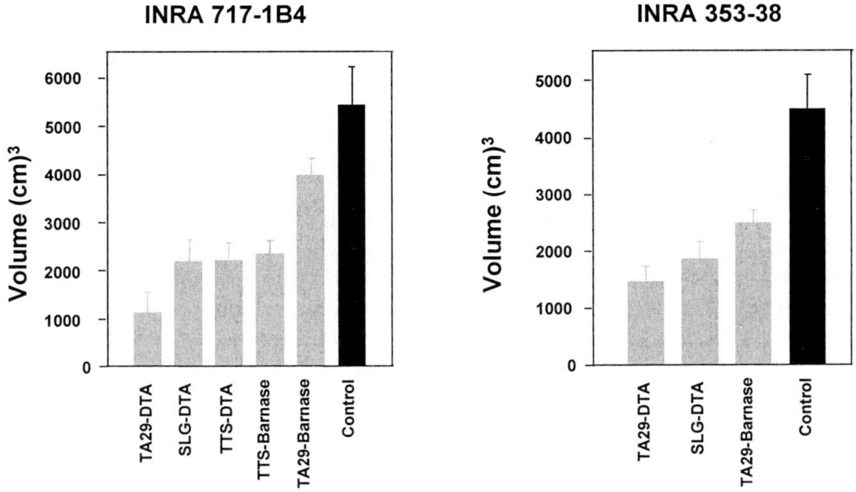

Figure 1. Effects of various sterility-inducing transgenes on growth. INRA 717-1B4 is a female *P. tremula* x *P. alba* clone; 353-38 is a male *P. tremula* x *P. tremuloides*.

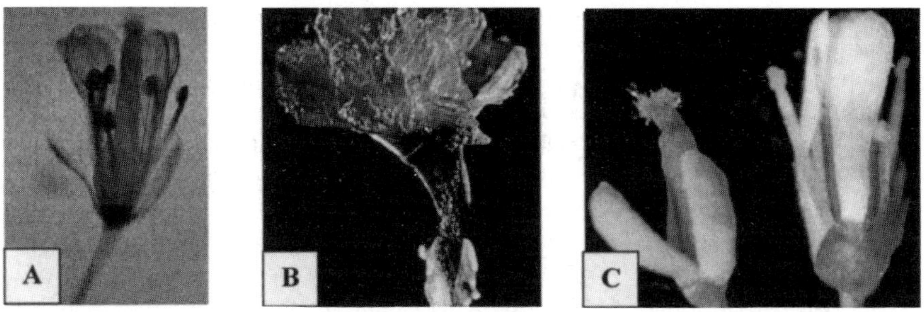

Figure 2. Specificity of PTD promoter. PTD::GUS expression in an *Arabidopsis* flower (A) and early-flowering (co-transformed with 35S::*LFY*) aspen (B). Dark zones indicate GUS expression. Effects of PTD::*DTA* on flower development in *Arabidopsis* (C); a transgenic flower is shown on the left, a non-transgenic control on the right.

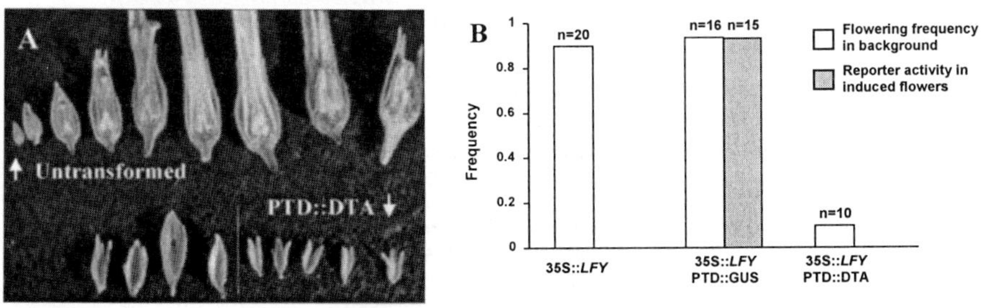

Figure 3. Effects of PTD::*DTA* on flower development in tobacco (A) and aspen (B). The upper panel in (A) shows normal flower development. Florets from transgenic tobacco plants consisted of only sepals [lower panel in (A)]. The 35S::*LFY* construct was used to induce flowering in aspen.

from forming on poplar co-transformed with 35S::*LFY* (Fig. 3B). Expression of PTD::*DTA* had no significant effects on growth in tobacco (data not shown).

DOMINANT NEGATIVE MUTATIONS

The approach taken for generating our DNM constructs is shown in Fig. 4. A key feature of these constructs is the promoter used to drive expression of the DNM transgene. For strong inhibition, the DNM protein should be present at a much higher level than that of the native protein. We are using hybrid promoters composed of two copies of the enhancer element from the 35S promoter (e35S) coupled to either the *ACTIN2* (*ACT2*) or *ACTIN11* (*ACT11*) promoters from *Arabidopsis*. The *ACT* promoters show strong expression in meristems, young growing tissues, and floral tissues. We have verified that the *2e35S/ACT2::GUS* and *2e35S/ACT11::GUS* constructs function in transgenic poplar, tobacco and *Arabidopsis*.

Guided by a study of an *AG* DNM [9], we altered poplar and *Arabidopsis* MADS-box (a motif common to many floral homeotic genes) cDNAs via polymerase chain reaction (PCR) to encode proteins that lack the C-terminal domains (Fig. 4). Constructs containing full-length coding regions were also produced to provide positive controls for analysis of transgenic phenotypes. All constructs have been introduced into *Arabidopsis* via *in planta* transformation and co-transformed with 35S::*LFY* into poplar. Fifteen constructs containing *Arabidopsis* transgenes are currently under evaluation in transgenic *Arabidopsis*; five each for the *Arabidopsis* genes *AP1*, *AP3*, and *AG*. Though not a MADS-box gene, *PTLF* (the *P. trichocarpa* homolog of the *Arabidopsis* *LFY* gene) also has a modular structure [21]. Alignment of all known *LFY* homologs revealed a total of six putative protein domains, based on amino acid conservation (Fig. 4). However, the actual functions of these domains are unknown, so analysis of *LFY*

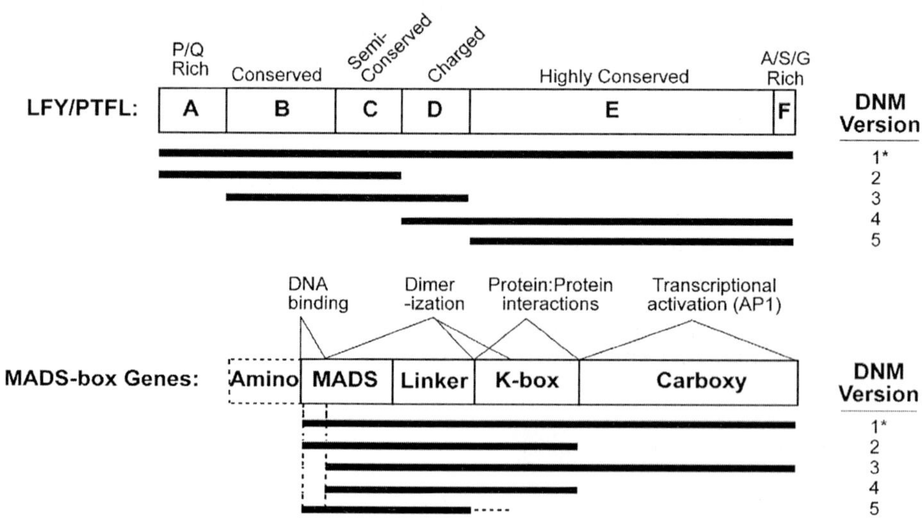

Figure 4. Domain structure and deletions used to generate dominant negative mutant versions of genes being tested in transgenic *Arabidopsis*, tobacco, or poplar. Constructs denoted with an asterisk are full-length versions.

genes lacking various domains was undertaken. Five *LFY* DNM transgenes have been introduced into *Arabidopsis* and tobacco; results from this will guide development of one or more *PTLF* DNMs for use in poplar.

In total, 27 different kinds of DNM and control constructs have been introduced into *Arabidopsis*. Preliminary results suggest that the hybrid promoter used in our first-generation DNM constructs may not be exhibiting the desired expression pattern. Thus, we are now beginning to assemble another set of DNM constructs that will utilize a different promoter.

RNA INTERFERENCE

We have just begun to experiment with several options for suppressing flowering via RNAi of single and multiple floral homeotic genes. In these studies, we are using an early-flowering genotype of *P. alba* that allows us to test transgenes for their effects on flowering in a short time interval. We are focusing on several genes isolated in our laboratory, including poplar homologs to the *Arabidopsis* genes *LFY* (*PTLF*), *AG* (*PTAG*), *APETALA1* (*PTAP1*), and *APETALA3* (*PTD*).

Following studies by Chuang and Meyerowitz (2000) [13] with *AG*, the MADS-box region will be excluded from our constructs in order to avoid suppression of non-target MADS-box genes. Untranslated regions (UTRs) will also be excluded from the RNAi transgenes so that UTR-specific probes can be used to distinguish native gene and transgene expression while analyzing transgenic plants. An incomplete version of *PTLF* will be employed to ensure that it cannot encode a functional protein.

Constructs will be produced to suppress: 1) *PTLF*, 2) *PTAP1-1/PTAP1-2*, 3) *PTAG1/PTAG2*, and 4) *PTD*. While the first three are likely to give both male and female sterility, the phenotype of *PTD* is difficult to predict. *PTD* most resembles a B-function gene in sequence (required for stamen and petal development in most angiosperms); however, in poplar it is also expressed strongly in female flower primordia. We have characterized two *PTAP1*s and the two *PTAG*s [22]. Because each pair shares approximately 90% nucleotide identity in coding regions outside the MADS box, it is highly likely that an RNAi transgene-containing sequence from one of the pair (*e.g.*, *PTAP1-1*) will result in suppression of both (*e.g.*, *PTAP1-1* and *PTAP1-2*). RNAi and cosuppression studies both suggest that this level of nucleotide identity is sufficient for cross-suppression, which has been observed for sequences with as little as 84% identity [11]. If this turns out to be untrue, we will make constructs containing sequences from both genes within each pair.

Because of functional redundancy, suppression of more than one floral regulatory gene is likely to be necessary to achieve complete sterility. Thus, we will also generate constructs that are designed to suppress the following pairs of genes: 1) *PTLF/PTAP1*, 2) *PTLF/PTAG*, 3) *PTAP1/PTAG*. One construct is designed to suppress three genes: *PTLF/PTAP1/PTAG*.

EXPRESSION STABILITY

Stable transgene expression over the lifetime of a tree, and in its vegetative propagules, is critical for all engineered sterility systems. We have therefore begun investigating the stability of transgene expression using reporter genes.

The vector to be used for studying stability will consist of two reporter genes, green fluorescent protein (GFP) and a herbicide resistance gene, which were selected for economy and speed of assay. Glufosinate is a contact herbicide that inhibits glutamine synthetase. Glufosinate resistance is conferred by the *bar* gene from *Streptomyces hygroscopicus*, which encodes phosphinothricin acetyltransferase [23]. Both genes can be used for scanning entire plants for sectors that have undergone gene silencing--one using a UV light source (GFP) and the other using low concentrations of glufosinate-containing herbicide, which acts in a non-systemic manner.

Because gene silencing may be very different with a native versus a foreign promoter, we are using a different promoter with each reporter gene. We have chosen the *rbc*s promoter to drive expression of the *bar* gene because it exhibits strong leaf expression. An *Arabidopsis rbcS* promoter fused to *bar* has been shown to confer high levels of glufosinate resistance in poplar [24], a result that we have repeated with transgenic poplars generated in our laboratory and grown in local field trials (unpubl. data). The GFP reporter gene will be driven by the 35S promoter, which has been used widely in transgenic poplars and is known to yield high levels of expression in leaves.

Poplar lines containing these reporter-gene constructs will be subjected to various stresses and grown in the laboratory and the field for several years to observe expression.

FLOWERING-TIME GENES

We are also identifying genes that regulate the transition from the vegetative to the reproductive phases in trees. We are trying to determine the degree of correspondence between the genetic control of phase transition in poplar, which has a juvenile period of four to six years, and that of the herbaceous annual plant *Arabidopsis*, which initiates flowering after a juvenile period that is only weeks in length. Both genera belong to the same clade of the eudicots [25], and the *Populus* genome is small [26], facilitating gene-to-gene comparisons. In addition, we are trying to relate vernalization, the induction of flowering by cold treatment, to the chilling temperatures needed to end dormancy in temperate zone trees, and to determine whether changes in DNA methylation are involved in post-dormancy changes in gene expression.

We are identifying candidate genes for study by isolating genes from poplar based on sequence homology with known flowering-time genes in *Arabidopsis* and other annual plants. To evaluate whether these candidate genes have important regulatory roles, we are taking advantage of a few characteristics that have established *Populus* as a model system for genetic and molecular analyses of woody plants. First, the ease of vegetative propagation, and the fidelity of juvenile characteristics in vegetative propagules, provides a continuous age gradient of a single genotype through the first year of flowering. This facilitates intensive study of quantitative and cell-specific changes in gene expression in relation to phase transition. Moreover, the ease with which poplars can be transformed has made them the tree taxa of choice for transgenic studies worldwide [27-30]. This allows direct functional tests of the roles of specific genes via RNAi and overexpression.

We have already conducted expression analysis on several candidate genes in relation to maturation and seasonal changes in expression. We collected various tissues at different seasonal times (Fig. 5) from one female and one male *P. trichocarpa* x *P. deltoides* genotype. Ramets of each clone were represented in a continuous age gradient of one to six years (*i.e.*, they had been through one to six growing seasons when we began our collections). For both genotypes, inflorescences were first initiated at age four (*i.e.*, during their fifth growing season). In total, more than 60 RNA samples

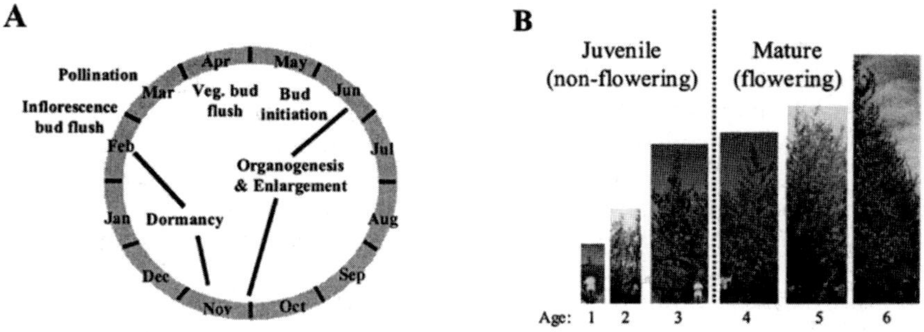

Figure 5. Seasonal cycle for *P. trichocarpa* x *P. deltoidies* floral development in western Oregon (A). An age gradient of the female poplar clone from which tissues were collected (B).

were isolated. For accurate quantification of the modest expression levels shown by most of these genes, real-time quantitative RT-PCR was employed.

The vegetative expression levels of *PTLF* differed significantly between juvenile and mature ramets in vegetative buds initiated during the current season (shoot apical meristem leaf primordia, and bud scales were removed) (Fig. 6). This difference persisted, although to a lesser degree, into autumn, but all ramets showed a uniformly low level of *PTLF* transcript in post-dormancy vegetative buds. The expression of *P. trichocarpa ID1-LIKE5* (*PTID1L5*) was markedly upregulated in newly expanding shoots (shoot apical meristem, leaves, internode) from mature ramets that would soon initiate inflorescences (Fig. 6).

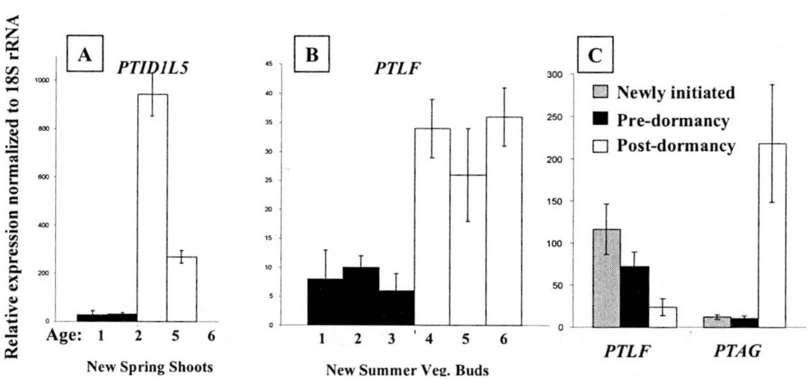

Figure 6. Variation in gene expression with age in a male poplar clone for poplar homologs of *INDETERMINATE1* (A) and *LEAFY* (B). Darkened bars indicate juvenile trees, open bars indicate sexually active trees. (C) Seasonal variation in vegetative bud gene expression in a mature female ramet for the poplar *LEAFY* gene (left) and *AGAMOUS* gene (right) at five years of age. Bars show one standard deviation based on three replicate measurements.

That *PTLF* vegetative bud expression was highest in the newly initiated vegetative buds of mature trees during long days is consistent with expectations based on the expression characteristics of *LFY*. Similarly, the expression pattern for *PTID1L5* corresponds well with that of the maize gene *ID1* [31]. *PTAG2* vegetative bud expression did not correlate with maturation, but was instead expressed at surprisingly high levels in ramets of all ages. It also exhibited a striking pattern of seasonal variation, in contrast to that seen with *PTLF* in mature ramets (Fig. 6). *PTAG2* was expressed at low levels in newly initiated vegetative buds and pre-dormancy, autumn vegetative buds. However, it was dramatically upregulated in post-dormancy vegetative buds. This result is particularly interesting given that the methylation and expression level of *AG* is altered in *Arabidopsis* lines with an overall decrease in methylation, and that the *CURLY-LEAF* gene acts to prevent *AG* vegetative expression [32]. The *PTAG2* pattern suggests the possibility that dormancy causes a transient change in its epigenetic regulation.

GENOMICS

Another way in which we will identify genes whose expression changes with maturation is via hybridization of poplar expressed sequence tag (EST) microarrays. With the aid of our collaborators in Sweden and France, we will use various RNA collections to screen microarrays for additional genes whose expression is correlated with maturation, and confirm expression patterns of selected genes using real-time PCR. Based on sequence homologies and expression patterns, we will begin to study the function of several genes via transformation using RNAi suppression and over-expression. All constructs will be tested in the early-flowering *P. alba* clone.

ACKNOWLEDGEMENTS

This work was funded by the Tree Genetic Engineering Research Cooperative (http://www.fsl.orst.edu/tgerc/index.htm), U.S. Dept. of Energy (Bioenergy Feedstock Development Program, grant #85X-ST807V), National Science Foundation I/UCRC Program (grant number 9980423-EEC), Consortium for Plant Biotechnology Research (grant number OR22072-78), Agenda2020 (grant number FC0797ID13552), and a grant from the Monsanto Company. The authors would also like to express their gratitude to Caiping Ma, Pearce Smithwick, Vicky Hollenbeck, Sarah Dye, Shuping Cheng, and Jace Carson for their help with the work described herein.

REFERENCES

1. Strauss, S.H., W.H. Rottmann, A.M. Brunner & L.A Sheppard, Genetic engineering of reproductive sterility in forest trees, *Mol. Breed.*, **1995**, 1, 5-26.
2. Brunner, A.M., R. Mohamad, R. Meilan, L.A. Sheppard, W.H. Rottmann & S.H. Strauss, Genetic engineering of sexual sterility in shade trees, *Journal of Arboriculture*, **1998**, 24(5), 263-273.
3. Skinner, J.S., R. Meilan, A.M. Brunner & S.H. Strauss, Options for Genetic Engineering of Floral Sterility in Forest Trees, In: *Molecular Biology of Woody Plants*, S.M. Jain & S.C. Minocha (eds.), Kluwer Academic Publishers, Dordrecht, The Netherlands, **2000**, pp. 135-153.

4. Eis, S., E.H. Garman & L.F. Ebell, Relation between cone production and diameter increment of Douglas-fir [*Pseudotsuga menziesii* (Mirb.) Franco], Grand fir [*Abies grandis* (Dougl.) Lindl.], and western white pine [*Pinus monticola* (Dougl.)]. *Can. J. Bot.*, **1965**, 43, 1553-1559.
5. Tappeiner, J.C., Effect of cone production on branch, needle and xylem ring growth of Sierra Nevada Douglas-fir, *For. Sci.*, **1969**, 15, 171-74.
6. Teich, A.H., Growth reduction due to cone crops on precocious white spruce provenances, *Environ. Canada Bi-monthly Res. Notes*, **1975**, 31, 6.
7. Mariani, C., M. DeBeuckeleer, J. Truettner, J. Leemans & R.B. Goldberg, Induction of male sterility in plants by a chimaeric ribonuclease gene, Nature, **1990**, 347, 737-741.
8. Espeseth, A.S., A.L. Darrow & E. Linney, Signal transduction systems: Dominant negative strategies and mechanisms, *Mol. Cell. Diff.*, **1993**, 1, 111-161.
9. Mizukami, Y., H. Huang, M. Tudor, Y. Hu & H. Ma, Functional domains of the floral regulator AGAMOUS: Characterization of the DNA binding domain and analysis of dominant negative mutations, *Plant Cell*, **1996**, 8, 831-845.
10. Fire, A., RNA-triggered gene silencing, *Trends Genet.*, **1999**, 15, 358-363.
11. Bosher, J.M. & M. Labouesse, RNA interference: Genetic wand and genetic watchdog, *Nature Cell Biol.*, **2000**, 2, E31-E36.
12. Waterhouse, P.M., M.W. Graham & M.-B. Wang, Virus resistance and gene silencing in plants can be induced by simultaneous expression of sense and antisense RNA, *Proc. Natl. Acad. Sci. USA*, **1998**, 95, 13959-13964.
13. Chuang, C.-F. & E.M. Meyerowitz, Specific and heritable genetic interference by double-stranded RNA in *Arabidopsis* thaliana, *Proc. Natl. Acad. Sci. USA*, **2000**, 97, 4985-4990.
14. Koltunow, A.M., J. Truettner, K.H. Cox, M. Wallroth & R.B. Goldberg, Different temporal and spatial gene expression patterns occur during anther development, *Plant Cell*, **1990**, 2, 1201-1224.
15. Hackett, R.M., M.J. Lawrence & C.H. Franklin, A *Brassica* S-locus related gene promoter directs expression in both pollen and pistil of tobacco, *Plant J.*, **1992**, 2, 613-617.
16. Wang, H., H.M. Wu & A.Y. Cheng, Development and pollination regulated accumulation and glycosylation of a stylar transmitting tissue-specific proline-rich protein, *Plant Cell*, **1993**, 5, 1639-1650.
17. Greenfield, L., M.J. Bjorn, G. Horn, D. Fong, G.A. Buck, R.J. Collier & D.A. Kaplan, Nucleotide sequence of the structural gene for diphtheria toxin carried by corynebacteriophage β, *Proc. Natl. Acad. Sci. USA*, **1983**, 80, 6853-6857.
18. Hartley, R.W., Barnase and barstar: Expression of its cloned inhibitor permits expression of a cloned ribonuclease, *J. Mol. Biol.*, **1988**, 202, 913-915.
19. Sheppard, L.A., A.M. Brunner, K.V. Krutovskii, W.H. Rottmann, J.S. Skinner, S.S. Vollmer & S.H. Strauss, A *DEFICIENS* homolog from the dioecious tree *Populus trichocarpa* is expressed in both female and male floral meristems of its two-whorled, unisexual flowers, *Plant Physiol.*, **2000**, 124, 627-639.
20. Weigel, D. & O. Nilsson, A developmental switch sufficient for flowering initiation in diverse plants, *Nature*, **1995**, 12, 495-500.
21. Rottmann, W.H., R. Meilan, L.A. Sheppard, A.M. Brunner, J.S. Skinner, C. Ma, S. Cheng, L. Jouanin, G. Pillate & S.H. Strauss, Diverse effects of overexpression of *LEAFY* and *PTLF*, the poplar homolog of *LEAFY/FLORICAULA*, in transgenic poplar (*Populus trichocarpa*) and *Arabidopsis*, *Plant J.*, **2000**, 22, 235-245.

22. Brunner, A.M., W.H. Rottmann, L.A. Sheppard, K. Krutovskii, S.P. DiFazio, S. Leonardi & S.H. Strauss, Structure and expression of duplicate *AGAMOUS* orthologs in poplar, *Plant Mol. Biol.*, **2000**, 44 (5), 619-634.
23. Riemenschneider, D.E., Genetic engineering of horticultural and forestry crops for herbicide tolerance, In: *Biotechnology of Ornamental Plants*, R.L. Geneve, J.E. Preece & S.A. Merkle (eds.), CAB International, **1997**, pp. 367-384.
24. DeBlock, M.D., Factors influencing the tissue culture and the Agrobacterium tumefaciens-mediated transformation of hybrid aspen and poplar clones, *Plant Physiol.*, **1990**, 93, 1110-1116.
25. Soltis, P.S., D.E. Soltis & M.W. Chase, Angiosperm phylogeny inferred from multiple genes as a tool for comparative biology, *Nature*, **1999**, 402, 402-403.
26. Bradshaw Jr., H.D., Case history in genetics of long-lived plants: Molecular approaches to domestication of a fast-growing forest tree: *Populus*, In: *Molecular Dissection of Complex Traits*, A.H. Paterson (ed.), CRC Press, NY, **1998**, pp. 219-228.
27. Tsai C-J., G.K. Podila & V.L. Chiang, *Agrobacterium*-mediated transformation of quaking aspen (*Populus tremuloides*) and regeneration of transgenic plants, *Plant Cell Rep.*, **1994**, 14, 94-97.
28. Tzfira, T., C.S. Jensen, W. Wang, A. Zuker, B. Vincour, A. Altman & A. Vainstein, Transgenic *Populus tremula*: A step-by-step protocol for its *Agrobacterium*-mediated transformation, *Plant. Mol. Biol. Rep.*, **1997**, 15, 219-235.
29. Kim, M.-S., N.B. Klopfenstein & Y.W. Chun, *Agrobacterium*-mediated transformation of *Populus* species. In: *Micropropagation, Genetic Engineering, and Molecular Biology of Populus*, N.B. Klopfenstein, Y.W. Chun, M.-S. Kim & M.R. Ahuja (eds.), Gen. Tech. Rep. RM-GTR-297. U.S. Dept. of Agric. Forest Service, Fort Collins, CO, **1997**, pp. 51-59.
30. Han, K.-H., R. Meilan, C. Ma & S.H. Strauss, An *Agrobacterium* transformation protocol effective in a variety of cottonwood hybrids (genus *Populus*), *Plant Cell Rep.*, **2000**, 19, 315-320.
31. Colasanti, J., Z. Yuan & V. Sundaresan, The indeterminate gene encodes a zinc finger protein and regulates a leaf-generated signal required for the transition to flowering in maize, *Cell*, **1998**, 93, 593-603.
32. Goodrich, J., P. Puangsomlee, M. Martin, D. Long, E.M. Meyerowitz & G. Couland, A polycomb-group gene regulates homeotic gene expression in *Arabidopsis*, *Nature*, **1997**, 386, 44-50.

Molecular Breeding of Woody Plants
N. Morohoshi and A. Komamine, editors.

POSSIBLE APPROACHES FOR STUDYING THREE DIMENSIONAL STRUCTURE OF LIGNIN

Noritsugu Terashima

Emeritus Professor of Nagoya University
2-610 Uedayama, Tenpaku, Nagoya, 468-0001, Japan

ABSTRACT

Information on three dimensional structure of lignin in the cell walls is essential for better understanding of chemical and biological properties of wood. An advanced structural model should provide following information on structural heterogeneities and macromolecular properties of lignin in the cell walls, (1) distribution and frequencies of different kinds of C_9 units, inter-unit bonds including stereochemistry, functional groups, lignin-polysaccharide bonds within a lignin macromolecule and within the cell walls; (2) higher order structure and size of lignin macromolecule; and (3) assembly of lignin and polysaccharides in different morphological regions of cell walls and in different kinds of cells. An effective approach is nondestructive observation of growing process of lignin macromolecule during formation of cell wall from an early stage to a last stage. Nondestructive observation of lignin formation process and its structural analysis were carried out by radio-and stable isotope tracer techniques combined with light microscopy and nuclear magnetic resonance spectroscopy. Combination of mild degradation analyses with nondestructive analysis is also effective approach. A 3D model can be proposed tentatively based on the information on (1~3) obtained by the nondestructive analyses and the information obtained by electron microscopy and other various destructive analyses. For further revision of the model, more improvements of the approaches to obtain more reliable information on (1~3) are required.

KEYWORDS

Lignin, 3D structure, nondestructive analysis, isotope tracer method, ^{13}C-NMR

INTRODUCTION

Physical, chemical and biological properties of wood depend largely on the properties of major cell wall polymers, and their assembly in the cell walls. Therefore, three dimensional (3D) structure of lignin macromolecule, and assembly of lignin, cellulose and hemicellulose in the cell walls are essential information for successful development and application of biotechnology in wood science and technology. Various 2D structural models have been proposed for softwood and hardwood lignins in the past [1-6]. These models provide information on average frequencies of different kinds of C_6-C_3 units, frequencies of various types of inter-unit bonds and functional groups, and the information is not sufficient to understand the chemical properties and behavior of lignin in the cell walls. The 2D models were proposed based on the information obtained mainly by degradation analysis. However, the total yield of degradation products is less than a half of lignin in the cell wall. A large part of products were put aside from detailed structural investigation due to difficulties in elucidation of their structures by traditional methods of organic chemistry. As a result, the proposed models differ considerably from one another in structure depending on the researcher and approaches employed [1-6].

In order to circumvent those problems involved in degradation analyses, various nondestructive approaches such as UV- and Raman microspectroscopy, electron microscopy and NMR spectroscopy have been applied to lignified wood tissue or isolated lignins. However the use of those methods is not satisfactory for determination of 3D structure of macromolecular lignin in the cell wall. Improved radio- and stable isotope tracer methods can be employed as another nondestructive approaches.

Lignin is thought to be formed by random polymerization of monolignols in the cell walls. However, the lignification of cell wall occurs in a biochemically regulated manner to form structurally ordered lignin macromolecules [7,8]. Therefore, detailed observation of the growing process of lignin macromolecule by nondestructive approaches should give necessary information for studying 3D structure of protolignin.

Difficulties in elucidation of 3D structure of lignin

In the study of lignin structure, researchers encounter following difficulties.
(a) There is considerable difference in lignin structure between softwood, hardwood and grasses. This is partly due to difference in composition of different types of the cells. While, it is not easy to collect different types of cells by separation from wood tissue without causing any modification of lignin structure.
(b) During the lignification of cell walls, different types of monolignols are supplied to polymerization site depending on the stage of the cell wall differentiation. As a result, the structure of lignin is not uniform with respect to morphological regions of the cell walls [7-10]. While, it is not easy to separate lignified cell walls into middle lamella (ML), primary wall (PW) and secondary wall (SW) completely.
(c) During formation of lignin macromolecule, different types of inter-unit bonds are prevailed depending on the stage of the polymerization. Thus, structure of lignin is not uniform within a single macromolecule. Destructive analysis lose this information.
(d) During lignification, chemical and physical bonds are formed between lignin and polysaccharides. It is impossible to isolate whole lignin from the wood cell wall without degrading 3D macromolecular structure. Milled wood lignin (MWL) has been often used for characterization of lignin. However, the yield of MWL is less than a half of lignin in the cell wall, and MWL is derived mainly from secondary wall lignin [11-13], which is different in structure from CML (ML and PW) lignin.

Necessary information for elucidation of 3D structure of lignin

An advanced 3D structural model should provide information on structural heterogeneities and macromolecular properties of lignin in the cell walls as follows.
(1) Total frequencies and heterogeneous distribution of various types of C_6-C_3 units (*p*-hydroxyphenyl-, guaiacyl- and syringylpropane units) within a lignin macromolecule and in different cell wall layers (ML, PW and SW).
(2) Total frequencies and distribution of inter-unit bonds, functional groups and lignin-polysaccharide bonds within a lignin macromolecules and within a cell wall
(3) Stereochemistry of side chain carbons of C_6-C_3 units
(4) Higher order structure, shape and size of lignin macromolecule

(5) Assembly of lignin, hemicellulose and cellulose in the cell walls
(6) Difference in (1) ~ (5) with respect to cell type and plant species.

RESULTS AND DISCUSSION

Effective approaches to clear up the difficulties in structural study

To clear up above difficulties, various nondestructive approaches have been employed. UV microspectroscopy [9] is effectively used for obtaining above information (1), and Raman microspectroscopy [14] for (5). To obtain information (4) and (5), electron microscopy has been employed [15] in combination with rapid freeze fracture technique[16-18]. Another promising approach for (1) ~ (3) is application of isotope tracer methods including dual radio-labeling technique, microautoradiography [19-22], solid state NMR and solution NMR combined with specific ^{13}C-enrichment technique [23-26]. The side chain carbons, Cα, Cβ, Cγ, ring-C4 and C5 of guaiacyl lignin are specifically enriched with ^{13}C [23,26], and difference NMR spectrum is determined between spectra of specifically ^{13}C-enriched lignin and unenriched lignin. The difference spectrum provides definite information on (1)~(3) [24-27]. Information on (3) has been also obtained by degradation analysis [28,29]. By integrating the results of nondestructive observation of the assembly process of polysaccharides and lignin from the early stage to the last stage of cell wall differentiation, it is possible to obtain most of the necessary information for building a structural model of lignin [7,8,27]. Useful information on the sequence of inter-unit bonds is obtained by combining results obtained by tracer method with those by mild destructive analysis such as thioacidolysis [27].

Proposed 3D structural model of softwood lignin

Most softwood composed mainly of tracheid, and major part of lignin is guaiacyl lignin containing a few percent of syringyl and *p*-hydroxyphenyl lignin. So long as lignin in softwood is dealt with, above difficulty (a) to obtain necessary information (6) can be put aside of consideration. The MWL prepared from ginkgo wood resembles those prepared from spruce and pine woods very close in structure [27]. Therefore, the experimental results obtained by the use of ginkgo can be combined with those obtained by the use of pine and spruce for building a structural model of softwood lignin.

Table 1 shows frequencies of inter-unit bonds and functional groups in SW lignin estimated mainly by radio- and stable isotope tracer methods. Thioacidolysis cleaves β-O-4' bond selectively, and gives dimeric products when the dimer were connected by β-O-4' bond on both sides or one side at the end of macromolecule. Therefore, the fact that β-β dimers are not obtained by thioacidolysis indicates that all β-β substructures must be connected to adjacent unit by bonds such as 5-5' which is resistant to thioacidolysis. The results of tracer experiments also suggest that bulk type polymerization occurs in the early stage followed by end-wise polymerization in the later stage of formation of lignin macromolecule [7,8]. Minor structures, such as dibenzodioxocin [30], may be formed in the end-wise polymerization stage.

It is possible to combine all above information into a 3D structural model for SW lignin in softwood tracheids [27]. It is noted that the average frequencies of inter-unit bonds and functional groups estimated by tracer methods in Table 1 are not so much different from those in 2D model proposed by Sakakibara [5] in which results of many researchers by various analytical methods are averaged and combined into a model.

Table 1. Frequencies of inter-unit bonds and functional groups / 100 C_9 units for SW lignin estimated tentatively by tracer method, and yield of thioacidolysis products from secondary wall fraction of ginkgo

	Structural model	Estimation by tracer method		Thioacidolysis product (mol %)
β-O-4'	55			23.3 (G monomer)
G-CH(OH)-CH(OH)CH$_2$OH	2	2	} 59 ±1.5	
β-O-C(CH$_2$OH)H-CHO	2			
β-1'(α-O-α')	4			
β-1'(α-OH)	2	} 6-8		1.3 (Dimer)
β-6'	2		} 35 ±1.5	0.4 (Dimer)
β-β'	8	8-10		0.0 (Dimer)
β-5'	16	15-17		1.6 (Dimer)
5-5'	18		} 41	1.4 (Dimer)
4-O-5'	7			0.4 (Dimer)
α-Carbonyl	3		3 ±1.5	
α-O-4'	5	(5)		
α-O-polysaccharides	10			
G-CH=CH-CH$_2$OH	4		4 ±1.5	
G-CH=CH-CHO	3		3 ±1.5	
G-CH$_2$CH$_2$CH$_2$OH	4	4		

CONCLUSION AND FUTURE PROSPECTS

A 3D structural model can be tentatively proposed for SW lignin in softwood based on mainly nondestructive analyses supplemented by mild destructive analyses. The average frequencies of inter-unit bonds and functional groups in the proposed model are not much different from one of the 2D models [5] proposed mainly by destructive approaches. At least another one model for CML lignin must be proposed for better understanding of softwood properties. Technical improvements are necessary in the future especially on following points. (1) Nondestructive analyses of lignin in CML of softwood; (2) Nondestructive analysis of lignin in other types of cells such as ray cells; (3) To achieve above targets, new technical developments must be made for effective separation of cell wall layers, and different types of cells without affecting structure of lignin; (4) Nondestructive analyses of bonds between lignin and polysaccharides.

For structural study of lignin in anatomically complicated hardwood, developments of techniques on above points are very important to obtain necessary information (6).

Among various nondestructive methods, isotope tracer methods provide useful and definite information which is not obtainable by any other methods. Especially, ^{13}C-tracer method can be used as a powerful tool for tracing structural change under any reaction conditions such as pulping, bleaching or biodegradation.

ACKNOWLEDGEMENTS

The author is grateful to following researchers for their help and cooperation in the papers related to this presentation, Drs. K. Fukushima, K. Takabe, J. Nakashima, Y. Xie, R. Atalla, J. Ralph, S. Ralph, L. Landucci, D. VanderHart, D. Robert, C. Lapierre,

B. Monties, U. Westermark, J. Hafrén, D. Evtuguin, C. Pascoal Neto. The author wishes to thank following institutes and funds for financial support, Dept. of Energy, USDA Forest Product Laboratory, USA; Skogsindustrins Forskningsfond, and STFI, Sweden; Ministry of Science & Technology, and University of Aveiro, Portugal.

REFERENCES

1. K. Freudenberg, 'The Constitution and biosynthesis of lignin', In: *Constitution and Biosynthesis of Lignin.* K. Freudenberg & A.C. Neish (eds.), **1968**, Springer-Verlag, Berlin, pp. 82-101.
2. Y.Z. Lai & K. V. Sarkanen, Isolation and structural studies. In: *Lignins, occurrence, formation, structure and reactions*, K. V. Sarkanen & C. H. Ludwig (eds.), Wiley Interscience, New York, **1971**, pp. 165-240.
3. H. Nimz, 'Beech lignin - Proposal of a constitutional scheme', *Angew. Chem.* Internat. Edit., **1974**, 13, 313-321.
4. E. Adler, 'Lignin chemistry-past, present and future', *Wood Sci. Technol.*, **1977**, 11, 169-218.
5. A. Sakakibara, 'A structural model of softwood lignin', *Wood Sci. Technol.*, **1980**, 14, 89-100.
6. W.G. Grasser & H. R. Grasser, 'The evaluation of lignin's chemical structure by experimental and computer simulation techniques'. *Paperi ja Puu.* **1981**, 63, 71-83.
7. N. Terashima, K. Fukushima, L. He & K. Takabe, 'Comprehensive model of lignified plant cell wall', In :*Forage Cell Wall Structure and Digestibility* : Jung, H.G., D.R Buxton, R.D. Hatfield & J. Ralph (eds.), Am. Soc. Agr., Madison, WI, **1993**, pp. 247-270.
8. N. Terashima, J. Nakashima, K. Takabe, 'Proposed structure of protolignin in the cell walls'. In: '*Lignin and Lignan Biosynthesis*', N.G. Lewis & S. Sarkanen (eds.), ACS Symp. Series, 697. Am. Chem. Soc., Washington DC., **1998**, pp. 180-193.
9. B.J. Fergus & D.A.I. Goring, 'The distribution of lignin in birch wood as determined by ultraviolet microscopy', *Holzforschung*, **1970**, 24, 118-124.
10. H-L. Hardell, G.L. Leary, M. Stoll & U. Westermark, 'Variations in lignin structure in defined morphological parts of spruce', *Svensk Papperstidn.*, **1980**, 83, 44-49.
11. P. Whiting & D.A.I. Goring, 'The morphological origin of milled wood lignin', *Svensk Papperstidn.*, **1981**, 84, R120-R122.
12. A. Maurer & D. Fengel, 'On the origin of milled wood lignin', *Holforschung* ,**1992** 46, 417-423; 471-475.
13. N. Terashima, K. Fukushima & T. Imai, 'Morphological origin of milled wood lignin studied by radiotracer method', *Holzforschung*, **1992**, 46, 271-275.
14. Atalla, R.H. & U.P. Agarwal, 'Raman microprobe evidence for lignin orientation in the cell wall of native woody tissue', *Science*, **1985**, 227, 636-638.
15. K. Ruel, F. Barnoud & D.A.I. Goring, 'Lamellation in the S2 layer of softwood tracheids as demonstrated by scanning electron microscopy'. 1978, *Wood Sci. Technol.* **12**, 287-291.
16. J. Hafrén, T. Fujino & T. Itoh, 'Changes in the cell wall architecture of differentiating tracheids of *Pinus thunbergii* during lignification', *Plant Cell Physiol.*, **1999**, 40, 532-541.
17. J. Hafrén, T. Fujino, T. Itoh, U. Westermark & N. Terashima, 'Ultrastructural change in the compound middle lamella of *Pinus thunbergii* during lignification and lignin removal', *Holzforschung*, **2000**, 54, 234-240.

18. J. Nakashima, T. Mizuno, K. Takabe, M. Fujita & H. Saiki, 'Direct visualization of of lignifying secondary wall thickenings in *Zinnia elegans* cells in culture' ,*Plant Cell Physiol.*, **1997**, 38(7),818-827.
19. M. Fujita & H. Harada, 'Autoradiographic investigation of cell wall development . II. Tritiated phenylalanine and ferulic acid assimilation in relation to lignification', *Mokuzai Gakkaishi*, **1979**, 25, 84-94.
20. N. Terashima, K. Fukushima, Y. Sano & K. Takabe, 'Heterogeneity in formation of lignin X. Visualization of lignification process in differentiating xylem of pine by microautoradiography', *Holzforschung*, **1988**, 42, 347-350.
21. K. Takabe, K. Fukazawa & H. Harada, 'Deposition of cell wall components in conifer tracheids', In: *Plant Cell Wall Polymers, Biogenesis and Biodegradation*, N.G. Lewis & M.G. Paice (eds.), ACS Symp. Series 399. Am. Chem. Soc., Washington DC. **1989**, pp. 47-66.
22.N. Terashima & K. Fukushuma, 'Biogenesis and structure of macromolecular lignin in the cell wall of tree xylem as studied by microautoradiography', In: *Plant Cell Wall Polymers, Biogenesis and Biodegradation*, N.G. Lewis & M.G. Paice (eds.), ACS Symp. Series 399. Am. Chem. Soc.,Washington DC. **1989**, pp. 160-168.
23.N. Terashima, Y. Seguchi & D. Robert, 'Selective ^{13}C-enrichment of side chain carbon of guaiacyl lignin in pine', *Holzforschung*, **1991**, 45(Suppl.), 35-39.
24. N. Terashima, R.H. Atalla & D. L. VanderHart, 'Solid state NMR spectroscopy of specifically ^{13}C-enriched lignin in wheat straw from coniferin', *Phytochemistry*, **1997**, 46, 863-970.
25. N. Terashima, J. Hafrén, U. Westermark & D.L. VanderHart, 'Nondestructive Analysis of Lignin Structure by NMR Spectroscopy of Specifically ^{13}C-Enriched Lignins 1: Solid State Study of Ginkgo Wood', *Holzforschung*, **2001**,In press.
26. N. Terashima, D. Evtuguin & C. P. Neto, 'An improved ^{13}C-tracer method as an analytical tool in lignin chemistry – Specific ^{13}C-enrichment of aromatic carbons in lignin' – In:*Proceedings of Post-Symposium Workshop of the 11th Int. Symp. Wood and Pulping Chem*, Grenoble, **2001**
27 N. Terashima, J. Hafrén, U. Westermark, Y. Xie, K. Fukushima & D.L. VanderHart, 'Proposed 3D structural model for softwood lignin', In: *Proceedings of 10th Int. Symp. Wood and Pulping Chem*, Yokohama, **1999**, Vol. 1, pp.106-109.
28. Y. Matsumoto, K. Minami, A.Ishizu, 'Structural study on lignin by ozonation --The erythro and threo ratio of the β-O-4 structure indicates how lignin polymerizes--', *Mokuzai Gakkaishi*, **1993**, 39, 734-736.
29. N. Habu, Y. Matsumoto, A. Ishizu & J. Nakano, 'Configurational study of phenylcoumaran type structure in lignin by ozonation', *Mokuzai Gakkaishi*, **1988**, 34, 732-738.
30. G. Brunow, I. Kilpeläinen, J. Sipilä, K. Syrjänen, K. Karhunen, H. Setälä & P. Rummakko, 'Oxidative coupling of phenols and the biosynthesis of lignin', In: '*Lignin and Lignan Biosynthesis*', N.G. Lewis & S. Sarkanen (eds.), ACS Symposium Series, 697. Am. Chem. Soc., Washington DC., **1998**, pp. 131-147.

Molecular Breeding of Woody Plants
N. Morohoshi and A. Komamine, editors.

INVOLVEMENT OF PEROXIDASE AND HYDROGEN PEROXIDE IN THE METABOLISM OF β-THUJAPLICIN IN FUNGAL ELICITOR-TREATED *CUPRESSUS LUSITANICA* SUSPENSION CULTURES

Jian Zhao [1,2], Kokki Sakai [1]

[1] *Laboratory of Forest Chemistry and Biochemistry, Faculty of Agriculture, Kyushu University, Fukuoka 812-8581 Japan;*

[2] *Author for correspondence, Institute of Materia Medica, Chinese Academy of Medical Sciences & Peking Union Medical College, Beijing, 100050, China*

ABSTRACT

The β-thujaplicin in *Cupressus lusitanica* cell cultures can be stimulated to a high level by a fungal elicitor on day 3 to 6 after elicitation, and then decreased. This pattern should be a combined result of β-thujaplicin biosynthesis and degradation, which simultaneously occur in the elicited cell cultures. *In vitro* studies revealed that a horseradish peroxidase could oxidize about 80 % of β-thujaplicin in the presence of H_2O_2. The culture medium containing some secreted peroxidases also oxidized a part of β-thujaplicin. Peroxidase activity in the medium and in the culture cells increased to a high level after elicitation, which was consistent with the decrease of β-thujaplicin accumulation. Further studies showed that H_2O_2 production in the culture medium and NADPH-dependent superoxide anion synthase activity in the cells exhibited a transient peak at 20-40 min after fungal elicitation. Both a NADPH oxidase specific inhibitor diphenylene iodonium (DPI) and peroxidase specific inhibitor KCN inhibited O_2^- and H_2O_2 *in vitro* production by NADPH-dependent superoxide anion synthase. These results suggest that the oxidative burst also occurred in fungal elicited *C. lusitanica* cell cultures, and the origin of O_2^- and H_2O_2 production in the cell cultures includes both NADPH oxdase and peroxidases. Addition of DPI or catalase to the cell cultures inhibited a part of elicitor-induced β-thujaplicin production, while exogenous H_2O_2 stimulated β-thujaplicin production, suggesting that H_2O_2 production induced by fungal elicitor is an important reason for β-thujaplicin production. These results suggest that peroxidases and H_2O_2 are profoundly involved in β-thujaplicin biosynthesis and biodegradation in fungal elicited *C. lusitanica* cell cultures.

KEY WORDS

Cupressus lusitanica, peroxidase, NADPH oxidase, β-thujaplicin, elicitor, hydrogen peroxide, biosynthesis and biodegradation

INTRODUCTION

β-Thujaplicin (Hinokitiol) is a tropolone compound with a seven-membered carbon ring and an isopropyl side chain. It has a broad spectrum of antimicrobial activity and therefore is widely used in cosmetics, clinic products and other areas [1,2]. Because β-thujaplicin is mainly contained in the heartwood of some Cupressaceae trees in a low content [2], its production by plant cell culture was introduced as an alternative source. At the same time, the plant cell cultures provide a good experimental system for studying metabolism of this novel tropolone since there are some difficulties when studying this heartwood component in intact plant. The *de novo* β-thujaplicin production in *C. lusitanica* cell cultures can be stimulated by fungal elicitor or methyl jasmonate [3,4]. Like most of other plant secondary metabolites, β-thujaplicin accumulation in the cell cultures increases to the maximum on day 3-6 after elicitor treatment, and then decreases to a low level [3,4]. This typical accumulation pattern should result from β-thujaplicin biosynthesis and biodegradation, which simultaneously occur in the cell cultures. Therefore besides the attempts to improve β-thujaplicin production, it would be also very important to study how β-thujaplicin is degraded or transformed into other compounds in the cell cultures. From the chemistry point of view, β-thujaplicin could be unstable under light because it can accept light energy and C1-C2 bond is broken to produce some compounds. But in dark culture conditions, this slow light-catalyzed degradation should not function as a main mode. On the contrary, the biological factors most probably play a major role in β-thujaplicin degradation.

Peroxidases are some oxidoredctases that catalyze the oxidation of a diverse group of organic compounds using hydrogen peroxide as the ultimate electron acceptor [5]. Peroxidases have been suggested to be involved in various metabolisms, such as auxin and indole alkaloid matabolism, flavone metabolism, biosynthesis of cell wall and lignin [5,6,7,8]. Peroxidases are also involved in plant defense responses and other physiological processes. Peroxidase-dependent H_2O_2 production has been found in a larger body of plant species [8,9]. Apoplastic peroxidase-dependent oxidative burst is recently observed in some plant species when exposed to pathogens or fungal elicitors [10, 11,12]. These novel functions of peroxidases have attracted more attentions. Since peroxdases are universal enzymes with multiple functions, and also a previous study revealed that β-thujaplicin and its iron chelate showed strong antioxidant activities [13], we tried peroxidase to oxidize β-thujaplicin. We found that horseradish peroxidase can oxide β-thujaplicin by about 80 %. This phenomenon stimulated our interests on peroxidase-catalyzed β-thujaplicin biodegradation in the cell cultures and related aspects, since these studies could be of great importance for the regulation of secondary metabolism. However, so far there is no report on these aspects.

MATERIALS & METHODS

Plant cell cultures and treatment profiles

The *Cupressus lusitanica* suspension cultures from callus was established as described previously [4]. About 2.5 g of fresh cells was inoculated into 20 ml production medium [3] in 100-ml flasks and incubated on a rotary shaker (110 rpm) at 23 ± 2 °C in the dark. For the time-course study, an autoclaved yeast elicitor (1mg /ml), or water (control) was added to the cell cultures and the cell cultures were collected in intervals for analysis of β-thujaplicin, hydrogen peroxide and enzyme activity assay. Catalase and superoxide dismutase were obtained from Sigma and dissolved in water with 5 % glycerol. Diphenylene iodonium (DPI) was from Wako Pure Chemicals and prepared in 0.05 % DMSO solution. These reagents were added to the cell cultures, respectively, together with fungal elicitor. The cell cultures were collected after 24 h of incubation.

In vitro biotransformation of β-thujaplicin

The horseradish peroxidase (HRP) was obtained from BDH (England) and prepared in 50 mM Na-phosphate buffer (pH 6.0). β-Thujaplicin was dissolved in 50 % of methanol solution. Transformation solutions were the different combinations of β-thujaplicin, phosphate buffer, HRP and 2 mM H_2O_2. The transformation of β-thujaplicin by culture medium was similar to that by HRP. The cell-free culture media were prepared from *C. lusitanica* cell cultures treated with fungal elicitor for 0, 4, 6 and 8 days, respectively, by filtration under vacuum. These cell-free culture media were directly combined with β-thujaplicin or 2 mM H_2O_2. All reactions were incubated at 30 °C in the dark for 12 h, then the unreacted β-thujaplicin was extracted twice with ethyl estate and determined by HPLC. An equally mixed medium, boiled for 5 min, was used as the negative control.

Determination of H_2O_2 in the culture medium

The H_2O_2 production in culture medium was assayed according to Mithofer *et al.*[14]. After elicitor treatment for different time, 1 ml of the supernatant medium was collected for H_2O_2 production assay. H_2O_2 concentration in the medium was determined by measuring increase of absorbance at 450 nm that resulted from the endogenous peroxidase-catalyzed oxidation of the exogenous *o*-dianisidine. Concentration of H_2O_2 was calculated from a standard curve obtained by incubating variable amounts of H_2O_2 ranging from 6 to 60 μM with 5 U of horseradish peroxidase and 50 μM *o*-dianisidine.

Enzyme extraction and activity assay

The cells were collected and frozen immediately, then were homogenized in liquid nitrogen into powder with an extraction buffer containing 0.1 M Tris-HCl (pH 7.2), 1 % PVP (w/v), 5 mM $MgCl_2$, 0.1 % Triton X-100 (v/v) and 10 % glycerol. The homogenate was centrifuged at 13000 rpm (4°C) for 20 min, and the supernatant was used as crude enzymes. Superoxide anion synthase activity was assayed according to the NBT-NADPH method of Mithofer *et al.* [14] with slight modifications. Peroxidase activity was assayed according to Chance and Maehly [15]. The protein content was determined with the Bradford method using bovine serum albumin as standard. To assay peroxidase activity of the culture medium, the medium was filtrated and centrifuged at 12000 rpm (4°C) for 10 min. The supernatant was used to assay peroxidase activity.

Extraction and determination of β-thujaplicin

Extraction and determination of β-thujaplicin were carried out using HPLC as previously described [4]. Vanillin was used as an internal standard. Biomass was expressed on fresh weight basis. All data were generated from triplicate independent experiments. Statistical analysis was carried out using the Student's t-test.

RESULTS & DISCUSSION

β-Thujaplicin accumulation in elicited *C. lusitanica* suspension cultures

As shown in Fig. 1, *C. lusitanica* cell cultures rapidly produced β-thujaplicin after

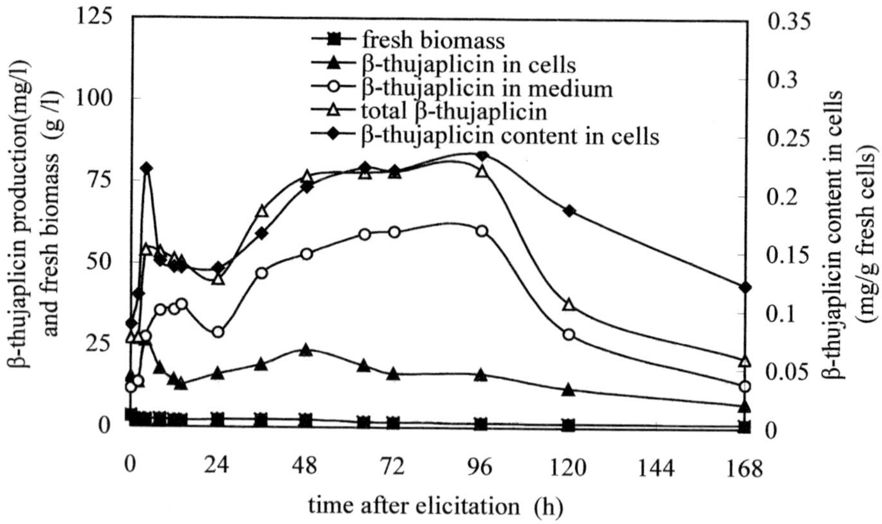

Figure 1. Time-course of β-thujaplicin accumulation in *C. lusitanica* cell cultures as a function of fungal elicitation. The cell cultures were incubated in a production medium and treated with 1mg/ml of a yeast elicitor.

treated with fungal elicitor. A low and transient β-thujaplicin peak was observed at about 8 h after elicitation. Then the β-thujaplicin production rapidly increased and reached the maximum on day 4-5 after elicitation, and then decreased to a low level. A large portion of β-thujaplicin production was released into the medium upon the fungal elicitation. These changes in β-thujaplicin accumulation reflect some biochemical processes occurring in the elicitor treated cell cultures. It is proposed that β-thujaplicin biosynthesis were activated and dominant at earlier stage of elicitation; and then the biotransformation or biodegradation of β-thujaplicin began to function at late stages of elicitation because β-thujaplicin in high concentration is also toxic to plant cell itself.

In vitro transformation of β-thujaplicin by peroxidase

The results presented in Table 1 show that horseradish peroxidase can transform 80 % of total β-thujaplicin in the presence of H_2O_2. Horseradish peroxidase or H_2O_2 alone almost cannot transform β-thujaplicin. It was a typical peroxidase activity that oxidizes β-thujaplicin. The oxidation products of β-thujaplicin were still under investigation because of some difficulties in identifying these reaction products. GC-MS analysis showed that there were some smear peaks at low m/z regions. It is proposed that the oxidation of β-thujaplicin by peroxidase may be like chain reactions that produce more than one final product.

Because the peroxidases and β-thujaplicin could accumulate in the medium to high levels after elicitation, we tried the culture medium as a peroxidase source to transform β-thujaplicin. Table 2 shows that the cell-free culture media prepared from non-elicited or elicited-cell cultures for different time were able to transform β-thujaplicin although the transformation efficiency was not high. The transformation abilities of the elicited culture media increased with elicitation time, which was in agreement with increased peroxidase activity and the decrease of β-thujaplicin accumulation in the culture medium at the corresponding time.

Table 1. *In vitro* transformation of β-thujaplicin by horseradish peroxidase

Horseradish Peroxidase (EU)	H_2O_2 (m M)	β-thujaplicin		
		Added (mg)	recovered (mg)	Transformed (%)
–	–	5	5.0 ± 0.02	0
82	2	5	1.0 ± 0.01	80
–	2	5	4.9 ± 0.04	12
82	–	5	4.8 ± 0.03	14
82	2	–	0 ± 0.03	–

Notes: three ml of reaction solutions contained 50 mM Na-phosphorate buffer (pH 6.0) and different combinations of β-thujaplicin, H_2O_2 and HRP. After incubation at 30°C in the dark for 12 h, the unreacted β-thujaplicin was immediately extracted with ethyl acetate for analysis with HPLC.

Table 2. *In vitro* transformation of β-thujaplicin by *C. lusitanica* culture medium treated or without treated with a fungal elicitor

Culture medium (peroxidase activity, ΔOD_{470} mg^{-1} protein.min^{-1})	H_2O_2 (m M)	β-thujaplicin		
		Added (mg)	Recovered (mg)	Transformed (%)
denatured (0)	2	5	5.2 ± 0.03	0
non-elicited (45.2)	2	5	2.19 ± 0.02	58
4 days after elicitation (10.2)	2	5	3.85 ± 0.02	26
6 days after elicitation (13.0)	2	5	3.53 ± 0.03	32
8 days after elicitation (17.8)	2	5	2.75 ± 0.02	47
denatured (0)	2	–	0.05 ± 0.01	–

Notes: the culture media were collected by filtration under vacuum on different day after elicitation. The reaction mixtures contained culture medium, H_2O_2 and β-thujaplicin. A denatured medium by boiling was used as control. Other conditions were the same as the horseradish peroxidase transformation.

Time-course of peroxidase activity in elicited *C. lusitanica* cell cultures

As shown in Fig. 2, the culture medium kept a high level of peroxidase activity in normal *C. lusitanica* cell cultures. But after elicitation, peroxidase activity dramatically decreased at first and then rapidly increased to the maximum after 5 days of elicitation, and kept this level for a long time. Peroxidese activity in the culture cells rapidly increased to an extremely high level on day 3-4 after fungal elicitation and kept the high level. Comparison of time-course changes in peroxidase activity and β-thujaplicin production shows that changes of peroxidase activity were consistent with the decrease of β-thujaplicin accumulation.

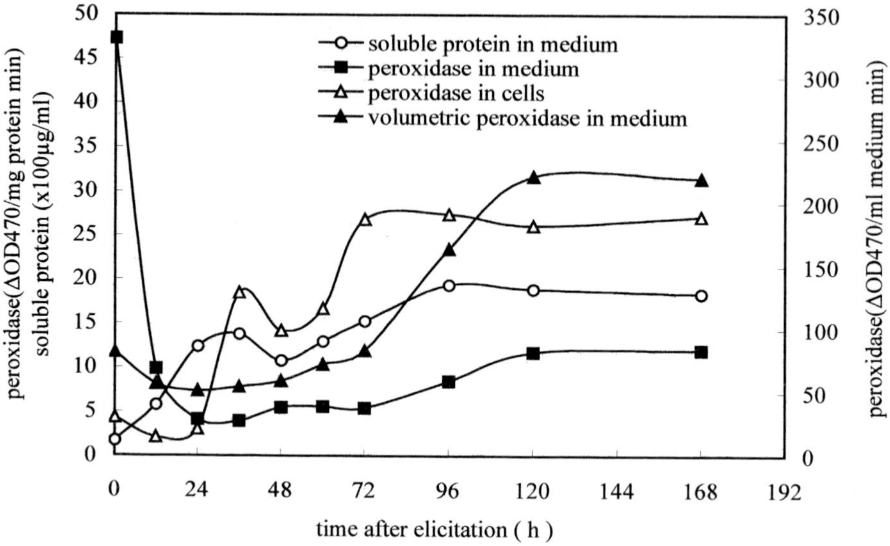

Figure 2. Kinetics of peroxidase activity in elicitor-treated *C. lusitanica* cells and the culture medium. Peroxidase activity was assayed using guaiacol method as described in "Materials and Methods" section.

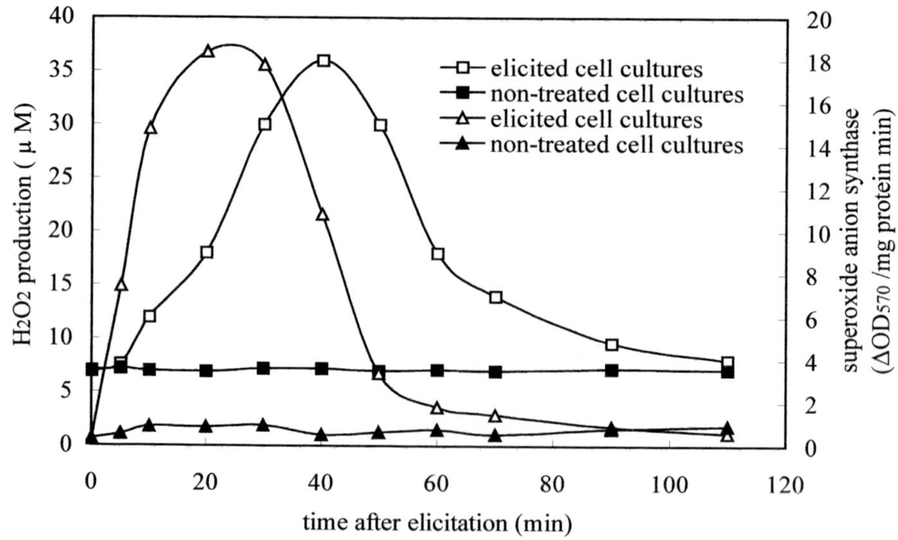

Figure 3. Yeast elicitor-induced H_2O_2 and transient superoxide anion sythase activity in *C. lusitanica* cell cultures. After yeast elicitation, H_2O_2 production (□, ■) of the culture medium was determined with an *in vivo o*-dianisidine oxidation method using endogenous peroxidase in the cell cultures. Superoxide anion synthase(Δ,▲)was assayed with a NADPH-NBT method as described in "Materials and Methods" section.

H_2O_2 production in elicitor treated *C. lusitanica* cell cultures

As an ultimate electron acceptor for peroxidase activity, H_2O_2 is produced in plant cells as a byproduct of electron transport system [12]. A rapid H_2O_2 production in fungal elicitor-treated plant cell cultures is often observed and plays important roles in plant defense responses [14,16]. A transient overproduction of H_2O_2 was observed in fungal elicitor-treated *C. lusitanica* cell cultures after 30 min of elicitation (Fig. 3), while non-elicited cell cultures only produced a basic level of H_2O_2. The NADPH oxidase, a supposed source for O_2^- and H_2O_2 overproduction [14,16], was induced by fungal elicitor (Fig.3). This NADPH-dependent superoxide anoin synthase activity was activated to a high level after fungal elicitation, which was corresponding to H_2O_2 production. These results suggest that the oxidative burst occurs in the elicited *C. lusitanica* cell cultures.

H_2O_2 stimulated β-thujaplicin production in elicited *C. lusitanica* cell cultures

When *C. lusitanica* cell cultures were pre-treated with DPI (an inhibitor of NADPH oxidase) or catalase (a H_2O_2-scavenging enzyme) 20 min prior to fungal elicitation, obvious inhibitory effects on elicitor-induced β-thujaplicin production were observed. Direct addition of H_2O_2 to the cell cultures, besides causing a hypersensitive cell death, the increased β-thujaplicin production was also observed after 24h of treatment (Fig. 4). All these results argued that the elicitor-induced β-thujaplicin production is at least partially mediated by fungal elicitor-induced H_2O_2 production, which support that H_2O_2 production is an important signal for secondary metabolite accumulation [17].

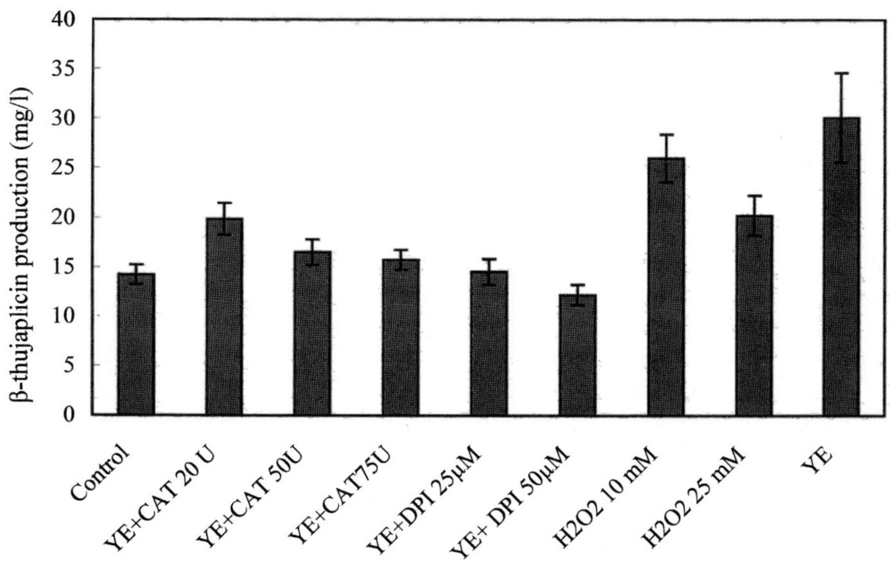

Figure 4. Effects of H_2O_2 on β-thujaplicin production in the fungal elicited cell cultures. H_2O_2 was solely added to the cell cultures. DPI or catalase was added to the cell cultures before yeast elicitation (YE+DPI, YE+CAT). The yeast elicitor (YE) and water (Control) was added to the cell cultures as positive and negative control. After 24 h of incubation, the cell cultures were harvested.

Table 3. *In vitro* superoxide anion and hydrogen peroxide synthesis by crude enzyme extracts from fungal elicited *C. lusitanica* cells

Inhibitor or enzyme additions	O_2^- production (Δ OD $_{570}$ mg^{-1} protein min^{-1}) (% of the control)	H_2O_2 production (nmol mg^{-1} protein min^{-1}) (% of the control)
Control	9.92 (100)	200 (100)
DPI (5 μM)	6.34 (64)	177 (86)
DPI (100 μM)	4.20 (42)	156 (78)
Imidozale (5 mM)	7.06 (71)	176 (88)
Imidozale (10 mM)	8.57 (86)	181 (91)
KCN (0.2 mM)	6.6 (66)	124 (62)
KCN (0.5 mM)	6.42 (65)	117 (59)
Superoxidae dismutase (10U/ml)	10.28 (103)	250 (125)
Superoxide dismutase (20U/ ml)	18.5 (186)	293 (146)
Catalyse (20 U/ ml)	8.77 (87)	55 (27)
Catalyse (50 U/ ml)	7.56 (76)	35 (18)

Notes: superoxide anion synthase activity was assayed at 25°C in 1 ml of reaction solution containing 50 mM Na-phosphate buffer (pH 7.0), 2 mM $MgCl_2$, 0.02 % Triton X-100 (v/v), 0.2 mM NBT, 20 μM NADPH, 50 μl of crude enzyme, and various reagents. Reactions were started by addition of NADPH. Superoxide anion production was determined by increase of absorbance at 570 nm over the period of 5 min. For H_2O_2 production assay, after 5 min of reaction without NBT, 100 μl of 100 mM HCl solution was added to each reaction solution to eliminate the KCN or other substances interfering the direct H_2O_2 assay, after 1h the samples were taken for H_2O_2 production assay by using horseradish peroxidase-catalyzed *o*-dianisidine oxidation.

NADPH-dependent apoplastic peroxidase is also responsible for elicitor-induced H_2O_2 production

Besides the NADPH oxidase as a source of the oxidative burst, apoplastic peroxidases may also play a role in the elicitor-induced oxidative burst[18, 19]. We therefore further investigated the oxidative burst generator in elicited *C. lusitanic* by studying the properties of superoxide anion synthase activity *in vitro*. Table 3 shows that NADPH-dependent superoxide anion synthase activity was sensitive both to DPI and to KCN, suggesting that superoxide anion synthase also included peroxidases. Moreover, another NADPH oxidase inhibitor imidozale (5 mM) only caused about 30 % inhibition on O_2^- production and 20 % inhibition on H_2O_2 production. Catalase slightly influenced O_2^- production and almost totally inhibited H_2O_2 production, and superoxide dismutase slightly stimulated O_2^- accumulation and strong enhanced H_2O_2 production.

DISCUSSION

The present results suggest peroxidases in the cell cultures could oxidize β-thujaplicin produced by fungal elicitation, which seems in agreement with its antioxidant activity. This property is mainly due to its neighboring hydroxyl and carbonyl groups. The oxidation site of β-thujaplicin by peroxidases may be the hydroxyl group. β-Thujaplicin ferrous and copper chelates in *C. lusitanica* cell cultures may not substantially affect its antioxidant activity and the peroxidase-catalyzed oxidation of β-thujaplicin. Therefore, peroxidases may play an important role in transformation of β-thujaplicin. In addition, our lab also found methylation is another important way to transform β-thujaplicin in *C. lusitanica* cell cultures to detoxify high concentration of β-thujaplicin [20], although the mechanisms for transformation of β-thujaplicin by peroxidase and β-thujaplicin O-methyltransferase in *C. lusitanica* cell cultures remain to be clarified.

It is known that several sources for the generation of reactive oxygen species exist in plant cells, including plasma membrane localized NAD(P)H oxidases, apoplastic peroxidases, amine oxidases and oxalate oxidases and protoplastic sources from mitochondria, chloroplasts and peroxisomes [12]. A membrane-bound NAD(P)H oxidase is popularly accepted as a oxidative burst generator [14,16,18,19]. However, an increasing body of evidence suggests that apoplastic peroxidase also functions as a source of the oxidative burst in some plant species [11,12,18,19]. Here our results suggest that in elicited *C. lusitanica* cell culture, NAD(P)H oxidase and peroxidases may simultaneously function in the generation of the oxidative burst. That adds another evidence for the existence of multiple systems for generation of active oxygen species in plant cells [18,19].

Our studies demonstrated that peroxidases and H_2O_2 are profoundly involved in the fungal elicitor-induced β-thujaplicin biosynthesis and biodegradation in *C. lusitanica* cell cultures. At earlier stages of elicitation, the NADPH-dependent apoplastic peroxidases are activated to catalyze the O_2^- and H_2O_2 production, which further stimulated β-thujaplicin production; at later stages of elicitation, peroxidases excreted to the culture medium catalyze β-thujaplicin oxidation and transform some β-thujaplicin using basic level of H_2O_2 in the medium. The biosynthesis and degradation of β-thujaplicin in the cell cultures are related to defense responses, from this point of view, the fungal elicited *C. lusitanica* cell cultures produce β-thujaplicin against pathogens. However, overproduction of β-thujaplicin is also toxic to cells themselves, and therefore they developed some strategies, including oxidation and methylation, to transform some parts of β-thujaplicin. It is revealed that the peroxidase isoforms responsible for H_2O_2 production and β-thujaplicin oxidation was different in R_f values (data not shown). Further study on β-thujaplicin oxidation by peroxidases and characterization of peroxidases responsible for β-thujaplicin oxidation and H_2O_2 production may lead a more profound understanding on these important processes.

ACKNOLEDGEMENT:

This work was supported by a postdoctor fellowship and grant (No. 12099345) from Japan Society for the Promotion of Science (JSPS), which are gratefully acknowledged. This work was also partially supported by the scientific research fund (No.11876040) of the Japanese Ministry of Education, Science and Culture.

REFERENCES:

1. T. J. Trust & R. W. Coombs, 'Antibacterial activity of β-thujaplicin', *Can. J. Microbiol.*, **1973**, 19, 1341-1347.
2. T. Okabe, K. Saito & Y. Otomo, 'Antimicrobial activity and safety of Hinokitiol', *Fragrance J.*, **1989**, 17, 74-79.
3. R. Itose & K. Sakai, 'Improved culture condition s for the production of β-thujaplicin

by suspension cell cultures of *Cupressus lusitanica', Plant Biotechnol.* **1997**, 14, 163-167.

4. J. Zhao, K. Fujita, J. Yamada & K. Sakai, 'Improved β-thujaplicin production in *Cupressus lusitanica* suspension cultures by fungal elicitor and methyl jasmonate', *Appl. Microbiol. Biotechnol.*, **2001**, 55: in press.
5. J. H. Dawson, 'Probing structure-fuction relations in heme-containing oxygenasses and peroxidase', Science, **1988**, 240, 433-439.
6. H. M. Awad, M. G. Boersma, J. Vervoort & I. M. Rietjens, 'Peroxidase-catalyzed formation of quercetin quinone methide-glutathione adducts', *Arch. Biochem. Biophys.*, **2000**, 378, 224-233.
7. M. Quiroga, C. Guerrero, M. A. Botella, A. Barcelo, I. Amaya, M. I. Medina, F. J. Alonso, S. E. de Forchetti, H. Tigier & V. Valpuesta, 'A tomato peroxidase involved in the synthesis of lignin and suberin', *Plant Physiol.*, **2000**, 122,1119-1127.
8. G. G. Gross, C. Janse & E. F. Elstner, 'Involvement of malate, monophenols and the superoxide radical in hydrogen peroxide formation by isolated cell wall from horseradish (*Armoracia lepathifolia* Gilib)', *Planta,* **1977**, 136, 271-276.
9. B. Halliwell 'Lignin synthesis: the generation of hydrogen peroxide and superoxide by horseradish peroxidase and its stimulation by manganese (II) and phenols', *Planta,* **1978**, 140, 81-88.
10. G. P. Bolwell, D. R. Davies, C. Gerrish, C. K. Auh & T. M. Murphy, 'Comparative biochemistry of the oxidative burst produced by rose and French bean cells reveals two distinct mechanisms', *Plant Physiol.,* **1998**, 116,1379-1385.
11. A.C. Allan & R. Fluhr, 'Two distinct sources of elicited reactive oxygen species in tobacco epidermal cells', *Plant Cell,* **1997**, 9, 1559-1572.
12. G. P. Bolwell & P. Wojtaszek, 'Mechanism for the generation of reactive oxygen species in plant defence-a broad perspective', *Physiol. Mol. Plant Path.*, **1997**, 51, 347-366.
13. T. Yamaguchi, K.Fujita & K.Sakai, 'Biological activity of extracts from *Cupressus lusitanica* cell culture', *J. Wood Sci.,* **1999**, 45, 170-173.
14. A. Mithofer, A. Daxberger, D. Fromhold-Treu & J. Ebel 'Involvement of an NAD(P)H oxidase in the elicitor-inducible oxidative burst of soybean', *Phytochem.* **1997,** 45,1101-1107.
15. B. Chance & A. C. Maehly, 'Assay of catalase and peroxidase', In: *Methods in Enzymology*, Vol. II, L. Packer (ed.), Academic press, NewYork, **1955**, pp 764-775.
16. N. Doke & Y. Miura, 'In vitro activation of NADPH-dependent O_2^- generating system in a plasma membrane-rich fraction of potato tube tissues by treatment with an elicitor from Phytophthora infestans or with digitonin', *Physiol. Mol. Plant Path.,* **1995**, 46,17-28.
17. T. Jabs, M. Tschöpe, C. Colling, K. Hahlbrock & D.Scheel, 'Elicitor-stimulated ion fluxes and O_2^- from the oxidative burst are essential components in triggering defense gene activation and phytoalexin synthesis in parsley', *Proc. Natl. Acad. Sci.USA,* **1997**, 94, 4800-4805.
18. A. K. Papadakis & K. A. Roubelakis-Angelakis, 'The generation of active oxygen species differs in tobacco and grapevine mesophyll protoplasts', *Plant Physiol.,* **1999**, 121,197-205.
19. C. Martinez, J. L. Montillet, E. Bresson, J. P. Agnel, G. H. Dai, J. F. Daniel, J. P. Geiger & M. Nicole, 'Apoplastic peroxidase generates superoxide anions in cells of cotton cotyledons undergoing the hypersensitive reaction to *Xanthomonas campestris* pv malvacearum Race 18', *Mol. Plant-Microbe Inter.,* **1998**,11, 1038-1047.
20. J. Yamada, K. Fujita & K. Sakai, A factor controlling β-thujaplicin production in suspension cultures of *Cupressu lusitanica.* A processing of IWBS, **2001**, in press.

Molecular Breeding of Woody Plants
N. Morohoshi and A. Komamine, editors.

A FACTOR CONTROLLING ß-THUJAPLICIN PRODUCTION IN SUSPENSION CULTURE OF *CUPRESSUS LUSITANICA*

Junko Yamada, Koki Fujita, Kokki Sakai

Department of Forestry and Forest Products Sciences
Graduate School of Bioresource and Bioenvironmental Sciences
Kyushu University, Fukuoka 812-8581

ABSTRACT

Suspension cell culture of Cupressus lusitanaca produce ß-thujaplicin, a heart wood tropolone. The product inhibition in the ß-thujaplicin biosynthesis was investigated in the present paper. Initial cell density at the beginning in the culture did not affect the ß-thujaplicin accumulation level, and the initial addition of ß-thujaplicin to the culture suppressed its *de novo* production. When ß-thujaplicin accumulation was reached at a certain level (ca. 40mg/l) in the *C. lusitanica* culture medium, the culture seemed to cease ß-thujaplicin production. However, the culture regained ß-thujaplicin productivity if the ß-thujaplicin-containing medium was exchanged for new one. The formation of ß-thujaplicin methyl ether was also observed. These results suggest that ß-thujaplicin synthesis was regulated by the product feed back mechanism in this cell line and the excess accumulation ß-thujaplicin would be relieved by the conversion to its methyl ether.

KEYWORDS

ß-thujaplicin, methylation, cell culture, *Cupressaceae*, product inhibition

INTRODUCTION

Plant cell cultures provide an attractive route to obtain highly valuable plant-derived products, such as flavors, fragrances, alkaloids, pigments and pharmaceuticals that are expensive to synthesize chemically and that naturally occur only at low concentrations. Recently, there have been some significant successes for the production of valuable plant second metabolites, but in most cases the yields of secondary metabolites are too low for commercial production [1,2].

ß-Thujaplicin, known as hinokitiol in Japan, is responsible for the durability and resistance of tree against insect attacks and fungal decays, which are characteristic of the heart wood of most species in the Cupressaceae family [3]. Its broad antimicrobial spectrum led to its wide use in cosmetics, clinical products and other areas. Most of this compound is provided from sawdust of *Thujopsis dolabrata* in Japan, but its content is very low (0.02% on sawdust basis) [4]. Therefore, plant cell culture is reasonable way for commercial production of ß-thujaplicin, since its demand is expected to increase.

From such a view point, we have investigated the production of ß-thujaplicin with cell cultures. We have already reported that a callus line induced from *Cupressus lusitanica* (Mexican cypress) had a high productivity of ß-thujaplicin and this productivity was stimulated by an elicitor addition [5, 6]. Some of the biological activities were demonstrated in an ethyl acetate extract from elicitor-treated *C. lusitanica* suspension cell culture [8], suggesting that the extract becomes a valuable bioactive source. In a previous paper [7], it was shown that sufficient supply of nutrients was necessary, but Fe (II) had an inhibitory effect on cell growth in a suspension culture of *C. lusitanica* cell. However, we also reported that insufficient inorganic nutrients and excess Fe(II) were advantageous for the ß-thujaplicin production [7]. Though several different cell cultures from different *Cupressaceae* species has been derived for ß-thujaplicin production [9, 10], the establishment of mass production system is still undergoing. It is usually difficult to make industrial manipulation tree culture cells because of their slow growth speed and sensitivity to culture stress [11]. During optimization of culture conditions such as nutrients, elicitor methods etc, we found the possibility of product inhibition in ß-thujaplicin biosynthesis.

Therefore, in this study, we investigated this regulating factor of ß-thujaplicin production for improving the productivity.

MATERIALS & METHODS

Cell cultures and culture conditions

Callus cultures of *C. lusitanica* have been maintained on Gamborg B5 medium [13] supplemented with 20 g/l sucrose, 0.01 μM BAP, 10 μM NAA and 2.7 g/l Gel-rite at pH 5.5 for more than 10 years as described previously [7]. The callus cells of *C. lusitanica* were transferred to suspension cultures and maintained in IS-1 medium [7], which is a modified Gamborg B5 medium containing 0.01 mM Fe(II), one tenth of that in the original B5 liquid medium, at 25°C in the dark on a rotary shaker (70 rpm).

ß-Thujaplicin production and determination

C. lusitanica cells grown in a 500ml flask with 120ml IS-1 medium [7] were separated from medium by filtering samples through Miracloth with suction and a part of residue was transferred to a 50ml flask with 10ml IS-2 medium containing 0.25 mM Fe(II) and major inorganic nutrients at 0.1 strength of the original B5 medium [7]. To stimulate ß-thujaplicin production, 1ml of an elicitor solution (18g/l) [5] was added to this culture. The cells were incubated at 25°C in the dark on a rotary shaker (70 rpm) for 6 days. After the incubation, cells separated by Miracloth were homogenized in a mortar with a pestle. Homogenated cell and medium were separately extracted twice with ethyl acetate. ß-Thujaplicin content in the extract was determined by the Endo's method [14].

Purification and identification of ß-thujaplicin methyl ether

Crude diethyl ether extract containing ß-thujaplicin methyl ether was prepared from 10L medium on day 3 after elicitation. The ethyl extract was concentrated and extracted with 1N HCl aq. The acidic aqueous phase obtained was neutralized and extracted with diethyl ether. ß-Thujaplicin methyl ether dissolved in this diethyl ether layer was purified with silica gel chromatography and identified by GC-MS and NMR analysis.

RESULTS & DISCUSSION

The effect of cell density on ß-thujaplicin production in the cell culture was investigated. As shown in figure 1, all cell densities tested showed the same level of ß-thujaplicin production after 5 day incubations. This result suggests that the cells can perceive the concentration of ß-thujaplicin in the medium and control its production depending upon the perception.

In order to prove this idea, the effect of initial addition of ß-thujaplicin to the culture on final level of ß-thujaplicin accumulation was investigated. As shown in figure 2, total concentrations of ß-thujaplicin in the medium were at the same level regardless of the initial doses of ß-thujaplicin. That is, the initial addition of ß-thujaplicin to the culture suppressed its *de novo* production. These results suggest that ß-thujaplicin synthesis was regulated by the product feed back mechanism in this cell line. Because ß-thujaplicin accumulation in the cells did not reduce with the initial addition of ß-thujaplicin, we suppose that ß-thujaplicin production depends on the concentration of ß-thujaplicin in the medium.

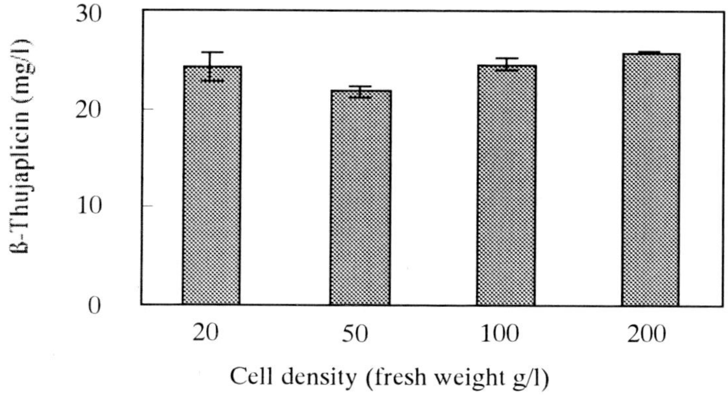

Figure 1. The effect of cell density on ß-thujaplicin production in suspension culture of *C. lusitanaca* cells.

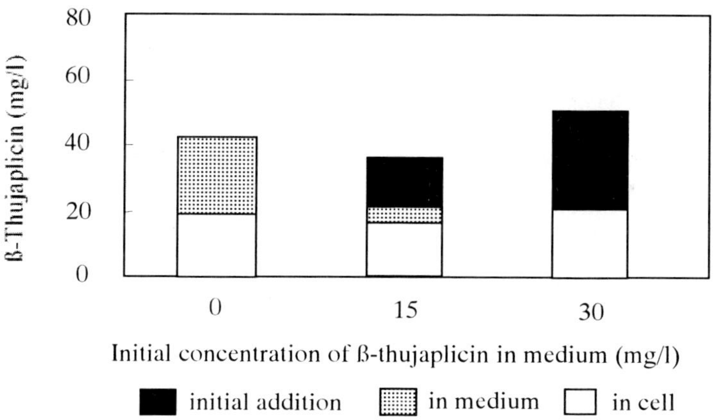

Figure 2. The effect of initial ß-thujaplicin concentration on ß-thujaplicin production in suspension culture of *C. lusitanaca* cells.

From the above discussion, it is expected that removal of ß-thujaplicin from the *C. lusitanica* culture system would stimulate its further production and its excretion to the medium. Therefore, medium exchange was performed to decrease ß-thujaplicin concentration in the medium, and hence to improve total ß-thujaplicin productivity. As shown in figure 3, without exchange of medium, the accumulation of ß-thujaplicin in the medium was limited to ca. 0.3 mg/flask. But, after the medium was changed to fresh one on day 3, the cell further produced and excreted ca. 0.3 mg/flask of ß-thujaplicin from day 3 to day 6. Therefore, medium exchange increased ß-thujaplicin productivity. This result also supports that ß-thujaplicin production depends on the concentration of ß-thujaplicin in medium as mentioned above.

In many cases, ß-thujaplicin concentration in this cell line showed maximum level on days 3-5, then decreased in several days. On the other hand, because an unknown peak was observed on HPLC analysis of ß-thujaplicin production, we purified and identified a metabolite that was corresponding to the peak. As a result, this unknown compound was ß-thujaplicin methyl ether. Only one of two possible ß-thujaplicin methyl ether isomers (6-isoproyl-2-methoxy-2,4,6-cycloheptatrien-1-one) was selectively biosynthesized in the cell culture. The time course of ß-thujaplicin methyl ether concentration was increased during reducing ß-thujaplicin concentration (figure 4). Therefore it seems that excess ß-thujaplicin is controlled by converting to the methyl ether.

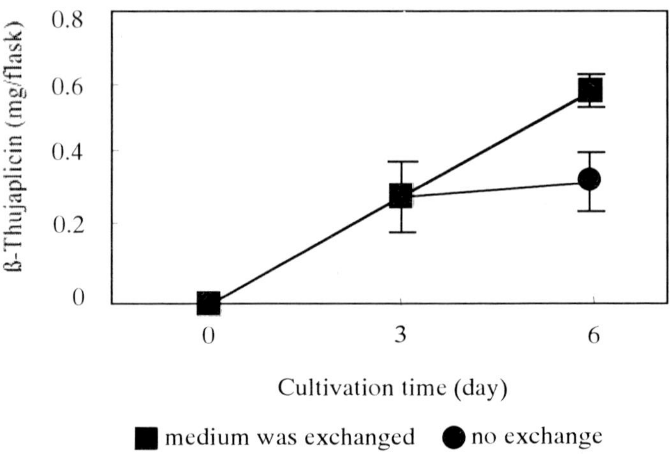

Figure 3. The effect of medium exchange on day 3 on ß-thujaplicin production in medium in suspension culture of *C. lusitanaca* cells. (ß-Thujaplicin concentration on day 6 with medium exchange is the sum total of the product for 0-3 and 3-6 days.)

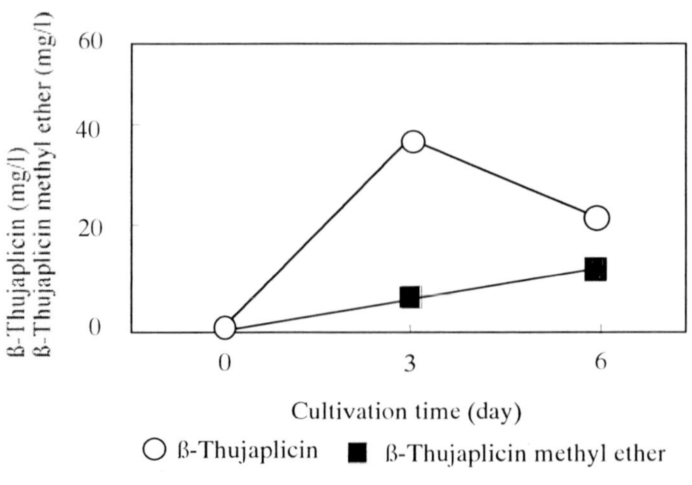

Figure 4. Time course of ß-thujaplicin and ß-thujaplicin methyl ether concentrations in medium of suspension culture of *C. lusitanaca* cells.

CONCLUSIONS

ß-Thujaplicin synthesis in *C. lusitanica* cell culture was regulated by the product feed back mechanism. Its methylation was also likely to control concentration of ß-thujaplicin in the medium. Therefore, removal of ß-thujaplicin from medium will be useful to establish a more effective culture system for the ß-thujaplicin production.

REFERENCE

1. F. Discosmo, M. Misawa, Plant cell culture secondary metabolism, (CRC press, USA) **1996**.
2. A. Komamine, Production of useful plant metabolites by plant cell culture technology, (in Japanese, CMC, Japan) **2000**.
3. S. Dev. Natural Products of Woody Plants, Vol. II, Rowe, J. W., Springer-Verlag, **1989**, p. 711.
4. T. Okabe, Aomorihibanofushigi, (in Japanese, Aomorihibakenkyukai) **1990**.
5. K. Sakai, K. Kusaba, Y. Tsutsumi, T. Shiraishi, 'Secondary Metabolites in Cell Culture of Woody Plants III, Formation of ß-thujaplicin in *Cupressus lusitanica* callus culture treated with fungal elicitors', Mokuzai Gakkaishi, **1994**, 40, 1-5.
6. S. Inada, Y. Tsutsumi, K. Sakai, 'Elicitor of the ß-Thujaplicin Accumulation in Callus Cultures of *Cupressus lusitanaca.'*, J. Fac. Agr., Kyushu Univ., **1993**, 38, 119-126.
7. R. Itose and K. Sakai, 'Improved culture conditions for the production of ß-thujaplicin by suspension cell culture of *Cupressus lusitanaca'*, Plant Biotechnol., **1997**, 14 (3), 163-167.
8. T. Yamaguchi, K. Fujita, K. Sakai, 'Biological activity of extracts from *Cupressus lusitanaca* cell culture', J. Wood Sci., **1999**, 45:170-173.
9. R. Fujii, K. Ozaki, M. Ono and H. Watanabe, 'Hinokitiol Production in Suspension Cells of Thujopsis *dolabrata* var. *hondai* Makino', Plant Tissue Cul. Lett., **1995**, 12 (1), 55-61.
10. M. Ono, T. Asai, and H. Watanabe, 'Hinokitiol Production in a suspension Culture of *Calocedrus formosana* Florin', Biosci. Biotechnol. Biochem., **1998**, 62(9), 1653-1659.
11. K. Sanada, A. Kawaguchi, T. Furuya, K. Ishihara, N. Nakajima. H. Hamada, 'High production of ß-Thujaplicin with Thuja dolabrata var. hondai cells in a semi-continuous culture system', J. Mol. Catalysis B: Enzymatic, **2000**, 11, 59-61.
12. W. Zhang and S. Furusaki, 'Regulation of anthocyanin synthesis in suspension cultures of strawberry cell by pH', Biotechnology Lett., **1997**, 19 (11), 1057-1061.
13. O. L. Gamborg, R. D. Miller, K. Ojima, 'Nutrient requirements of suspension culture of soybean root cells', Exp. Cell Res., **1968**, 50, 151-156.
14. M. Endo, T. Mizutani. M. Matsumura, M. Moriyasu, M. Ichimaru, A. Kato and Y. Hashimoto, 'High-performance liquid chromatographic determination of hinokitiol in cosmetics by the formation of difuluoroborone', J. Chromatography, **1988**, 455, 430-433.

Molecular Breeding of Woody Plants
N. Morohoshi and A. Komamine, editors.

ENDOGENOUS PLANT HORMONES IN PROTOPLASTS OF EMBRYOGENIC CELLS OF CONIFERS

Hamako Sasamoto[1], & Shinjiro Ogita[2]

1 Cell Manipulation Laboratory, Forestry and Forest Products Research Institute, P.O.Box 16, Tsukuba, Ibaraki 305-8687, Japan

2 Research Association for Biotechnology, Nara Institute of Science and Technology, 8916-5, Takayama, Ikoma, Nara, 630-0101, Japan

ABSTRACT

Two different methods were used to examine the regulatory factors for cell division and subsequent plant regeneration in protoplast cultures of several Japanese conifers. One method involves the small-scale analysis of endogenous levels of plant hormones in protoplasts. The other is the protoplast culture system itself, using multi-well culture plates, to determine the optimum concentrations and optimum combination of plant growth regulators, *e.g.* 2,4-dichlorophenoxyacetic acid, naphthaleneacetic acid, benzyladenine, gibberellic acid and abscisic acid. Both methods have been successfully used for plant regeneration in the protoplast cultures of broad-leaved trees, *e.g.Populus* and *Betula*. Endogenous levels of abscisic acid were very high on the basis of cell number in recalcitrant protoplasts of cotyledons and hypocotyls of *Chamaecyparis obtusa,* compared to those of embryogenic cells of *Larix leptolepis* and *Cryptomeria japonica.* Endogenous gibberellins were detected in the latter embryogenic cells. As exogenously supplied gibberellic acid increased colony formation in the former and non-embryogenic callus of *Cryptomeria*, part of the recalcitrancy could be related to the high content of abscisic acid and low content of gibberellins.

KEYWORDS

Conifers, plant hormones, protoplast culture, somatic embryo

INTRODUCTION

In cell culture of forest tree species, conifers have proven to be quite recalcitrant for efficient plant regeneration, which is the key technology for further application of genetic engineering [1]. We first developed two different methods to examine the regulatory factors for cell division and differentiation in broad-leaved trees. One is the small-scale analysis of endogenous levels of plant hormones in protoplasts. The other is the protoplast culture system using multi-well culture plate to determine the optimum concentrations and optimum combination of plant growth regulators. The former supplies experimental data for design of the latter. Such methods have been successfully applied for plant regeneration in protoplast cultures of *Populus alba* [1-3] and *Betula platyphylla* [4,5]. High levels of endogenous gibberellins [3] or abscisic acid (ABA) [5] are inhibitory for regeneration from protoplasts of long-subcultured seed-derived poplar suspension cells [3], or of birch leaves [5]. Then, optimum concentrations of ABA or phenyl-urea type cytokinin, 4-CPPU, were found to be key factors for each protoplast culture [1-5].

In conifer somatic embryogenesis, usage of high concentration of ABA is reported in tissue culture for development of mature somatic embryos from embryogenic cells, after induction and subculture of them with auxins and cytokinins [6,7]. Endogenous levels of plant hormones including ABA were measured in the developmental stages of somatic embryos of *Larix* [8] and *Picea* [9,10], however, there are no reports dealing with protoplasts of them.

In this report, we applied the above two methods to the embryogenic cells in liquid culture of a plant regenerable Japanese conifer, *Larix leptolepis*, for finding key factors in protoplast-to-plant regeneration. In addition, the data were compared with cells of recalcitrant Japanese conifers, *e.g.* embryogenic[11] and non-embryogenic cells of *Cryptomeria japonica* and *Chamaecyparis obtusa* [12,13].

MATERIALS & METHODS

Materials

A. Seedlings of Chamaecyparis obtusa

Seed of *Chamaecyparis obtusa* were sterilized with 2% of Na-hypochlorite solution and washed with autoclaved water several times and planted on 5 ml of solid Murashige and Skoog's medium (MS) [14] (0.8% agar) in a test tube. They were grown for 3 weeks to one month before protoplast isolation.

B. Induction and subculture of embryogenic cells of Larix leptolepis

Induction and subculture of embryogenic cells of *Larix leptolepis* was performed as described by Ogita et al. (1997) [6, 7]. They were subcultured in a liquid modified CD medium containing 7μM of 2,4-dichlorophenoxyacetic acid (2,4-D) and 3μM of benzyladenine (BA), 600mg/l glutamine, 0.09M sucrose (mCD), on a rotary shaker (G-10, New Brunswick Scientific) at 100 rpm speed.

C. Induction and subculture of embryogenic cells and non-embryogenic callus of Cryptomeria japonica

Embryogenic cells and non-embryogenic callus of *Cryptomeria japonica* were induced from a seed on 1μM of 2,4-D-containing medium as reported by Ogita et al.[11]. The former was subcultured in liquid medium of the same composition as the embryogenic cells of *Larix leptolepis.* The latter was subcultured in the solid medium containing 1 μM of 2,4-D and 0.2% of Gellan Gum (Wako Pure Chemical Industries, Ltd., Osaka).

Protoplast isolation and culture

A. Chamaecyparis obtusa

Protoplasts were isolated from the cotyledons and hypocotyls of seedlings of *Chamaecyparis obtusa* separately and cultured as described by Sasamoto et al. [12,13]. The tissues were excised in 0.6M mannitol solution and protoplasts were isolated with 1% Cellulase RS and 0.25% Pectolyase Y-23 in 0.6M mannitol solution for 4-5 hrs at 23-25°C in static condition. After passing through the 94µm nylon mesh, they were washed with 0.6M of mannitol solution by centrifugation at 300rpm (10*g*; cotyledon) or 800rpm (100*g*, hypocotyl). A portion of the protoplasts was cultured in 0.3ml (24 well, Falcon No.3047) or 0.05ml (96 well, Falcon No.3075)) of medium in multiwell culture plate at cell density of 10^3–$2x10^4$/ml. The basic culture medium was MS containing 1µM or 10 µM of naphthalene acetic acid (NAA), 0.09M sucrose and 0.6M mannitol at pH5.8. Effects of concentrations of BA, gibberellic acid (GA_3) and abscisic acid (ABA) were tested. Sterilized water was dispensed between wells to maintain the humidity, and the plate was sealed with Parafilm R. Protoplasts were cultured in the dark at 28°C using a CO_2 incubator with a chiller (Hirasawa CPD-170S, Tokyo) without the supply of CO_2. All of the above procedures were performed under sterile conditions.

B. Embryogenic cells of Larix leptolepis and Cryptomeria japonica

Protoplasts were isolated from the embryogenic cells by incubation with 1% Cellulase RS and 0.25% Pectolyase Y-23 in mannitol solution of 0.4M (*Larix*) or of 0.6M (*Cryptomeria*) for 4-8 hrs on a rotary shaker at a speed of 80rpm. 2% Cellulase RS (*Larix*) or combination of 1% each of Cellulase R-10 and Hemicellulase (*Cryptomeria*) was also used. After passing through nylon mesh (40 to 94 µm) they were washed three times with mannitol solution by centrifugation at 100 *g* for 4 min. A portion of the protoplasts was cultured in multi-well culture plate at cell density of 0.5-$10x10^4$/ml as in the *Chamaecyparis*. The basic culture media were ammonium nitrate-free MS (*Larix*) or MS (*Cryptomeria*) containing 10 µM each of 2,4-D and BA. Effects of varying concentrations of medium components were tested, *e.g.* mannitol (0.04 to 0.6M), sucrose (up to 0.31M) and BA (0µM).

C. Non-embryogenic cells of Cryptomeria japonica

Protoplasts were isolated from yellow-green callus of *Cryptomeria japonica* by incubation with 1% Cellulase RS and 0.25% Pectolyase Y-23 in 0.6M mannitol solution for 4-8 hrs under static conditions. They were purified by centrifugation using density gradient of 0.6M sucrose and 0.6M mannitol solution. The basic culture medium was MS containing 10 µM of 2,4-D and 0.6M mannitol. Effects of BA, ABA and GA_3 were tested.

Small-scale analysis of ABA and gibberellins (GAs)

A. Extraction and partition:

The protoplasts were extracted with 0.5ml of 80% methanol or ethanol at 4°C for 18hrs. After centrifugation at 1600g, the precipitate was re-extracted for 4hrs with 0.3ml of cold 80% methanol. After further centrifugation, the combined extracts were dried in vacuo at 38°C using a Vacuum centrifugal vaporizer (CVE-100, EYELA, Tokyo, Japan) with a cold trap (Unitrap UT-50, EYELA). The residual precipitate was dried at 80°C, 18hr for dry weight measurement. After addition of 0.6ml or 0.5 ml of H_2O to the extract, pH was adjusted to 2.5 with 1-5 µl of 6N HCl and extracted three times with 1/3 volume of methylene chloride (ABA) or ethyl acetate (GAs). Multiflex tip (# MU553008, Funakoshi, Tokyo) was used for fractionation. Combined solvent fraction was dried in vacuo and stored at -85°C before purification by micro-HPLC. When cited in the text, the fraction was further purified by successive partition with 0.5M phosphate buffer, pH 8.0 and acidic ethyl acetate, and dried by the addition of 100mg of sodium sulfate.

B. Purification by Micro-HPLC

Frozen extracts were dissolved with 100µl of 80% ethanol and filtered (0.22µm Millex-GV4, Millipore). Fifty µl portion was analyzed using micro-HPLC (MicroPrep, Unipoint 233XL, Gilson, M & S Instruments Trading Inc. Japan) with automatic injection and fractionation system, and C18 reversed phase column (Betasil C18, 4.6x50 mm, Keystone, USA). Gradient of 30min (ABA), or 35min (GAs), from 10% acetonitrile to 100% (ABA) or to 60% (GAs) in 0.1% acetic acid at flow rate of 0.3ml/min was used. Each one min fraction around standard ABA (A-1012, Sigma) position, was evaporated to dryness and stored at –80°C before ELISA test. For GAs, each one min fraction from 5 min to 35 min was used for micro-bioassay after dissolving with 10 µl of 50% acetone.

C. ELISA test

ABA was re-eluted from combined two min fractions with 200µl of 25mM Tris buffered saline (0.15M NaCl, 1 mM of $MgCl_2$, pH 7.4). ELISA test (ABA assay kit, Sigma) was performed as instructed by the company. Optical density was detected by using Flying-Spot Scanner with micro plate system (CS-9000, Shimadzu, Kyoto). ABA amount was calculated from at least six different concentration data.

D. Micro-bioassay

Modified micro-drop bioassay using dwarf rice, Tan-ginbozu, was used for gibberellin detection [15]. Length of second leaf sheath was measured after 4 days of culture of treatment at 25°C.

RESULTS & DISCUSSION

Endogenous levels of ABA

ABA contents of protoplasts of embryogenic cells of *Larix leptolepis* and *Cryptomeria japonica*, and cotyledons and hypocotyl protoplasts of *Chamaecyparis obtusa*, respectively are shown in Table 1. Very high value (93 pmoles / $6x10^6$ protoplasts in cotyledons and 52 pmoles / $6x10^6$ protoplasts in hypocotyl) was obtained with *Chamaecyparis*. Compared to them, much less ABA was detected in leaf protoplasts of *Populus alba* [3,5], whose diameter was around 20 µm. As the sizes of protoplasts of *Larix, Cryptomeria* and *Chamaecyparis* varied (20-50 µm in diameter), the calculated concentrations of ABA in them were reduced as shown in Table 2.

Table 1. Endogenous levels of ABA in various protoplasts.

	pmoles	
	/ $6x10^6$ protoplasts	/ mg dry wt.
Larix EC	6.8*	0.3
Cryptomeria EC	24.1	2.1
Chamaecyparis cotyledon	92.6	1.8
Chamaecyaris hypocotyl	52.2	0.52
Populus alba leaf	0.94**	0.37

*, **: Average of three (*) and two (**) independently isolated samples.

Table 2. Calculated concentrations of ABA in protoplasts of different diameters.

	Calculated concentrations (μM)		
Diameter	20μm	30μm	50μm
Larix EC	0.27	0.08	0.02
Cryptomeria EC	0.96	0.28	0.06
Chamaecyparis cotyledon		1.09	0.23
Chamaecyaris hypocotyl		0.61	0.13
Populus alba leaf	0.037		

Endogenous levels of GAs

Fig. 1 shows the profiles of endogenous GAs in protoplasts of embryogenic cells of

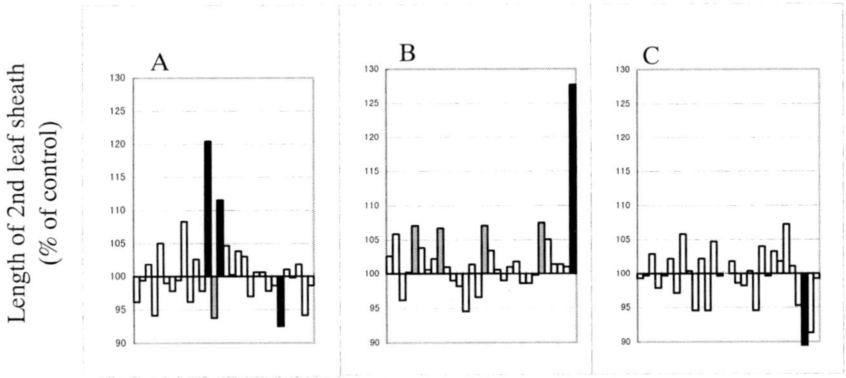

Figure 1. Patterns of bioassay of gibberellins in protoplasts of embryogenic cells, of *Larix leptolepis* (A) and *Cryptomeria japonica* (B), and cotyledon protoplasts of *Chamaecyparis obtusa* (C). Black column: $p<0.05$. Brown column: $p<0.1$.

Table 3. Endogenous levels of gibberellins in conifer protoplasts, calculated by adding the statistically significant values of GA3 equivalent.

	pmoles / $6x10^6$ protoplasts	pmoles / mg dry wt.	Concentrations in a protoplast
Larix EC ($4x10^6$*)	1.36	0.054	0.016 μM (30μm d.)
Cryptomeria EC ($6x10^6$)	1.55	0.108	0.018 μM (30μm d.)
Chamaecyparis cotyledon (10^6)	-**	-	
Chamaecyaris hypocotyl (10^6)	-	-	
Populus alba leaf (10^7)	0.31	0.12	0.012 μM (20 μm d.)

* : Number of protoplasts analyzed. **-: not detected.

Larix leptolepis (1A), *Cryptomeria japonica* (1B) and cotyledons of *Chamaecyparis obtusa* (1C), respectively. Hypocotyl protoplasts of *Chamaecyparis obtusa* showed a similar or more inhibitory pattern. Number of protoplasts used was 10^6 for *Chamaecyparis*, and $4x10^6$ for *Larix*, $6x10^6$ for *Cryptomeria*. GA3 equivalent values ($p>0.01$) were added and concentration in single protoplasts was roughly calculated (Table 3). Both embryogenic cells contained at least several times higher content of GAs of GA3 equivalent on the basis of cell number. Their diameter sizes varied from 20 to 50 μm, while the diameter of *Populus* was around 20 μm. Such data were obtained with one step purification with acidic ethyl acetate. As both protoplasts of *Chamaecyparis* contain large amount of ABA, their patterns of GAs at 15-18min around might reflect the inhibitory activity of it.

Content of GAs in leaf protoplast of *Populus alba* (Table 3) was lower compared to the value obtained using TLC [3]. Further purification steps of partition with 0.5M phosphate buffer, pH 8.0 and successive partition with acidic ethyl acetate, before HPLC was effective for reduction of inhibitory substances in leaf protoplasts of *Populus alba*. However, only slightly higher content (0.50 pmoles / $6x10^6$ protoplasts) was obtained.

The purpose of measurement of endogenous plant hormones in this study is not the finding nor purification of each specific gibberellin, but to know the bulk profile or order of concentration of it in a protoplast, which could be related to its reactivity to exogenous plant growth regulators. A simple and quick method is preferred. Thus TLC [3,5] can be used for such purpose, although elution of ABA or GAs from the TLC bed is time-consuming. Avoiding contamination of large amount of exogenous hormones used in culture is a point which is common to both TLC and HPLC [16].

Effects of exogenous plant growth regulators in protoplast cultures of conifers

A. Protoplasts of Chamaecyparis obtusa [12,13]

As shown in Fig. 2, at optimal auxin condition (1 μM of NAA), ABA was inhibitory to cell divisions, especially at the highest concentration, i.e.10μM in protoplast culture of cotyledons of *Chamaecyparis obtusa*. Colony formation was promoted by the addition of 0.1 to 1 μM of GA3. Better results were obtained with hypocotyl protoplasts by the addition of a low concentration of BA[12,13]. Optimal cell density was $1-2x10^3$ /ml.

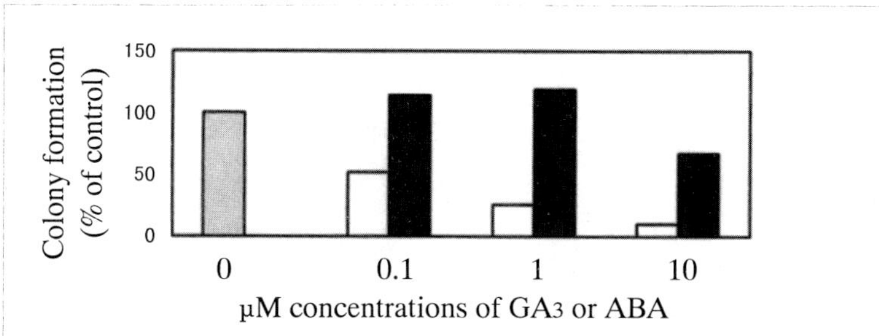

Figure 2. Effects of ABA and GA3 on colony formation in protoplast culture of cotyledons of *Chamaecyparis obtusa*. Medium was MS containing 1μM of NAA and each concentration of GA3 (black columns) or ABA (white columns).

B. Embryogenic cells of Larix and Cryptomeria

Proliferation of embryogenic cell culture from protoplast cultures of both species was obtained (Fig.3). The same combination of auxin (10 μM of 2,4-D) and cytokinin (10 μM of BA) was successful for controlled cell divisions and differentiation of early somatic embryos from protoplast cultures of embryogenic cells of both species. MS salts without NH_4NO_3 was preferred as the basic medium composition by *Larix* and MS for *Cryptomeria.* In *Larix*, as shown in Fig. 4, a high concentration of sucrose was promotive for head cell divisions. At 3 % sucrose, suspensor cells elongated, and at 9 % sucrose, only small cells divided. However, at 6 % sucrose, both structures were observed. 0.2M or more of mannitol was inhibitory at 9 % sucrose. Optimal cell density for efficient cell divisions was $5\text{-}10\times10^4$/ml in *Larix.* In *Cryptomeria*, reduction of mannitol to 0.3M was effective for induction of embryogenic culture from protoplast culture.

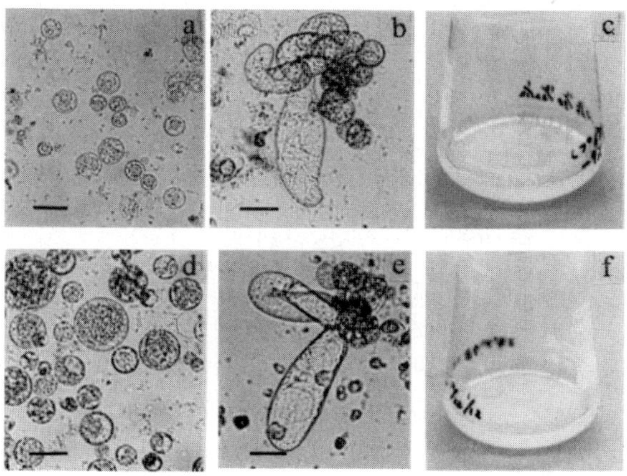

Figure 3. Proliferation of embryogenic cell clusters in protoplast cultures originating from embryogenic cells of *Larix leptolepis* (a-c) and *Cryptomeria japonica* (d-f). 3b,3e: embryogenic cell clusters obtained after 11days of culture. Bar=50μm.

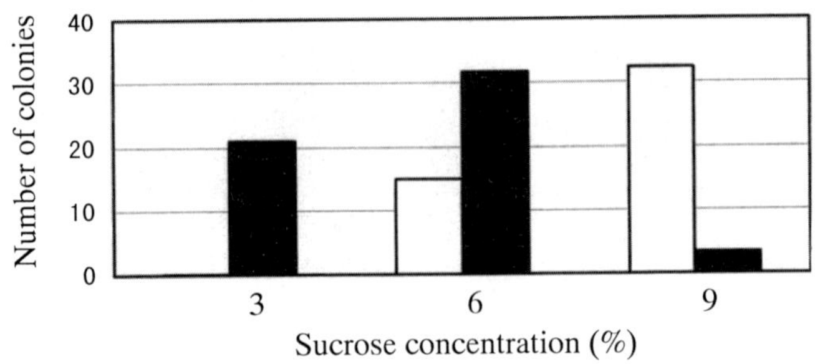

Figure 4. Effects of sucrose and mannitol concentrations on colony formation in the protoplast culture of embryogenic cells of *Larix leptolepis*. After 26days of culture with (black columns) or without (white column) 0.2M of mannitol.

C. Non-embryogenic callus of Cryptomeria

Protoplasts were isolated from non-embryogenic callus of *Cryptomeria japonica* by 1% of Cellulase RS and 0.25% of Pectolyase in 0.6M mannitol solution. The debris and tracheary elements were separated by the gradient of sucrose and mannitol. Though numbers of purified protoplasts were low, there was no difficulty in surveying of the culture conditions for this material, as low cell density was optimum ($3x10^4$/ml). In the protoplast culture of *Cryptomeria*, 1 and 10 μM of NAA or 2,4-D in both mCD medium or MS medium containing 0.6M mannitol was sufficient for colony proliferation. Effects of BA, ABA and GA3 on early cell enlargement of protoplasts, which preceded cell divisions, were examined (Fig. 5). BA and ABA were inhibitory at the various concentrations tested. In contrast, GA3 was promotive for cell divisions. Such promotive effect of GA3 was also detected in protoplasts of non-embryogenic cells of *Chamaecyparis* (see section A). Callus proliferation was obtained from the colonies developed in auxin medium with or without GA3 (Fig. 6).

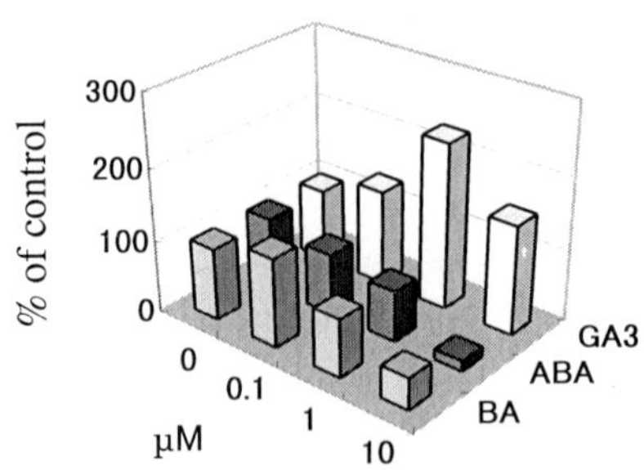

Figure 5. Effects of BA, ABA and GA3 on cell enlargement in protoplast culture of non-embryogenic callus of *Cryptomeria japonica*.

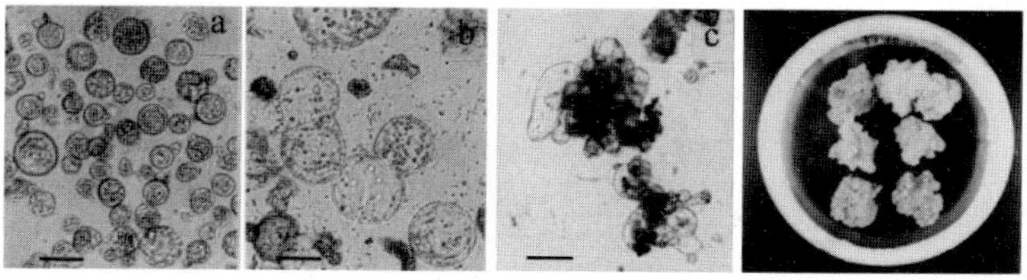

Figure 6. Proliferation of callus in the protoplast culture originating from non-embryogenic callus of *Cryptomeria japonica*. 6b: cell divisions after 10 days of culture. Bar=50µm (a, b) and 250µm (c).

CONCLUSIONS

Measurement of levels of plant hormones, ABA and gibberellins using micro-purification and assay steps was performed in protoplasts of embryogenic cells of Japanese conifers, e.g. *Larix leptolepis* and *Cryptomeria japonica* and non-embryogenic cells of *Chamaecyparis*. On the basis of numbers of protoplasts, very high level of ABA was found in non-embryogenic cells of *Chamaecyparis obtusa*. In contrast, content of GAs was low in it. Both protoplasts of non-embryogneic cells, e.g. *Chamaecyparis* and *Cryptomeria*, could be cultured by the addition of auxin (NAA or 2,4-D) alone. Low concentration of GA_3 increased cell enlargement and cell divisions of protoplasts in *Cryptomeria* and *Chamaecyparis*, respectively.

In addition, in the embryogenic cells of *Larix leptolepis*, both GAs and ABA levels were moderately high. As they divided rapidly when cultured in 10 µM of 2,4-D and BA, regulation of cell divisions with other factors, *e.g.* sucrose and mannitol was necessary for prolonged colony proliferation and differentiation to early somatic embryos.

Combination of auxin and cytokinin is used for many successful protoplast cultures, however, for recalcitrant conifers, endogenous and, therefore, exogenous ABA and GAs could be key factors for protoplast-to-plant regeneration. Therefore, data of endogenous levels of ABA and GAs by micro-analysis could aid in experimental design for further adjustment of culture conditions in protoplast cultures, using multi-well culture plates, for plant regeneration of embryogenic and possibly non-embryogenic cells of Japanese conifers.

ACKNOWLEDGEMENTS

Authors are grateful to Prof. Trevor A. Thorpe of the University of Calgary for critically reading this manuscript.

REFERENCES

1. H. Sasamoto, Y. Wakita, S.Yokota & N. Yoshizawa, Large electro-fused protoplasts of *Populus alba* selected by a micromanipulator: Techniques and some characteristics of cells and their regenerants. *J. For. Res*. **2000**, 5, 265-270.
2. H. Sasamoto, Y. Hosoi, K. Ishii, T. Sato & A. Saito, Factors affecting the formation of callus from leaf protoplasts of *Populus alba*, *J.Jpn.For.Soc.* **1989**, 71, 449-55.

3. H. Sasamoto, Y. Hosoi & M. Koshioka, Endogenous levels of plant hormones may affect the culture conditions of Poplar protoplasts to regenerate plants. In: *Current Issues in Plant Molecular and Cellular Biology,* M.Terzi et.al. (eds)., Kluwer Academic.Netherlands **1995**, pp481-486
4. Y. Wakita, H. Sasamoto, S. Yokota & N. Yoshizawa, Plantlet regeneration from mesophyll protoplasts of *Betula platyphylla* var.japonica. *Plant Cell Reports* **1996,** 16, 50-53.
5. H. Sasamoto, Y. Wakita & S. Ogita, The relation between endogenous levels of plant hormones and regeneration efficiency of *Betula* leaf protoplasts. *Abst. 16th. Int. Conference of Plant Growth Substances,* **1998,** p92.
6. S. Ogita, T. Kubo & M. Fushitani, Anatomical characteristics in early embryogenesis from immature embryo of *Larix leptolepis*. *For. Resources Environ.* **1997**, 35, 45-51.
7. S. Ogita. H. Sasamoto & T. Kubo, Control of the development of somatic embryo of Japanese conifers by the density of embryogenic cells in liquid culture. In: *Trasnplant Production in the 21st Century,* C.Kubota and C. Chun (eds.), Kluwer Academic, Netherlands, **2000**, pp209-214.
8. I. Jourdain, M-A. Lalu & P. Label, Hormonal changes during growth of somatic embryogenic
masses in hybrid larch. *Plant Physiol. Biochem*. **1997**, 35(9), 741-749.
9. L. Kong & E.C. Yeung, Effects of silver nitrate and polyethylene glycol on white spruce (Picea glauca) somatic embryo development: enhancing cotyledonary embryo formation and endogenous ABA content. *Physiol. Plant.* **1995,** 93, 298-304.
10. D.J.Carrier, E.J.Kendall, C.A.Bock, J.E.Cunningham & D.I.Dunstan, Water content, lipid deposition, and (+)-abscisic acid content in developing white spruce seeds. *J. Exp. Bot*. 50, **1999,** 1359-1364.
11. S. Ogita, H. Ishikawa, T. Kubo & H. Sasamoto, Somatic embryogenesis from immature and mature zygotic embryos of *Cryptomeria japonica* I: Embryogenic cell induction and its mrphological characteristics. *J. Wood Sci.* **1997**, 45, 87-91.
12. H. Sasamoto, A. Kondo, Y. Hosoi, H. Maki & K. Odani, Callus regeneration from cotyledon protoplasts of *Chamaecyparis obtusa* (Hinoki Cypress). *In Vitro Cell. and Dev. Biol.Plant* **1992**, 28, 132-136.
13. H. Sasamoto, Y. Hosoi, H. Maki, Protoplast cultures of conifers, *Chamaecyparis obtusa* and *Picea abies*. In: *Plant Tissue Culture and Gene Manipulation for Breeding and Formation of Phytochemicals*, K.Oono et.al.(eds)., NIAR/Japan **1992**, pp211-218
14. T. Murashige & F. Skoog, A revised medium for rapid growth and bioassay with tobacco tissue cultures. *Physiol. Plant*. **1962,** 15, 473-497
15. T. Nishijima & N.Katsura, A modified micro-drop bioassay using dwarf rice for detection of femtomol quantities of gibberellins. *Plant Cell Physiol.* **1989,** 30(5), 623-627.
16. L. River & A. Crozier (eds.), Principles and Practice of Plant Hormone Analysis Vol.1, Academic Press, **1987**, pps167.

Molecular Breeding of Woody Plants
N. Morohoshi and A. Komamine, editors.

EFFICIENT PLANT REGENERATION OF *LARIX KAEMPFERI*

Shinjiro Ogita [1,]* & Hamako Sasamoto [2]

[1] *Research Association for Biotechnology (NAIST), 8916-5, Takayama, Ikoma, Nara, 630-0101, Japan*

[2] *Cell Manipulation Laboratory, Forestry & Forest Products Research Institute, Ibaraki, P.O.Box 16, Tsukuba Norin Kenkyu Danch-nai, 305-8687, Japan*

ABSTRACT

Embryogenic tissues of Japanese larch (*Larix kaempferi* Lamb. = *L. leptolepis* Gord.) which were induced and maintained on the modified Campbell and Durzan (mCD) medium with 7 μM of 2,4-dichlorophenoxyacetic acid (2,4-D), 3 μM of N^6-benzyladenine (BA), consisted of distinctive aggregates composed of small cytoplasm-rich cells and elongated suspensor-like cells. The small aggregates showed competence for both proliferation and differentiation with dynamic morphogenic changes. We selected two embryogenic cell lines with different morphologies and studied the effects of abscisic acid (ABA) on maturation and germination of somatic embryos from these lines. Our studies indicated that the ABA requirement for maturation varied in relation to cell line morphology. Based on these results, an efficient embryogenic suspension cell line was developed with reproducible morphological and developmental characteristics. We found that in high cell density cultures, although embryogenic cells continued to proliferate, no further development occurred. In low cell density cultures, development of distinctive cell aggregates into somatic embryos was promoted. Thus, by manipulating the culture cell density, an efficient production of uniform somatic embryos of Japanese larch is possible.

KEYWORDS

Cell density, embryogenic tissue, *Larix kaempferi* = *Larix leptolepis*, morphology, somatic embryogenesis, suspension culture

INTRODUCTION

In order to utilize plant tissue culture techniques for mass micropropagation of selected trees and for production of new superior strains of trees by molecular breeding, development of reliable and high-efficient tissue culture techniques are highly recommended. Somatic embryogenesis is an excellent technique which allows mass production of genetically identical plants from a single selected explant. In recent years, many successes have been reported using the somatic embryogenic technique in the micropropagation of conifers, especially *Picea* and *Larix* species[1]. In the case of Japanese larch (*Larix kaempferi* Lamb. = *L. leptolepis* Gord.), an efficient plant regeneration method through somatic embryogenesis is not available, although initiation of embryogenic culture has been studied by von Aderkas et al. [2].

In previous communications, we reported the initiation and maintenance of embryogenic tissues, as well as embryo maturation and plant recovery from these tissues in Japanese larch[3-5]. We found that the embryogenic tissues of Japanese larch were consisted of distinctive aggregates composed of small dense, cytoplasm-rich cells and elongated suspensor- like cells. These aggregates showed competence for both proliferation and differentiation with dynamic morphogenic changes. In order to create a reliable and an efficient production system, it is essential that we have a good understanding of the morphogenic changes occurring within the embryogenic aggregates and how these changes are regulated. An efficient embryogenic suspension cell line has recently been developed for a Japanese larch with reproducible morphological and developmental characteristics[6]. In this report, we discuss the current production and some problems related to somatic embryogenesis in this system.

MATERIALS & METHODS

The embryogenic tissues of Japanese larch were induced from mature and immature zygotic embryos[3,4]. The induction medium was the modified Campbell and Durzan (mCD) medium (the concentration of NH_4NO_3 was reduced to 400 mg l^{-1} from 800 mg l^{-1} of the original[7], and 600 mg l^{-1} of glutamine was added) supplemented with 7 μM of 2,4-D, 3 μM BA and 30 g l^{-1} sucrose. The pH of the medium was adjusted to 5.6 before solidifying with 2 g l^{-1} gellan gum. The stock embryogenic tissues were maintained using the same medium at 4- to 6-week intervals at 25 °C, with a 16-h photoperiod under fluorescent illumination (60 $\mu Em^{-2}.s^{-1}$). Liquid suspensions of the embryogenic tissues were also generated. The stock embryogenic tissues were cultured in 10 ml liquid medium of the same composition as the induction-maintenance medium using a 100 ml conical flask. Furthermore, cell density was measured and adjusted using a value of packed cell volume (PCV). The PCV was determined by centrifuging cultures in conical tubes at 500 rpm for 3 minutes. The flasks were placed on a rotatory shaker with a speed of 100 rpm. The suspension methods were detailed in previous reports[5, 8]. For maturation, mCD medium containing ABA was applied as described by Ogita et al. (1999c)[5]. For germination of matured somatic embryos, the hormone-free mCD medium was used. Morphological characteristics of embryogenic tissues were observed using an inverted fluorescence microscope under U- or B-excitation fluorescence light after staining with a 0.0002% solution of 4'-6-diamidino-2-phenylindole (DAPI)[4].

RESULTS & DISCUSSIONS

Somatic embryogenesis of Japanese larch

Embryogenic tissue induction & maintenance

In general, various plant organs and tissues can be used as a source of explants for the initiation of embryogenic cultures. In the case of conifers, zygotic embryos are preferred[1]. Initially, we attempted to induce embryogenic tissue of Japanese larch using mature zygotic embryos excised from stored seeds. However, we soon realized that the frequency of initiation was extremely low and stable proliferation was unable to achieve[3]. Therefore, immature embryos were tested and this explant source proved to be more effective in embryogenic tissue production[9].

Whitish or whitish-yellow tissues were induced from immature embryos on the solid mCD medium containing 7 μM of 2,4-D and 3 μM of BA during the initial culture of 4 weeks. At this point, since the amount of tissues was small and distinct color differences were not present; it was difficult to determine whether embryogenic tissues were present. All tissues initiated were then transferred to fresh medium of the same composition. Four weeks after the transfer, i.e. 8 weeks after culture initiation, different forms of tissues could be found. Embryogenic tissues had a white and friable appearance (Fig. 1 a). Non-embryogenic tissues were also present and they were characterized by having yellow, green or gray color and the tissues appeared to be more compact in appearance. The non-embryogenic tissues tended not to proliferate and they gradually turned brown and dead. Additional morphological characteristics of the tissues could be identified by DAPI staining and fluorescence microscopy. Many small cell clumps consisted of distinctive aggregates of small dense cells (embryonal regions) and elongated cells (suspensor regions) were present (Fig. 1 b). Strong blue fluorescence could be observed after staining with DAPI, indicated that the embryonal regions were cytoplasmic with large nuclei (Fig.1 c). Chromosomes in mitotic phases were also observed as shown in Fig. 1d. These cell aggregates had the typical structure of conifer embryogenic cell tissues, as shown by Attree and Fowke (1991)[10], Bonga et al. (1997)[11] and Ogita et al. (1997)[4]. In contrast, dense cell aggregates with suspensor-like cells could not be found in the non-embryogenic tissues. The embryogenic tissues of Japanese larch have been maintained on the same mCD medium for more than 4 years by subculturing once every 4 - 6 weeks and have shown continued competence for both proliferation and differentiation.

Maturation of somatic embryos & plant recovery

Two embryogenic cell lines characterized by different morphology and color were obtained from the stock embryogenic tissues of Japanese larch. One of the cell lines had only small embryogenic aggregates and was white in color, and the other line had large embryogenic aggregates and was red in color. As shown in our previous paper[4], a high concentration (50 μM) of ABA was effective in the development of the embryogenic aggregates of the white cell line into mature somatic embryos while a low concentration (0.1 μM) was sufficient for embryo maturation in the red cell line. After maturation of somatic embryos, no difference was found in plant recovery potency between white and red embryogenic cell lines of different morphology. Mature somatic embryos were taken from the both the white and the red embryogenic tissues and placed on the hormone-free mCD medium. Regenerated plants were observed on and after 4 - 6 weeks of culture as shown in Figs. 1 e and f.

Somatic embryogenesis of conifers has 4 main phases, i.e. (1) initiation and maintenance of embryogenic tissues, (2) maturation of somatic embryos, (3) germination of somatic embryos, and (4) plant regeneration and naturalization in soil. Many researchers who work in the area of somatic embryogenesis are sometimes too eager to form final products, i.e. somatic seedlings. Therefore, in normal practice, once an optimal concentration of ABA is determined for a given species, the amount of ABA used would be 'fixed' for any future experiments. Recently, a well documented study which relates to somatic embryogenesis and plant regeneration of Japanese larch in solid culture condition is reported by Kim et al. (1999)[12]. They found that only one embryogenic cell line could proliferate in subsequent culture and it was selected from a total of 294 cell lines. They also indicated that there were no clear optimal concentration for ABA for embryo maturation and there was no clear differences with respect to the duration of ABA exposure in Japanese larch. Moreover, our data from two cell lines of Japanese larch clearly showed a necessity to optimize the concentrations of plant growth regulators used in relation to embryogenic tissue formation. In order to establish an efficient culture system through somatic embryogenesis in conifer species including Japanese larch, the conditions for

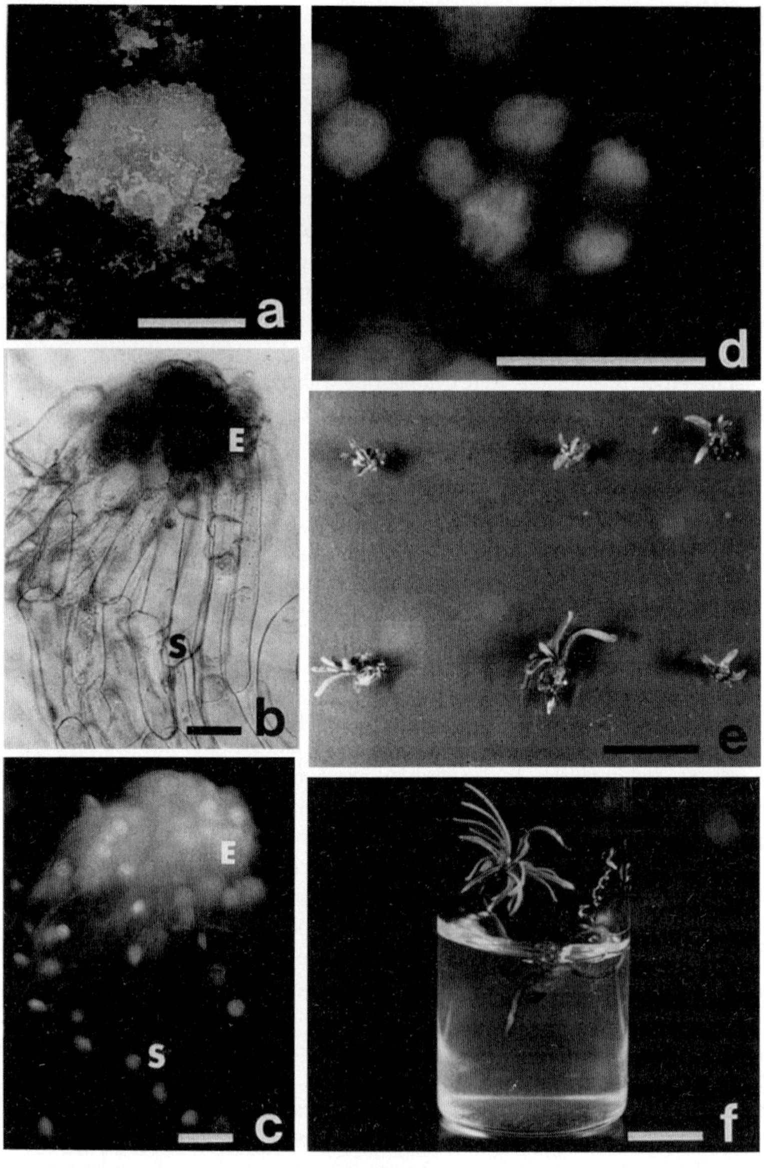

Figure 1. Somatic embryogenesis in Japanese larch.
(a) Embryogenic tissues derived from immature embryos. E, embryonal region; S, suspensor region. (b) Embryogenic tissues are characterized by having many small aggregates of cells. (c) DAPI staining shows in blue fluorescence from the embryonal region indicating that the cells are cytoplasmic with large nuclei.
(d) Chromosomes of embryogenic cells in mitotic stages.
(e) Germinated seedlings in petri dish.
(f) A somatic plantlet regenerated from a somatic embryo.
Scale bars in (a), (e) and (f) represent 1 cm.
Scale bars in (b), (c) and (d) represent 50 μm.

embryogenic tissue induction and maintenance should be carefully optimized. This first phase is the least studied step but it is also the most critical step for the subsequent phases that follows, i.e. maturation and germination of somatic embryos and plant recovery. Based on our own observations, we focus our attention on developing a liquid suspension system for embryogenic tissue proliferation for Japanese larch.

A potential application of embryogenic suspension culture

As indicated above, the conditions for somatic embryogenesis on solid medium from the initiation of embryogenic tissues to plant recovery, were established for Japanese larch. However, continual development within the embryogenic tissues was not homogeneous. For example, uneven proliferation of embryogenic tissues could be found even in the maturation medium containing ABA and differences of developmental stages of somatic embryos could be seen on the solid medium. The heterogeneous cell populations within an embryogenic tissue may be the cause of the varied response observed. Therefore, in order to avoid the heterogenous response as seen from solid medium, we have developed a liquid suspension culture method that allows the generation and maintenance of embryogenic tissues with reproducible morphological and developmental characteristics.

The cell density used in a given culture system can have dramatic effects on cell proliferation[13]. In Japanese larch, cell density also played an important role in regulating the developmental pattern of the embryogenic tissue. When the maintenance cultures were transferred into medium with different cell densities, the embryogenic tissue continued to proliferate and dramatic morphological changes could be seen[6]. We found that in high cell density culture of early stage embryogenic cells (0.5 - 1 ml PCV/ 10 ml culture medium), although the cells continued to proliferate, no further development occurred. In low cell density culture (0.01 - 0.1 ml PVC/ 10 ml culture medium), development of distinctive cell aggregates with well organized embryonal head regions and elongated suspensor cells was promoted. Similar cell density effects were also observed in other Japanese conifer suspension cells[6].

One of the potential applications of the observed cell density effect was performed as shown in Fig. 2. By culturing small spherical aggregates at high cell densities (preferably 0.5 - 1 ml PCV/ 10 ml culture medium, 1 ml PCV of the suspension culture have an average of 1.5 - 2 x 10^4 of the cell aggregates) approximately every 3 - 4 weeks, they continued to proliferate while maintaining their morphology, i.e. small cytoplasm-rich cells together with the rounded, vacuolated cells (Fig. 2a). The final PCV values increased 4 - 6 times per one culture period, and the number of cell aggregates per PCV was similar to the initial density at the beginning of the culture cycle. The non-destructive observation using an inverted microscope enabled the determination of growth characteristics of the small spherical aggregates.

Once the cell density was lowered to 0.01 - 0.1 ml PVC/ 10 ml culture medium, further uniform development of embryogenic tissues occurred after 3 - 4 weeks of culture (Fig. 2b). Furthermore, most of the embryogenic tissues with well organized embryonal heads and suspensors, i.e. pro-embryos, could mature when transferred to a liquid medium containing ABA (Fig. 2c). We also noted that a further reduction of pro-embryo density to approximately 150 - 200 pro-embryos per 10 ml of maturation medium enabled almost all pro-embryos to develop into mature embryos. If the relative cell and proembryo densities were higher than that described above, uniform development of somatic embryos would be gradually arrested in our experimental condition.

Based on the experiment described above, we made a trial calculation of number of somatic

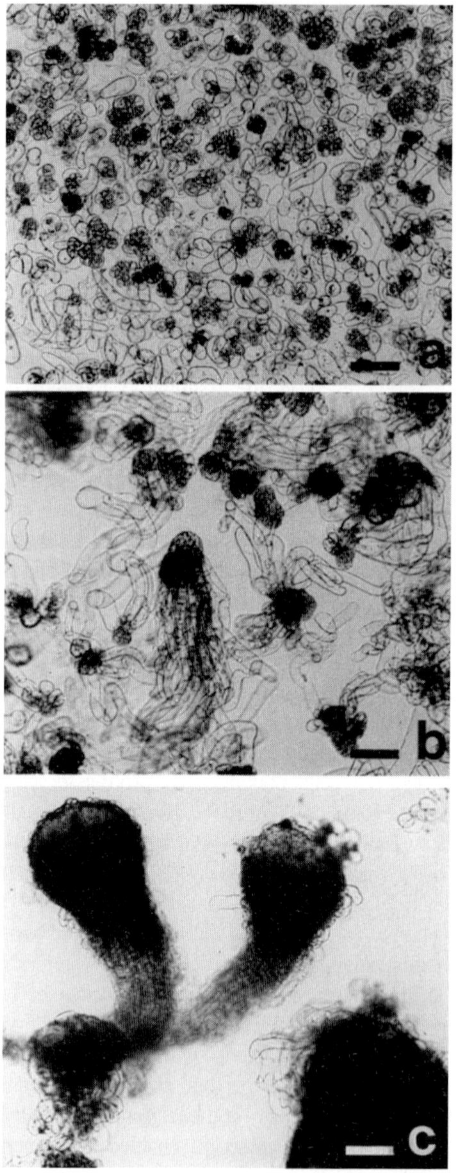

Figure 2. An application of the high - low cell density suspension culture method for the production of uniform somatic embryos in Japanese larch.
(a) Small spherical aggregates maintained in a high-density culture.
(b) In a low-density culture, embryogenic tissues have well organized embryonal regions and an elongated suspensor regions.
(c) Somatic embryos developed in a liquid maturation medium containing ABA.
Scale bars represent 200 μm.

embryos formed by the high-low cell density suspension culture method. Approximately 1.5 - 2 x 10^4 of mature embryos would be obtained per unit explant, i.e. one ml PCV of high-cell-density suspension cultures which contains approximately 1.5 - 2 x 10^4 of small spherical cell aggregates. Our calculation showed an efficiency of the cell density effect, although that is a laboratory experimental scale, there is room for further improvement and for scale-up production of somatic embryos in Japanese larch.

In the latest study, for example, the effects of bioreactor configuration on the growth and maturation of somatic embryos was investigated and an efficacy of the bioreactor system was indicated in *Picea sitchensis*[14]. In this report, the authors also indicate that the morphology of the cell lines used influenced the effect of different bioreactor configuration on the growth and maturation of somatic embryos. Our current work which elucidated in part the morphogenetic changes of embryogenic tissues during the initiation and maintenance phases would contribute to the development of large scale propagation of somatic embryos of Japanese larch using a bioreactor system, in the near future.

ACKNOWLEDGEMENTS

The authors gratefully acknowledge Prof. Edward C. Yeung of the University of Calgary for critical reviews of this manuscript.

REFERENCES

1. D. I. Dunstan, T. E. Tautorus & T. A. Thorpe, Somatic embryogenesis in woody plants, In: *In vitro Embryogenesis in Plants*, T. A. Thorpe (ed.), Kluwer Academic Publishers, Dordrecht, **1995**, pp. 471-538.
2. P. von Aderkas, K. Klimaszewska & J. M. Bonga, Diploid and haploid embryogenesis in *Larix leptolepis*, *L. decidua*, and their hybrids, *Can. J. For. Res.*, **1990**, 20, 9-14.
3. S. Ogita, T. Kubo & M. Fushitani, Histological and morphological characteristics of embryogenesis from mature embryos callus in *Larix leptolepis*, In: *The 4th Plant Tissue Culture Symposium,* The Japanese Association for Plant Tissue Culture, **1994**, pp. 80-81 (in Japanese).
4. S. Ogita, T. Kubo & M. Fushitani, Anatomical characteristics in early embryogenesis from immature embryo of *Larix leptolepis*, *Forest Resources Environ.*, **1997**, 35, 45-51.
5. S. Ogita, H. Sasamoto & T. Kubo, Maturation and plant recovery from embryogenic cells of Japanese Larch: Effect of abscisic acid in relation to their morphology, *J. For. Res.*, **1999c**, 4, 241-244.
6. S. Ogita, H. Sasamoto & T. Kubo, Control of the development of somatic embryo of Japanese conifers by the density of embryogenic cells in liquid culture, In: *Transplant Production in the 21st Century*, C. Kubota & C. Chun (eds.), Kluwer Academic Publishers, Dordrecht, **2000**, pp. 209-214.
7. R. A. Campbell & D. J. Durzan, Induction of multiple buds and needles in tissue cultures of *Picea glauca*, *Can. J. Bot.*, **1975**, 53, 1652-1657.
8. S. Ogita, H. Sasamoto & T. Kubo, Selection and micro-culture of single embryogenic cell clusters in Japanese conifers: *Picea jezoensis*, *Larix leptolepis* and *Cryptomeria japonica*, *In Vitro Cell. Devel. Biol.-Plant*, **1999d**, 35, 428-431.
9. J. M. Bonga, K. Klimaszewska, M. A. Lelu & P. von Aderkas, Somatic embryogenesis in *Larix*, In: *Somatic Embryogenesis in Woody Plants, Vol. 3*, S. Jain, P. Gupta & R. Newton

(eds.), Kluwer Academic Publishers, Dordrecht, **1995**, pp. 315-339.

10. S. M. Attree & L. C. Fowke, Micropropagation through somatic embryogenesis in conifers, In: *Biotechnology in Agriculture and Forestry, Vol.17, High-Tech and Micropropagation I*, Y. P. S. Bajaj (ed), Springer-Verlag, Berlin, **1991**, pp. 53-70.
11. J. M. Bonga, P. von Aderkas & K. Klimaszewska, Cytology of in vitro cultured tissues of forest trees, In: *Cytogenetic Studies of Forest Trees and Shrub Species*, Z. Borzan & S. E. Schlarbaum (eds.), Croatian Forests, Inc., Zagreb, **1997**, pp. 325-343.
12. Y. W. Kim, Y. Youn, E. R. Noh & J. C. Kim, Somatic embryogenesis and plant regeneration from immature zygotic embryos of Japanese larch (*Larix leptolepis*), *Plant Cell, Tiss. Org. Cult.*, **1999**, 55, 95-101.
13. S. S. Bhojwani & M. K. Razdan, Chapter 4, Cell culture, In: *Plant Tissue Culture: Theory and practice, a revised edition*, Elsevier, Amsterdam, **1996**, pp. 63-93.
14. B. Ingram & F. Mavituna, Effect of bioreactor configuration on the growth and maturation of *Picea sitchensis* somatic embryo cultures, *Plant Cell, Tiss. Org. Cult.*, **2000**,1,87-96.

Molecular Breeding of Woody Plants
N. Morohoshi and A. Komamine, editors.

SOMATIC EMBRYOGENESIS OF JAPANESE CONIFERS

Katsuaki Ishii* , Emilio Maruyama & Yoshihisa Hosoi

Tissue Culture Laboratory, Bio-resources Technology Division, Forestry and Forest Products Research Institute, Ibaraki 305-8687, Japan

ABSTRACT

The present work focused on improving somatic embryogenesis of the Japanese conifer species by studying the factors affecting culture condition. They include *Chamaecyparis obtusa* Sieb. et Zucc., *Chamaecyparis picifera* Sieb. et Zucc., *Cryptomeria japonica* D. Don, *Pinus densiflora* Sieb. et Zucc. and *Pinus thunbergii* Parl. When the zygotic embryos in the explants were either proembryos or early embryos, initiation of somatic embryogenesis was successful. The medium significantly affected induction success, the medium with auxin (3 or 5 or 10 uM 2,4-dichlorophenoxyacetic acid) and cytokinin(1 or 3 uM 6-benzylaminopurine) being better than the medium with cytokinin alone. Maltose instead of sucrose was better for the maturation of the somatic embryos. Sawara cypress (*Chamaecyparis picifera*) plantlets regenerated from somatic embryo and grown in a greenhouse were planted out in the field to compare with the control seedlings. Somatic embryogenesis and plant regeneration from shoots cultures of Hinoki cypress (*Chamaecyparis obtusa*) *in vitro* was also successful.

KEYWORDS

Hinoki cypress, *Chamaecyparis obtusa*, Sawara cypress, *Chamaecyparis picifera* ,Sugi cedar, *Cryptomeria japonica* , Japanese red pine, *Pinus densiflora*, Japanese black pine, *Pinus thunbergii* ,somatic embryo, maturation, germination, regeneration

INTRODUCTION

Somatic embryogenesis is ideal procedure not only for effective propagation of plus trees but also target tissue for genetic transformation. Up to now, there are many somatic embryogenic regeneration with conifers[1]. However, information about somatic embryogenesis in Japanese conifers is still scarce[2-7]. In almost all gymnosperm woody species, zygotic embryos are the frequently used explants for initiation of somatic embryogenesis[8]. However, regeneration of plants are usually difficult or at very low frequency except *Larix* species[9,10]. Micropropagation of Hinoki cypress (*Chamaecyparis obtusa* Sieb. et Zucc.) using shoot culture was reported[11]. Somatic embryo induction and plantlet regeneration at low rate was reported with Hinoki cypress only among *Chamaecyparis* species[6]. We here succeeded in the production of relatively many plantlets from somatic embryo culture from immature zygotic embryos of Hinoki cypress (*Chamaecyparis obtusa* Sieb. et Zucc.) and Sawara cypress(*Chamaecyparis picifera* Sieb. et Zucc.). The wood quality of Hinoki cyprress is superior among conifers in Japan, however it grows slower in its young stage [12]. It also has high adaptability to dry and semi-dry planting site [13]. Plant regeneration through somatic embryogenesis of Hinoki cyprress is important for genetic engineering and somatic hybridization breeding to create resistant hybrid for disease.

Somatic embryogenesis of Sugi cedar(*Cryptomeria japonica*), Japanese red pine(*Pinus densiflora*) and Japanese black pine(*Pinus thunbergii*) were also described.

MATERIALS & METHODS

1. Somatic embryo induction and proliferation

The aim of the experiments started in 1995 was to obtain an overall view of somatic embryogenesis (from initiation to regeneration) in Japanese conifers like Black Pine, Hinoki cypress , Red pine, Sawara cypress and Sugi Cedar.

Somtic embryogensis from shoots cultured *in vitro* was also used in Hinoki cypress. Those shoots were induced from a juvenile seedling germinated in vitro. That material was collected from natural forest of Hinoki cypress in Nagaoka county in Kochi prefecture. Leafy shoot was cultured on the half strength LP medium [14] containing 5 uM BAP and 10 uM NAA or 2,4-D.

Immature zygotic embryos were collected from end of June to early August from 1995 to 2000.Media listed in the Table 1 were used for somatic embryo induction and further cultivation. Culture condition was first under dark on the agar solidified medium at 25 °C. For further proliferation, culture condition for immature zygotic embryos were under dark in the 100 ml flask containing 30 ml liquid culture at the 70 r.p.m. rotary shaker. For cell proliferation test, agar solidified medium in the test tubes (18mm x 180mm) was also used.

Open-pollinated cones were cleaned with ethanol and dissected to remove the seeds. Then, removed seeds were disinfected with 1-3% (w/v available chlorine) sodium hypochlorite solution for 15-30 min and subsequently rinsed with sterile distilled water. The excised explants were cultured in SM1 (Standard Embryonic Tissue Capture medium[15], MS (Murashige and Skoog's medium[16], or LPm (modified from Quorin and Lepoivre's medium[14] containing basal salts reduced to half concentration from the standard, 10 g/l sucrose and 500 mg/l L-glutamine. Media were supplemented with 0-30μM 2,4-D plus 0-10μM BAP (Table 1). The pH of the media was adjusted to 5.8 prior to autoclaving for 15 min at 121°C.

Embryogenic tissues were subcultured at 2 weeks intervals on rotary shakers at 100 rpm, in 100 ml flasks containing 30-40 ml of LPm, MSm (modified Murashige and Skoog's medium; basal salts were reduced at half concentration from the standard, and replacing NH_4NO_3 with 1,000 mg/l L-glutamine), or EMm (modified Embryo Maturation medium[7] containing basal salts reduced to half concentration from the standard, concentration of KCl was reduced to 75 mg/l, and casein hydrolysate and all aminoacids were replaced with 1,500 mg/l L-glutamine.

2. Somatic embryo maturation

Induced somatic embryos were transplanted on the half-strength LP medium containing 7.57μM ABA and 90 mM Sucrose. Maltose was also used in the maturation medium. At this stage, natural type ABA((s)(+)-ABA) and synthetic type ABA ((R)(-)-ABA) were compared for their effects on the maturation of the somatic embryos of Hinoki cyprress.

One hundred to 500 mg FW of proliferated embryogenic cell masses suspended in 2-3 ml were plated on filter paper disks (70 mm in diameter) over 90 x 20 mm petri dishes containing SM3 (Embryo Development medium)[15] (Table 1) and cultured for 6 to 12 weeks.

Culture condition was under dark on the agar solidified medium at 25 °C.

3. Plant regeneration from mature somatic embryos

For germination from the somatic embryos, half-strength LP medium containing 5g/l activated charcoal was used. In the other case, CD medium[17] containing 0.0027 μM NAA was also used for plant regeneration from somatic embryos. Culture condition was under

Table 1. Culture conditions for somatic embryogenesis of Japanese conifers

Species	Explant	Induction	Proliferation	Maturation	Germination	Acclimatization
Cryptomeria japonica (Sugi)	Megagametophyte	SM1, LPm (25℃, darkness) Sucrose 10 g/l 2,4-D 0-10 μM BAP 0-5 μM AC 0-2 g/l Gelrite 2-3 g/l or D. Bacto Agar 6-8 g/l	LPm, MSm, EMm (25℃, darkness) Sucrose 10-30 g/l 2,4-D 1-10 μM BAP 0-5 μM Rotary shaker (100 rpm)	SM3 (25℃, darkness) Sucrose 60 g/l or Maltose 60 g/l PEG 75 g/l or Sorbitol 20 g/l ABA 50 mg/l Gelrite 4-5 g/l	1/2 LP (25℃, lighted) Sucrose 20 g/l AC 5 g/l Agar 11.5-12.5 g/l	Vermiculite Hyponex 0.1% or Nm
Chamaecyparis obtusa (Hinoki)	Megagametophyte	SM1, LPm (25℃, darkness) Sucrose 10 g/l 2,4-D 0-10 μM BAP 0-5 μM AC 0-2 g/l Gelrite 2-3 g/l or D. Bacto Agar 6-8 g/l	LPm, MSm, EMm (25℃, darkness) Sucrose 10-30 g/l 2,4-D 1-10 μM BAP 0-5 μM Rotary shaker (100 rpm)	SM3 (25℃, darkness) Maltose 60 g/l PEG 150 g/l AC 2 g/l ABA 100 mg/l Gelrite 4-5 g/l	1/2 LP (25℃, lighted) Sucrose 10-30 g/l AC 0-5 g/l Agar 8-12.5 g/l	Vermiculite : Perlite (1 : 1) Hyponex 0.1% or Nm
Chamaecyparis pisifera (Sawara)	Whole seed	MS, 1/2 MS (25℃, darkness) Sucrose 30 g/l 2,4-D 0-30 μM BAP 0-10 μM	LPm, MSm, EMm (25℃, darkness) Sucrose 10-30 g/l 2,4-D 1-10 μM BAP 0-5 μM Rotary shaker (100 rpm)	SM3, EM (25℃, darkness) Maltose 30-60 g/l PEG 50-150 g/l AC 2 g/l ABA 100 mg/l Gelrite 3-5 g/l	1/2 LP (25℃, lighted) Sucrose 10-30 g/l AC 0-5 g/l Agar 8-12.5 g/l	Vermiculite : Perlite (1 : 1) Hyponex 0.1% or Nm
Pinus densiflora (Akamatsu)	Megagametophyte	SM1, LPm (25℃, darkness) Sucrose 10 g/l 2,4-D 0-10 μM BAP 0-5 μM AC 0-2 g/l Gelrite 2-3 g/l or D. Bacto Agar 6-8 g/l	LPm, MSm, EMm (25℃, darkness) Sucrose 10-30 g/l 2,4-D 1-10 μM BAP 0-5 μM Rotary shaker (100 rpm)	SM3 (25℃, darkness) Sucrose 30 g/l or Maltose 30-60 g/l PEG 0-150 g/l AC 2 g/l ABA 50-100 mg/l Gelrite 3-10 g/l	1/2 LP (25℃, lighted) Sucrose 5-30 g/l AC 0-5 g/l Agar 8-12.5 g/l	Vermiculite : Perlite (1 : 1) Hyponex 0.1% or Nm
Pinus thunberghii (Kuromatsu)	Megagametophyte	SM1, LPm (25℃, darkness) Sucrose 10 g/l 2,4-D 0-10 μM BAP 0-5 μM AC 0-2 g/l Gelrite 2-3 g/l or D. Bacto Agar 6-8 g/l	LPm, MSm, EMm (25℃, darkness) Sucrose 10-30 g/l 2,4-D 1-10 μM BAP 0-5 μM Rotary shaker (100 rpm)	SM3 (25℃, darkness) Sucrose 30 g/l or Maltose 30-60 g/l PEG 0-150 g/l AC 2 g/l ABA 50-100 mg/l Gelrite 3-10 g/l	1/2 LP (25℃, lighted) Sucrose 5-30 g/l AC 0-5 g/l Agar 8-12.5 g/l	Vermiculite : Perlite (1 : 1) Hyponex 0.1% or Nm

fluorescent light of 60 to 70 μES^{-1} on the agar solidified medium at 25 °C.

The number of embryo germinated and converted to plantlets was recorded after 4-12 weeks.

4. Habituation of regenerated plants

Somatic seedlings were habituated in the vermiculite or vermiculite + perlite soil under high humidity for 2 weeks then gradually uncovered the cover over the pot. Developed emblings were transferred into pots and acclimatized in plastic boxes inside the growth cabinet (Table 1).

5. Field performance of Sawara cypress

Emblings of habituated Sawara cypress were planted out to the field with the control seedlings. Height Growth was measured after one growing season.

RESULTS

From shoots of in vitro cultured plantlets of Hinoki cypress

Cell clusters induced from shoot cultured *in vitro* was first pale greenish, however after several subculturing on the somatic embryo induction medium under dark condition, it turns whitish and translucent cell clusters were obtained. After twice subculture at the 2 months interval, 500 mg fresh cell clusters were subcultured to the maturation medium. After 2 months, there appeared the embryo heads at the end of the suspensor. When comparing the different type of ABA at the same concentration in the maturation media, number of matured somatic embryos were greater in the case of natural type ABA. Matured somatic embryos with cell clusters were subcultured first to the half-strength LP containing activated charcoal where many mature somatic embryos became brown and died. However some mature embryos germinated. Further subculture to the germination medium of CD containing NAA then half-strength medium containing activated charcoal produced the rooted converted seedlings from mature embryos(Fig. 1).

From immature zygotic embryos

Proembryonic type cell clusters were obtained from immature zygotic embryos of Hinoki cypress as shown in the Fig. 2. It can be subcultured in the same induction medium at least 6 months at the 3 weeks interval. At the same time, using the agar solidified medium, the different rate of cell proliferation from immature zygotic embryo at different sampling period was checked. The best proliferation period was from early July to mid July in Hinoki cypress. However, the cell clusters obtained at those period was not necessarily the embryogenic character. For embryogenesis, a little bit early period from end of June to the early July was the best.

The highest response was achieved with materials collected between late of June and beginning of July in general with Japanese conifers. Medium was not a critical factor for embryogenic cell induction. The initiation of embryonal masses was observed after 2-6 weeks of culture on media supplemented or not with exogenous plant growth regulators and the initiation frequencies varied according to the species. Somatic embryogenesis was initiated at relatively high frequencies in Sugi cedar, Hinoki cypress and Sawara cypress, and at very low frequencies in the species of Pinus (Table 2).

Table 2. Induction of embryogenic cells in Japanese conifers

Species	Seed collection date	Induction media	2,4-D (μM)	BAP (μM)	Embryogenic cell induction (%)
Cryptomeria japonica	July 04	SM1	0	0	10
(Sugi)		LPm	10	5	11
Chamaecyparis obtusa	July 05	SM1	0	0	20
(Hinoki)		LPm	10	5	15
Chamaecyparis pisifera	July 02	MS	10	0.1	15
(Sawara)					
Pinus densiflora	July 02	SM1	0	0	1
(Akamatsu)		LPm	10	5	1
Pinus thunberghii	June 26	SM1	0	0	1
(Kuromatsu)		LPm	10	5	3

Proliferation of embryogenic cells was possible in medium containing a combination of 2,4-D plus BAP (Table 1). Since proliferation is slow on solid media, these media supported cell growth longer than liquid media did. Thus, embryogenic cells were subcultured at intervals of 4 weeks for maintenance on solid media, and 2 weeks for rapid proliferation in liquid media. Embryogenic cell lines have been maintained and proliferated for more than 2 years.

Maturation of somatic embryos of Japanese conifers

Cotyledonary embryos were obtained mostly on maturation media containing Maltose (50-60 g/l), ABA (50-100 μM), PEG (50-150 g/l), and AC (0-2 g/l). Embryogenic cells on media without ABA, generally, did not develop beyond the embryo stage 1. The addition of PEG into ABA-media stimulated the cotyledonary embryo production, suppressing the precocious germination and the secondary embryogenesis. The number of somatic embryos was greater improved (excepting Sugi) in presence of AC (2 g/l).

Germination and plant conversion

High and medium plant conversion rate were obtained in Sawara cypress and Hinoki cypress somatic embryos (92 and 37%, respectively). Germination was initiated 1-2 weeks after transferring to germination media, and about 4 weeks after transferring almost all embryos germinated and regenerated plants. Germination and conversion of somatic embryos in both *Chamaecyparis* species were achieved without pregerminative treatments. Regenerated emblings were acclimatized successfully and transferred to a glasshouse before transplanting to the field.

In contrast the percentage of germinants and emblings obtained in Sugi cedar and *Pinus*

Figure 1. Plantlets regenerated through somatic embryo induced from shoot cultured *in vitro*

Figure 2. Somatic embryos induced from immature embryos of Hinoki cypress

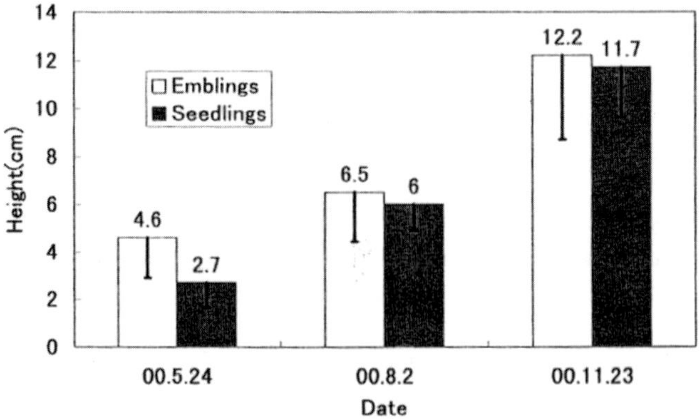

Figure 3. Field Growth of emblings and seedlings of Sawara cypress during one growing season

species was very low. Attempts to enhance germination in these species using partial drying and cool drying treatments were tried but without a significant improvement in the number of germinants. Most of the nongerminated embryos developed cotyledons which turned green under light conditions and although some of them showed epycotyl growth and later plant conversion with adventitious root formation, most of them turned brown and then died. Further improvement of the maturation and plant conversion steps is necessary.

From the proembryogenic tissue, the mature somatic embryos were induced and then those embryos were germinated on the germination medium *in vitro*. After habituation, somatic seedling were obtained in the pot.

Field performance

Sawara cypress from somatic embryos grows equally well with control seedlings during the one growing season as shown in Figure 3.

DISCUSSION

The present study is the first English written progressive report on the regeneration and habituation of somatic embryo plants in Hinoki cypress and other Japanese conifers. In conifers, explants including immature zygotic embryos have been the most successful for initiating embryogenic cultures. The optimum stage of immature zygotic embryo development for the initiation of embryogenic cultures in conifers has been the precotyledonary stage, when fertilization has occurred but the cotyledons have not yet been formed[18]. In the present study, the window for initiation for embryogenic potent cells in 1997 was a period from end of June to early July. Natural type ABA was much more êffective for somatic embryo induction, because synthetic type ABA is supposed to contain only half concentration of active ABA.

Even though the regeneration rate was quite low, this is the second report where the embryogenic cells was induced and regenerated from shoot explants not from immature embryo in the case of conifers. First report was from Larix species . Regeneration of somatic embryos from mature stage of conifers must be studied in the future for getting the genetic gain of selected plus trees.

As is the case for other conifers, further improvement of the maturation and conversion steps of somatic embryogenesis is necessary for application of the method to practical forestry.

CONCLUSIONS

Somatic embryogenesis from immature zygotic embryos and shoot of Hinoki cypress cultured *in vitro* was achieved using different culture media for induction, maturation and germination steps. Somatic embryos from both explants sources converted to the plantlets. However, the regeneration rate was less than 1 %, and further improvement study was needed. Other Japanese conifers like Black pine, Red pine, Sawara cypress and Sugi cedar were successfully regenerated through somatic embryos from immature embryos.

REFERENCES

1. E.I. Hay, & P.J. Charrest, Somatic embryo germination and desiccation tolerance in conifers, In: *Somatic Embryogenesis in Woody Plants*. Jain, S.M. *et al.* (ed), 547 pp. Kluwer Academic Publishers, Dordrecht, 1999, 61-96.
2. K. Ishii, Somatic embryo formation and plantlet regeneration through embryogenic callus from mature zygotic embryos of *Picea jezoensis* var. *hondoensis*, *Picea jezoensis*,

and *Picea glehnii, J. Jpn. For. Soc.*, 1991, 73, 24-28.

3. K. Ishii & E. Maruyama, Somatic embryogenesis and plant regeneration in tissue culture of *Chamaecyparis obtusa*, (In Japanese) *Trans 109th Annu Meet Jpn For Soc.*, 1998,337-338.
4. K. Ishii & E. Maruyama, In vitro culture of Japanese black pine (*Pinus thnbergii*), In: *Transplant Production in the 21st Century*, C. Kubota & C. Chen (eds.), Kluwer Academic Publishers, Dordrecht, 2000, pp 205-208.
5. S. Ogita, H. Ishikawa, K. Kubo & H. Sasamoto, Somatic embryogenesis from immature and mature zygotic embryos of *Cryptomeria japonica* 1: Embryogenic cell induction and its morphological characteristics. *J. Wood Sci.*,1999,45,87-91.
6. S. Ogita, H. Sasamoto & T. Kubo, Maturation and plant recovery from embryogenic cells of Japanese larch: Effects of abscisic acid in relation to their morphology, *Journal of Forest Research*, 1999, 4, 241-244.
7. E. Maruyama, T. Tanaka, Y. Hosoi, K. Ishii & N. Morohoshi, Embryogenic Cell Culture, Protoplast Regeneration, Cryopreservation, Biolistic Gene Transfer and PlantRegeneration in Japanese Cedar (*Cryptomeria japonica* D. Don). *Plant Biotechnology*, 2000,17, 281-296.
8. K. Raemakers, E. Jacobsen & R. Visser, Proliferative somatic embryogenesis in woody species. In: *Somatic Embryogensis in Woody Plants*. Jain, S.M. *et al.* (ed), 547pp, Kluwer Academic Publishers, Dordrecht, 1999,29-59.
9. M.A. Lelu, K. Klimaszewska & P.J. Charest, Somatic embryogenesis from immature and mature zygotic embryos and from cotyledons and needles from somatic plantlets of *Larix.*, *Can. J. For. Res.* 1994a ,20,100-106.
10. M.A. Lelu, C. Bastien, K. Klimaszewska & P.J. Charest, An improved method for somatic plantlet production in hybrid larch (*Larix x leptoeuropaea*): part 2. Control of germination and plantlet development. *Plant Cell Tiss. Org. Cult.* 1994b,36,117-127.
11. K. Ishii & T. Sato, *In vitro* plantlet regeneration in some species of Japanese conifers. *Proc. of the 6th Internatl. Comgr. of SABRAO*,1989,869-872.
12. N. Fukuhara, Meiotic observation in the pollen mother cell of interspecific hybrid between *Chamaecyparis obtusa* and *C. picifera*. J. Jap. For.Soc. 1978,60,437-441.
13. T. Maeta, Effects of gamma-rays irradiation on inerspecific hybridization between *Chamaecyparis obtusa* S. *et* Z. and *C. picifera* S. *et* Z. *Hoshasen Ikusyujo Kenkyu Hokoku*, 1982,5,1-87.
14. J. Aitken-Chrisstie & T.A. Thorpe, Clonal propagation: Gymnosperms, In: *Cell Culture and Somatic Cell Genetics of Plants* (Vasil, I.K. ed.) Academic Press, Inc. Orlando, San Diego 1984,1, pp.82-95.
15. D.R. Smith, Growth medium. *United States Patent* ,1996, # 5, 565,355.
16. T. Murashige & F. Skoog, A revised medium for rapid growth and bioassays with tobacco cultures. *Physiol. Plant.*, 1962,15, 473-497.
17. R.A. Campbell & D.J. Durzan, Inducton of multiple buds and needles in tissue culture of *Picea glauca*, *Can. J. Bot.* 1975,53, 1652-1657.
18. H. Haggman, A. Jokela, J. Krjnakova, A. Kauppi, K. Niemi & T. Aronen, Somatic embryogenesis of scots pine: cold treatment and characteristics of explants affecting induction, *Journal of Experimental Botany*,1999,50,1769-1778.

Molecular Breeding of Woody Plants
N. Morohoshi and A. Komamine, editors.

APPLICATION OF SOMATIC EMBRYOGENESIS TO TREE IMPROVEMENT IN CONIFERS

David R. Cyr *, Stephen M. Attree, Yousry A. El-Kassaby, David D. Ellis, Dan R. Polonenko & Ben C.S. Sutton

CellFor Inc., PO Box 133, Brentwood Bay, BC, Canada V8M 1R3

ABSTRACT

Somatic embryogenesis (SE) offers advantages in tree breeding due to the genetic gain improvements that can be realized through selection and production of elite (clonal) lines. Additionally, it is a platform through which value-added traits can be introduced via genetic engineering.

At CellFor considerable effort has been directed towards the improvement of SE protocols for coniferous species, with significant focus placed on development *in vitro*. Recent progress includes rapid bulk-up of cultures via liquid culture, production of desiccation-tolerant embryos in bioreactors and direct delivery of somatic seed to the nursery. Additionally, transgenic spruce, loblolly pine and radiata pine have been developed via SE, with the targets including enhanced cellulose synthesis, lignin modification and control of key developmental pathways

As a tool for tree improvement, conifer SE circumvents the infrastructure and time constraints that are inherent to production of improved seed via seed orchards. The application of SE in this context necessitates the development of a deployment strategy that addresses genetic diversity and the genetic gain achieved. Approaches, based on the principles of population genetics, have been developed to optimize both the initial selection strategy and subsequent deployment of clonal lines.

Since 1995, CellFor and its predecessor companies have implemented commercial programs for spruce (interior; Sitka), pine (loblolly; radiata; slash) and Douglas-fir. Annual volume production in excess of 1 million seedlings has been achieved.

KEYWORDS

Bioreactor, Clonal Forestry, Conifer, Genetics, Nursery, Somatic Embryogenesis, Transformation

INTRODUCTION

Conifer somatic embryogenesis (SE), since first reported in 1985 for spruce[1,2], has been extended to greater than 40 species and hybrids[3]. SE has become an important component of conifer tree improvement programs due to its capacity for providing a broad genetic base for selection, insuring long-term genetic stability of juvenile germplasm (cryopreservation), and enabling sustained production of plants[4,5]. Tree improvement targets for SE include growth, form, wood quality, and resistance to insects or disease, with deployment strategy aimed at maximizing gain of selected lines on productive or high index plantation sites. Gains, in the relative short-term, are expected based on traditional breeding targets, whereas, in the longer term larger gains are expected from genetic engineering as the technology comes of age.

By the late 90's, applied SE programs were in place for at least 9 commercial conifer species[3]. These programs, for the most part, have focused on feasibility studies and trial establishment. Operational production of seedlings, in the case of spruce, has been demonstrated[5,6,7]. The current status of the commercial development of conifer SE at CellFor Inc. will be discussed in this paper.

CONIFER SOMATIC EMBRYOGENESIS TECHNOLOGY

Research and development efforts have been directed towards increasing the yield and quality of culture bulk-up, somatic embryo production and conversion to plants. Emphasis is placed on optimization of through-put, and integration of the SE system. In general, for any given species, this entails the development and optimization of protocols on a research line subset, validation of improvements using an operationally relevant population of lines and subsequent integration into high-volume production (technology transfer).

Maintenance culture (proliferation) is commonly performed through frequent subculture of embryonal suspensor masses (ESM) on semi-solid media using auxins and cytokinins[3]. However, liquid cultures offer a greater opportunity to develop a maintenance system that can be streamlined into production of mature somatic embryos. The liquid approach, not only facilitates the rapid bulk-up of maintenance tissue and is scalable, but also enables the effective modulation of development temporally prior to initiation of the maturation process. The current system, initially developed on spruce (Attree, unpublished), has been extended to Douglas-fir, and most recently to loblolly pine. In the case of Douglas-fir, scale-testing has demonstrated long-term stability of maintenance cultures, predictability within and among genotypes, and bulk-up capabilities which supercede that of maintenance culture on semi-solid medium.

The development of the liquid system, as implied previously, is not independent of the targets for improving embryo quality; namely, desiccation tolerance and high frequency vigorous germination. In simple terms, the liquid culture steps facilitate increased embryo yield and synchronous development of high-quality embryos during maturation. Abscisic acid (ABA) and water stress have been identified as significant factors in maturation. In concert with high levels of ABA, water-stress can be applied using high gel-strength or non-permeating osmotica such as polyethylene glycol[8]. The latter of these water-stress approaches exhibits the greatest potential for scale-up, as it is amenable to a liquid-based system.

Studies were conducted to examine the interaction between ABA and PEG during spruce maturation (Attree, unpublished). Preliminary results indicated that optimal ABA and water potential levels are different for early versus late maturation, with increases in the intensity of these applications promoting development and inhibiting precocious development. Further optimization of these two parameters led to a continuous reduction in moisture content to a level similar to that observed in conifer seeds undergoing natural development. It was found that increased ABA could promote reductions in moisture content, however, this was more effectively modulated using a combination of water stress and ABA. Maturation using these principles produced embryos capable of desiccation at 4°C and ambient relative humidity (45%). With this approach long-term storage of embryos (several years) and subsequent regeneration of vigorous plants has been demonstrated. More recently, long-term storage has been achieved for an operational population of Douglas-fir lines (~ 15 lines).

Bioreactors[9] and embryo purification are important components of the *in vitro* SE process. Bioreactors are a format suitable for automation. These were initially developed for spruce and are currently being applied for embryo production from Douglas-fir and loblolly pine. One bioreactor is considered as the equivalent of at least 30 maturation Petri-plates, and in Douglas-fir, yields that dramatically exceed the productivity of Petri-plates (i.e., 6,000 mature embryos per bioreactor) have been demonstrated. Pilot production using bioreactors was implemented for Douglas-fir during the year 2000. Embryo purification techniques, when applied in association with bioreactor production, have proven to be an effective tool for harvest, enrichment and desiccation of embryos. Using this approach several thousand desiccated embryos can be collected and stored in an area less than that of a 5.5-cm filter paper.

Although efforts have gone into optimization of in vitro germination, the ultimate aim is to eliminate this step from the operational SE process. In this context, several artificial seed approaches have been proposed most commonly involving some form of encapsulation[3]. Regardless of the approach, the objective is to enable mechanical sowing into conventional growing environments. Additionally, desiccation is essential to year-round production, their storage allowing mechanical sowing in a short window.

Specific systems have been developed for encapsulated embryos, however, high costs and mechanical limitations have to date prevented them from moving beyond the prototype stage and automated commercial systems are not currently available. An additional challenge with encapsulation the recapitulation of the role of the nutritive tissue (e.g., megagametophyte). It has been concluded that significant research is still required before this type of approach would have practical application[10].

The CellFor approach has been to develop an alternative and flexible system for sowing and germination that eliminates most of the previously encountered issues. The result has been an approach that allows somatic embryos to be sown by conventional seeders into a range of container formats. The planting substrates used are soil-less media commonly used by horticultural industry. Moreover, the sowing steps can be carried out in commercial greenhouse environments without the need for a sterile environment or specialized equipment. Proof of concept has been achieved with spruce, loblolly and radiata pine and Douglas-fir, with embryos achieving germination 5-7 days after sowing and conversion after 2-3 weeks. Pilot operational production of Douglas-fir using this technology is scheduled for 2001.

GENETICS OF CLONAL FORESTRY

Seed orchards and vegetative propagation are options for delivering genetic gain to plantation forestry. For seed orchards there are two areas of limitation. Firstly, maximal genetic gain can only occur if the orchard performs as a closed and perfect population. However, seed orchard research has indicated this is affected by sub-optimal reproductive synchrony, varying proportion of gamete production among genotypes, pollen contamination and orchard site maladapation. Secondly, although controlled pollinated (CP) seed can be produced from orchards, it is costly and the volume of seed per annum is finite.

Vegetative propagation can enhance orchard outputs through the development of clonal forestry. The advantages for forest productivity is the incremental gain that can be realized through selection and mass propagation of elite individuals (clonal lines). Additionally, clonal lines can be established from small quantities of elite seed, and operational deployment from newly selected individuals can be achieved without the cost and time delay involved in generating a seed orchard. In tree improvement additional benefits can be garnered through improved forest management and end product utilization, particularly in the case of clonal forestry.

The most prevalent of the clonal propagation approaches are rooted-cuttings: common in Australia, Brazil, Chile, New Zealand and South Africa. Seedling production entails the establishment of mother plants and, in some cases, limited serial propagation. The disadvantages of this approach include the time to reach production, a maximal multiplication rate, physiological aging of donor plants, and the inability to store germplasm for future propagation. These factors increase the cost per gain delivered as compared to what can be achieved with effective tissue culture propagation. Although benefits are realized from cuttings in some short-rotation hardwood species, in conifers this is more often than not limiting for clonal forestry. SE can overcome these limitations by virtue of its capacity for the establishment of embryogenic cultures from a wide array of gentoypes, amenability to long-term frozen storage, and the capacity for sustained embryo production[10]. SE, as an alternative vegetative propagation system, delivers not only the benefits of controlled pollinated families but also the substantial additional value of clonal forestry (increased yield, quality and uniformity).

Regardless of the approach, genetic gain and genetic diversity must be optimized. In the case of SE, where multiple copies per line and lines per family are produced, a method of estimating genetic diversity that takes into account the breeding strategy is required for deployment. By choosing an approach, on the basis of potential genetic gain and predicted effective population size, a forester can customize levels of gain and genetic diversity. Maximum potential gain, and particularly genetic diversity, is defined by the number of parents, their genetic relationship and the mating design. Where clonal forestry is practised, genetic factors include within line (multiple copies), among lines/within cross (full-sib) and among lines for common parent (half-sib) scenarios. The genetic advances from SE, as compared to seed orchard operations, is illustrated in Figure 1.

APPLICATIONS OF CONIFER SE IN CLONAL FORESTRY

Clonal Selection: The starting point for SE commercial production is the clonal establishment phase. The specific design of a clonal selection program depends on the pedigreed families available, level of gain expected and identification of target traits. Efforts on interior spruce are an example. In its conception, the program was designed to establish 1,000 clonal lines via SE using 30 to 40 top ranked families[11]. These families had been selected from 173 plus-trees based on 10 and 15-year progeny trial data for weevil-resistance and height, respectively. Subsequent execution of this plan resulted in the delivery of more than 1,300 lines from 36 top-ranked families to clonal trials from 1996 to 1998[12]. The deployment strategy included the establishment of 6 to 8 copies on each of 3 or more sites selected for yield and high weevil-hazard. Prior to this program, approximately 250 lines from families varying in rank had been installed in trials during 1994 to 1995. Initial selection of 30 to 50 operational lines expected to begin in 1999, was based on a selection intensity of 5%[20]. Using a 'Best' family approach, a first-generation population of was constructed in 1997 (i.e., 5 lines x 12 top families)[11]. This was estimated to provide 60 % gain in weevil-resistance and 11% gain in height over the baseline population. Gains after clonal selection, by definition are expected to be substantially higher; it has been calculated that these can be 40 to 60% as compared to 13% for seed orchards (El-Kassaby and Sutton, unpublished).

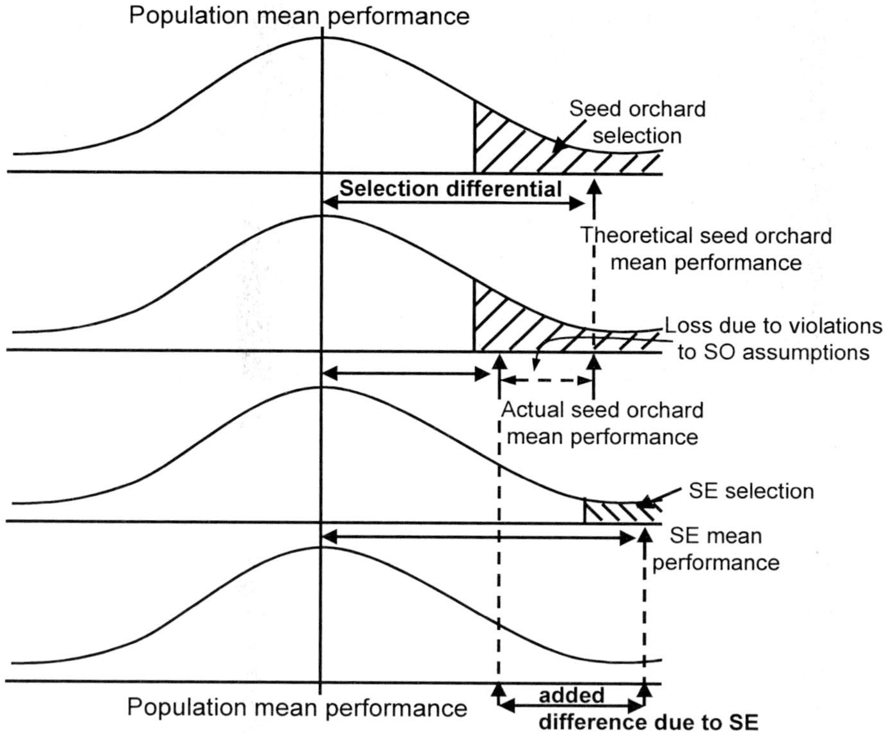

Figure 1. Selection and production of clones.

The steps of clonal selection include line initiation (induction), cryopreservation and production of plants for clonal trials. Typically, embryogenic cultures are established from excised embryo (i.e., spruce) and immature seed (i.e., pines and Douglas-fir) explants, with 200 to 2000 seed cultured per family. For large projects, seed volumes of 20 to 24K may be handled in a 3 to 6 week window. Establishment of lines generally occurs within 1 to 8 weeks, with cryopreservation completed in 3 to 6 months. Production of plants (e.g., 10 to 200 copies per line) are delivered to the customer at 12-18 months; the timing of the delivery varies according to the species, region, trial schedule and the product format. For interior spruce, *in vitro* plants were delivered to the nursery within 9 to 10 months of receipt of the donor seed. Clonal trials were installed 13 to 14 months thereafter. In this program, close to 95% of the cryopreserved lines produced embryos and 80% of these met target numbers and stock quality for field trials.

CellFor Inc. and its predecessor companies have conducted clonal selection programs since 1990. Much of the current focus is on the development of programs for species with shorter rotation ages and advanced breeding, and in regions where plantation forestry is part of the forest estate. To date, greater than 3,000 clonal lines from 7 species have been placed in trials (Table 1).

Table 1. Summary of clonal selection programs.

Species	Region	Selection Target (s)	Trial Year	Trial Lines (Families)	Projected* Trials: New Families/ Lines (Trial Year)
Interior Spruce	Canada	weevil resistance	94-98	1,455 (51)	na
Sitka Spruce	Canada	weevil resistance	97-00	378 (25)	na
Douglas-fir	Canada, USA	yield & form	na	na	12 / 200 (03)
Loblolly Pine	Chile, USA	yield	98-01	429 (14)	7 / 180 (02)
Radiata Pine	Australia, Chile New Zealand	yield, form & wood traits	98-01	1,000 (37)	na
Slash Pine	USA	fusiform rust resistance	01	103 (8)	2 / 75 (02)
Pinus patula	South Africa	yield & blight & *Fusarium* resistance	00	63 (9)	na
			Total:	3,418 (144)	21 / 455

*programs under development for delivery in 2003-04 not included.

Production: Two production phases typically occur during the commercialization SE, and are defined by operational activities occurring before and after the availability of the first clonal selection results for a given species and program. The first phase can be referred to as pilot-scale commercial production; this can deliver clonal material based on two scenarios for line selection. The first scenario is based on performance data collected from R&D assessments of embryo production, *in vitro* germination and nursery trials, whereas, the second is based on a specified number of elite families and lines per family, typically with balanced genetic diversity (i.e., family forestry). By definition, pilot production is carried out at a semi-operational level (i.e., 100,000 to 500,000 per annum).

The primary benefit of initiating pilot-scale operational production and testing at an early stage is to enable demonstrations of the substantial genetic gains that family forestry can provide to the customer. Additionally, early pilot-scale production and testing facilitate the evaluation of somatic seedlings under operational nursery and field conditions, adapting the product to the nursery growing regimes of customers, and assessing the performance of clones as clonal block plantings on a variety of sites.

Commercial production per se would commence upon the customers' selection of their preferred elite clones based on the results of clonal field tests. Given the dynamics and long-term nature of this selection process, the clonal composition of the production population would evolve as new R&D and field performance data are collected. This highlights the advantage of using somatic embryogenesis versus other propagation approaches, in that the ability to cryopreserve lines facilitates integration of alternate clones into the production population. Thus the clone bank performs as a flexible source of proven high-value genetic material.

Production of interior spruce somatic embryos was initiated in 1992 with 12,000 somatic seedlings delivered the following spring. This production volume doubled in 1994, and then tripled in 1995. By 1999, the cumulative total for delivered interior spruce somatic seedlings had increased to 1 million (Table 2). With the implementation of standard manufacturing principles, production and delivery of 1.5 million spruce somatic seedlings was achieved in during 2000. Recently, operational seedling numbers have been produced for Douglas-fir and loblolly pine.

Table 2. Summary of SE Production

Species	Year		Seedling Volume
Interior Spruce	94-99		1,000,000
Sitka Spruce	98		50,000
White Spruce	00		1,450,000
Loblolly Pine	99-01		200,000
Douglas-fir	00-01		500,000
		Total:	**3,200,000**

GENETIC ENGINEERING

As discussed previously, an added benefit of SE is the development of a tissue culture system where regeneration from single cells is possible. Therefore, coupled with the extensive development of a commercial SE program, CellFor has an active research program in the genetic engineering of value-added traits in conifers. In-house transformation systems are in place for interior spruce, Sitka spruce, Douglas fir, loblolly pine and radiata pine, with our primary emphasis in the pines. The goal is to evaluate the integration of transgenics into the SE program such that when elite superior genotypes are identified through our ongoing field testing programs, individual value added traits can be incorporated, where applicable, into superior lines.

Current programs with transformation focus on the modification and enhancement of wood properties, transgene expression and the development of strategies for the deployment of transgenic trees such that the environmental impacts, both hypothetical and real, are minimized. With the modification of wood properties, work is proceeding in two areas: 1) modification of lignin composition, such that the lignin is easier to remove during pulping, and 2) the effect of increasing the levels of cellulose precursor(s) (i.e., UDP-glucose) on cellulose formation. While transgenic pines have been regenerated from both projects, most of the results to date are derived using tobacco and poplar model systems. In both species, significant changes in both lignin and biomass accumulation have been achieved. Pulping trials are currently underway to characterize the effect of these modifications on pulping rates. Another area of emphasis is gene expression. Specifically, this is aimed at directed towards expression of transgenes to the cambial region as well as cells and/or tissues associated with reproductive structures. The latter studies are aimed at the manipulation of reproductive organs to limit transgene spread through pollen or seed.

SUMMARY

Conifer somatic embryogenesis, as a propagation technology for clonal forestry, has come into maturity, as evidenced by CellFor's efforts over the last decade. To date, programs for selection of elite lines are in place in 6 countries spanning 3 continents, with expansion of these activities to other regions currently under development. Within the next few years, the first results of clonal selection will be available; concurrently, integration of new scale-up technologies into the commercial operations will occur. As a result, volume production will increase by a magnitude from current levels in the next few years and the level of gain delivered via SE will increase. Additional gains are anticipated as genetic engineering outputs are integrated into the operational production system.

REFERENCES

1. V. Chalupa, Somatic embryogenesis and plantlet regeneration from cultured immature and mature embryos of *Picea abies* (L.) Karst., *Commun. Inst. For. Cech.*, **1985**, 14, 57-63.
2. I. Hakman, L.C. Fowke, S. von Arnold & T. Eriksson, The development of somatic embryogenesis in tissue cultures initiated from immature embryos of *Picea abies* (Norway spruce), *Plant Sci.*, **1985**, 38, 53-9.
3. D.R. Cyr, Seed substitutes from the laboratory. In: *Seed Technology*, J.D. Bewley & M. Black (eds.), CRC Press LLC Academic Press, Boca Raton FL, **2000,** pp 326-58.
4. D.R. Cyr, Cryopreservation of embryogenic cultures of conifers and its application to clonal forestry. In: *Somatic Embryogenesis in Woody Plants Vol. 4*, S.M. Jain, P.K. Gupta & RJ Newton (eds.), Kluwer Academic, Boston MA, **1999**, pp 239-61.
5. S.C. Grossnickle, D.R. Cyr & D.R. Polonenko, Somatic embryogenesis tissue culture for the propagation of conifer seedlings: a technology comes of age. *Tree Plant. Notes*, **1996**, 47, 48-57.
6. B.C.S. Sutton & D.R. Polonenko, Commercialization of plant somatic embryogenesis In: *Somatic Embryogenesis in Woody Plants Vol. 4*, S.M. Jain, P.K. Gupta & RJ Newton (eds.), Kluwer Academic, Boston MA, **1999**, pp 263 -91.
7. B.C.S. Sutton, D. Polonenko, D. Cyr & S.C. Grossnickle, Commercialization of somatic embryogenesis in forestry, *Biotech. Intl.*, **1999**, 2, 254-60.
8. S.M. Attree, & L.C. Fowke, Desiccated conifer somatic embryos, *US Patent*, **1995**, 5,464,769.
9. S.M. Attree, M.K. Pomeroy, & L.C. Fowke, Production of vigorous, desiccation tolerant white spruce (*Picea glauca*-Moench. Voss.) synthetic seeds in a bioreactor, *Plant Cell Rep.* , **1994**, 13, 601-06.
10. W.C. Carlson & J.E. Hartle, Manufactured seeds of woody plants, In: *Somatic Embryogensis in Woody Plants: Volume 1-History, Molecular and Biochemical Aspects, and Applications,* S.M. Jain, P.K. Gupta & R.J. Newton (eds.), Kluwer Academic, London, **1995**, pp 253-64.
11. B.C.S. Sutton, S.C. Grossnickle, D.R. Roberts, J.H. Russell & G. Kiss, Somatic embryogenesis and tree improvement in interior spruce. *For. J.*, **1993**, 91, 34-8.
12. D.R. Cyr, Cryopreservation: Roles in clonal propagation and germplasm conservation of conifers, In: *Cryopreservation of Tropical Plant Germplasm-Current Research Progress and Applications, JIRCAS International Agriculture Series No. 8*. F. Engelmann & H. Takagi (eds), Proc. JIRCAS/IPGRI Joint International Workshop 20-23 Oct., 1998, Tsukuba, Japan, **2000**, pp 261-68.

Molecular Breeding of Woody Plants
N. Morohoshi and A. Komamine, editors.

SOMATIC EMBRYOGENESIS AND PLANTLET REGENERATION IN *PINUS ARMANDII* VAR. *AMAMIANA*

Yoshihisa Hosoi* and Katsuaki Ishii

Division of Bio-resources Technology, Forestry and Forest Products Research Institute, P.O.Box 16,Tsukuba Norin Kenkyu Danchi-nai, Ibaraki, 305-8687 JAPAN

ABSTRACT

Embryogenic cell suspensions were induced from a mature seed of *Pinus armandii* var. *amamiana* known as an endangered species. MS liquid medium supplemented with 10 μ M 2,4-D and 3% sucrose was used for initiation of the suspensions. The suspensions were incubated in the dark at 24℃. Obtained suspensions were transferred to NH_4NO_3-free MS liquid medium supplemented with 10 μ M 2,4-D, 3mM L-glutamine and 3% sucrose and subcultured every 2 weeks. In order to develop somatic embryos, the suspensions were transferred into NH_4NO_3-free MS media supplemented with 3% sucrose, 100 μ M ABA, 0-20% PEG (MW4,000), 0-20mM L-glutamine and 0.2% activated charcoal. After 1-2 weeks of incubation the suspension cultures were transferred onto NH_4NO_3-free media solidified with 0.4% gellan gum containing 100 μ M ABA, 0.2% activated charcoal, 0-10% PEG and 0-20mM L-glutamine. The cultures were incubated under a 16h light / 8h dark photoperiod. After 1-2 momths of culture, differentiation of embryos progressed and cotyledonary embryos were obtained. These embryos were transferred on NH_4NO_3-free solid MS media containing 0-10mM L-glutamine and incubated at 24℃ and under a 16h light / dark photoperiod. After 2-3weeks some plantlets with roots and green cotyledons were obtained. Plantlets were transplanted into vermiculite containing modified MS liquid medium in 200ml culture flasks. This is the first report of somatic embryogenesis and plantlet regeneration in *P. armandii* var. *amamiana.*

KEYWORDS

Pinus armandii var. *amamiana*, endangered species, somatic embryogenesis, MS-Murashige and Skoog medium

INTRODUCTION

Pinus armandii var. *amamiana* belongs to the subgenus *Strobus* (a.k.a. *Haploxylon*) in the genus *Pinus*. This species is endemic in Japan and found only on the islands of Yakushima and Tanegashima in Kagoshima prefecture. The regeneration rate of *P. armandii* var. *amamiana* in its natural habitat is quite low. Propagation through seeds is unreliable due to poor germination and death of young seedlings under natural habitats. Mature trees have also been affected by typhoons and wilt disease. Hence, this species is recorded as an endangered plant in the Red Data Book of Japanese plants [1]. Tissue culture could be an important tool both for preservation and multiplication of such an endangered species. Somatic embryos can also be used for the long-term storage of germplasm, thereby conserving rare and endangered species [2-4]. Embryogenic cultures of conifers were first established from immature zygotic embryo explants of *Picea abies* L.[5, 6] Since 1985, researchers have made a great deal of progress in the development of somatic embryogenesis technology for conifer reforestation. However, *Pinus* species have been regarded as highly recalcitrant to somatic embryogenesis[4, 7]. There has been no information on *in vitro* culture of *P. armandii* var. *amamiana* to date.

The objectives of this study were to investigate conditions for somatic embryogenesis and the feasibility of converting these somatic embryos of *P. armandii* var. *amamiana* to plantlets.

MATERIALS & METHODS

Initiation and maintenance of embryogenic cell suspensions

Mature seeds were collected from open-pollinated cones of P. *armandii* var. *amamiana* and stored at 4℃ for a year. After removing the seed coats, seeds were surface disinfected for 1min in 70% ethanol, then 45min in 3% sodium hypochloride with one drop of detargent, followed by three rinses in sterile ultrapure water. Each embryo with female gametophyte tissue was transversally bisected, and cultured in liquid MS medium supplemented with 10μM 2,4-D in a 100ml culture flask . The flasks were put on a rotary shaker at 100 rpm and maintained at 24℃ in the dark.

After 1 month of culture, the suspensions derived from single embryo were transferred into 50ml of MS (with or without NH_4NO_3) liquid media containing 10μM 2,4-D and 3mM L-glutamine in petri dishes (90×20mm). The suspensions were incubated stationarily and kept at 24℃ in the dark. The suspensions were subcultured by transferring 1ml of suspension into 50ml fresh medium at two-week-interval.

Somatic embryo maturation

The suspension cultures were transferred into 50ml NH_4NO_3-free liquid MS media containing 100μM abscisic acid (ABA), 0.2% activated charcoal, 0-20% polyethylene glycol (PEG: MW4,000) and 0-20mM L-glutamine. The cultures were incubated in 90×20mm petri dishes at 24℃ in the dark.

After 1-2 week of incubation, the suspensions were transferred onto filter paper disks (Advantec No.2, 70mm in diameter), which were put on 50ml of maturation media poured onto 90×20mm petri dishes. The media comprised of NH_4NO_3-free MS supplemented with 100μM ABA, 0.2% activated charcoal, 0-10% PEG, 0-20mM L-glutamine, 3% sucrose and 0.4% gellan gum. The cultures were maintained at 24℃ in the dark.

Embryo germination and development

After further 1-2 months, cotyledonary somatic embryos were plated on filter disks over germination media (NH_4NO_3-free MS containing 0-10mM L-glutamine, 3% sucrose and 0.4% gellan gum). The cultures were maintained at 24℃, 16h photoperiod and 5,000 lx.

Obtained seedlings were then transferred to sterile vermiculite moistened with NH_4NO_3-free MS containing 3mM L-glutamine and 3% sucrose in 200ml flasks for further elongation in a growth chamber maintained at 24℃, 16h photoperiod and 5,000 lx.

RESULTS & DISCUSSION

Initiation of embryogenic cell suspensions from mature zygotic embryo

60 open-pollinated seeds were tested for initiation of embryogenic cell suspensions. Except for contaminated seeds (5%), calli were initiated from the mature zygotic embryo explants of all seeds. The calli developed in liquid medium but they were non-embryogenic. Although many free single cells were initiated from megagametophytes, cell division did not occur at all.

In two embryos, embryogenic cell suspensions, consisting of an embryonic region of cells with dense cytoplasm, attached to elongated, vacuolated cells, were observed 1 month after culture (Fig. 1). Following experiments, single embryogenic cell line was employed.

Maintenance and proliferation of embryogenic cell suspensions

Maintenance of the embryogenic cell line has been done by transferring cultures in liquid medium every 2 weeks into fresh medium because most cells turned brown if the transfer period was longer than 3-4 wks. The cell line has been maintained its embryogenic character, by showing embryogenic heads with attached suspensors for the last 15 months. The differences on growth of suspensions between two media (see materials & methods) were not clear. For next experiments NH_4NO_3-free MS medium containing 10μM 2,4-D, 3% sucrose and 3mM L-glutamine was used as subculture medium.

Maturation and germination of somatic embryo

Development of somatic embryos to the cotyledonary stage was observed after 1-2 months on filter disks over solid media (Fig. 2).

The use of PEG in combination with ABA has been reported as a common method for stimulating somatic embryo maturation in many conifers. In present experiments the effect of PEG as an osmotic agent was not clear. Further study is needed. The supplement of L-glutamine into media enhanced embryo maturation and prevented somatic embryos from browning.

126 cotyledonary somatic embryos were transferred to germination media. After 2-3 weeks, 73 (58%) cotyledonary embryos failed to develop and died. 53 (42%) cotyledonary embryos turned green and showed cotyledon expansion and hypocotyl elongation. Furthermore, 3 embryos underwent complete germination showing root development (Fig. 3). Obtained seedlings were transferred to vermiculite as described above (Fig. 4). Other 50 embryos had precocious radicles, which failed to elongate. These embryos were transferred onto rooting medium (MS medium supplemented with 3μM IBA and 0.8% agar) after cutting back the radicles. After transfer on the rooting medium, 12 embryos developed needles soon but rooting has not been observed yet (Fig. 5).

CONCLUSIONS

Up to now, no studies had explored cell or tissue culture of *P. armandii* var. *amamiana*. To our knowledge, this is the first report of somatic embryogenesis from mature zygotic embryos in *P. armandii* var. *amamiana*. In conclusion, the present study has demonstrated production of somatic embryos and their subsequent development into plantlets. However, the rate of regeneration from embryogenic suspensions was low. Further optimization of culture conditions is required for embryo maturation and plantlet formation.

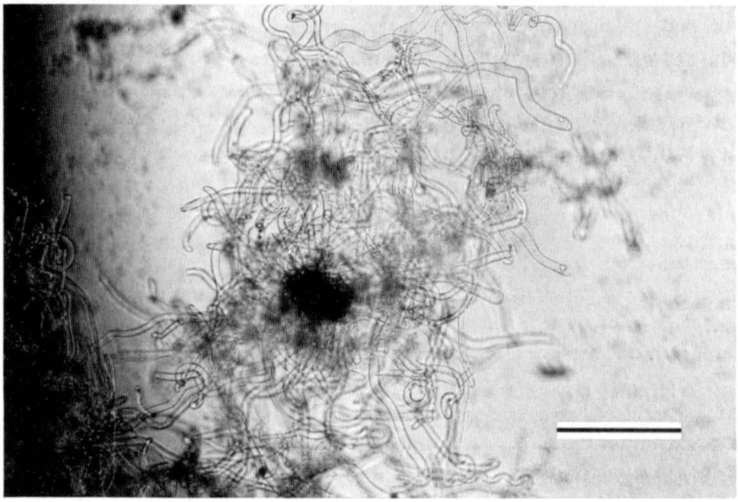

Figure 1. *P. armandii* var. *amamiana* embryogenic suspension culture composed of immature embryos with long suspensions. Bar = 200 μm

Figure 2. Differentiation of cotyledonary somatic embryo on maturation medium. Bar=1mm

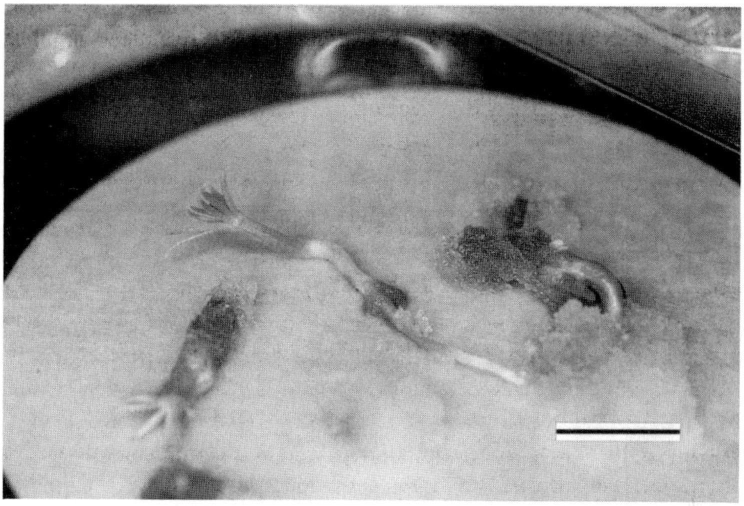

Figure 3. Root elongation of germinating somatic embryo. Bar=1cm

Figure 4. Germinated plantlet transferred to vermiculite. Bar=2cm

Figure 5. Normal shoot formation and elongation from somatic embryo after cutting back the precocious radicle. Bar=1cm

REFERENCES

1. Environment Agency of Japan, (ed.), '*Pinus armandii* var. *amamiana*' , In: *Threatened wildlife of Japan - Red Data Book 2nd ed.* -, Japan Wildlife Research Center, **2000**, 8, p. 271.
2. F. S. Catharine, N. B. Jones & van S. Johannes, 'Optimization of a working cryopreservation protocol for *Pinus patula* embryogenic tissue' , *In Vitro Cell. Dev. Biol.-Plant*, **2000**, 36, 366-369.
3. S. M. Attree & L. C. Fowke, 'Embryogeny of gymnosperms: advances in synthetic seed technology of conifers' *Plant Cell Tiss. Org. Cult.,* **1993**, 35, 1-35.
4. T. E. Tautorus, L. C. Fowke & D. I. Dunstan, 'Somatic embryogenesis in conifers' , *Can. J. Bot.,* **1991**, 69,1873-1899.
5. V. Chalupa, 'Somatic embryogenesis and plantlet regeneration from cultured immature and mature embryos of *Picea abies* (L.) Karst' , *Communi. Inst. For. Cech.,* **1985**, 14, 57-63.
6. I. Hakman & S. von Arnold, 'Plantlet regeneration through somatic embryogenesis in *Picea abies* (Norway spruce)' , *J. Plant Physiol.*, **1985**, 121, 149-158.
7. R. Ljiljana, A. Carmen, F. F. Mario & R. Roberto, 'Somatic embryogenic tissue establishment from mature *Pinus nigra* Arn. ssp. *salzmanni* embryos' , *In Vitro Cell. Dev. Biol.-Plant*, **1999**, 35, 206-209.

Molecular Breeding of Woody Plants
N. Morohoshi and A. Komamine, editors.

PLANT REGENERATION FROM SOMATIC EMBRYOS IN *PINUS THUNBERGII* (JAPANESE BLACK PINE) AND *PINUS DENSIFLORA* (JAPANESE RED PINE)

Toru TANIGUCHI

Forest Tree Breeding Center, Ministry of Agriculture, Forestry and Fisheries, Ibaraki 319-1301 Japan

ABSTRACT

Embryogenic tissues were initiated from megagametophyte explants containing pre-cotyledonary embryos. The explants were collected from juvenile cones of two *Pinus* species, *Pinus thunbergii* and *P. densiflora*, in mid and late July. The medium used for initiation was modified DCR medium (inorganic nitrate salts were the same as in DCR but the other components were the same as in MS) containing 1.5g/l glutamine, 10 μM 2,4-D, 5μM BA, 30g/l sucrose and 8g/l agar. The initiation frequencies in *Pinus thunbergii* and *P. densiflora* were 8.0 % (31 out of 389 explants) and 2.9 % (3 out of 105 explants), respectively. Embryogenic tissues were subcultured on modified DCR medium containing 1.5g/l glutamine, 2 μM 2,4-D, 1 μM BA, 30g/l sucrose, and 8g/l agar or 4 g/l gelrite. After one year of the initial culture, eleven *P. thunbergii* and three *P. densiflora* embryogenic lines were established. Matured somatic embryos were obtained by using Smith embryo development medium containing 50 μM ABA, 75 g/l PEG, 60 g/l sucrose and 5.5 g/l gelrite. Average numbers of mature somatic embryos per dish were 32 and 22 in best cell lines of *P. thunbergii* and *P. densiflora*, respectively. The dishes sealed with permeable paper tape benefited for germination of somatic embryos compared with parafilm. Among modified GD medium Smith embryo germination medium, modified DCR medium, and LP medium, the best performance for germination was carried out by the modified GD medium containing 10 g/l sucrose and 5 g/l gelrite. Germination frequency of selected cell lines of both species was ca. 28 %. Regenerated plantlets have been transferred to soil.

KEYWORDS

Pinus thunbergii, *Pinus densiflora,* somatic embryos, plant regeneration

INTRODUCTION

Both of *Pinus thunbergii* PARL. (Japanese black pine) and *P. densiflora* SIEB. et ZUCC (Japanese red pine) are important forest tree species in Japan. Disease induced by pine wood nematode has been very serious problem of these pines. Therefore, resistant clones of these pines to pine wood nematode were selected and seed orchards were established using the resistant clones.

In future, to create the resistant trees to disease or insect by gene transformation and to propagate efficiently the preferable trait trees are expected. In order to realize these, plant regeneration system by somatic embryogenesis is useful. Since plantlet production through somatic embryogenesis of *Pinus* species was reported in *P. lambertiana* [1], plant regeneration studies by embryogenesis were reported in some species (for instance in *P.*

taeda [2], *P. radiata* [3], *P. strobus* [4] and *P. nigra* [5]). In the case of *P. thunbergii* and *P. densiflora*, reports of plant regeneration through somatic embryogenesis are scarce. In this study, immature seeds collected from resistant trees to pine wood nematode of *P. thunbergii* and *P. densiflora* were used as explants, and somatic embryogenesis and plant regeneration were shown.

MATERIALS & METHODS

Plant materials

Open pollinated and immature cones of *P. thunbergii* were collected from two clones in the clonal bank of Forest Tree Breeding Center located in Juo (Ibaraki, Japan) on 29 June, 15 July and 22 July, 1999. In the case of *P. densiflora*, cones were collected from two clones on 22 July, 1999. All of these clones were resistant clones to pine wood nematode. Immature seeds were extracted from the cones and then surface disinfected for 1 min in 70 % (vol/vol) ethanol, followed by 15 min in 7.5 % (vol/vol) hydrogen peroxide, and rinsed three times in sterile deionized water. The immature seeds were dissected, and explants (intact megagametophytes with zygotic embryos) were extracted from the seeds.

Embryogenic tissue induction and maintenance

Initiation media used in this study were MS medium [6] and mDCR medium (modified from DCR medium [7]). Inorganic nitrate salts, i.e. ammonium nitrate (NH_4NO_3), potassium nitrate (KNO_3) and calcium nitrate tetrahydrate ($Ca(NO_3)_2{\cdot}4H_2O$) in the mDCR medium were the same as in DCR medium but the other components were the same as in MS medium. Both media were containing 1.5 g/l L-glutamine, 10 μM 2,4-dichlorophenoxyacetic acid (2,4-D), 5 μM N6-benzyladenine (BA), 30 g/l sucrose, and 8 g/l agar. Five explants were cultured in each petri dish. All cultures were transferred onto fresh medium of same components with three-week intervals. Nine weeks after culturing on initiation media, embryogenic tissues were transferred onto maintenance medium which was consisting of mDCR medium components, but 2,4-D and BA were reduced to 2 μM and 1 μM, respectively. This medium was gelled by 4 g/l gellan gum instead of agar. Embryogenic tissues were transferred onto maintenance medium with monthly intervals. Cultures were incubated in the dark at 25°C.

Maturation of somatic embryos

Eight months after initial culture, embryogenic tissues were transferred onto maturation media. Maturation media were Smith embryo development medium [8] containing 7.5 % polyethylene glycol, 60 g/l sucrose and 5.5 g/l gellan gum. In order to investigate the effect of abscisic acid (ABA) on somatic embryo maturation, ABA (mixed isomers, Sigma) was added to maturation media at four different concentrations (0, 1, 10, 50 μM). Six embryogenic tissue mass, each approximately 50 mg (fresh weight), were cultured on each dish in darkness at 25°C.

Germination and soil establishment

Cotyledonary somatic embryos were removed from maturation media after 40-50 days of culture on maturation media, and transferred onto germination media. Culture condition for germination was a 16-h photoperiod and temperature of 25°C.

At first step, to investigate the effects of permeability of dishes on germination two kinds of sealing tapes, parafilm and permeable adhesive paper tape were used to seal the lids of dishes. Smith embryo germination medium [8] (SM-ger) containing 10 g/l sucrose was used at this step. In order to check the water vapor permeability, petri dishes containing medium without embryos were weighted at 4-5 day intervals for 40 days.

At next step, to seek the preferred medium for germination, five kinds of media were investigated to germinate the somatic embryos: mGD [9] (modified from GD [10]), SM-ger [8], mDCR (describe above), half-strength mDCR, and LP [11]. All of these media were containing 10 g/l sucrose and 5 g/l gelrite.

Some of germinated plantlets developed roots and needles were transferred into bottles containing vermiculite fertilized with 0.1 % (v/v) Hyponex. After two months plantlets were transferred into pots containing soil mixture (peatmoss and vermiculite) and grown in green house. For initial 10 days pots were covered with translucent polyethylene to maintain high humidity. Afterwards the polyethylene covers were removed and the plantlets watered daily.

RESULTS & DISCUSSION

Embryogenic tissue induction and maintenance

As shown in Table 1, better performance for initiation was carried out by the mDCR medium that was containing lower concentration of ammonium than MS medium. In both *Pinus* species, embryogenic tissues were initiated from explants collected in mid or late July when the developmental stage of zygotic embryos were pre-cotyledonary. In late June, when zygotic embryo could not recognized under stereomicroscope, embryogenic tissue was not induced. *P. thunbergii* might be more responsible for embryogenic tissue initiation than *P. densiflora*. Moreover, initiation rates were different among family. In *P. thunbergii*, K2 family was more responsible than K1 family (Table 1). In *P. densiflora*, embryogenic tissues were initiated only in A1 family (Table 1). Totally, thirty-three and four embryogenic tissues were induced in *P. thunbergii* and *P. densiflora*, respectively, and after one year of initiation eleven *P.*

Table 1. Effects of medium and collection date on initiation of embryogenic tissue (ET) in two *Pinus* species.

Species	Mother tree*	Date of collection	mDCR Explants	mDCR With ET		MS Explants	MS With ET	
Pinus thunbergii	K1	29-Jun	100	0	(0.0)	100	0	(0.0)
	K1	15-Jul	101	1	(1.0)	100	1	(1.0)
	K1	22-Jul	100	0	(0.0)	100	0	(0.0)
	K2	29-Jun	100	0	(0.0)	100	0	(0.0)
	K2	15-Jul	98	17	(17.3)	98	0	(0.0)
	K2	22-Jul	90	13	(14.4)	90	1	(1.1)
Pinus densiflora	A1	22-Jul	70	3	(4.3)	70	1	(1.4)
	A2	22-Jul	35	0	(0.0)	35	0	(0.0)

* Immature and open pollinated cones were collected from these mother trees.
Values in parenthesis are percentages.
The composition of mDCR (modified from DCR medium) is written in the text.

thunbergii and three *P. densiflora* embryogenic lines were remained.

Maturation of somatic embryos

After two weeks of culture on maturation media, cylindrical embryonal head structure was developed at the top of suspensor in both species. Within five weeks mature embryos with cotyledons were developed. As shown in Table 2, in *P. thunbergii* cell lines K105 and K202 and *P. densiflora* cell line A102, best result was obtained by the medium containing 50 μM of ABA. Lower concentration of ABA, 10 μM was preferred in *P. thunbergii* cell line K106. A few numbers of maturated embryos developed in cell lines K201 (*P. thunbergii*) and A102 (*P. densiflora*). Average numbers of mature somatic embryos per dish were 32 (K105) and 22 (A101) in best cell lines of *P. thunbergii* and *P. densiflora*, respectively.

These results shown above were obtained from the embryogenic tissues, which were maintained for eight months. When the older embryogenic tissues, which were maintained for 18 months, were used for maturation study, the number of maturated embryos decreased, even if best cell lines, K105 (*P. thunbergii*) or A101 (*P. densiflora*) were used. These decreases of the ability of maturation were serious problem.

Germination and soil establishment

In both species the dishes sealed with paper tape were better for germination (that is both of radicle extension and needle development) than with parafilm. Higher frequencies of germination (Table 3) and better growth of plantlets were obtained on paper tape sealed dishes than on parafilm sealed dishes. Weight decreases of dish sealed with paper tape and with parafilm were 315 mg/day and 22 mg/day, respectively. From these results, water vapor permeability was thought to be preferred for germination of somatic embryos of pine. Resemble results were obtained in somatic embryo germination of Japanese larch (unpublished data).

Among five media tested, mGD medium was best for germination of both pine somatic embryos (Table 4). The germination frequency on mGD medium in both species was ca. 28 %. This mGD medium was used for embryo culture for adventitious bud induction in *P. thunbergii* and *P. densiflora* [12].

Before acclimatization of both species, new needles were developed but epicotyl was very short. After acclimatization the epicotyl of regenerated plantlet started to elongate. Plantlets were acclimatized successfully under conditions described above.

Table 2. Effect of ABA concentration on maturation of somatic embryos in two *Pinus* species. Average numbers of maturated somatic embryos per dish are shown.

Species	Cell line	ABA (10^{-6}M) 0	1	10	50
	K201	0.0	1.5	3.0	0.5
Pinus	K202	0.0	5.0	0.0	22.0
thunbergii	K105	2.5	15.0	15.0	32.0
	K106	1.0	3.7	21.0	6.0
Pinus	A101	2.3	12.3	12.3	22.0
densiflora	A102	0.0	0.0	1.0	1.0

Table 3. Effects of tapes sealed the lids of dishes on germination of somatic embryos in two *Pinus* species.

Species*	Type of tapes sealed dishes	No. of embryos cultured	No. of germinated embryos	Germination frequency (%)
Pinus thunbergii	Parafilm	32	2	6.3
	Paper tape	51	6	11.8
Pinus densiflora	Parafilm	30	2	6.7
	Paper tape	30	6	20.0

*Three (K202, K105 and K106) and one (A101) cell lines were used in *P. thunbergii* and *P. densiflora* , respectively.
The germination medium was SM-ger.

Table 4. Effect of media on germination of somatic embryos in two *Pinus* species.

Species and cell line	Media	No. of embryos cultured	No. of germinated embryos	Germination frequency (%)
Pinus densiflora A101	mGD	70	20	28.6
	SM-ger	80	20	25.0
	mDCR	30	0	0.0
	1/2mDCR	30	0	0.0
	LP	40	0	0.0
Pinus thunbergii K105	mGD	50	14	28.0
	SM-ger	50	5	10.0
	mDCR	30	0	0.0
	1/2mDCR	30	0	0.0
	LP	30	0	0.0

The lids of dishes were sealed with permeable paper tape.

CONCLUSIONS

As shown above, plants were regenerated from somatic embryos in both of *P. thunbergii* and *P. densiflora*. However there are some problems, low initiation rate, particularly in *P. densiflora,* decrease or loss of maturation ability after long subculture period of embryogenic cells, and low germination rate. To solve these problems and use the somatic embryogenesis system in genetic transformation, improvement of protocol is needed. By screening of families, seeking of high potential cell lines for embryogenesis is also important.

REFERENCES

1. P. K. Gupta & D. J. Durzan, Somatic polyembryogenesis from callus of mature suger pine embryos, Bio/Technology, **1986**, 4, 643-645.
2. P. K. Gupta & D. J. Durzan, Biotechnology of somatic polyembryogenesis and

plantlet regeneration in loblolly pine, Bio/Technology, **1987**, 5, 147-151.

3. J. Aitken-Christie, K. Gough, D. J. Maddocks, M. R. Sigley, V. A. Hodder, F. Burger & P. C. S. Carter, Embryogenesis of radiata pine, In: *Proceedings of International Wood Biotechnology Symposium*, Tokyo, **1994**, pp91-98.
4. E. Garin, N. Isabel & A. Plourde, Screening of large numbers of seed famillies of *Pinus strobus* L. for somatic embryogenesis from immature and mature zygotic embryos, Plant Cell Rep., **1998**, 18, 37-43.
5. T. Salajova, J. Salaj & A. Kormutak, Initiation of embryogenic tissues and plantlet regeneration from somatic embryos of *Pinus nigra* Arn., Plant Sci., **1999**, 145, 33-40.
6. T. Murashige & F. Skoog, A revised medium for rapid growth and bioassays with tobacco tissue cultures, Physiol. Plant., **1962**, 15, 473-497.
7. P. K. Gupta & D. J. Durzan, Shoot multiplication from mature trees of Douglas-fir (*Pseudotsuga menziesii*) and Sugar pine (*Pinus lambertiana*), Plant Cell Rep., **1985**, 4, 177-179.
8. D. R. Smith, Growth medium, US Patent, No. 5565355, October **1996**.
9. H. E. Sommer, C. L. Brown & P. P. Kormanik, Differentiation of plantlets in longleaf pine (*Pinus palustris* MILL.) tissue culture in vitro, Bot. Gaz., **1975**, 136, 196-200.
10. P. M. Gresshoff & C. H. Doy, Development and differentiation of haploid *Lycopersicon esculentum* (Tomato), Planta, **1972**, 107, 161-170.
11. J. Aitken-Christie & T. A. Thorpe, Clonal propagation: Gymnosperms, In: Cell culture and somatic cell genetics of plant, Vol. 1, I. K. Vasil (Ed.), Academic Press, San Diego, **1984**, pp. 82-95.
12. M. Okamura & T. Kondo, Manual for tissue culture of pine, Bull. National For. Tree Breed. Center, **1995**, 13, 139-143.

Molecular Breeding of Woody Plants
N. Morohoshi and A. Komamine, editors.

CONCEPTS AND BACKGROUND OF PHOTOAUTOTROPHIC MICROPROPAGATION

Chieri Kubota

*Department of Bioproduction Science, Faculty of Horticulture, Chiba University
Matsudo, Chiba 271-8510, Japan*

ABSTRACT

Micropropagation is a method to produce genetically identical plantlets by using tissue culture techniques. Recent research revealed that chlorophyllous plantlets in vitro had high photosynthetic ability but that their net photosynthetic rates were restricted by the in-vitro environmental conditions, mainly the low CO_2 concentration during photoperiod. Photoautotrophic micropropagation refers to micropropagation with no exogenous organic components (sugar, vitamins, etc.) added to the medium, and it has been developed along with the development of techniques of in-vitro environmental control. CO_2 concentration, photosynthetic photon flux, relative humidity, and air current speed in the vessel are some of the most important environmental factors affecting plantlet growth and development; controlling these factors requires knowledge and techniques of greenhouse and horticultural engineering as well as the knowledge of physiology of in-vitro plantlets. Photoautotrophic micropropagation has many advantages with respect to improvement of plantlet physiology (biological aspect) and operation/management in the production process (engineering aspect), and it results in reduction of production costs and improvement in quality of plantlets. Feasibility of photoautotrophic micropropagation has been reportedly shown in both herbaceous and woody plant species. Photoautotrophic micropropagation will give a breakthrough in large-scale production of genetically identical, pathogen-free plantlets with vigorous growth and better overall quality and therefore, has a great potential to be introduced in transplant production and biotechnology research.

KEYWORDS

Carbon dioxide enrichment, in-vitro environmental control, large-scale culture system, photosynthesis, PPF, sugar-free

Micropropagation, an in-vitro vegetative propagation method using pathogen free propagules, has been considered significant in agriculture and forestry for producing pathogen free stock plants or genetically superior clones that cannot be propagated by seeds or whose propagation efficiency is low in conventional vegetative propagation. However, the use of micropropagation is still limited to small number of species (mostly those with high commercial values such as landscape and ornamental materials). Much research has been done to enhance the multiplication ratio, improve the uniformity and enhance the growth of plantlets in vitro, mostly focusing on manipulation of medium components and growth regulators. Photoautotrophic micropropagation [1] is a unique approach of producing in-vitro quality plantlets with high photosynthetic and growth rates by controlling in-vitro environmental conditions.

ENVIRONMENTAL CONTROL IN MICROPROPAGATION

A tissue culture vessel is a system for growing small plants under aseptic conditions. The vessels are made of glass or clear plastics to allow light transmission into the vessel. In this sense, tissue culture vessels can be considered as miniature greenhouses. The difference between a tissue culture vessel and a greenhouse as a system for plant production is that greenhouses are systems for controlling the environment (temperature, radiation, photoperiod, etc.), while, conventionally, tissue culture vessels are not. Consequently, tissue culture vessels must be placed in a growth room with environmental control. Nevertheless, environmental conditions surrounding plantlets are not directly controlled in tissue culture. There has been much research related to environmental control in greenhouse crop production, while not much research for in-vitro environmental control. Greenhouse environmental control contributes to improved growth and quality of plants. Can in-vitro environmental control be of similar benefit and improve the growth and quality of plantlets? Plants fix atmospheric carbon. Can plantlets in vitro grow without sugar in the medium as a carbon source? Concepts of photoautotrophic micropropagation and methods of in-vitro environmental control have been developed through research initiated from such simple questions universally asked by anyone having worked in micropropagation.

1. Definition of Photoautotrophic micropropagation

Photoautotrophic micropropagation often refers to micropropagation with no sugar added to the medium. Narrowly defined, photoautotrophy is a nutritional status where living organisms grow without any additional exogenous organic components as nutrients. When defined in this narrow sense, media in photoautotrophic micropropagation should exclude all organic components. As in hydroponics, media for photoautotrophic micropropagation consist exclusively of inorganic components. Vitamins, growth regulators, and gelling agent should not be added to the medium in photoautotrophic micropropagation. Instead of gelling agents, porous substances like vermiculite should be employed as supporting materials in photoautotrophic micropropagation.

Ideally, photoautotrophic micropropagation should be distingished from sugar-free micropropagation. However, while we will define photoautotrophy as the plant nutritional status where only endogenous carbohydrate is used as the energy source, for all practical purposes photoautotrophic micropropagation often refers to micropropagation with no sugar added to the medium. Sugars and other carbohydrates may be significant components of agar and other gelling agents, but perhaps it is reasonable not to consider it as an exogenous carbohydrate source in the practical definition of photoautotrophic micropropagation.

There are often misconceptions or misusage of the terminology for the nutritional statuses of in-vitro cultures. Any chlorophyllous cells/tissues/organs/plantlets (chlorophyllous cultures) growing under conventional conditions using sugar-containing medium grow photomixotrophically. Photomixotrophy is the nutritional type where living organisms use not only endogenous but exogenous carbohydrates as an energy source. Thus regardless of the degree of dependence on sugar in the medium, chlorophyllous cultures growing on media containing sugar should be considered as photomixotrophic. Heterotrophy is the nutritional type where exogenous (medium) carbohydrate is the sole source of energy. Any cultures without chlorophyll are heterotrophic and grow heterotrophically.

For successful photoautotrophic micropropagation, understanding in-vitro environment and basics of environmental control is critical. For plants growing photoautotrophically, promotion of photosynthesis is the primary way to enhance the growth rate of the plantlets. To promote in-vitro photosynthetic rates, it is necessary to know the status of environmental conditions (for example, air temperature and CO_2 concentration) in the vessels and how to maintain them in optimum ranges for maximizing net photosynthetic rates of the plantlets. Lack of understanding of the in-vitro environment or of the interaction of plantlets and the in-vitro and ex-vitro environments, makes it more difficult to improve the micropropagation system by applying the photoautotrophic micropropagation method.

Successful photoautotrophic micropropagation also requires knowledge of when (or at which stage) cultures should transit from photomixotrophic into photoautotrophic status. Developmental and operational stages for conventional micropropagation are generally classified into four or five stages. For photoautotrophic micropropagation, the number of classified stages may be less than in conventional photomixotrophic micropropagation, since multiplication and rooting stages are often combined as one stage in photoautotrophic micropropagation by reproducing photosynthetically active, leafy nodal cuttings to be used as explants. Therefore, theoretically, only the introduction/initiation stage of the culture (Stage I) must be under heterotrophic/photomixotrophic conditions where virus (or pathogen) free cultures are established by culturing meristematic tissue. Once chlorophyllous organs able to conduct photosynthesis are developed, the cultures are ready to move on to photoautotrophic micropropagation conditions. The acclimatization stage is often eliminated when plantlets are grown under optimal photoautotrophic conditions. Thus, a photoautotrophic system could exclusively consist of two stages, initiation (Stage I) and multiplication/rooting (Stage II), while the conventional, photomixotrophic micropropagation requires four stages, initiation (Stage I) , multiplication (Stage II), rooting/preparation (Stage III), and acclimatization (Stage IV) (Fig. 1).

2. Characteristics of in-vitro environmental conditions in the conventional photomixotrophic micropropagation.

In-vitro aerial conditions are affected by physical properties of the vessels, environmental conditions outside the vessel (inside the culture room) and plantlets (photosynthesis, transpiration, etc.) and generally characterized as 1) low CO_2 concentration during photoperiod, 2) high CO_2 concentration during dark period, 3) low water vapor pressure deficit (high relative humidity), 5) low air current speed and 4) low photosynthetic photon flux (PPF). As shown in Fujiwara et al. (1987) [2] and other reports, the typical diurnal change in CO_2 concentration in a conventional culture vessel containing chlorophyllous plantlets is characterized with a linear increase during dark period followed by a sharp decrease within a few hours after onset of photoperiod to reach as low as CO_2 compensation point. This decrease in CO_2 concentration clearly show that the chlorophyll plantlets in vitro retain high photosynthetic ability but the low CO_2 concentration, caused by the limited ventilation of the vessel, forces the plantlets grow photomixotrophically. Enhancing ventilation is therefore the first important key to success in photoautotrophic micropropagation.

3. Important techniques in photoautotrophic micropropagation

(1) Ventilation of the vessels

Natural ventilation is a type of gas exchange of conventional culture vessels. Unless the vessel is specially designed to achieve a complete seal against air exchange, almost all vessels conventionally used in micropropagation have some air exchange by natural

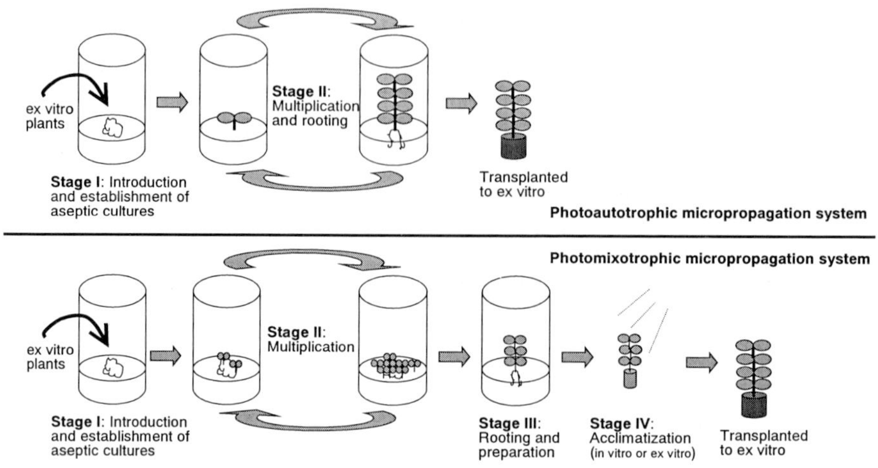

Figure 1. Photoautotrophic and photomixotrophic (conventional) micropropagation systems (Kozai and Kubota, 2001).

Figure 2. Tomato plantlets cultured (left) photoautotrophically under high PPF, vessel ventilation rate and CO_2 concentration without explant (nodal cuttings) leaf removal, and photomixotrophically under low PPF, vessel ventilation rate and CO_2 concentration with (right, the conventional method) and without (center) explant leaf removal.

ventilation through the gaps of contact surfaces of vessels, lids, and sealing tapes. The driving force of the natural ventilation is the difference in air pressure inside and outside the vessel, caused by the difference in air temperature (density) inside and outside the vessel and/or the velocity and current pattern of the air surrounding the vessels. Therefore, the shape of the vessel, orientation of the lids and vents, and air current environment around the vessels will affect the number of air exchanges of naturally ventilated vessels. Air current speed around a vessel was experimentally confirmed to enhance air exchange of the vessel [3]. When the air inside and outside a vessel is stagnant, the difference in air pressure between the inside and outside of the vessel is zero. Under this condition, the driving force of gas diffusion is a partial gas pressure difference.

Enhanced ventilation of a vessel is desired, when attempting to improve the aerial microenvironment inside the vessel. One simple way to enhance the natural ventilation of a vessel is to use lids having gas permeable filters, or to replace the vessels with vessels having improved ventilation properties. Fig. 2 shows tomato plantlets growing photomixotrophically and photoautotrophically in the vessels with enhanced natural ventilation [4]. The CO_2 concentration inside the vessel can be increased further by increasing the ventilation rate of the vessel and/or increasing the CO_2 concentration inside the culture room (CO_2 enrichment).

Optimization of the aerial environment for maximizing photoautotrophic growth of plantlets requires that the ventilation rate should be adjustable according to the magnitude of net photosynthetic rate of the plantlets inside the vessel. A forced ventilation system is therefore more suitable than a natural ventilation system for adjusting ventilation rates by manipulating air flow rates through the vessel, and therefore has been used as an optimized controlling system for in-vitro plantlets [5-9].

(2) CO_2 enrichment

Carbon dioxide enrichment is effective for increasing CO_2 concentration inside the vessel with enhanced natural ventilation rate. With natural ventilation, CO_2 concentration inside the vessel during photoperiod cannot be increased to a level higher than that outside the vessel. CO_2 concentration in the culture room can be measured and controlled using an infrared CO_2 analyzer/controller (IRGA) with a solenoid valve, liquid CO_2 container and connecting tubes. Such a setup is widely used for CO_2 enrichment in the greenhouse and it can be introduced into a tissue culture room. The diurnal changes in CO_2 concentration can be recorded by an analogue or digital recorder connected to the IRGA. For efficient usage of enriched CO_2 gas to the culture room, it is important to keep the culture room as airtight as possible. For more information, see Kozai (1991) [1] and Jeong et al. (1993) [10].

(3) High photosynthetic photon flux (PPF)

Photosynthetic photon flux is around 50 μmol m^{-2} s^{-1} under conventional culture conditions. Increasing PPF to a range between 100 to 300 μmol m^{-2} s^{-1} generally increases the net photosynthetic and growth rates of the photoautotrophically cultured plantlets, with sufficient CO_2 supply in the vessel.

The PPF levels can be selected to provide the maximum net photosynthetic rate. Another way is to select the level of PPF that results in the greatest efficiency of converting PAR energy to plantlet dry mass (chemical energy).

(4) Porous supporting materials

Use of porous supporting materials improve the root zone environment and thus enhance rooting. This is especially beneficial for the woody plant species that are difficult to root under conventional photomixotrophic conditions. Rooted plantlets generally have high survival percentage upon transferring to the greenhouse. We should notice that without sugar in the medium and with porous supporting materials and liquid medium, the system is more like a small hydroponic system, referred to as "microponics" [11]. Improvement of root zone environmental control in photoautotrophic micropropagation may well benefit from the expertise and techniques available from hydroponics.

Replacing conventional agar gel with porous materials significantly affects the root zone environment and therefore the anatomical characteristics of roots. Root system of sweetpotato developed in Florialite® (vermiculite and cellulose fiber mixture) with liquid medium has numerous lateral roots, compared with the other supporting materials. The well-developed root system helps nutrient and water uptake and promotes overall growth of the plantlets [12]. Plantlets grown in such porous supporting materials survive well in ex-vitro environmental conditions. High survival percentage of *Eucalyptus* plantlets ex vitro was highly correlated with the enhanced root development due to improved root zone environment [13]. Similar results, in terms of correlation of improved root system, enhanced growth and high survival percentage were observed with other crops including acacia (*Acacia mangium*) [14], coffee (*Coffea arabusta*) [15], mangosteen (*Garcinia mangostana*; Eramayanti et al., 1999), and sweetpotato (*Ipomoea batatas* [12]. Woody plants are generally difficult to root, and therefore, a supporting material such as vermiculite or Florialite® would be beneficial to those crops.

(5) Leafy explants

Criteria for selecting usable explants in photoautotrophic micropropagation must be different from photomixotrophic micropropagation. In photoautotrophic micropropagation, leafy explants are used, while explant leaf removal is a common practice in conventional photomixotrophic micropropagation. Photosynthetic ability of explants is critical, especially regarding the initial growth of plantlets. The enhanced net photosynthetic rates of explants could stimulate new root and axillary shoot development from the explants. Leaf area is, therefore, an important quality variable of explants. Potato explants with larger leaves give greater initial net photosynthetic rates per plantlet, and therefore provide greater growth rates [16].

ADVANTAGES AND DISADVANTAGES OF PHOTOAUTOTROPHIC MICROPROPAGATION

Photoautotrophic micropropagation has many advantages with respect to improvement of plantlet physiology (biological aspect) and operation/management in the production process (engineering aspect).

<u>Promotion of growth and photosynthesis</u>

Net photosynthetic rate, and thus the growth rate of in-vitro plantlets, is often enhanced when the plantlets are cultured photoautotrophically under a properly controlled environment, compared with those cultured conventionally. The positive effects of controlling environment are observed on growth of both photoautotrophic and photomixotrophic plantlets. The growth promotion is caused by the environmental conditions controlled to favor photosynthesis. Presence or absence of sugar in the medium affects the net photosynthetic rate of the plantlets. It has been observed that plantlets had higher net photosynthetic rates under photoautotrophic rather than under photomixotrophic conditions, perhaps due to enhanced RuBisCO activities as explained by Desjardins et al. (1995) [17].

High survival percentage / smooth transition to ex-vitro environment

Photoautotrophically cultured plantlets are grown under environmental conditions that enhance net photosynthetic and transpiration rates. High photosynthetic rate, normal anatomical structure and functional stomata [18] contribute to enhanced survival percentages upon transfer to the ex-vitro environment.

Elimination of morphological and physiological disorders

A good example of the morphological and physiological disorders that we often observe in conventional micropropagation but not in photoautotrophic micropropagation is hyperhydricity (vitrification). Hyperhydricity is reportedly caused by physical and chemical factors including high relative humidity and ethylene concentration inside the vessel, conditions, which do not exist in a photoautotrophic system.

Little loss of plantlets due to contamination

An advantage of the absence of sugar in the medium is the reduced chance of microbial contamination. Because of lower plantlet loss, production costs will be reduced by the reduction of cultures that the production manager may need to schedule for production as compensation for contamination loss. The conceptual separation of microbial contamination and pathogen infection is possible in photoautotrophic micropropagation since, to some extent, microbes can be accepted as long as they do not overgrow the cultures or are not pathogen. The advantage of lower plantlet loss due to contamination emphasizes the advantages in engineering (operational and management related) aspects described below.

Advantages of engineering aspects include:

Flexibility in the design of the vessel (larger vessels)

In photoautotrophic micropropagation, selection of vessel size and material is flexible, whereas it is limited due to risk of contamination in conventional micropropagation. Based on the concept of pathogen-free plant propagation, it is not necessary to grow plantlets under aseptic conditions as long as pathogens are excluded. Thus, the surface sterilization method for vessels might be simpler (or less expensive) than for the conventional micropropagation.

Automation

Since vessel size is not a limitation in the photoautotrophic micropropagation system, automation for handling plantlets (cutting and transferring into new vessels) can be introduced. Furthermore, automatic plug transplanters are now commercially available, and as such, technologies developed for plug seedling production can be introduced with simple modifications to fit the micropropagation-based plug transplant production system. However, successful use of automation in the in-vitro and ex-vitro stages of micropropagation demands uniformity of cultures and plantlets. Photoautotrophic micropropagation with environmental control will contribute to producing the uniform plantlets that can be handled by robots.

Simplification of the micropropagation system

Due to the reduced probability of contamination in photoautotrophic systems, micropropagation systems can be simplified, especially those employing larger vessels. Additionally, the multiplication stage can be combined with the rooting stage by using leafy nodal cuttings as explants (Fig. 1). Furthermore, the acclimatization stage necessary in conventional micropropagation can be often eliminated or shortened.

The following items are often considered as disadvantages of photoautotrophic micropropagation:

Relative complexity of techniques and knowledge required for controlling in-vitro environment

Several industries have successfully introduced photoautotrophic micropropagation (i.e., Long, 1997 [19]) but application is still limited. One of the reasons for this is that photoautotrophic micropropagation requires the use of specific technology and expertise to control the in-vitro environment. Without fully understanding the relationship between the physiology of the in-vitro plantlets, in-vitro environment, ex-vitro environment, and the physical or structural characteristics of the culture vessels, it will be almost impossible to optimize the photoautotrophic micropropagation system, maximizing the growth of the plantlets while maintaining high quality, yet minimizing inputs of energy and resources. The complexity of controlling the photoautotrophic micropropagation system may contribute to the failure of the establishment of successful photoautotrophic micropropagation.

Expense for lighting, CO_2 enrichment, and cooling

Photoautotrophic micropropagation often requires increased levels of PPF, and concentration of CO_2 available to in-vitro plantlets. The former can be achieved by increasing the number of lamps installed per unit shelf area, although improvement of reflection to obtain greater efficiency and other alterations of lighting methods can significantly increase the PPF received by plantlets inside the vessel. By improving the efficiency of providing CO_2 enrichment, lighting and cooling, increases in costs for photoautotrophic micropropagation can be minimized.

Limitation of application to multiplication systems using multiple buds/shoots

Multiple bud/shoot formation is sometimes expected in conventional Stage II (multiplication stage) micropropagation (such as herbaceous perennials) and it is induced primarily by adding growth regulators. High multiple bud/shoot induction has not yet been achieved in photoautotrophic micropropagation (without sugar but with growth regulators as in Erturk and Walker (2000) [20]) as compared to that in photomixotrophic micropropagation and this is often pointed out as a limitation when applying photoautotrophic micropropagation to the multiplication stage using conventional multiple shoots as explants (propagules). Comparison of the quality of explants when obtained by photoautotrophic micropropagation and conventional photomixotrophic micropropagation is necessary, since the conventional photomixotrophic micropropagation often produces many buds/shoots with low quality (too small in size, hyperhydrated, etc.), while photoautotrophic micropropagation produces fewer but better quality buds/shoots. Further understanding of physiological mechanisms and environmental limitations of enhancing bud development is necessary for wider application of photoautotrophic micropropagation.

APPLICATIONS OF PHOTOAUTOTROPHIC MICROPROPAGATION

Vegetative propagation can be most effectively done under controlled environment. Therefore, photoautotrophic micropropagation will replace the conventional vegetative propagation method, provided that the production cost per propagule is comparable with that in the conventional vegetative propagation in greenhouse. Propagation efficiency and costs for maintaining stock plants are variable, being affected by the weather conditions in greenhouse vegetative propagation. Thus, vegetative propagation under a controlled and pathogen free environment (i.e., photoautotrophic micropropagation) can be feasible. The production costs can be reduced in photoautotrophic micropropagation, but energy and mass balance in photoautotrophic micropropagation systems need to be analyzed for further improvement of the production efficiencies.

Potential applications of techniques and methods developed for photoautotrophic micropropagation include using photoautotrophic micropropagation as a research tool of testing phenotypes (such as specific resistances) of genetically engineered plantlets under aseptic, controlled environment conditions as indicated by Kirdmanee and Mosaleeyanon (2000) [21]. Considering the high level of regulation against testing genetically modified plants under natural environmental conditions, the photoautotrophic micropropagation method could provide the advantage of inducing phenotype expression under conditions similar to those of a greenhouse, but in a highly contained situation.

Another potential application will be somatic embryogenesis, which is a key technology for mass production of elite clones and has been introduced commercially, producing transplants for planting in clonal forestry. One of the challenges preventing wider application of somatic embryos is low percent germination of somatic embryos and conversion to plantlets. Photoautotrophic micropropagation may contribute favorably in this area since somatic embryos have chlorophyll in their cotyledons and/or newly emerged true leaves, in which active photosynthesis can be expected (Afreen, this book). Long (1997) [19] also suggested the possibility of using photoautotrophic methods for improving germination of somatic embryos.

CONCLUSION

Photoautotrophic micropropagation is an advanced plant production techniques that emerged as an integration of biology and engineering for practical application. Such integration would be necessary for the future development of transplant production systems and the outcomes of research and development in photoautotrophic micropropagation will contribute to improvement and problem-solving in future agriculture, forestry and horticultural production systems.

REFERENCES

1. T. Kozai, 'Autotrophic micropropagation', In: Biotechnology in Agriculture and Forestry 17: High-Tech and Micropropagation I, Y.P.S. Bajaj (ed.), **1991**, Springer-Verlag, N.Y., pp. 313-343.
2. K. Fujiwara, T. Kozai T & I. Watanabe, 'Measurements of carbon dioxide gas concentration in closed vessels containing tissue cultured plantlets and estimates of net photosynthetic rates of the plantlets', *J. Agr. Meteorol*, **1987**, 43, 21-30. (in Japanese)
3. Y. Ibaraki, Y. Iida & K. Kurata, 'Effects of air currents on gas exchange of culture vessels', *Acta Hort.*, **1992**, 319, 221-224.
4. C. Kubota, N. Kakizaki, T. Kozai, K. Kasahara & J. Nemoto, 'Growth and net photosynthetic rate of tomato plantlets during photoautotrophic and photomixotrophic micropropagation', *HortScience*, **2001**, 36, 49-52.
5. K. Fujiwara, T. Kozai & I. Watanabe, 'Development of a photoautotrophic tissue culture system for shoot and/or plantlets at rooting and acclimatization stages', *Acta Hort.*, **1988**, 230, 153-158.
6. C. Kubota & T. Kozai, 'Growth and net photosynthetic rate of *Solanum tuberosum* in vitro under forced ventilation', *HortScience*, **1992**, 27, 1312-1314.
7. S.M.A. Zobayed, C. Kubota & T. Kozai, 'Development of a forced ventilation micropropagation system for large-scale photoautotrophic culture and its utilization in sweetpotato', *In Vitro Cell. Dev. Biol.-Plant*, **1999**, 35, 350-355.
8. H. Erturk & P.N. Walker, 'Effects of light, carbon dioxide, and hormone levels on

transformation to photoautotrophy of sugarcane shoots in micropropagation', *Transactions of the ASAE*, **2000**, 43, 147-151.
9. J. Heo, S.B. Wilson & T. Kozai, 'A forced ventilation micropropagation system for photoautotrophic production of sweetpotato plug plantlets in a scale-up culture vessel: I. growth and uniformity', *HortTechnology*, **2001**, 11, 90-94.
10. B.R. Jeong, K. Fujiwara & T. Kozai, 'Carbon dioxide enrichment in autotrophic micropropagation: Methods and advantages', *HortTechnology*, **1993**, 3, 332-334.
11. E.J. Hahn & Y.B. Lee, 'A new method on mass-production of micropropagated chrysanthemum plants using microponic system in plant factory', *Acta Hort.*, **1996**, 440, 527-532.
12. F. Afreen-Zobayed, S.A.M. Zobayed, C. Kubota & T. Kozai, 'Supporting material affects the growth and development of in vitro sweet potato plantlets cultured photoautotrophically', *In Vitro Cell. Dev. Biol. Plant*, **1999**, 35, 470-474.
13. C. Kirdmanee, Y. Kitaya & T. Kozai, 'Effects of CO_2 enrichment and supporting material in vitro on photoautotrophic growth of *Eucalyptus* plantlets in vitro and ex vitro', *In Vitro Cell. Dev. Biol. Plant*, **1995**, 31, 144-149.
14. T.M. Ermayanti, M. Imelda, T. Tajuddin, C. Kubota & T. Kozai, 'Growth promotion by controlling the in vitro environment in the micropropagation of tropical plant species', *Proc. of Intl. Workshop on Conservation and Sustainable Use of Tropical Bioresources. Nov. 9-10, 1998. Tokyo, Japan*, **1999**, pp. 10-25.
15. Q.T. Nguyen, T. Kozai, G. Niu & U.V. Nguyen, 'Photosynthetic characteristics of coffee (*Coffea arabusta*) plantlets in vitro in response to different CO_2 concentrations and light intensities', *Plant Cell, Tissue Organ Cult.*, **1999**, 55, 133-139.
16. Y. Miyashita, Y. Kitaya, C. Kubota & T. Kozai, 'Photoautotrophic growth of potato plantlets as affected by explant leaf area, fresh weight and stem length', *Scientia Horticulturae*, **1996**, 65, 199-202.
17. Y. Desjardins, C. Hdider & J. De Rick, 'Carbon nutrition in vitro – regulation and manipulation of carbon assimilation in micropropagated systems', In: Automation and Environmental Control in Plant Tissue Culture, J. Aitken-Christie, T. Kozai, and M.L. Smith (eds.), **1995**, Kluwer Academic Publishers, The Netherlands, pp. 441-471.
18. S.M.A. Zobayed, F. Afreen-Zobayed, C. Kubota & T. Kozai, 'Stomatal characteristics and leaf anatomy of potato plantlets cultured in vitro under photoautotrophic and photomixotrophic conditions', *In Vitro Cell. Dev. Plant*, **1999**, 35, 183-188.
19. R.D. Long, 'Photoautotrophic micropropagation – a strategy for contamination control?' In: Pathogen and Microbial Contamination Management in Micropropagation., A.C. Cassells (ed.), **1997**, Kluwer Academic Publishers, The Netherlands, pp. 267-278.
20. H. Erturk & P.N. Walker, 'Effects of rooting period, clump size, and growth medium on sugarcane plantlets in micropropagation during and after transformation to photoautotrophy', *Transactions of ASAE*, **2000**, 43, 499-504.
21. C. Kirdmanee & K. Mosaleeyanon, 'Environmental engineering for transplant production', In: Transplant Production in the 21st Century, C. Kubota and C. Chun (eds.), **2000**, Kluwer Academic Publishers, The Netherlands, pp. 78-82.
22. T. Kozai & C. Kubota, 'Photoautotrophic Culture', In: Plant Propagation by Tissue Culture, A.C. Cassells (ed.), **2001**, (in press).

Molecular Breeding of Woody Plants
N. Morohoshi and A. Komamine, editors.

PHOTOAUTOTROPHIC MICROPROPAGATION OF TROPICAL AND SUBTROPICAL WOODY PLANTS

Quynh T. Nguyen[1,2*] & Toyoki Kozai[2]

[1] *Department of Plant Cell Technology, Institute of Tropical Biology, NCST-VN, 1 Mac Dinh Chi Street, Dist. 1, Hochiminh City, Vietnam.*

[2] *Faculty of Horticulture, Chiba University, Matsudo 648, Chiba 271-8510, Japan.*

ABSTRACT

Photoautotrophic micropropagation, recently applied to several plant species, has shown that the growth, development and quality of these plants are significantly affected by *in vitro* environmental factors. The photoautotrophic micropropagation approaches for tropical and subtropical woody plants are discussed, including fruit species (mangosteen), industrial plant species (coffee), and forest trees (eucalyptus, acacia, paulownia, neem, gmelina). The photosynthesis and growth of *in vitro* plants in air-diffusive, small culture vessels were remarkably increased when cultured photoautotrophically. Improving environmental factors, such as photosynthetic photon flux (PPF), number of air exchanges of the culture vessel, CO_2 concentration inside the culture vessel and air porosity of supporting material, improved significantly the *in vitro* plant quality. Under photoautotrophic condition, rooting *in vitro* observed anatomically showed a similarity to that of vegetative propagation as described in greenhouse. The application of large culture vessels in photoautotrophic micropropagation is characterized by a direct flush of a particular air mixture into the vessel using an air pump. CO_2 and other gaseous concentrations can be controlled by an airflow rate meter and an infrared CO_2 analyzer. Growth of *in vitro* plants of paulownia and coffee was significantly greater in the forced ventilation system compared with those in the natural ventilation system.

KEYWORDS

Forced ventilation, hyperhydricity, microbial contamination, natural ventilation, primary roots.

INTRODUCTION

Demands for woody transplants have been rising considerably for aforestation, reforestation, and environmental conservation, for pulp, paper and furniture industries and for energy production. Micropropagation used for transplant production is an advanced technology of vegetative propagation to produce a large number of genetically identical and pathogen-free transplants of several valuable plant species within a certain time period [1]. However, its commercial use is still limited due to several problems such as relatively low growth rate of *in vitro* plants, significant loss of these plants by microbial contamination, hyperhydricity, excessive exudation of explants (especially woody perennials, etc.), poor rooting, low survival percentage of *in vitro* plants during the *ex vitro* stage, and high labor costs due to intensive manual labor works [2]. Most of the above reasons are on account of the conventional micropropagation method, directly or indirectly. In the conventional micropropagation, i.e. *in vitro* plants growing heterotrophically or photomixotrophically, sucrose exists in the culture medium as a main carbon source. Thus, photosynthetic organs of plantlets *in vitro* are said to be underdeveloped resulting in low net photosynthetic rate of these plants [3].

Intensive research has shown that most chlorophyllous *in vitro* plants can grow photoautotrophically, and the growth of these plants can be significantly promoted by a proper control of environmental factors [4]. Environmental control in photoautotrophic micropropagation is mainly characterized by the reduction or elimination of sucrose, vitamins and plant growth regulators in the culture medium, the increase in CO_2 concentration in the culture vessel (by enhancement of vessel ventilation or CO_2 concentration of the culture room), the increase in photoperiod and photosynthetic photon flux (PPF), the decrease in relative humidity in the culture vessel and the use of air-porous supporting materials instead of gelling agents. By using photoautotrophic micropropagation method, the loss of *in vitro* plants due to microbial contamination or hyperhydricity in the *in vitro* stage is significantly reduced. Thus, *in vitro* plants will grow rapidly and vigorously during acclimatization stage and, consequently, high quality transplants can be obtained at lower production costs [5].

In this article we demonstrate experimental results of photoautotrophic micropropagation and emphasize the advantages of this approach over the conventional, photomixotrophic approach for tropical and subtropical woody plants species such as acacia (*Acacia mangium*), coffee (*Coffea arabusta*), eucalyptus (*Eucalyptus camaldulensis*), gmelina (*Gmelina arborea* Roxb.), mangosteen (*Garcinia mangostana*), neem (*Azadirachta indica*) and paulownia (*Paulownia fortunei*). We also discuss the advantages of air-porous supporting materials and large culture vessels with forced ventilation over gelling agents and small culture vessels with natural ventilation.

NATURAL VENTILATION SYSTEMS USING SMALL CULTURE VESSELS AND ITS APPLICATION

In photoautotrophic micropropagation using small (or conventional) culture vessels like Magenta boxes (GA-7 type) with an air volume of 370 ml, gas permeable filter discs are attached to the vessel lid or vessel sides for enhancing natural ventilation. Therefore, the CO_2 concentration inside the sterile vessels will increase with the increase of CO_2 concentration in the surrounding atmosphere. As the ventilation of the culture vessel is promoted, the CO_2 concentration inside the culture vessel during the photoperiod is kept higher to increase the net photosynthetic rate and the relative humidity is kept lower to increase the transpiration rate than those in the vessel without gas permeable filter discs.

Paulownia

Single nodal cuttings of *in vitro Paulownia fortunei* plants with two opposing leaves for each cutting were cultured on half strength MS [6] medium without plant growth regulator. On day 30, paulownia plants grew more vigorously under photoautotrophic condition than under photomixotrophic condition, especially when supporting materials with high air porosity, such as vermiculite and Florialite (a mixture of cellulose fibers and vermiculite, Nisshinbo Industries Inc., Japan) were used. There was no significant difference in multiplication ratio when plants grew under photomixotrophic or photoautotrophic condition (data not shown). Paulownia plants cultured phototautotrophically on half strength MS agar medium without plant growth regulator in 800 mL Phytacon column-type vessels (Sigma Co., USA) demonstrated a significant increase in the dry weight and shoot length under long photoperiod and large number of air exchanges [7]. Under CO_2 enriched and high PPF condition, the photoautotrophic growth of paulownia plants on the vermiculite was significantly promoted on day 28 (Fig. 1). High CO_2 concentration and high PPF enhanced the development of lateral roots and normal vascular systems. Paulownia plants cultured under CO_2 enriched and high PPF condition in the *in vitro* stage continued increasing their growth significantly during 15 days in the *ex vitro* stage (data not shown).

Acacia [8]

Apical shoots of *Acacia mangium* grown in MS medium containing 1 mg l^{-1} IBA, 7 g l^{-1} agar and 30 g l^{-1} sucrose were used as explants. Effects of the presence (30 g l^{-1}) or absence of sugar and growth regulator (IBA, 1 mg l^{-1}) in the medium, number of air exchanges of the culture vessel (0.6 or 6.7 h^{-1}), CO_2 concentration in the culture room (400 or 1500 μmol mol^{-1}) and types of supporting materials (agar or Florialite) on the growth *in vitro* and survival percentage *ex vitro* of acacia plants under photosynthetic photon flux of 150 μmol m^{-2} s^{-1} were examined. The combination of high CO_2 concentration (CO_2 enrichment) and high number of air exchanges of the culture vessel increased the fresh and dry mass of the plants, while presence/absence of sugar and growth regulator did not have any significant effects (Table 1) [8]. The plants cultured in a conventional way (with sugar and IBA containing agar medium in the culture vessel and with low number of air exchanges under low CO_2 concentration in the culture room) gave the lowest fresh and dry mass, and no root formation, while rooting of the plants was enhanced in other treatments (Table 1) [8].

Figure 1. Photoautotrophic growth of *in vitro* paulownia plants as affected by high/low CO_2 concentration and PPF on day 28.

Table 1. Fresh (FW) and dry weight (DW), and percent of rooting of *in vitro* acacia plants cultured for 28 days. Means ± SD are shown [8].

Treatments					FW (mg/plant)	DW (mg/plant)	% rooting
Sucrose conc. ($g\ l^{-1}$)	Growth regulators	CO_2 conc. ($\mu mol\ mol^{-1}$)	No. of air exchanges (h^{-1})	Supporting material			
30	Yes[z]	1500	6.7	Florialite	222 ± 89	40 ± 21	94
30	No	1500	6.7	Florialite	155 ± 21	32 ± 1	81
0	Yes	1500	6.7	Florialite	410 ± 145	62 ± 24	100
0	No	1500	6.7	Florialite	298 ± 4	41 ± 14	82
30	Yes	400	0.7	Florialite	152 ± 15	22 ± 2	38
30	No	400	0.7	Florialite	116[y]	17[y]	75
0	Yes	400	0.7	Florialite	110 ± 33	18 ± 4	82
0	No	400	0.7	Florialite	139 ± 38	20 ± 4	46
30 (control)	No	400	0.7	Agar	100[y]	14[y]	0
ANALYSIS OF VARIANCE[x]							
Sucrose concentration					NS	NS	NS
Growth regulator					NS	NS	NS
CO_2 enrichment and No. of air exchanges (h^{-1})					**	**	NS

[z] Medium contained 1 mg l^{-1} IBA. [y] Only one replication for treatment with 30 g l^{-1} sucrose, without growth regulator and CO_2 of 400 mmol mol^{-1}. [x]ANOVA (Analysis of variance) was applied for 9 treatments (except for the control treatment). NS, nonsignificant; **, significant at $p \leq 0.01$.

Eucalyptus [9]

Eucalyptus camaldulensis shoots (2.2 cm long) were cultured *in vitro* photoautotrophically in Magenta culture vessels (V = 370 ml) for 6 weeks under CO_2-nonenriched (400 $\mu mol\ mol^{-1}$ in the culture room) or CO_2-enriched (1200 $\mu mol\ mol^{-1}$ in the culture room) conditions. Each vessel contained different types of supporting materials (agar, Gelrite, plastic net or vermiculite). Gas permeable filter discs (Milli-Seal, Millipore Japan, Tokyo) were attached on sidewalls of the vessel. CO_2 enrichment significantly increased growth (dry weight and number of primary roots) of plants *in vitro* regardless of the type of supporting materials (Table 2) [9]. The growth *in vitro* was greatest in the vermiculite, followed by the plastic net, Gelrite, and agar (in descending order) under either CO_2-nonenriched or CO_2-enriched condition. The growth of plants *ex vitro* was the highest and percent damaged leaves/roots was the lowest in the vermiculite under the CO_2-enriched condition (data not shown). The extensive root system with many secondary roots was produced *in vitro* in the vermiculite.

Neem

In vitro Azadirachta indica single node cuttings were cultured on agar or Florialite medium with the presence (20 g l^{-1}) or absence of sucrose in a Magenta vessel having low (0.15 h^{-1}) or high (3.5 h^{-1}) number of air exchanges at CO_2 concentration of 450 $\mu mol\ mol^{-1}$ and PPF of 100 $\mu mol\ m^{-2}\ s^{-1}$. Dry weight and leaf fresh weight of *in vitro* neem plants significantly increased under photoautotrophic condition compared with photomixotrophic condition (Fig. 2).

Neem plants grown in the Florialite developed significantly longer roots than those in the agar (data not shown). Under photoautotrophic condition, the growth of *in vitro* neem plants increased with the increase in PPF (data not shown). At a higher PPF the neem shoot elongation was suppressed, whereas the root elongation was promoted significantly (Fig. 3).

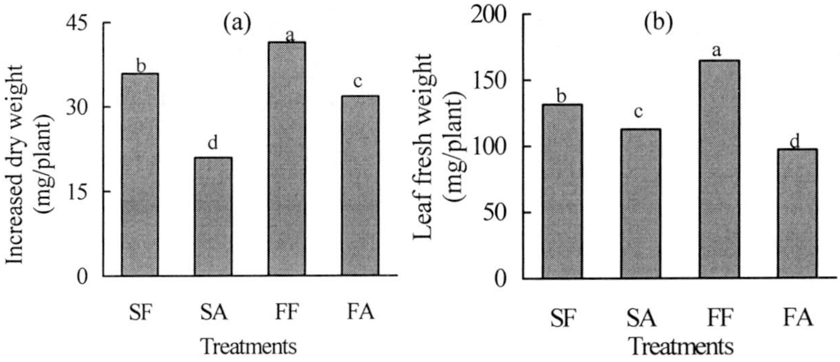

Figure 2. (a) Increased dry weight and (b) leaf fresh weight of *in vitro* neem plants on day 28. For treatment codes, S and F in the left represent for the presence and absence of sucrose, respectively. F and A in the right represent for Florialite and agar medium, respectively. Different letters on each column show a significant difference at the 5% level by LSD test.

Table 2. Effects of CO_2 enrichment and supporting material *in vitro* on dry weight, leaf area, number of primary roots per plant (NoR), and length of primary roots of eucalyptus plants after 6 weeks *in vitro* [9].

CO_2 condition	Supporting material	Dry weight (mg/plant)	Leaf area (cm^2/plant)	NoR	Length of primary roots (mm/plant)
Non-enriched	Agar	45	8	1	27
(400 $\mu mol\ mol^{-1}$)	Gelrite	49	9	1	30
	Plastic net	64	11	4	39
	Vermiculite	82	12	5	42
Enriched	Agar	54	9	2	32
(1200 $\mu mol\ mol^{-1}$)	Gelrite	62	10	2	37
	Plastic net	76	12	5	45
	Vermiculite	103	13	6	49
$LSD_{P=0.05}$		8	3	1	11
ANALYSIS OF VARIANCE[z]					
CO_2 condition (C)		**	NS	*	*
Supporting material (S)		**	*	**	**
C x S		NS	NS	NS	NS

[z] NS, *, ** Nonsignificant and significant at $p \leq 0.05$ and 0.01, respectively.

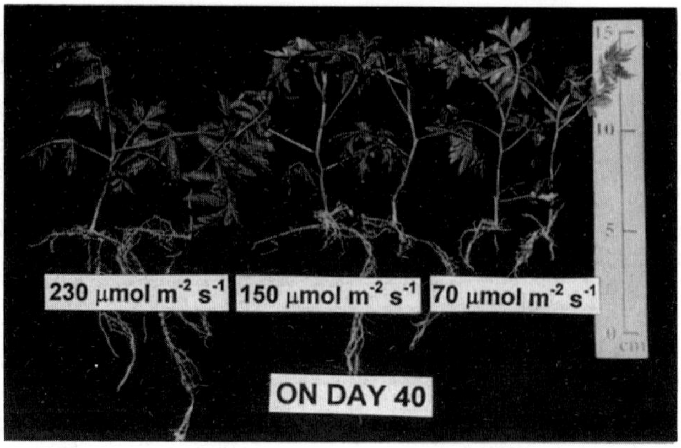

Figure 3. Photoautotrophic growth of *in vitro* neem plants as affected by PPF levels on day 40

Mangosteen [8]

Mangosteen (*Garcinia mangostana* L.) apical shoots excised from *in vitro* germinated seedlings were cultured on the vermiculite medium with presence (30 g l^{-1}) or absence of sucrose and growth regulator (10 mg l^{-1} of 2-ip and 1 mg l^{-1} of IBA) in the culture vessel with high (4.4 h^{-1}) number of air exchanges at a PPF of 110 μmol m^{-2} s^{-1} under high CO_2 concentration (1300 μmol mol^{-1} in the culture room). For the control (conventional) treatment, the number of air exchanges of the culture vessel was 0.1 h^{-1}, and the agar medium containing 30 g l^{-1} of sucrose, 10 mg l^{-1} of 2-ip and 1 mg l^{-1} of IBA was used. This treatment was put under a non-enriched CO_2 condition (Table 3) [8].

Table 3. Dry weight (DW), percent rooting, number of leaves per plant (NoL) of mangosteen plants, cultured for 30 days. Means ± SD are shown [8].

Treatments					DW	%	NoL
Sucrose conc. (g l^{-1})	Growth regulators	CO_2 conc. (μmol mol^{-1})	No. of air exchanges (h^{-1})	Substrate	(g/plant)	rooting	
30	Yes[z]	1300	4.4	Vermiculite	0.30 ± 0.13	40	6 ± 1.04[NS y]
30	No	1300	4.4	Vermiculite	0.23 ± 0.18	20	4 ± 0.9[NS]
0	Yes	1300	4.4	Vermiculite	0.23 ± 0.10	20	6 ± 0.9[*]
0	No	1300	4.4	Vermiculite	0.20 ± 0.11	40	4 ± 0.9[NS]
30 (control)	Yes	400	0.1	Agar	0.21 ± 0.11	0	5 ± 0.9
ANALYSIS OF VARIANCE[x]							
Sucrose concentration					NS	NS	NS
Growth regulator					NS	NS	**

[z] Medium contained 10 mg l^{-1} 2-ip and 1 mg l^{-1} IBA.

[y] NS, *: Nonsignificantly or significantly different from the control treatment at $p \leq 0.05$ according to *t*-test, respectively.

[x] ANOVA was applied for 4 treatments (except for the control treatment). NS, nonsignificant; **, significant at $p \leq 0.01$.

Fresh and dry weights of the plants were not significantly different among treatments on day 30. Addition of the growth regulator in the medium increased the number of leaves. Twenty to forty percent of shoots exhibited root induction in the treatments with high number of air exchanges and vermiculite as supporting material, either with or without sugar/growth regulator. On the other hand, in the control treatment with low number of air exchanges and sugar containing agar medium, no rooting was observed (Table 3) [8]. The CO_2 concentration inside the culture vessel was the lowest, and the net photosynthetic rate per leaf area in the control treatment was about 10 % of those in the treatments with high number of air exchanges (data not shown).

Coffee

Single nodal cuttings of *Coffea arabusta* plants were cultured *in vitro* under PPF of 75 $\mu mol\ m^{-2}\ s^{-1}$ and an ambient CO_2 concentration (400 $\mu mol\ mol^{-1}$). Shoot length and leaf area of coffee plants were significantly greater on the medium with 0 $g\ l^{-1}$ sucrose than with 20 $g\ l^{-1}$ sucrose (data not shown). In addition, no callus formation was observed at the shoot base of plants cultured on the sugar-free medium, whereas all plants grown on sugar containing medium were found having callus [10]. High CO_2 concentration significantly increased the fresh weight, shoot length, leaf area and net photosynthetic rate of *in vitro* coffee plants cultured on the sugar-free medium. However, high PPF did not increase the growth of coffee plants significantly (Table 4). Increased dry weight and shoot length of coffee plants developed on Florialite were also significantly greater than those cultured on agar (data not shown).

Gmelina

The growth of single node *Gmelina arborea* Roxb. cuttings cultured photoautotrophically increased with the increase in number of air exchanges of the culture vessel and on the Florialite medium under ambient CO_2 concentration of 500 $\mu mol\ mol^{-1}$ and PPF of 150 $\mu mol\ m^{-2}\ s^{-1}$. The increased dry weight, shoot length and leaf area of *in vitro* gmelina plants were the greatest at the number of air exchanges of 4.2 h^{-1} compared with those of 2 or 3.5 h^{-1} on day 28 (data not shown). The air-porous supporting material such as Florialite improved root formation resulting in lower shoot/root dry weight ratios of all treatments using Florialite compared with those using agar (Fig. 4). The development of lateral roots and normal vascular systems were enhanced by the use of high ventilation of the culture vessel and Florialite in the medium.

FORCED VENTILATION SYSTEMS USING LARGE CULTURE VESSELS AND ITS APPLICATION

In the forced ventilation method, a particular gas mixture is flushed directly through the culture vessel and, thus, increase CO_2 concentration and ventilation rate inside the vessel. This method helps to control and maintain CO_2 concentration inside the culture vessel at an optimal level relatively precisely by use of an airflow rate meter or mass flow controller and an infrared CO_2 analyzer. In photoautotrophic micropropagtion using a large culture box/chamber, many forced ventilation systems were developed having advantages over natural ventilation systems [5]. The uniform growth, shorter multiplication/rooting cycle and high survival percentage are major beneficial impacts of this system.

Table 4. Effects of high/low CO_2 concentrations and PPFs on the increased dry weight (DW), leaf area (LA), shoot length (SL) and net photosynthetic rate (P_n) of *in vitro* coffee plants cultured photoautotrophically on day 45.

Treatment codes	DW (mg)	LA (cm^2)	SL (mm)	P_n (μmol h^{-1}/plant)
HE	$109.3^{ab\ z}$	64.9^{ab}	17^{b}	14
LE	132.1^{a}	81.3^{a}	31^{a}	15
HN	61.8^{c}	43.1^{bc}	8^{c}	5
LN	41.0^{c}	24.4^{c}	10^{c}	4

[z] a, b, c within a column show a significant difference at the 5% level by LSD test. As symbols for treament codes, H and L in the left represent for PPF at 350 and 150 μmol m^{-2} s^{-1}, respectively. E and N in the right represent for CO_2 concentration at 1450 and 450 μmol mol^{-1}, respectively.

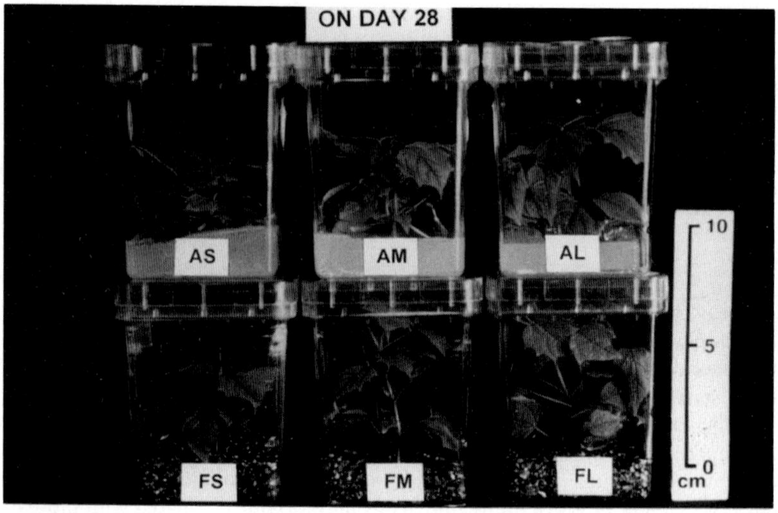

Figure 4. Growth of *G. arborea* plants as affected by number of air exchanges of the vessel and supporting materials on day 28. For treatment codes, A and F in the left represent for agar and Florialite, respectively. S, M and L in the right represent for the number of air exchanges of 2, 3.5 and 4.2 h^{-1}, respectively.

A forced ventilation system for photoautotrophic growth of *Paulownia fortunei* using a large box (V = 16.8 L) was developed (Nguyen *et al.*, unpublished). The growth of *in vitro* paulownia plants was significantly increased in the photoautotrophic, forced ventilation compared with that in the photomixotrophic, natural ventilation (Fig. 4). The net photosynthetic rate increased with the increase in the air flow rate over the culture period of 28 days (data not shown). Zobayed *et al.* [11] recently demonstrated an efficient scaled-up vessel for mass propagation of eucalyptus with an automatic supply of nutrient. Nguyen *et al.* [12] developed a forced ventilation system for enhancing the growth of *in vitro* coffee (*C. arabusta*) plants inside a large box (V = 11.2 L). Coffee plants significantly increased fresh weight, shoot length and leaf area as compared with those in natural ventilation (Table 5).

Table 5. Increased fresh weight (IFW), shoot length (SL) and leaf area (LA) of coffee plants cultured *in vitro* for 40 days as affected by ventilation method and PPF.

Treatments					
Ventilation method	Number of air exchanges (h^{-1})	PPF[z] ($\mu mol\ m^{-2}\ s^{-1}$)	IFW (mg/plant)	SL (mm/plant)	LA (cm^2/plant)
Natural (control)	3.9	150	338	11	14
Forced	2.7	250	260**	16* [y]	15[NS]
Forced	2.7	150	116**	13*	12*
Forced	5.9	250	604**	27**	23**
Forced	5.9	150	569**	36**	24**
ANOVA[x]					
Ventilation (A)			**	**	**
PPF (B)			*	**	NS
A x B			**	**	*

[z] PPF from day 16 to day 40 of the *in vitro* period.

[y] NS, *, **: nonsignificantly or significantly different from the control treatment at $p \leq 0.05$, respectively, according to *t*-test.

[x] ANOVA was applied for 4 treatments with forced ventilation method. NS, *, **: nonsignificant or significant at $p \leq 0.05$ or 0.01, respectively.

Figure 5. Growth of *P. fortunei* plants as affected by ventilation methods on day 28.

CONCLUSION

Photoautotrophic micropropagation has significantly improved the growth, quality and multiplication of *in vitro* woody transplants. Research on a large-scale, photoautotrophic mass propagation system with forced ventilation being conducted successfully in several laboratories will increase the importance role of this method in the micropropagation industry in the near future.

REFERENCES

1. J. Aitken-Christie, T. Kozai & M.A.L. Smith, *Automation and Environmental Control in Plant Tissue Culture*, Kluwer Academic Publishers, Dordrecht, The Netherlands, **1995**, 574 p.
2. T. Kozai, Photoautotrophic micropropagation, *In vitro Cell Dev. Biol.*, **1991**, 27, 47-51.
3. J. Pospisilova, J. Solarova & J. Catsky, Photosynthetic responses to stress during *in vitro* cultivation, *Photosynthetica*, **1992**, 26, 3-18.
4. K. Fujiwara & T. Kozai, Physical microenvironment and its effects. In: *Automation and Environmental Control in Plant Tissue Culture*, J. Aitken-Christie, T. Kozai & M.A.L. Smith (eds.), Kluwer Academic Publishers, Dordrecht, The Netherlands, **1995**, pp. 319-369.
5. T. Kozai, C. Kubota, S. Zobayed, Q.T. Nguyen, F. Afreen-Zobayed & J. Heo, Developing a mass propagation system for woody plants, In: *Proc. of the 12th Toyota Conference: Challenge of Plant and Agricultural Sciences to the Crisis of Biosphere on the Earth in the 21st Century*, K. Watanabe & A. Komamine (eds.), Eurekah.com., Texas, U.S.A., **2000**, pp. 289-302.
6. T. Murashige & F. Skoog, A revised medium for rapid growth and bioassays with tobacco tissue cultures, *Physiol. Plant*, **1962**, 15, 473-479.
7. Q.T. Nguyen, T. Kozai & J. Heo, Enhanced growth of *in vitro* plants in photoautotrophic micropropagation with natural and forced ventilation systems, In: *Proc. of the Intl. Symposium on Transplant Production in Closed System for Solving the Global Issues on Environmental Conservation, Food, Resources and Energy: Transplant Production in the 21st Century*, C. Kubota & C. Chun (eds.), Kluwer Academic Publishers, Dordrecht, The Netherlands, **2000**, pp. 246-251.
8. T. M. Ermayanti, M. Imelda, T. Tajuddin, C. Kubota & T. Kozai, Growth promotion by controlling the *in vitro* environment in the micropropagation of tropical plant species, In: *Proc. of Intl. Workshop on Conservation and Sustainable Use of Tropical Bioresources*, Nov. 9-10, 1998, Tokyo, Japan, **1999**, pp. 10-25.
9. C. Kirdmanee, Y. Kitaya & T. Kozai, Effects of CO_2 enrichment and supporting material *in vitro* on photoautotrophic growth of Eucalyptus plantlets *in vitro* and *ex vitro, In vitro Cell Dev. Biol.-Plant*, **1995**, 31, 144-149.
10. Q.T. Nguyen, T. Kozai, K.L. Nguyen & U.V. Nguyen, Photoautotrophic micropropagation of tropical plants, In: *Proc. of the 9th Intl. Congr. of IAPTC: Plant biotechnology and in vitro biology in the 21st century*, A. Altman, M. Ziv & S. Izhar (eds.), Kluwer Academic Publishers, Dordrecht, The Netherlands, **1999**, pp. 659-662.
11. S.M.A. Zobayed, F. Afreen, C. Kubota & T. Kozai, Mass propagation of Eucalyptus camaldulensis in a scaled-up vessel under *in vitro* photoautotrophic condition, *Ann. of Bot.*, **2000**, 85, 587-592.
12. Q.T. Nguyen, T. Kozai, J. Heo & D.X. Thai, Photoautotrophic growth response of *in vitro* coffee plantlets to ventilation methods and photosynthetic photon fluxes under carbon dioxide enriched condition, *Plant Cell Tiss. Org. Cult.*, **2001**, in submission.

Molecular Breeding of Woody Plants
N. Morohoshi and A. Komamine, editors.

LARGE-SCALE PHOTOAUTOTROPHIC MICROPROPAGATION IN A SCALED-UP VESSEL

S.M.A. Zobayed*[1], F. Afreen[1], C. Kubota[2] and T. Kozai[2]

[1]*Department of Biological Sciences, University of Hull; Hull; HU6 7RX, UK*
[2]*Department of Bioproduction Sciences, Chiba University; Matsudo, Chiba 271-8510, Japan*

ABSTRACT

A scaled-up culture vessel for the large-scale micropropagation has been designed and the method for the large-scale micropropagation of *Eucalyptus* under photoautotrophic condition (sugar-free medium) using this scaled-up culture vessel with forced ventilation (SV-treatment) is described. Growth, multiplication, epicuticular wax content, stomatal function of plantlets and transpiration rate and *ex vitro* survival of the plants grown under SV-treatment were studied and compared with those of the plantlets grown in a Magenta-type vessel with natural ventilation and under photomixotrophic condition (30 g l^{-1} sucrose in the medium). Results revealed that fresh and dry mass in plantlets grown under SV-treatment were significantly higher than those of plantlets from Magenta type vessel. Multiplication was enhanced and normal stomatal closing and opening were observed in the leaves of plantlets grown in scaled-up vessel. Epicuticular leaf-wax content was significantly higher compared with that of the leaves grown in Magenta type vessel. The anatomical study showed well-organized palisade and spongy mesophyll layers for leaves grown in SV-treatment. Importantly after transplanting *ex vitro*, plantlets survived well without any specialized *ex vitro* acclimatization treatment. This was because they were better able to control transpiration and thus lost less water and showed no signs of wilting. In contrast, plantlets cultured in Magenta type vessel had open stomata which were not functional, showed a higher transpiration rate and uncontrolled and rapid water loss immediately after transferring *ex vitro*. Minimization of large-scale plantlet production cost by using scaled-up vessel is also discussed in this article.

KEY WORDS

Ex vitro, forced ventilation, photoautotrophic, stomata, wax content.

INTRODUCTION

Micropropagation is an obvious method for rapid propagation of genetically superior and pathogen-free plantlet propagation. However the commercial use of conventionally micropropagated plantlets are often undermined by their poor physiological quality and growth; substantial percentage of plantlets cannot survive or cannot grow fast and vigorously after transplanting *ex vitro*. For the large-scale micropropagation the use of small culture vessel (volume <0.5 L) has a number of disadvantages, for instance, it can enhance the labour cost significantly. It has been predicted that by using large vessel labour cost can be reduced at least by 50% [1]. In contrast, for the large-scale micropropagation small vessels require more shelf space and thus can enhance the production cost. Moreover, in the small vessel with natural ventilation, planting density is a limiting factor, and thus the increase of planting density can seriously reduce the growth of the plantlets. In large vessel under forced ventilation, it is possible to increase the planting density without any significant reduction in the plant growth. [2]

By attaching gas permeable filter discs (natural ventilation) on the lid or on the wall of the culture vessels, the air exchange can be increased. However, with natural ventilation, the CO_2 and other gaseous concentrations in the culture vessel are interrelated with a number of factors such as the metabolic activity of the plants in vitro, the size and leaf area of the plants, number of air exchanges of the culture vessel etc. Thus the gaseous concentration in the culture vessel with natural ventilation are often unpredictable and uncontrollable; it is also difficult to measure the gaseous concentration in the vessel continuously. [1]

In this article we have discussed the possible use of large vessels for the mass propagation of *in vitro*-grown plantlets. We have described one model design of a large vessel with forced ventilation and nutrient supply system. The growth and physiology of *Eucalyptus* plantlets grown in this vessel are also described.

LARGE VESSELS UNDER PHOTOAUTOTROPHIC MICROPROPAGATION

One of the major advantages of the photoautotrophic micropropagation system is that it makes it possible to use large (scaled-up) culture vessels with minimum risk of microbial contamination. To reduce the production cost of the micropropagated plantlets, it is necessary to develop a scaled-up culture vessel in which hundreds of plantlets can be propagated simultaneously with uniform growth. Recently, there has been much interest to propagate plantlets in scaled-up culture vessels for large-scale micropropagation. Fujiwara *et al.,* [3] developed a large vessel (volume 20 L) with forced CO_2 supply system, which can be used to culture shoots and/or plantlets photoautotrophically during the rooting and acclimatization stages. Roche *et al.* [4] developed a commercial-scale photoautotrophic micropropagation system for potato microplants where 100 nodal explants were cultured under natural ventilation in a stainless steel tray containing a block of polyurethane foam (85 x 300 x 25 mm) and enclosed with a polyethylene sleeve. Kubota and Kozai [5] grew potato plantlets under forced ventilation in a large vessel (2.6 L) containing a multi-cell tray. Recently, Heo and Kozai [6] developed a similar type of system using an even larger culture vessel (volume 13 L); in this system, 20 sweet potato plantlets were cultured photoautotrophically. However, one disadvantage of these large culture vessels is that there is commonly a variation in growth of the cultured plantlets mainly due to the ununiform distributions of CO_2 concentration and other environmental factors in the culture headspace.

Zobayed *et al.* [7] developed a scaled-up culture vessel (volume 3.5 L) in which air distribution pipes for forced ventilation are installed to provide an air current pattern giving uniform distributions of CO_2 concentration and relative humidity as well as those of air current speeds, and thus the uniform plant growth. However, in a larger vessel (*c.* 20 L), the air distribution pipes were unable to supply CO_2 enriched air uniformly. This was because of the significant pressure drop (and thus the flow rate) in the pipe as the distance from the source (air pump) increased. The system was later modified (see Zobayed *et al.* [1] and below) and an uniform distribution of CO_2 and plant growth was achieved.

Control of environmental conditions and nutrient in the culture vessel can significantly improve the growth and quality of the cultured plantlets. To enhance the air exchange in the scaled-up vessel, Roche *et al.* [4] achieved natural ventilation by enclosing the vessel in a low density polyethylene sleeve with pin hole perforation. On the other hand forced ventilation was used by Kubota and Kozai [5] and Heo and Kozai. [6] In our previous study we found that forced ventilation has many advantages over natural ventilation (see Zobayed *et al.* [7,8]). One of the major advantages of the forced ventilation system is that the air exchange rate of the vessel can be controlled as the CO_2 concentration in the headspace throughout the culture period. Relative humidity in the culture headspace is also possible to control by increasing or decreasing the flow rate or by controlling the ambient relative humidity.

Chlorophylous plants use CO_2 and water in the presence of light to synthesize organic compounds. Usually, due to the continuous flushing of unsaturated air with water vapor by the forced ventilation system, the relative humidity in the headspace is reduced compared with that of the conventional system and thus the nutrient medium may dry out rather quickly. For example, Buddendorf-Joosten and Woltering [9] found it necessary to control the external relative humidity in their system involving a throughflow of air over the vessels capped with diffusive membranes. However, high relative humidity may have deleterious effect on cultured plantlets. Thus for long-term culture under forced ventilation, a continuous supply of nutrient solution throughout the culture period is essential. Use of nutrient supply system can make it possible to control the pH, composition and volume of the nutrient solution in the culture vessel. By circulating the nutrient solution, the root zone is possible to immerse temporarily in the nutrient solution, which ensure the availability of oxygen around the root and thus improve the root quality and the growth.

DESIGN OF A SCALED-UP VESSEL

A model system has been developed (see Zobayed *et al.* [2]) for the large-scale micropropagation of plantlets. Each vessel was 610 mm long, 310 mm wide and 105 mm high with a volume of *c.* 20 L. and consisted of two chambers. The lower chamber was narrow (internal height 2 mm) and was used as an air distribution chamber; the upper one was used for culturing the plantlets. Four pipes (length 25 mm; internal diameter 1.5 mm) were fitted vertically to the bottom of the lower (air distribution) chamber. Each of these acted as an air inlet pipe and was directly connected to an air pump (Non Noise S200, Artem Co. Ltd., Japan) by means of flexible tubing via a flow rate meter. One filter disc (50 mm diameter; 0.5 μm pore diameter; Millipore, Japan) was placed in between the air pump and the culture vessel in order to prevent microbes entering into the vessels. Several narrow tubes (0.5 mm inner diameter; 3.5 mm length) were fitted vertically in between the upper and lower chamber, each of which opened to the headspace of the culture vessel and directed the air flow from the air distribution chamber to the culture vessel headspace. The outlet tubes (1.5 mm inner diameter; 10 mm length) were located on the two opposite side-walls of the culture chamber, 10 mm from the top. Filter discs (50 mm diameter; 0.5 μm pore diameter) were attached to each of these outlet tubes on the outside of the culture vessel. The culture vessel contained an autoclavable plug tray (Minoru Sangyo Co. Ltd.

Japan) with 500 cells. A nutrient reservoir (2.5 L volume) was connected to the culture vessel via a flexible tubing (2 mm inner diameter). The bottom of the nutrient reservoir was 60 mm below that of the culture vessel to facilitate periodic drainage of the nutrient solution from the culture vessel.

The culture vessel was surface sterilized by wiping three times with sodium hypochlorite solution (2% w:v) followed by rinsing with sterile water. The whole system was placed in a growth room with enriched CO_2 concentration (1100 - 1200 μmol mol^{-1}). However for commercial use, the inlets pipe of the system could be directly connected to gas cylinder or to gas mixing apparatus.

GROWTH, PHYSIOLOGY AND SURVIVAL OF *EUCALYPTUS* PLANTLETS CULTURED IN A SCALED-UP VESSEL

Nodal cuttings of *Eucalyptus* were cultured in this scaled-up vessel (SV-treatment) under photoautototrophic condition (without any sugar in the medium and with a PPF of 100 μmol mol^{-1} and under CO_2 enrichment) and growth, physiology and survival *ex vitro* of the plantlets after 32 days of culture were studied and compared with those of the plantlets grown in Magenta type vessels under photomixotrophic condition.

Growth

The *in vitro* environment of the scaled-up vessel attached with a forced ventilation system is favorable to plant growth and the plantlets are more normal physiologically and morphologically and acclimatize more easily in the *ex vitro*.

Growth of *Eucalyptus* plantlets was enhanced compared with that of the plantlets grown in a Magenta type vessel. Despite the higher planting density (*c.* 2.2x that of the control), the plantlets grown under SV-treatment developed more favorable shoots and roots than those grown in the control (Table 1). The fresh mass of leaves and roots per plantlet was 281 mg (2.5x that of the control) and 168 mg (2.2x) respectively in SV-treatment. The leaf area was 2.5x that of the control. After 28 days of culture, the number of unfolded leaves was 9.3 per plantlet in the SV-treatment, compared with 8.3 in control.

Table 1. Growth of *Eucalyptus* plantlets grown in a scale-up vessel (SV treatment) under photoautotrophic conditions and in Magenta-type vessel (control) under photomixotrophic conditions.

Treat-ments	Leaf area (cm^2)	No. of leaves	Stem length (cm)	Fresh mass (mg)			Dry mass (mg)		
				Leaves	Stem	Root	Leaves	Stem	Root
SV	20.8±3*	9.3±0.7*	6±0.9*	281±38**	84.1±13*	168±17**	61.1±9.6**	13.7±2.1**	19.9±2.7**
Control	8.3±0.1	8.0±0.2	3.5±0.3	114±7	42.1±3.1	77±4.3	13.4±1.0	5.8±0.3	8.2±0.3

Each value represents mean±SE of 20-100 replicates. Mean with asterisks are significantly different from those of control by 't test' at $P = 0.01$ or 0.05.

The dry mass was significantly greater in plantlets grown under SV-treatment, leaf and root dry mass were 4.6- and 2.4x those of the control, respectively. The stems of SV-plantlets also showed significantly dry mass compared to those of the control (Table 1). Greater dry mass of plantlet is a desirable character leading to a higher quality transplant.

Stomatal characteristics

In the SV-treatment, stomata opened during the light period and closed in the dark (functional stomata). In control (Magenta type vessel with natural ventilation), many stomata remained widely open in both light and dark periods, indicating abnormal functioning of stomata. Large stomatal area was also noticed (Table 2); in the dark, the leaf area occupied by stomata was 25% and 18% for the control and the SV-treatment, respectively. The length of stomata on the leaves of control was 1.7X (66 μm) and the width was nearly double (39 μm) that of stomata in SV-leaves (Table 2). The density of stomata was relatively high in the leaves subjected to the SV-treatment, and was 1.3X that of the control (Table 2).

Leaf anatomy

Plantlets from SV-treatment had leaves that were thicker (723 μm) than the control ones and with well-organized palisade and spongy mesophyll layers (Table 2). These layers were 1.6X and 2.4X thicker, respectively, than those of the control. The epidermal cells were well developed and the average depth was 42 μm in the SV-treatment compared with 28.4 μm for the irregular cells of the control. Large air spaces were obvious in the leaves of control especially in the spongy mesophyll layer. In contrast, the SV-plantlets had cells that were more closely packed with smaller intercellular spaces.

Epicuticular waxes

The development of epicuticular wax is known to be advantageous for plantlets during the acclimatization period [10,11] for protection from desiccation. The reduced amount of epicuticular wax is directly correlated with substantially increased water loss of plantlets after transplanting [12,13]. In the present experiment, the leaves from the SV-treatment exhibited a high amount of epicuticular wax development (3.1 g m^{-2} leaf area; Table 2) compared with control (0.9 g m^{-2}).

Relative water loss and transpiration *ex vitro*

As shown in Fig. 1, the leaves from the SV-treatment exhibited relatively slow and controlled water loss immediately after exposure to the *ex vitro* conditions. Within 60 minutes of transplanting, only 10% water had been lost. The percent water loss of the leaves of the control increased rapidly and nearly 30% of the water had been lost within 60 min of transplanting (Fig. 1). Severe desiccation with irreversible tissue damage of wilted leaves was observed under the microscope. Usually these leaves died within a few days. In case of the SV-treatment, the stomata closed immediately after exposure to the dry atmosphere resulting in controlled water loss.

Table 2. Stomatal characteristics, leaf anatomy and epicuticular wax content of *Eucalyptus* leaves grown photoautotrophically in a scaled-up vessel under forced ventilation and photomixotrophically in a Magenta type vessel (control) for 28 d (After, Zobayed *et al.* [13]).

	SV-treatment	Control
Stomatal characteristics		
Stomatal density (number per mm^2)	338±13.3**	259±9.9
Percent of stomatal area per leaf area	18.2±0.9**	24.7±1.1
Stomatal length (μm)	39.1±6.0**	66.3±7.3
Stomatal width (μm)	21.4±2.0**	41.9±3.9
Leaf anatomy		
Leaf thickness (μm)	723.3±14.7**	421.2±21.1
Palisade layer thickness (μm)	242.1±6.1**	151.3±8.6
Spongy mesophyll layer thickness (μm)	405.3±29.4**	166.7±8.7
Upper epidermal cell thickness (μm)	42.1±2.1**	28.4±3.4
Epicuticular wax content (g m^{-2} leaf area)	3.1±0.6	0.9±0.1

For stomatal measurements, leaves were collected from dark. Each value represents mean±SE of 15 replicates from 5 different leaves. For leaf anatomy and epicuticular wax content each value represents mean±SE of 5 replicates from 5 different leaves. Mean with asterisks are significantly different from those of the control by 't test' at $P = 0.01$.

Transpiration rate was measured immediately after transplanting *ex vitro*. The rate was higher in the control than in the SV-treatment; transpiration rate in the control and the SV-treatment was 425 and 110 g m^{-2} leaf area h^{-1} respectively.

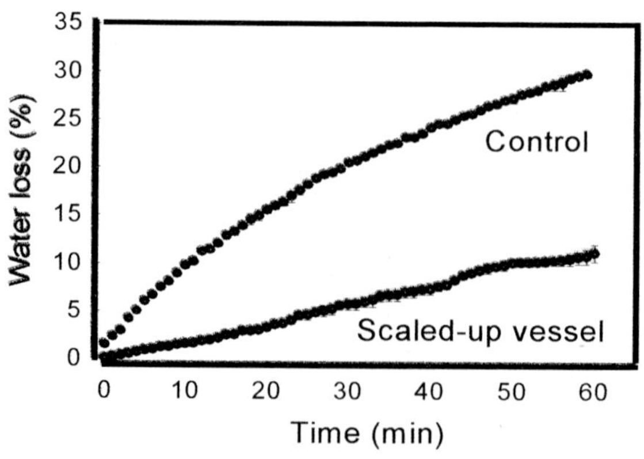

Fig. 1. The percent water loss per leaf area measured immediately after transplanting *ex vitro* for leaves of *Eucalyptus* grown photomixotrophically in a Magenta vessel (control) and photoautotrophically in a scaled-up vessel under forced ventilation. Each bar represents mean±SE of 10 replicates.

Transplantation and percent survival

When transplanted *ex vitro* (without any *ex vitro* acclimatization), 91% of the plantlets from the SV-treatment survived, while only 27% survived from the control. Plants, which survived from the control, had leaves that were permanently wilted. The new leaves usually developed within 7–15 days after transplanting and thus the growth was slower than that of the SV-treatment (data not shown). On the other hand, a few leaves were wilted when plantlets transferred from SV-treatment.

MINIMISING THE PRODUCTION COST IN MICROPROPAGATION

Commercial application of plant micropropagation is still limited by high production costs and damage and loss of plants during 'hardening' when micropropagated plants adapt to normal photoautotrophic conditions. Use of scaled-up vessel in the photoautotrophic micropropagation system can minimise the production cost of the micropropagated plants. This is explained below:

The variable part of the production cost (excluding the fixed cost) of a plantlet (C) can be roughly estimated by the following equation[8]:

$$\text{Cost per transplant, } C = \frac{C_e+C_l+C_m+C_a+C_n}{E*S_c*M_r*S_e}$$

where, E = number of explants cultured per production area; S_c = ratio of survived explants (without contamination) during the *in vitro* period; M_r = multiplication ratio or number of new shoots/nodes produced per explant; S_e = ratio of survived plantlets after transplanting *ex vitro*; C_e = electricity cost (electricity cost for the lamps and other equipments); C_l = labor cost during the *in vitro* and *ex vitro* periods; C_m = maintenance cost during the *in vitro* and *ex vitro* periods; C_a = costs to maintain the ambient environment in the culture room such as CO_2 concentration and relative humidity; C_n = cost of the nutrient solution, supporting materials and other chemicals. Each of these factors is discussed elaborately below.

1) Kozai *et al.* [1] estimated that the use of scaled-up culture vessels is expected to reduce labor costs by nearly 50%, as compared with those in the conventional micropropagation system.

2) By using scaled-up vessel with forced ventilation, the planting density can be increased without reduction of the dry mass per plantlet.[2] Therefore, the production per shelf-area can be increased and thus production cost per transplant can be reduced. Moreover, by using the scaled-up vessel, the multiplication and the acclimatization periods can be shortened, thus transplant production per year can be increased by increasing the number of production cycles, which should reduce the production cost further.

3) As the photoautotrophic micropropagation system does not require any sugar, it can significantly reduce the loss of plantlets *in vitro* due to microbial contamination and increase the ratio of survived explants (without contamination) during the *in vitro* period (S_c), which reduces the production cost.

4) By using scaled-up vessel under forced ventilation the ratio of survived plantlets after transplanting in *ex vitro* (S_e) can be increased significantly.[8] This will also contribute

in the reduction of production cost. Moreover, the growth of plants is faster *ex vitro* than those of the conventionally grown plants [14] which can directly reduce the maintenance cost (C_m).

5) To culture in a scaled-up vessel under photoautotrophic micropropagation system, higher photosynthetic photon flux is required compared with that of the conventional photomixotrophic system, which can slightly increase the electricity cost for lighting and air conditioning. However, in a culture room, 60-70% of photosynthetically active radiation (PAR) emitted from lamps is absorbed by the walls, shelves, vessels, and only 30-40% of it is absorbed by the plantlets *in vitro,* on the average. The ratio of PAR received by the plantlets and emitted by the lamps can easily be increased by using reflector sheets in the inner wall of the culture shelves and by using transparent material for sealing the vessels. [15]

6) In the photoautotrophic micropropagation system the PPF must be higher (*ca.* 200 $\mu mol\ m^{-2}\ s^{-1}$) than in heterotrophic or photomixotrophic micropropagation system (*ca.* 100 $\mu mol\ m^{-2}\ s^{-1}$). Thus, one may argue that the high cost for electricity is due to high PPF in the photoautotrophic micropropagation system. Electricity consumption for lighting per marketable plantlet, E, is roughly calculated as follows:

$$E = k*[(P * L)/N]$$

where, k is proportional coefficient, P is PPF, L is time integral of lighting at the multiplication period and N is the number of marketable plantlets at shipping stage. Thus, shortening L and increasing N can compensate the increase in PPF. N can be increased by increasing the multiplication rate and by increasing the percent survival at the multiplication and acclimatization stages.

7) Setting the photoperiod at night and dark period during the daytime can minimize electricity cost for cooling. [15] In some countries the price of electricity during the nighttime is about 50 % of that at the daytime.

8) In the conventional micropropagation procedure, when the plants are cultured in a culture vessel, usually the sealing or capping of the vessel can cut off at least 50% of the light. [16] This can be overcome by using transparent capping system (as used in the scaled-up vessel) thus the electricity cost per plantlet for lighting can be reduced.

9) Fibrous and decomposable supporting material (such as Florialite or vermiculite) can be used instead of agar [17,18], which can improve the root growth and it is also possible to transfer the plantlets to the soil directly without any manual wash of the root zone. Thus possible root damage can be avoided which will save time and reduce labor cost. Moreover, automation is feasible with this type of supporting material.

10) One may debate that CO_2 enrichment is expensive for commercialization. However, it has been estimated that to maintain the CO_2 concentration at 1000 $\mu mol\ mol^{-1}$ in a 250 m^3 culture room with 200,000 plantlets for 16 h photoperiod, 130 kg of liquid CO_2 is required per 30 days [15]. Currently, the price of 130 kg of liquid CO_2 is about 200 US$. On the other hand, to propagate 200,000 plantlets in the conventional method, the amount of sucrose consumption is 60 kg (30 g l^{-1} sucrose; 10 ml medium required per plantlet), which costs much more than CO_2. Therefore, the enrichment of CO_2 is considered as an inexpensive procedure.

Combining all of the features discussed above, it can be said that the production of the plantlets under photoautotrophic condition especially in a scaled-up vessel can significantly reduce the production cost per plantlet.

CONCLUSIONS

The results indicated that for the large-scale micropropagation of *Eucalyptus* use of small vessel (Magenta type) can have inhibitory effects on growth and development, induce non-functional stomata and reduce the *ex vitro* survival. Use of scaled-up vessel can enhance growth and improve physiology of plantlets and survival percentage. Moreover, the production cost can be reduced significantly. In conclusion scaled-up vessel described above is suitable for the large-scale photoautotrophic micropropagation of woody and other horticultural plants.

BIBLIOGRAPHY

1. T. Kozai & C. Kubota, S. M. A. Zobayed, Q. T. Nguyen, F. Afreen, J. Heo, Developing a mass-propagation system of woody plants. In: *Challenge of Plant and Agriculture Sciences to the Crisis of Biosphere on the Earth in the 21st Century* (eds. K. Watanabe & A. Komamine) Chapter 28, Texas, USA, Landes Bioscience, **1999**, pp: 293-306.
2. S. M. A. Zobayed, F. Afreen, C. Kubota & T. Kozai, Mass propagation of *Eucalyptus camaldulensis in* a scaled-up vessel under *in vitro* photoautotrophic condition. *Ann. Bot.,* **2000a,** 85: 587-592.
3. K. Fujiwara, T. Kozai & I. Watanabe, Development of a photoautotrophic tissue culture system for shoots and/or plantlets at rooting and acclimatization stages. *Acta Hort.,* **1988**, 230, 153-158.
4. T.D. Roche, R.D. Long, A.J. Sayegh & M.J. Hennerty, Commercial scale photo-autotrophic micropropagation: applications in Irish agriculture, horticulture and forestry. *Acta Hort.,* **1996**, 440, 515-520.
5. C. Kubota, & T. Kozai, Growth and Net Photosynthetic rate of *Solanum tuberosum in vitro* under forced ventilation. *Hort Sci.,* **1992,** 27: 1312-1314.
6. J. Heo & T. Kozai, Forced ventilation micropropagation system for enhancing photosynthesis, growth and development of sweet potato plantlets. *Env. Cont. Biol.-Plant,* **1999,** 37:83-92.
7. S. M. A. Zobayed, C. Kubota & T. Kozai, Development of a forced ventilation micropropagation system for large-scale photoautotrophic culture and its utilization in sweet potato. *In Vitro Cell. Dev. Biol. – Plant.,* **1999a**, 35: 350-355.
8. S. M. A. Zobayed, F. Afreen & T. Kozai, Quality Biomass Production via Photoautotrophic Micropropagation. *Acta Hort.* **2000b.** 530: 377-386.
9. J. M. C. Buddendorf-Joosten & E. J. Woltering, Controlling the gaseous composition *in vitro*-description of a flow system and effects of the different gaseous components on *in vitro* growth of potato plantlets. *Sci. Hort.,* **1996,** 65: 11 - 23.
10. E. Sutter, & R. W. Langhans, Formation of epicuticular wax and its effect on water loss in cabbage plants regenerated from shoot-tip culture. Can. J. Bot.. **1982,** 60: 2896 - 2902.
11. B. W. W. Grout, Wax development on leaf surfaces on *Brassica oleracea* var. Currawong regenerated from meristem culture. *Plant Sci. Lett.* **1975**, 5 : 401 - 405.

12. J.E. Preece & E. G. Sutter, Acclimatization of micropropagated plants to the greenhouse and field. In: *Micropropagation – technology and application* (ed. P. C. Debergh & R. H. Zimmerman) Kluwer Academic Publishers, The Netherlands, **1991**, pp: 71-93.
13. S. M. A. Zobayed, F. Afreen & T. Kozai,. Physiology of *Eucalyptus* plantlets cultured photoautotrophically under forced ventilation. *In Vitro Cell. Dev. Biol. – Plant,* **2001**,(in press).
14. S. M. A. Zobayed, F. Afreen, C. Kubota & T. Kozai, Water control ability of *Ipomoea batatas* grown photoautotrophically under forced ventilation and photomixotrophically under natural ventilation. *Ann. Bot.,* **2000c** 85: 603-610.
15. T. Kozai, Autotrophic Micropropagation. In: *Biotechnology in Agriculture and Forestry,* Vol. 17, High-Tech and Micropropagation I (ed. By Y.P.S. Bajaj). Springer-Verlag Berlin Heidelberg. **1991**, pp: 313-343.
16. K. Fujiwara, T. Kozai & I. Watanabe, Fundamental studies on environments in plant tissue culture vessels. (3) Measurement of carbon dioxide gas concentration in closed vessels containing tissue cultured plantlets and estimates of net photosynthetic rates of plantlets. *J Agr Met.*, **1987,** 43 : 21-30.
17. F. Afreen, S. M. A. Zobayed, C. Kubota & T. Kozai, Supporting material affects the growth and development of *in vitro* sweet potato plantlets cultured photoautotrophically. *In Vitro Cel. and Dev. Biol.- Plant,* **1999,** 35: 470-474.
18. C. Kirdmanee, Y. Kitaya & T. Kozai, Effects of CO_2 enrichment and supporting material *in vitro* on photoautotrophic growth of *Eucalyptus* plantlets *in vitro* and *ex vitro*. *In Vitro Cell. Dev. Biol. – Plant.,* 1995, 31: 144-149.

Molecular Breeding of Woody Plants
N. Morohoshi and A. Komamine, editors.

MASS-PROPAGATION OF COFFEE FROM PHOTOAUTOTROPHIC SOMATIC EMBRYOS

Afreen F.* [1], Zobayed S.M.A.[1], and Kozai T. [2]

[1]*Department of Biological Sciences, University of Hull; Hull HU6 7RX, UK*
[2]*Department of Bioproduction Science, Chiba University, Chiba 271-8510, Japan*

ABSTRACT

The somatic embryos of *Coffea arabusta* were cultured and different stage embryos (torpedo, precotyledonary, cotyledonary and germinated) were selected. When these embryos were grown photoautotrophically (in sugar-free medium with CO_2 enrichment in the culture headspace and high photosynthetic photon flux) for 60 days the dry mass of each of the cotyledonary and germinated embryos increased by up to 10% and 50%, respectively. In contrast, the torpedo and precotyledonary stage embryos lost at least 20-25% of their initial dry mass. Thus we have suggested that cotyledonary stage embryos can be considered as the earliest stage, which can be cultured photoautotrophically in order to develop plantlets.

In this article we have also discussed about a specially designed large culture vessel with forced ventilation system for the mass-propagation of coffee plantlets. The cotyledonary stage embryos were selected and grown in this specially designed large vessel having temporarily root zone immersing system with the view of large-scale embryo-to-plantlet development under photoautotrophic condition. The growth and survival of the plantlets after 45 days of culture were studied and compared with those grown photoautotrophically in a RITA vessel (Vitr Pic, CIRAD) and in a Magenta-type vessel. Results revealed that fresh mass and dry mass of leaves and roots of plantlets developed in the specially designed large vessel were significantly enhanced compared with those of plantlets developed in a RITA vessel. The poorest growth of the plantlets was observed when grown in a Magenta-type vessel. The survival of the plantlets after transplanting *ex vitro* followed a similar pattern and was highest in the large vessel followed by that in RITA vessel and in Magenta-type vessel.

KEYWORDS

Coffea arabusta, CO_2, embryo-to-plantlet, forced ventilation, large vessel.

INTRODUCTION

Coffee is playing a major role in the economies of more than 50 countries mainly Latin American countries such as Colombia, Brazil, other Central American, African and Asian countries. Among the many species of coffee *Coffea arabica* and *C.*

canephora are the only two species which are consumed worldwide. *C. arabica*, a high-altitude grown tropical plant (500-2000m) is popular for producing high quality coffee with a low content of caffeine and a fine aroma, although this plant is susceptible to main fungal diseases such as leaf rust, coffee berry disease (CBD). On the contrary, *C. canephora*, a low-altitude grown tropical plant (0-800m) produces coffee with higher caffeine content but poorer in quality than *C. arabica*. However, this species is a worthwhile source of genes for resistance to leaf rust, CBD and nematodes [1]. These features led the plant breeders to combine the complementary characteristics of both *C. arabica* and *C. canephora* in a new species named *Coffea arabusta* [2].

Conventionally the new hybrid *C. arabusta* is multiplied asexually [3]. Vegetative propagation therefore plays an important role in production of coffee transplants. Cuttings are the most commonly used vegetative propagation procedure. However, due to the dimorphism of the vegetative axes the number of cuttings available from one plant is extremely limited and time consuming specially for large-scale production. To obtain genetically uniform large-scale transplant production Dublin [4] used microcuttings; however, the *in vitro* growth of microcuttings is slow [5].

Somatic embryogenesis, which offers the promise of an inexpensive, large-scale propagation method, is therefore considered as a unique alternative technique to overcome these problems. In general, the production cost of plant propagation via somatic embryogenesis is potentially cheaper than the microcuttings specially when bioreactors and automation procedures are used in the production process. The most promising application of somatic embryo is in the field of genetic engineering where by means of somatic embryos specific and directed changes are introduced into elite individuals. As embryos originate from single cells therefore plants derived from somatic embryos tend to be genetically alike [6].

Due to the prospects of somatic embryogenesis for application to the subject of plant micropropagation and its success and rapid progress we therefore decided to investigate the possibility of growing different stages (torpedo, precotyledonary, coyledonary and germinated) *C. arabusta* somatic embryos under photoautotrophic condition (PA).

In our previous study, we have described that cotyledonary and germinated somatic embryos have photosynthetic ability specially when the 14 days pretreatment of high photosynthetic photon flux (PPF: 100 μmol m^{-2} s^{-1}) was applied [7]. In this article we have reviewed the scaling-up of the culture system for the large-scale propagation of somatic embryos. We have described the development of different stage coffee somatic embryos under PA condition. For the large-scale embryo-to-plantlet development we have developed a new culture system and the growth *in vitro* and survival of the plantlets *ex vitro* were compared with those grown in RITA vessel and in Magenta-type vessel.

Scaling-up of culture for mass propagation via somatic embryogenesis

In the past somatic embryogenesis has been used for the improvement of several coffee species and genotypes [8,9,10], but the multiplication rate was very low for its large-scale application [3]. Other reports suggest that after 14 weeks of culturing coffee somatic embryos on gelled medium 60-65 somatic embryos were obtained per explant [3]. In order to increase the production rate and to further reduce the cost of plantlet production attention was drawn on the use of liquid medium. For the mass production of coffee the practice of Erlenmeyer flasks [11] and bioreactors [12] using the liquid medium became quite popular. For the growth of embryogenic cultures, applications of bioreactors have been described by other scientists [13,14]. Spin filter bioreactors for carrot, airlift bioreactors for alfalfa and sweet potato and vibro-stirring bioreactors for poinsettia have been successfully used for the growth of embryogenic cultures [15]. Over 60,000 carrot somatic embryos can be produced from one litre of suspension culture in a bioreactor [16].

In the year 1993 a new temporary immersion technique known as RITA was developed at CIRAD [17] in order to overcome the physiological and technical limitations encountered in bioreactors. This new technique has been used for the improvement of banana [17], coffee [18], Hevea [19] etc. somatic embryogenesis. The main advantage of this system is that it enables the entire regeneration phase i.e. development of plantlets from embryogenic cell suspension cultures to take place in the same vessel.

Importance of photoautotrophic micropropagation of somatic embryos

The development of photoautotrophic micropropagation system for the chlorophyllous plants has proved to be a particularly useful research tool and expanding area for the commercial production of plants *in vitro*. The key advantages of this system involves enhancement of growth of *in vitro* plantlets [20], reduction of microbial contamination which minimizes the risk of losses of plantlets [21], reduction of production cost [22], improvement of physiological characteristics of plantlets [23] which enables better acclimatization *ex vitro* [21]. Moreover, we anticipate that the ability of somatic embryos to grow photoautotrophically can simplify the embryo-to-plantlet development procedures, make possible the automation and thus contribute to reduce the production cost and improve the plant quality.

METHODS OF PHOTOAUTOTROPHIC MICROPROPAGATION OF COFFEE SOMATIC EMBRYOS

Plant materials and treatments

Establishment of the culture

Nodal cuttings of coffee plantlets (*Coffea arabusta*) were cultured in Magenta-type vessel containing hormone free MS [24] medium supplemented with 20 g l^{-1} sucrose. After 4 weeks of culture, regenerated leaves were collected and cut into pieces (10 x 5 mm) and were placed in a Magenta-type vessel containing modified MS [24] medium (for details see Afreen *et al.*[7]). Agar (8 g l^{-1}; Kanto Chemical Co., Japan) was used as gelling agent and 30 g l^{-1} sucrose was added in the medium. Cultures were placed at 23°C air temperature and under cool-white fluorescent lamps (National Co., Japan) providing 16 h d^{-1} photoperiod in a growth chamber. The ambient CO_2 concentration was 400 µmol mol^{-1} and the photosynthetic photon flux (PPF) was 30 µmol m^{-2} s^{-1} measured on the empty culture shelf. Somatic embryos were developed within 9-12 weeks of culture.

Experiment I: Growth of different stage somatic embryos under photoautotrophic conditions

Somatic embryos were established as described above and were placed under a high PPF of 100 µmol m^{-2} s^{-1} (pretreatment) for 14 days (for details see Afreen *et al.* [7]). Different stage somatic embryos (torpedo, precotyledonary, cotyledonary and germinated) were selected and were transferred to plastic petridishes (vol. 30 ml) containing MS [24] medium (10 ml per petridishes). For the photoautotrophic treatment, sucrose and vitamins were deducted from the formulation and the petridishes were placed in a CO_2 enriched growth chamber (1000 µmol mol^{-1}) under 100 µmol m^{-2} s^{-1}

PPF. To introduce natural ventilation, one Millipore filter membrane (Nihon Millipore Ltd.Yonezawa, Japan) with 0.45 μm pore diameter was attached covering the hole (10 mm diameter) of the lid of each of these petridishes (number of air exchanges = 5.0 h^{-1}, estimated according to Kozai *et al.* [25]). For the photomixotrophic treatment (PM), 20 g l^{-1} sucrose was added in the medium and the petridishes were placed in a growth chamber with a CO_2 concentration of 400 μmol mol^{-1} and were placed under 50 μmol m^{-2} s^{-1} PPF. Each petridish contained five similar size and stage embryos and five petridishes were prepared per treatment. For both the treatments ambient temperature and relative humidity were 25°C and 80-85%, respectively, during the photoperiod. Experiment was conducted for 60 days and growth (fresh and dry mass) of the somatic embryos was recorded.

***Experiment II:* Embryo-to-plantlet development using different kinds of vessels**

Somatic embryos were established as described earlier and were placed under a high PPF of 100 μmol m^{-2} s^{-1} for 14 days (for details see Afreen *et al.* [7]). In the first attempts to achieve embryo-to-plantlet development, 16 weeks old cotyledonary stage embryos (including 2 weeks pretreatment) were selected and were used as experimental material. The embryos were cultured under photoautotrophic conditions (in sugar-free medium with CO_2 enrichment in the culture headspace and high PPF) in three different types of vessels as followed:

i) Magenta-type vessel; ii) Rita (temporary immersion system); iii) a specially designed large culture vessel (for details see Afreen *et al.*[7]).

Magenta-type and Rita vessels were modified by attaching gas-permeable Millipore filter membrane (two membranes in Magenta and three in Rita vessels) with 0.45 μm pore diameter and covering the hole (10 mm diameter) of the lid of each of these vessels. The number of air exchanges were 3.0 h^{-1} in Magenta vessel and 2.6 h^{-1} in Rita vessel throughout the experiment. A mixture of vermiculite and paper pulp (as described by Afreen *et al.* [26]) was used as supporting medium in the Magenta-type and specially designed large vessels and for 'Rita vessel' liquid nutrient solution was used and the immersion frequency was 5 min/12h by connecting an air pump through an electric timer.

In the large vessel, forced ventilation was introduced by using an air pump connected to the vessel; the flow rates were initially 50 ml min^{-1} (number of air exchanges was 1.6 h^{-1}) and were gradually increased every 2 or 3 days to maintain the CO_2 concentration in the culture headspace in a range *ca.* 1000 μmol mol^{-1}, the maximum flow rate was 200 ml min^{-1} on day 45 (number of air exchanges was 5.8 h^{-1}). The root zone was temporarily immersed (immersion frequency was total 15 min. before drainage started/12 h) by the nutrient medium.

For all the treatments hormone free MS medium was used as a basal medium; sucrose, vitamins and amino acids were subtracted from the formulation to ensure the photoautotrophic conditions. Vessels were placed in a growth chamber with a PPF of 100 μmol m^{-2} s^{-1} during the 16 h photoperiod; ambient CO_2 concentration was 1000 μmol mol^{-1}, relative humidity 80-85% and the air temperature was 23°C.

Experiment was conducted for 45 days and the harvesting included fresh and dry mass of the plantlets developed from the embryos and percentage of rooting. Plantlets were transplanted in the greenhouse (average temperature 29±2°C; RH 60-70%) and the survival percentage was recorded after 7 days.

RESULTS AND DISCUSSION

Experiment I: Growth of different stage somatic embryo under photoautotrophic conditions

Considering the growth (fresh and dry mass) of different stage somatic embryos the overall greater performance occurred in the embryos grown in photomixotrophic (PM) treatment than that of photautotrophic (PA). As illustrated in Fig.1 (a & b), in PM treatment the total fresh and dry mass of torpedo (X1.9 and X2.3 respectively) and precotyledonary embryos were remarkably higher (X 1.7 and X 1.8 respectively) than those of the embryos grown in PA treatment. In PM treatment the total fresh and dry mass of cotyledonary (X 1.6) and germinated embryos (X 1.3 and X 1.1 respectively) were also substantially higher those of the embryos grown in PA treatment.

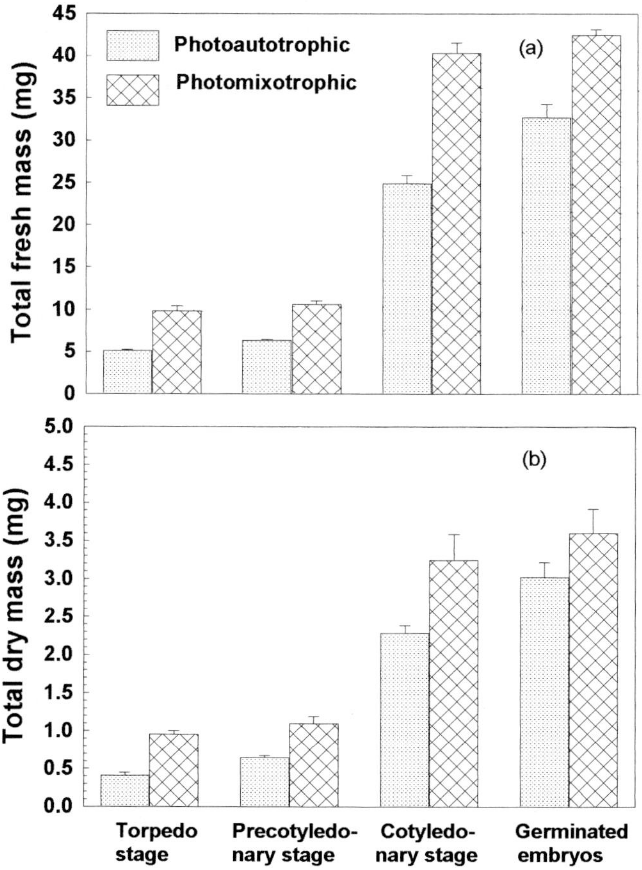

Fig. 1. Growth (a) fresh and (b) dry mass of different stage somatic embryos grown under photoautotrophic and photomixotrophic conditions for 60 days. Each bar represents mean ± SE of 7 replicates. Embryos were developed and grown under a PPF of 30 μmol m^{-2} s^{-1} for 14 weeks followed by 2 weeks pretreatment of high PPF (100 μmol m^{-2} s^{-1}).

In the present study it was also observed that after 60 days of culture, in the PA treatment, torpedo and precotyledonary stage embryos lost at least 25 and 20% respectively, of their initial dry mass (Fig. 2). The most likely reason for this loss could be that the low photosynthetic ability of the plant materials coupled with the continuous respiration, probably led the plantlets to depend completely on their own reserve food material. On the contrary, in the PM treatment, the dry mass of each of the torpedo and precotyledonary stage embryos increased by upto 190 and 200% respectively, of their initial dry mass. Under PA condition in the later stages i.e. cotyledonary and germinated embryos, the dry mass of each of the embryos was increased by up to 10% and 50%, respectively, of their initial dry mass (Fig. 2). The probable reason for the dry mass increment in the later stage under PA treatment could be that at the later stage embryos are capable of photosynthesizing more than those at the early stages [7]. These results confirm the above findings and are in agreement with our previous observation where the cotyledonary and germinated embryos showed stomatal development and scavenging of CO_2 (Afreen *et al.* [7]).

In both the treatments root formation was not observed in the torpedo or precotyledonary stage embryos. When grown in PM treatment 33% and 58% of the embryos produced root in the cotyledonary and germinated stages, respectively. Whereas under PA-treatment only 15% cotyledonary and 25% germinated embryos produced root.

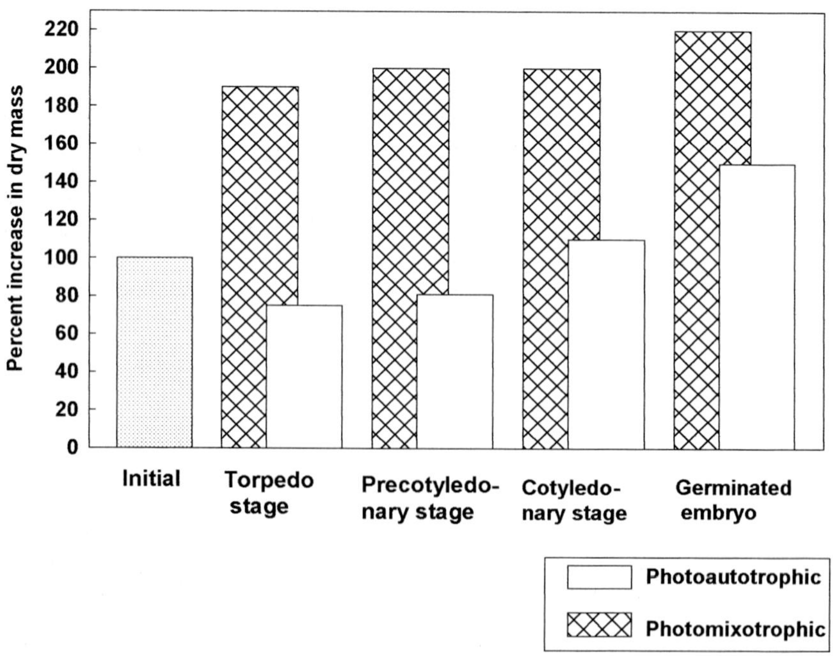

Fig. 2. Percent increase of dry mass (compared with that of the initial dry mass) of coffee plantlets developed photoautotrophically from different stage somatic embryos after 60 days of culture.

***Experiment II:* Embryo-to-plantlet development using different kinds of vessels**

Taking into account of all the parameters of growth and development within the three different types of culture vessels, it is evident that embryos grown in the large vessel produced more vigorous shoot and normal roots than those grown in Magenta-type vessel. The growth of the plantlets from RITA vessel being intermediate between the large and Magenta-type vessel (Fig. 3).

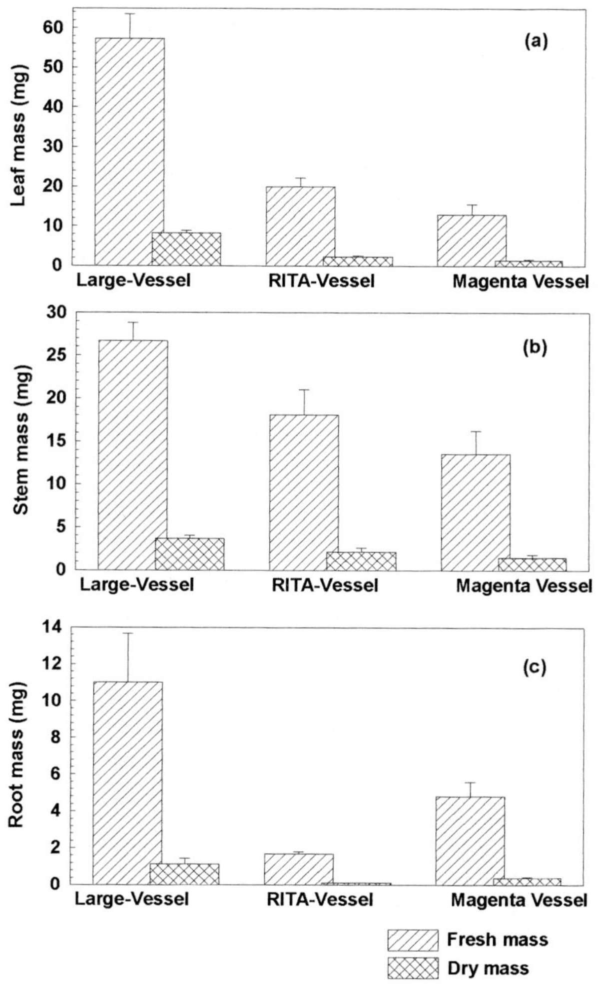

Fig. 3. The fresh and dry mass of a) leaf, b) stem and c) root of plants developed from cotyledonary stage embryos after 45 days of culture in specially designed large vessel (with temporarily root zone immersing system and forced ventilation), Rita vessel (with temporary immersion system and natural ventilation) and Magenta-type vessel (with natural ventilation). Each bar represents mean ± SE of 7 replicates.

The leaf fresh and dry mass of the plantlets from large vessel were significantly higher, being X2.8 and X3.4 respectively than those of the plantlets grown in Rita and X4.4 and X5.6 respectively than those of Magenta-type vessel (Fig. 3a). The dry mass of stem of the plantlets from large vessel was X1.1 and X3 respectively those of plantlets grown in Rita and Magenta-type vessel (Fig. 3b).

The most noticeable difference was observed in case of root growth. Plantlets cultured in large vessel exhibited the greatest fresh and dry mass being X6.5 and X12 those of plantlets grown in Rita vessel (Fig. 3c). It should be mentioned here that most of the plantlets in Rita did not produce any root, even the roots, which developed in few plantlets, remained very small. Plantlets cultured in Magenta-type vessel exhibited an intermediate root growth pattern (fresh and dry mass, 4.8 and 0.37 mg respectively) between those of large vessel and Rita. This result agrees with our previous finding Afreen *et al.* [26] where we have shown that the supporting medium Florialite was particularly beneficial to root growth.

The survival percentage *ex vitro* of the plants, which was recorded on Day 7 followed a similar pattern and was highest (98%) in the plantlets grown in large vessel followed by 61% and 30% survival of the plants from Rita and Magenta-type vessels respectively.

The above results pointed out that for the embryo-to-plantlet development under photoautotrophic condition, Magenta-type vessels exhibited the lowest growth regime; moreover, for large-scale production the use of small vessel has many disadvantages [20]. RITA vessels are normally used for the development of plantlets from embryogenic cell suspension cultures using sugar containing medium. This vessel is claimed to be suitable for embryo-to-plantlet development without handling the plant material [3]; however at the end of each phase the culture medium needs to be changed. In case of Rita, density of plant material is also a limiting factor. However, our research revealed that, for the embryo-to-plantlet development under photoautotrophic condition, the use of RITA vessel substantially reduced the growth compared to large-vessel.

As discussed by Gupta *et al.* [27], in the conventional system, for embryo-to-plantlet development following steps are necessary (i) embryo selection and transfer on the germination medium ii) germinated and rooted plantlet selection and transfer to soil and iii) acclimatization. Generally, in each of the above phases, cotyledonary, late cotyledonary or germinated somatic embryos are selected individually, in most cases by hand under the stereo microscope. The invention of machine vision [15] and image analysis [28] systems offer great potential for classifying and sorting embryos but the use is still limited. Selected embryos are transferred onto gelled medium for germination. After 6-10 weeks of germination, plantlets with epicotyl are selected by hand, transferred to soil and incubated in a greenhouse with frequent misting for acclimatization and growth. In somatic embryogenesis procedures aimed at mass production, these methods are still very time consuming and involve high labour costing. However, in our system, cotyledonary embryo selection is necessary which is done by hand, but once the embryos are transferred to the large vessel, germination, root development and acclimatization take place in the same vessel and without handling the plant material or changing the culture medium. Another advantage of our new system is that by increasing the number of cells the density limitations can be overcome.

CONCLUSIONS

The results provide us with two vital information, the first one is that cotyledonary stage embryo is the earliest stage which can be cultured photoautotrophically in order to develop coffee plantlets. The second one, from mass propagation viewpoint is that the use of large vessel consisting of the root zone immersing system. The major advantage of the system involves the production of quality plantlets, which ensures the rapid growth and development of plants after transplanting *ex vitro*. The results also give indications that the newly developed system should substantially reduce the cost of producing coffee plants by somatic embryogenesis.

REFERENCES

1. J. Bettencourt & C. J. J. R. Rodrigues, In: *Coffee*: *Agronomy*, R. J. Clarke and R. Macrae (eds.), Elsevier Applied Science Publishers, Amsterdam, **1988**, pp. 199-234.
2. J. Capot, 'L'amelioration du cafeier en Cote d'Ivoire – Les hybrides 'Arabusta' *Café Cacao Thé*, **1972**, 16, 3-17.
3. M. Berthouly & H. Etienne, In : *Somatic Embryogenesis in Woody Plants*, S. M. Jain, P. K. Gupta and R. J. Newton (eds.), Kluwer Academic Publishers, Dordrecht, The Netherlands, **1999**, 5, pp : 259-287.
4. P. Dublin, 'Multiplication vegetative *in vitro* de l Arabusta', *Café Cacao Thé*, **1980**, Vol. WWIV 4, 281-290.
5. P. Dublin, F. Enjalric, L. Lardet M. P. Carron, N. Trolinder & C. Pannetier, In: *Micropropagation - Technology and application*, P. C. Debergh and R. H. Zimmerman (eds.), Kluwer Academic Publishers, Dordrecht, The Netherlands, **1991**, pp 337-361.
6. G. L. Sita, In: *Somatic embryogenesis in Woody Plants*, S. M. Jain, P. K. Gupta and R. J. Newton (eds.), Kluwer Academic Publishers, Dordrecht, The Netherlands, **1999**, 5, pp. 95-111.
7. F. Afreen, S. M. A. Zobayed & T. Kozai, 'Photoautotrophic growth of *Coffea arabusta* somatic embryo', **2001,** in press.
8. E. S. Pierson, A. A. M. Van Lammeren, J. H. N. Schel & G. Staritsky, ' *In vitro* development of embroids from punched leaf discs of *Coffea canephora*', *Protoplasma*, **1983**, 115, 208-216.
9. T. Yasuda,.Y. Fujii & T. Yamaguchi, 'Embryogenic callus induction from *Coffea arabica* leaf explants by benzyladenine', *Plant Cell Physiol.*, **1985**, 26, 595-597.
10. M. Berthouly & N. Michaux-Ferriére, 'High frequency somatic embryogenesis in *Coffea canephora*: induction conditions and histological evolution', *Plant Cell Tiss. Org. Cult.*, **1996**, 44, 169-176.
11. G. Staritsky & G. A. M. Van Hassel, 'The synchronised mass propagation of *coffea canephora* in –vitro'. In : *9éme Colloque International sur le Café*, Paris, **1980**, pp. 597-602.
12. J. P. Ducos, A, Zamarripa, A. Eskes & V. Petiard, 'Production of somatic embryos of coffee in bioreactor'. In : *15th Colloq. Sci, Inter. Café*, Paris, **1993,** pp. 89-96.
13. R. Levin, V. Guba, B. Tal, S. Hirch, D. DeNola & I. K. Vasil, 'Automated plant tissue culture for mass propagation', *Bio/Technology G*, **1988**, 1035-1050.
14. W. Preil, P. Florek, V. Wix & A. Beck, 'Towards mass propagation by use of bioreactors', *Acta Horticult.*, **1988**, 226, 99-105

15. D. Cazzulino, H. Pederson & C. K. Chin, In: *Bioreactors and Image Analysis for scale-up and plant propagation*, I. K. Vasil (eds.), Academic Press, New York, **1990**, pp. 147-175.
16. P. V. Ammirato & D. J. Styer, In: *Biotechnology in Plant Science*, M. Zaitlin, P. Day and A. Hollaender (eds.), Academic Press, New York, **1985**, pp. 161-178.
17. D. Alvard, F. Cote & C. Teisson, 'Comparison of methods of liquid medium culture for banana propagation. Effects of temporary immersion of explants', *Plant Cell Tiss. Org. Cult.*, **1993**, 32, 55-60.
18. M. Berthouly, M. Dufour, D. Alvaro, C. Carasco, L. Alemanno & C. Teisson, 'Coffee micropropagation in liquid medium using temporary immersion technique'. In : *16éme Colloque*, Paris, **1995**, 2, pp. 514-519.
19. H. Etienne, M. Lartaud, N. Michaux-Ferriére, M. P. Carron, M. Berthouly & C. Teisson, 'Improvement of somatic embryogenesis in *Hevea brasilensis* (Mull. Arg.) using the temporary immersion technique'. *In vitro Cell Dev. Biol.* **1997**, 33, 81-87.
20. T. Kozai & C. Kubota, S. M. A. Zobayed, Q. T. Nguyen, F. Afreen & J. Heo, In: *Challenge of Plant and Agriculture Sciences to the Crisis of Biosphere on the Earth in the 21st Century*, K. Watanabe & A. Komamine (eds.), **1999,** Landes Bioscience, Texas, pp: 293-306.
21. T. Kozai & S. M. A. Zobayed, In*: Encyclopedia of Cell Technology*, R. Spier, (eds.), **2000,** John Wiley & Sons, Inc., USA, pp: 1-12.
22. S. M. A. Zobayed, F. Afreen & T. Kozai, 'Quality Biomass Production via Photoautotrophic Micropropagation', *Acta Hort.*, **2000**, 530: 377-386.
23. F. Afreen, S. M. A. Zobayed, C. Kubota & T. Kozai, In: *Transplant production in the 21st Century*, C. Kubota and C. Chun (eds.), **2000a**, Kluwer Academic Publishers, The Netherlands, pp: 238-245.
24. T. Murashige & F. Skoog, 'A revised medium for rapid growth and bioassays with tobacco tissue culture', *Physiol. Plant*, **1962,** 15, 473-497.
25. T. Kozai, K. Fujiwara & I. Watanabe, 'Fundamental studies on environments in plant tissue culture vessels. (2) Effects of stoppers and vessels on gas exchange rates between inside and outside of vessels closed with stoppers', *Journal of Agricultural Meteorology*, **1986**, 42**,** 119-127.
26. F. Afreen, S.M.A. Zobayed, C. Kubota, T. Kozai & O. Hasegawa, 'A combination of vermiculite and paper pulp supporting material for the photoautotrophic micropropagation of sweet potato', *Plant Sci.*, **2000b**, 225-231.
27. P. K. Gupta, R. Timmis & W. C. Carlson, In: *Advances in Development Biology and Biotechnology of Higher Plants*, W. Y. Soh, J. R. Liu and A. Komamine (eds.), **1993**, The Korean Society of P. T. C., Korea, pp. 18-37.
28. R. C. Harrell & D. J. Cantliffe, In: *Scale-up and Automation in Plant Propagation*, I. K. Vasil (eds.), **1991,** Academic Press, New York, pp. 179-195.

Molecular Breeding of Woody Plants
N. Morohoshi and A. Komamine, editors.

AUTOMATION IN SOMATIC EMBRYO PRODUCTION

Yasuomi Ibaraki

*Department of Biological Science, Faculty of Agriculture, Yamaguchi University
Yoshida 1677-1, Yamaguchi-shi, Yamaguchi 753-8515, Japan*

ABSTRACT

Automation enhances the advantages of using somatic embryogenesis for micropropagation. Automation techniques could contribute to the development of the practical somatic embryo production system from two aspects. First, for the established culture protocol, they can offer significant labor cost reductions. In addition, they can be used as effective tools for research on somatic embryogenesis. In this review, the techniques related to the automation in somatic embryo production were introduced and discussed.

KEYWORDS

Automation, bioreactor, image analysis, micropropagation, somatic embryogenesis

INTRODUCTION

Micropropagation techniques require intensive labor and this often limits their commercial application. Automation can reduce the labor required. Somatic embryogenesis is more suitable for automation and scale-up than micropropagation via organogenesis. In many cases, somatic embryos can be produced in liquid culture without the procedure of cutting plantlets into segments for proliferation [1]. Somatic embryos could be easier to handle since they are relatively small and uniform in size [2]. In addition, somatic embryos have a potential of long-term storage through cryopreservation or desiccation, which facilitates flexibility in scheduling production and transportation [2].

Automation can also contribute to an improvement in the productivity of somatic embryos linking to scale-up of culture. Automated sorting of somatic embryos is expected to achieve a high percentage of uniform embryos. Moreover, in an automated culture process using a bioreactor, the culture parameters, such as pH and dissolved oxygen concentration, could be better optimized by monitoring and controlling culture systems than would be the case when using flask batch culture.

In biological processes, such as somatic embryo production, some judgments for biological objects from experts are required. Therefore, for the automation of these processes, to handle information on the biological objects is needed. For example, in somatic embryo production, quality of embryogenic cultures as materials and embryos as products should be properly evaluated. The quality evaluation affects directly on embryo productivity. In addition, accurate judgment of contamination or physiological disorders in cultures such as hyperhydricity is essential for stabilizing productivity. It is expected not only to simply mechanize certain procedures, but also to develop a system to monitor and evaluate the processes for the automation of somatic embryo production. Image-analysis-based evaluation that can provide objective information will be helpful for improvement in reliability and/or reproducibility of culture processes, replacing expert judgements.

Automation techniques could be used as research tools for investigating somatic embryogenesis. Controllability of culture conditions in a bioreactor is advantageous for studies to optimize culture conditions. Bioreactors have been applied to investigate DO effects on somatic embryo development. The change in requirements for pH during carrot somatic embryo development was detected by using a bioreactor [3]. In addition, automated evaluation of somatic embryos can provide us with more information related to population dynamics during somatic embryogenesis. Image-analysis-based classification technique was used to investigate the effects of auxin transport inhibitor, 2,3,5-triodobenzoic acid [4], and semi-continuous cultures[5] on population dynamics and morphology of carrot somatic embryos

Although the detailed protocol largely depends on plant species, the production of somatic embryos generally involves three main processes: 1) induction, 2) maintenance of embryogenic cultures, and 2) development of embryos. Embryogenic cultures are induced from both non-embryogenic and embryogenic tissues under special growth regulator treatments. Induced embryogenic cultures are often maintained in suspension cultures because it has advantages of giving homogeneity and a higher efficiency of proliferation. For initiation of embryo development, it is required to change medium, especially to modify growth regulator concentrations. Processes for enhancing conversion and adaptability for delivery are also necessary for matured embryos, in order to use somatic embryos for transplant production including synthetic seed production. These include dehydration, growth regulator treatment, CO_2 treatment with lighting [6], and encapsulation. In terms of delivery, encapsulated embryos may be preferred because of ease of handling and the possibility for direct sowing into the soil.

Processes expected to be automated are labor intensive and require expert decisions as follows [7]:

1) Evaluation of embryogenic culture in maintenance,
2) Somatic embryo development, including treatment for initiation of somatic embryo development, and monitoring and control of the culture,
3) Harvest of the somatic embryos,
4) Pre-delivery process for enhancing conversion and preparing for delivery.

A fully automated system for embryo production has not yet been established. Techniques related to the automation of each process for somatic embryo production will be introduced in this presentation.

EVALUATION OF EMBRYOGENIC CULTURES IN MAINTENANCE

The productivity of somatic embryos largely depends on the quality of embryogenic cultures as a material [8]. The embryogenic potential could change with culture period and be affected by medium composition and environmental conditions. The evaluation of culture is crucial in this process. Selection of cultures based on appropriate evaluation of the embryogenic potential should be required. In addition, the evaluation of cultures would be useful for controlling the productivity of embryos. For example, to know the number of proembryogenic masses (PEMs) in inoculum for embryo development leads to the control of the number of somatic embryos produced. These quality evaluation processes often include visual inspections by experts, and therefore include subjective judgments. Sampling may be also required in the evaluation processes, which involves intensive labor. Therefore, it is desirable to have a method of quality evaluation that is simple, subjective, non-destructive and has the potential for automation. Image analysis techniques appear to be one of promising methods for the evaluation of cultures. Image analysis has a potential for non-invasively offering an objective index for visual information. In fact, Image analysis has been used as research tools in cell biology/plant tissue culture, which are a time-saving method to enhance and amplify the routine experimental measurement [9]. The simple and objective evaluation methods can help to gather empirical knowledge and to establish the stable culture protocol. Image analysis applications were reviewed in detail for plant cell culture and micropropagation[9] and for plant suspension culture[10].

Evaluating callus quality on solid media

In some cases, researchers judge callus for somatic embryogenesis suitable by visual inspection and select such callus for further proliferation. Indices often used for callus quality were color, compactness, and friability. Moreover, the term "nodular" was sometimes used for expressing embryogenic callus [11-12]. These indices are difficult to express numerically and are apt to be subjective.

The color information can be analyzed easily from images and is suitable for quantification. In addition, Olofsdotter (1993)[13] reported that it was possible to measure the compactness of each cell aggregate in relation to gray scale values. Texture analysis may be used for evaluating compactness and/or friability of callus if these properties can be evaluated based on the appearance of callus.

Information used for evaluating quality of suspension culture

The following information may be promising for use in image-analysis-based evaluation of embryogenic potential of suspension cultures; 1) amount of PEMs, 2) color, 3) growth rate (quantity), 4) cell aggregate size distribution, 5) macroscopic texture

Using microscopic observation, a PEM could be identified. In a number of systems studied to date, embryogenic cells shared similar structural features. They are small and highly cytoplasmic, and often have an accumulation of starch within the plastids [14]. On the other hand, non-embryogenic cells are large and vacuolated. Therefore, a PEM was selected with regard to its transparency and shape under microscopy. The amount of cell aggregates including these types of cells, i.e., PEMs, in suspension cultures may be one direct index for determining the embryogenic potential of culture. In a similar way, the amount in cultures of other proliferating embryogenic tissues as materials for embryo production such as embryonal suspensor masses (ESMs) and early globular embryos can be used for evaluation of cultures.

Culture growth characteristics are important indices for evaluating suspension culture[10]. Differences in growth characteristics between embryogenic and non-embryogenic cultures have been reported in maize suspension culture [15], in carrot suspension culture[16-17], and in *Ipomoea batatas* callus culture [18]. Ibaraki and Kurata (1997)[17] quantified embryogenic suspension cultures by image analysis of macroscopic images of the suspensions. It is also possible to measure the size of cell aggregates and to count the number directly using microscopic images to estimate the growth rates. Harrell *et al.* (1992)[19] estimated the growth of each size group cell aggregate from measurements of time course changes in cell aggregate size distributions.

Many methods for quantification in suspension culture were reported as well as image analysis. These methods have the potential to be used to evaluate cultures via an estimation of growth in automated culture processes [7]. These include optical sensors, turbidimeters, and electrical conductivity sensors.

Cell aggregate size distribution patterns in suspension culture vary significantly between cell lines and also as a consequence of culture age and culture conditions [20]. Several reports have shown differences in cell aggregate size distribution patterns between embryogenic and non-embryogenic suspension cultures [15,21]. Therefore, the cell aggregate size distribution pattern may be used as an index for the embryogenic potential of culture. The macroscopic visual appearance of suspension cultures may vary with the color distribution and size distribution of cell aggregates in the suspension cultures. It has been reported that the visual appearance of suspension cultures changes with the number of subcultures [22] or with variations in embryogenic potential [8,22]. Hence, the visual texture of a macroscopic image of a suspension culture may be used as an indicator of cultured cell quality [8].

Microscopic and macroscopic image analyses

Use of microscopic images has advantages of directly observing individually cells, cell aggregates, and tissues. However, this microscopic image analysis has difficulties in image acquisition [10], since sampling of the suspension for microscopic observation is necessary and it may be destructive with a risk of contamination and requiring intensive labour. In addition, there are questions whether the sample accurately represents population of the suspension culture. For promising ways, systems for acquiring cell images in the imaging cell compartment connected to a bioreactor have been reported[19,23]. As an alternative way, to use microscopic CCD (charge-coupled device) cameras with long working distance is promising for acquiring microscopic images [10]. These CCD cameras have a flexible scope head and enable us to get images through culture vessel walls.

As compared with microscopic images, macroscopic images, which viewed with normal or macro lens whose fields of view contain almost one whole culture, are more easily acquired from outside the culture vessel without sampling [10]. The spatial resolution of macroscopic images is several hundred micrometers per pixel and does not allow us to identify a small cell-aggregate. However, color analysis and/or texture analysis of macroscopic images of suspension cultures can provide us with information related to suspended cells and tissues. Macroscopic images were used for the evaluation of embryogenic potential of carrot embryogenic suspension cultures with texture analysis of the image [8].

PROCESS FOR SOMATIC EMBRYO DEVELOPMENT

Automation of initiating somatic embryo development

Generally, somatic embryo development is initiated by transferring embryogenic cultures to medium modified in growth regulator concentrations, in many cases medium without 2,4-D. Therefore, medium exchanging methods are essential for the automation of this process. The selection of PEMs is needed for controlling the productivity of somatic embryos in some cases. In addition, the population density of PEMs or embryos is very important factors for controlling the productivity of somatic embryos.

Examples of the promising automated systems for initiation of embryo development were reported in carrot [24-25] in *Capsicum annuum* [26].

Automated medium exchange

It is relatively easy to automatically exchange medium in bioreactor systems. The medium should be exchanged while maintaining aseptic conditions. Pumps whose heads are not in contact with the medium, such as peristaltic pumps, can be used to exchange medium during culture. Spin-filter bioreactors, in which the stirring central shaft contains a housing filter device, were reported to be promising for somatic embryo production because they can undergo medium exchange rapidly using the filter as a perfusion device [24].

Selection of PEMs

In order to improve synchrony in embryo development, a fixed cell aggregate size was often used to select PEMs as materials for embryo production. However, a large variation in the size of PEMs used for somatic embryogenesis has been reported. Large carrot PEMs were reported to produce multiple embryos [27-28], and this may cause asynchrony in the culture [29]. PEM size should be optimized for synchronization of embryo development and the embryo productivity [29]. A fixed cell aggregate size can be selected by sieving. Kurata and Futaya (1992)[25] reported an automated system for initiation of embryo development including an automated sieving.

Control of inoculated population density of PEMs or embryos

Several reports have shown that a specific PEM density is needed for somatic embryo development [24,26,27]. At the same time, development of embryos after the globular or heart-shaped stage is inhibited if the PEM population density is too high [28,30]. As a possible reason for the effect of high density, it was suggested that inhibitory conditioning factors released from embryos in the medium would increase with an increasing density [28]. Monitoring the number of PEMs and determining the amount of medium for regulating population density are useful for the control of somatic embryo production. Image analysis is promising for monitoring PEM/embryo density.

Bioreactor applications

For production of somatic embryos, the bioreactor system is one of the most promising ways of scaling-up the system. Use of bioreactors is advantageous for automation of processes linked to monitoring systems. In addition, by altering culture parameters, such as cell population density[28] and dissolved oxygen concentration[31], bioreactors could regulate somatic embryo development and improve synchronization.

Although there have been several reports for somatic embryo production in bioreactors, it does not seem to be a well-established technique[7]. The successful production of somatic embryos by bioreactors was reported in carrot[6,22,24], in sandalwood [32], and in *Eschscholtzia californica*[33]. However, failure in somatic embryogenesis or a reduction in productivity was also reported. As compared to culture in shaken flasks, somatic embryo production using bioreactors was limited in *Picea mariana* and *Picea glauca-engelmannii* [34], in alfalfa [35] and in *Ipmoea batatas* [36]. Moreover, some embryogenic cultures even if they can be maintained in liquid medium, simply cannot be grown as true suspension cultures [37]. It should be noted that a successful bioreactor system arrangement for one species might not be applicable to other species.

Inconsistencies in optimization of bioreactor types and culture parameters have been reported [7]. According to Vasil (1994)[38], the most difficult and intractable problems in the use of bioreactors for large-scale somatic embryo production are in the biology of the system and not in the engineering.

Performance of embryo production should be assessed, not with reference to the number of embryos, but according to the number of quality embryos. "Quality embryos" can be defined from the aim of embryo production and, in many cases, are defined as embryos with high conversion potential. Conversion should refer to development into a normal plantlet via germination, which has a potential for the survival in an ex vitro soil environment [37]. It is crucial that the term "conversion" be properly used so that full assessment of somatic embryo production protocols can be done [39].

Bioreactor application for embryo development of woody plants

In numerous conifer species, there are few reports on somatic embryo development and maturation from liquid cultures[40]. Therefore, conventional bioreactors, such as stirred tank reactors and air-lifted reactors, have not been successfully applied for somatic embryo development in these species.

Attree *et al.* (1994)[41] reported the production of vigorous embryos of *Picea glauca* in a bioreactor, which consisted of a culture chamber and medium reservoir. In the culture chamber immature somatic embryos (proembryos) were cultured on a flat absorbent pad above the surface of a liquid medium. Fresh medium was pumped into one end of the chamber with a peristaltic pump and the spent medium exited by gravity from the opposite end. Over a 7-week culture period, over 6,300 normal looking cotyledonary stage embryos were obtained from a single chamber without manual manipulation. The somatic embryos after drying converted into normal plantlets at a high frequency.

Pâques *et al.* (1995)[42] reported successful cotyledonary embryo production of *Picea abies* from EMSs using a bioreactor with polyurethane layers in liquid medium. The EMSs were immobilized in polyurethane layers that were placed vertically in liquid maturation medium, and were immersed intermittently in liquid medium.

Gupta and Timmis (1999)[40] reported the successful production of somatic embryos of Douglas fir in a bioreactor with pads. The medium was supplied semi-continuously from the lower surface of the pads to the developing embryos on the top. The medium was absorbed in the pads by capillary action, and after a few hours, was pumped out to the reservoir. This was repeated at regular intervals until mature cotyledonary embryos developed.

Etienne-Barry *et al.*(1999)[43] succeeded in *Coffea arabica* somatic embryo production by 1L Temporary immersion bioreactor (RITA®), and direct sowing of the embryos in soil.

HARVEST OF SOMATIC EMBRYOS

For obtaining high quality and homogeneous embryos, sorting embryos by using information on embryo quality, i.e., conversion potential, is an effective technique. To obtain high conversion rates, it has often been necessary to select embryos manually[22,42,44]. The procedure is labor intensive and is expected to be automated. The critical point to be considered in automation of this process is how to evaluate embryos. The promising information for evaluating embryo quality includes size, shape, density, color and growth rate [45]. Since the developmental stages of somatic embryos involve morphological changes, shape should be an index for somatic embryo selection at a suitable developmental stage for conversion. In conifer species, since careful selection of somatic embryos improved germination rates [44,46], the practice of selecting only normal-looking embryos for germination seems justified[47]. Therefore, shape information can be used for the evaluation of somatic embryo quality, i.e., conversion potential. In addition, size is often used as a feature for selecting embryos because embryo size varies according to embryo development, and selection according to size is easy to perform.

Image analysis methods have been applied to the evaluation of somatic embryo quality for this purpose because it has the potential for non-destructive quantification of these features.

Image-analysis-based evaluation of somatic embryos

The main features used in image analysis for embryo evaluation were related to the shape and size of somatic embryos. These features were reviewed by Ibaraki (1999)[45] in detail.

Using the size and shape related features, tools such as template matching techniques, neural networks, decision trees, discriminant analysis, k-nearest neighbor classification, fuzzy clustering and methods using lower and upper bounds or threshold values, were applied to embryo evaluation in the algorithms [7].

For recent study for woody plants, Zhang *et al.* (1999)[48] applied neural networks for a pattern recognition system for Douglas fir embryos that had four to seven cotyledons. The pattern recognition system used a decision tree and consisted of two neural classifiers. Ten size related features and Fourier features from image contour were used as inputs to the neural classifiers. In this system, 267 of 271 normal embryos and 22 of 32 abnormal embryos were judged correctly. In addition, 80% of embryos could be classified into the proper stage of the three developmental stages.

Embryo sorting

Image-analysis-based sorting

Embryo sorting systems combined with image analysis systems that have been proposed in previous studies are mainly of three types [45]. In one type of system, embryos were placed in an imaging area, analyzed, and selected embryos were picked up (e.g., Hämäläinen *et al.* (1993)[49]). The second type of system (called "flowing system" by Ibaraki (1999)[45]) had embryos flowing in liquid medium and sorted into the proper reservoirs by the medium flow [50-51]. The third type of system sorted gel beads containing embryos for synthetic seed production [52].

These systems are all prototypes and none have been commercialized. For the development of practical sorting systems, several improvements in sorting performance are essential. In addition, further research on embryo quality is necessary. In other words, it is important to know which embryos should be harvested for transplant production. Moreover, considering the physiological and economic aspects, it is necessary to further investigate at which stages embryos should be sorted [45].

Filtration

If harvestable somatic embryos could be selected based only on the size, filtration (sieving) would be a very simple and useful method for embryo sorting. Filtration has been used in practice for obtaining uniform-sized embryos[22,53,54]. However, it should be noted that at later stages, heart- and torpedo-shaped embryos often have the same diameter, but are of different length, and it is difficult to obtain sharp separation using filtration [55].

Encapsulation

Encapsulation can offer somatic embryos an advantage for delivery potential. Encapsulation reduces mechanical damage of embryos and may enhance the storage potential of embryos. For examples of automated encapsulation systems, Kirin Brewery Co. Ltd. (Japan) developed two types of machine based on a random encapsulation method [52].

CONCLUSIONS

Automation techniques are expected to contribute not only to reduction of labor, but also to an improvement in the productivity of somatic embryos by their use as research tools. Many improvements are still needed for practical automated somatic embryo production systems. They can be realized, not only from progress in techniques related to mechanization and monitoring, but also from a further comprehension of biological processes. It is expected that automation techniques contribute to the acquisition of knowledge of the processes.

REFERENCES

1. K. Kurata, In: *Automation and environmental control in plant tissue cultures,* J. Aitken-Christie, T. Kozai and M. A. L. Smith (Eds), 1995, Kluwer academic publishers, Dordrecht, pp.165-186
2. R. Cervelli & T. Senaratna, In: *Automation and environmental control in plant tissue cultures,* J. Aitken-Christie, T. Kozai and M. A. L. Smith (Eds.), 1995, Kluwer academic publishers, Dordrecht, pp.29-64
3. V. Jay, S. Genestier & J-C. Courduroux, 'Bioreactor studies of the effect of medium pH on carrot (*Daucus carota* L.) somatic embryogenesis', Plant Cell Tiss. Org. Cult., 1994, 36, 205-209
4. C-M. Chi, H. Vits, E. J. Staba, T. J. Cooke & W-S. Hu, 'Morphological kinetics and distribution in somatic embryo cultures', Biotechnol. Bioeng., 1994, 44, 368-378
5. C. Zhang, C-M. Chi, E. J. Staba, T. J. Cooke & W-S. Hu, 'Application of image analysis to fed-batch cultures of somatic embryos', In Vitro Cell. Dev. Biol., 1996, 32p, 190-198
6. N. Onishi, Y. Sakamoto & T. Hirosawa, 'Synthetic seeds as an application of mass production of somatic embryos', Plant Cell Tiss. Org. Cult., 1994, 39, 137-145
7. Y. Ibaraki & K. Kurara, 'Automation of somatic embryo production', *Submitted to* Plant Cell Tiss. Org. Cult.
8. Y. Ibaraki, Y. Kaneko & K. Kurata, 'Evaluation of embryogenic potential of cell suspension culture by texture analysis', Transactions of the ASAE, 1998, 41, 247-252

9. M. A. L. Smith, In: *Automation and environmental control in plant tissue cultures,* J. Aitken-Christie, T. Kozai T and M. A. L. Smith (Eds.), 1995, Kluwer academic publishers, Dordrecht, pp.145-163
10. Y. Ibaraki & K. Kurata, 'Application of image analysis to plant cell suspension cultures', Computer and Electronics in Agriculture, 2000, 30,193-203
11. S. Buyukalaca & F. Mavituna, 'Somatic embryogenesis and plant regeneration of pepper in liquid media', Plant Cell Tiss. Org. Cult., 1996, 46, 227-235
12. L. Alemanno, M. Berthouly & N. Mixchaux-Ferrière, Histology of somatic embryogenesis from floral tissues cocoa', Plant Cell Tiss. Org. Cult., 1996, 46, 187-194
13. M. Olofsdotter, 'Image processing: a non-destructive methods for measuring growth in cell and tissue culture', Plant Cell Rep., 1993, 12, 216-219
14. E. C. Yeung, In: *In Vitro Embryogenesis in Plants,* T. A. Thorpe (ed.), 1995, Kluwer Academic Publishers, Dordrecht, pp.205-247
15. S. Stirn, A. Hopstock & H. Lorz, 'Bioreactor cultures of embryogenic suspensions of barley (*Hordeum vulgare* L.) and maize (*Zea mays* L.)', J. Plant Physiol., 1994, 144, 209-214
16. S. M. Smith, & H. E. Street, 'The decline of embryogenic potential as callus and suspension cultures of carrot (*Daucus carota* L.) are serially subcultured', Ann. Bot., 1974, 38, 223-241
17. Y. Ibaraki & K. Kurata, 'Image analysis based quantification of cells in suspension cultures for producing somatic embryos', Environ. Control in Biol., 1997, 35, 63-70
18. Q. Zheng, A. P. Dessai & C. S. Parkash, 'Rapid and repetitive plant regeneration in sweetpotato via somatic embryogenesis', Plant Cell Rep., 1996, 15, 381-385
19. R. C. Harrell, M. Bieniek & D. J. Cantiffe, 'Noninvasive evaluation of somatic embryogenesis', Biotechnol. Bioeng., 1992, 39, 378-383
20. P. M. Kieran, P. F. MacLoughlin & D. M. Malone, 'Plant cell suspension cultures: some engineering considerations', J. Biotechnol., 1997, 59, 39-52
21. J. van Boxtel & M. Berthouly, 'High frequency somatic embryogenesis from coffee leaves', Plant Cell Tiss. Org. Cult., 1996, 44, 7-17
22. F. Molle, J. M. Dupuis, J. P. Ducos, A. Anselm, I. Crolus-Savidan, Y. Petiard & G. Freyssinet, In: *Synseeds*, K. Redenbaugh (ed.), 1993, CRC press, Boca Raton, pp.257-287
23. A. Grand d'Esnon, R. Chee, R. C. Harrell & D. J. Cantliffe, 'Qualitative and quantitative evaluation of liquid tissue cultures by artificial vision', Biofutur, 1989, 76, S3
24. P. V. Ammirato & D. J. Styer, In: *Biotechnology in Plant Science*, M. Zaitlin, P. Day & A. Hollaender (eds.), 1985, Academic Press Inc, London, pp.161-178
25. K. Kurata & Y. Futaya, 'Automation of selecting embryogenic callus by sieving', Acta Hort. , 1992, 319, 563-566
26. F. Marvituna & S. Buyukalaca, 'Somatic embryogenesis of pepper in bioreactors: a study of bioreactor type and oxygen-uptake rates', Appl. Microbiol. Biotechnol., 1996, 46, 327-333
27. W. Halperin, 'Population density effects on embryogenesis in carrot-cell cultures', Exp. Cell Res., 1967, 48, 170-173
28. K. Osuga & A. Komamine, 'Synchronization of somatic embryogenesis from carrot cells at high frequency as a basis for the mass production of embryos', Plant Cell Tiss. Org. Cult., 1994, 39, 125-135
29. Y. Ibaraki, R. Matsushima & K. Kurata, 'Analysis of morphological changes in carrot somatic embryogenesis by serial observation', Plant Cell Tiss. Org. Cult., 2000, 61, 9-14
30. O. Maes, P. Coutus-Thévenot, T. Jouenne, M. Boulay & J. Guern, 'Influence of extracellular proteins, proteases and protease inhibitors on grapevine somatic embryogenesis', Plant Cell Tiss. Org. Cult., 1997, 50, 97-105
31. T. Shimazu & K. Kurata, 'Improvement of synchronization on carrot somatic embryo culture by controlling dissolved oxygen concentration', Environ. Control in Biol., 1999, 37, 179-184.
32. V. A. Bapat, D. P. Fulzele, M. R. Heble & P. S. Rao, 'Production of sandalwood somatic embryos in bioreactors', Current Science, 1990, 59, 736-748
33. J. Archambault, R. D. Williams, L. Lavoie, M-F. Pépin & C. Chavarie, 'Production of somatic embryos in a helical ribbon impeller bioreactor', Biotechnol. Bioeng., 1994, 44, 930-943
34. T. E. Tautorus, M. M. Lulsdorf, S. I. Kikcio & D. I. Dunstan, 'Bioreactor culture of *Picea mariana* Mill. (black spruce) and the species complex *Picea glauca-engelmannii* (interior

spruce) somatic embryos', Appl. Microbiol. Biotechnol., 1992, 38, 46-51
35. T. H. H. Chen, B. G. Thompson & D. F. Gerson, '*In vitro* production of alfalfa somatic embryos in fermentation systems', J. Ferment. Technol., 1987, 65, 353-357
36. M. E. Bieniek, R. C. Harrell & D. J. Cantliffe, 'Enhancement of somatic embryogenesis of *Ipomoea batatas* in solid cultures and production of mature somatic embryos in liquid cultures for application to a bioreactor production system', Plant Cell Tiss. Org. Cult., 1995, 41, 1-8
37. S. A. Merkle, W. A. Parrott & B. S. Flinn, In: *In Vitro Embryogenesis in Plants,* T. A. Thorpe (ed.), 1995, Kluwer Academic Publishers, Dordrecht, pp.155-203
38. I. K. Vasil, 'Automation of plant propagation', Plant Cell Tiss. Org. Cult., 1994, 39, 105-108
39. K. Redenbaugh, In: *Synseeds,* K. Redenbaugh (ed.), 1993, CRC press, Boca Raton, pp.3-7
40. P. K. Gupta & R. Timmis, In: *Plant Biotechnology and In Vitro Biology in the 21st Century*, A. Altman *et al.* (eds.), 1999, Kluwer academic publishers, Dordrecht, pp.49-52
41. S. M. Attree, M. K. Pomeroy & L. C. Fowke, 'Production of vigorous, desiccation tolerant white spruce (*Picea glauca* [Moench.] Voss.) synthetic seeds in a reactor', Plant Cell Rep., 1994, 13, 601-606
42. M. Pâques, J. Bercetche & M. Palada, In: *Somatic embryogenesis in woody plants Vol.1,* S. Jain, P. Gupta & R. Newton (Eds.), 1995, Kluwer academic publishers, Dordrecht, pp.399-414
43. D. Etienne-Barry, B. Bertrand, N. Vasquez & H. Etienne, 'Direct sowing of *Coffea arabica* somatic embryos mass-produced in a bioreactors and regeneration of plants', Plant Cell Rep., 1999, 19, 111-117
44. P. K. Gupta, R. Timmis, K. Timmis, J. Grob, W. Carlson, E. Welty & C. Carpentar, 'Advances in conifer tree improvement through somatic embryogenesis', Proceedings of the 12th Toyata Conference, 1999, 309-315
45. Y. Ibaraki, In: *Somatic Embryogenesis in Woody Plants vol.4,* S. M. Jain, P. K. Gupta & R. J. Newton (Eds.), 1999, Kluwer Academic Publishers, Dordrecht, pp.169-188
46. G. W. Adams, M. G. Doiron, Y. S. Park, J. M. Bonga & P. J. Charest, 'Commercialization potential of somatic embryogenesis in black spruce tree improvement', For. Chron., 1994, 70, 593-598
47. E. I. Hay & P. J. Charest, In: *Somatic Embryogenesis in Woody Plants vol.4,* S. M. Jain, P. K. Gupta, R. J. Newton RJ (Eds.), 1999, Kluwer Academic Publishers, Dordrecht, pp.61-96
48. C. Zhang, R. Timmis & W-S. Hu, 'A neural network based pattern recognition system for somatic embryos of Douglas fir', Plant Cell, Tiss. Org. Cult., 1999, 56, 25-35
49. J. J. Hämäläinen, U. Kurten & V. Kauppinen, 'Classification of plant somatic embryos by computer vision', Biotechnol. Bioeng., 1993, 41, 35-42
50. R. C. Harrell, M. Bieniek, C. F. Hood, R. Munilla & D. J. Cantliffe, 'Automated, in vitro harvest of somatic embryos', Plant Cell, Tiss. Org. Cult., 1994, 39, 171-183
51. Y. Ibaraki, M. Fukakusa & K. Kurata, 'SOMES2: Image-analysis-based somatic embryo sorter', Current Plant Science and Biotechnology in Agriculture, 1995, 22, 675-680
52. Y. Sakamoto, T. Mashiko, A. Suzuki & H. Kawata, 'Development of encapsulation technology for synthetic seeds', Acta Hort., 1992, 319, 71-76
53. G. Giuliano, D. Rosellini & M Terzi, 'A new method for the purification of the different stages of carrot embryoids', Plant Cell Rep., 1983, 2, 216-218
54. D. L. Rodriguez, S. L. Kitto & K. M. Lomax, 'Mechanical purification of torpedo stage somatic embryos of *Daucus carota* L.', Plant Cell Tiss. Org. Cult., 1990, 23, 9-14
55. F. Lo Schiavo, 'A critical review of the procedures for embryo purification', Plant Mol. Biol. Rep., 1984, 2, 15-18

Molecular Breeding of Woody Plants
N. Morohoshi and A. Komamine, editors.

A CLOSED-TYPE TRANSPLANT PRODUCTION SYSTEM

Changhoo Chun* and Toyoki Kozai

*Department of Bioproduction Science,
Faculty of Horticulture, Chiba University,
Matsudo, Chiba 271-8510, Japan.*

ABSTRACT

Because photoautotrophic micropropagation systems do not need sugar in the culture media, they are thought to be much easier to scale up than photomixotrophic micropropagation systems. A closed, large-scale photoautotrophic micropropagation for transplant production system has been developed at Chiba University, Japan. New concepts and technologies were adapted to produce high quality transplants at minimum usage of resources, and as scheduled. Newly developed software for production management was used to enhance the efficiency of the transplant production system. Currently, virus-free transplants of sweetpotato (*Ipomoea batatas* (L.) Lam.) are vegetatively propagated and produced under sterilized conditions in this system. This system can also be used for production of transplants of any other species including woody plants with a minimum of modification.

KEYWORDS

Automation, micropropagation, photoautotrophic culture, scale-up, sweetpotato, woody plants.

INTRODUCTION

Transplant production based upon micropropagation has advantages over transplant production that uses seeds or cuttings, with respect to genetic and phenotypic uniformity and scheduled year-round production of disease-free or pathogen-free transplants[1]. Kozai[2] and, Nguyen and Kozai[3] reported systems developed to produce a large number of quality transplants at low cost, based upon the idea of photoautotrophic (no sugar in the culture medium) or photosynthetic micropropagation under high CO_2 concentration or CO_2 enrichment. With photoautotrophic micropropagation, it is possible to use large culture vessels with minimum risk of microbial contamination. In the large vessels used for the photoautotrophic micropropagation, forced ventilation has been found to have several advantages over natural ventilation[4]. Environmental control techniques have been developed to exclude any problems caused from scaling-up the micropropagation systems and the efficiency of the systems can be maximized as a result.

A scaled-up micropropagation system based on the closed system of Kozai et al.[5, 6] has been developed at the Matsudo campus of Chiba University, Japan[7]. This system is a research facility to study biological and engineering aspects of transplant production in closed systems and to test the feasibility of closed systems for production of transplants. Various technologies for scaling-up the micropropagation systems were applied to produce high quality transplants with the highest efficiency. In this system, the transplants are vegetatively propagated and produced under sterilized conditions. This system can be thought of as scaled-up culture vessels of a photoautotrophic micropropagation system[2]. High quality transplants, defined as transplants that are physiologically and morphologically superior and show vigorous growth after being transplanted into fields and greenhouses, can be produced regardless of weather conditions.

Although this system can produce transplants of any species, it is presently being used to produce virus-free transplants of sweetpotato (*Ipomoea batatas* (L.) Lam.). Sweetpotato is used as a fresh vegetable, a feed for animals and a raw material for producing biodegradable plastics and hydrogen gas, and vast amounts of quality transplants are expected to be needed in the near future[5].

For sweetpotato production, vine cuttings are conventionally used as transplants. Because the vine-cutting transplants do not have roots, almost 100% of them wilt after transplanting. Even though large portions of the transplants recover from the wilting in a few weeks, the delay of the early growth caused by the wilting after transplanting may decrease the final yield and quality of sweetpotato. In the newly developed system, however, sweetpotato transplants with root balls are produced in multi-cell trays for direct transplanting into fields[8]. In this case, transplants keep growing without wilting and a higher yield can be expected. Machines can be used for transplanting the sweetpotato transplants with root balls, which is difficult with vine-cutting transplants.

SCALE-UP OF PHOTOAUTOTROPHIC MICROPROPAGATION SYSTEMS

Micropropagation is an advanced vegetative propagation technology for producing a large number of genetically superior and pathogen-free transplants in a limited time and space. Recent studies[2-4] revealed that most chlorophyllous plants in vitro have the ability to grow photoautotrophically, and that the low CO_2 concentration in the air-tight culture vessel during the photoperiod is the main cause of the low net photosynthetic rate of plants in vitro. Also, the net photosynthetic rate of plants in vitro is considerably lower when cultured on sugar-containing medium than when cultured on sugar-free medium. Furthermore, we have shown that the photoautotrophic growth of chlorophyllous plants in vitro can be significantly promoted by increasing the CO_2 concentration and light intensity or photosynthetic photon flux (PPF), by decreasing the relative humidity in the culture vessel, and by the use of fibrous and/or porous supporting materials with high air porosity. By using a culture medium containing no sugar, the loss of plants in vitro due to microbial contamination can be significantly reduced. When a culture vessel with a high ventilation rate or high number of air exchanges is used, the relative humidity in the vessel is reduced. This reduction in relative humidity results in enhanced rooting and a high percent survival at the ex-vitro acclimatization stage, especially when porous supporting materials are used in vitro.

In many photoautotrophic micropropagation systems, gas-permeable filters are attached to the lid or sides of the culture vessels to enhance the natural ventilation of culture vessels, and thus to maintain higher CO_2 concentration in the vessels during the photoperiod. This increases the net photosynthetic rate and suppresses the relative humidity, which in turn increases the transpiration rate. On the other hand, in forced ventilation, air pumps or air compressors are used to flush a particular gas mixture directly through the culture vessel. In photoautotrophic micropropagation using large culture vessels, forced ventilation has several advantages over natural ventilation.

Fujiwara et al.[9] developed a large culture vessel (58 cm x 28 cm x 12 cm high) with a forced ventilation system for enhancing the photoautotrophic growth of strawberry (*Fragaria* x *ananassa* Duch.) explants and/or plantlets during the rooting and acclimatization stages. Kubota and Kozai[10] showed that the net photosynthetic rate and photoautotrophic growth of potato (*Solanum tuberosum* L.) plants cultured using a large

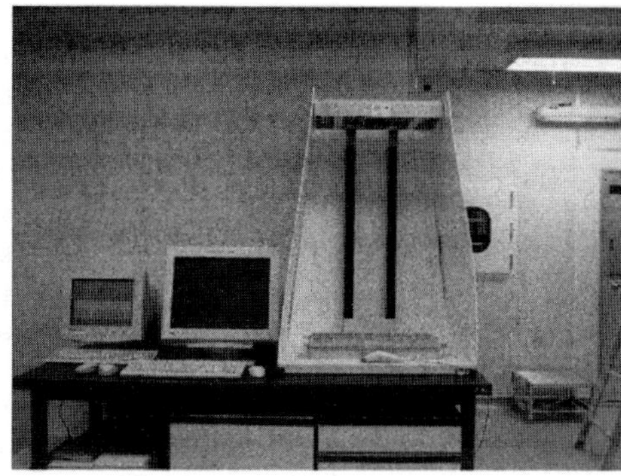

Figure 1. A transplant growth measuring system that consists of a stereo-eyed digital camera system, a bar-code scanner and a computer.

culture vessel with forced ventilation, containing a multi-cell tray with rock-wool cubes, were significantly greater than those cultured using a conventional (small) culture vessel with natural ventilation. Heo and Kozai[11] developed a forced ventilation micropropagation system with a culture vessel containing a multi-cell tray widely used for plug seedling production. The cells were filled with sterilized vermiculite or cellulose plugs. The photoautotrophic growth of sweetpotato plants cultured with this system were several times greater than the photomixotrophic growth of plants cultured with conventional or small culture vessels containing sugar and with natural ventilation. However, the growth in the culture vessel was not uniform, with larger plants near the air inlet and comparatively smaller plants near the air outlet. Zobayed et al.[12] developed large culture vessels with air distribution pipes for forced ventilation to provide an air current pattern which enables uniform distributions of CO_2 concentration and relative humidity as well as those of air current speeds, and thus uniform plant growth.

The idea of a forced ventilation micropropagation system can be further extended to an aseptic culture room. In this case, each aseptic culture room is considered as a large culture vessel that contains sterilized trays and plants. The closed transplant production system at Chiba University can be considered as a scaled-up culture vessel that adapts forced ventilation.

A CLOSED TRANSPLANT PRODUCTION SYSTEM AT CHIBA UNIVERSITY

A closed transplant production system at Chiba University was developed in 2000 to study biological and engineering aspects of transplant production in closed systems and to test the feasibility of closed systems for production of transplants[7]. In this system, which has a floor area of 500 square meters, sweetpotato transplants are vegetatively propagated and produced under sterilized conditions. The system consists of several spaces in a sterilized area and a non-sterilized area.

Sterilized area

The sterilized area consists of a transfer room, three production rooms (micropropagation room and production rooms 1 and 2) and a low-temperature storage room 2. To enter the sterilized area, workers must pass through air showers and materials must be sterilized and transferred though double-door systems that have UV lamps. To maintain a positive pressure in the sterilized area, air is supplied from outside through a ventilation system that has filters and air heaters. The amount of air supplied to the production rooms is kept at a minimum by maintaining the production rooms as closed as possible.

A. Transfer room

Plantlets grown in the production rooms, each with 4-6 unfolded leaves, are "transferred" into the transfer room. In this room, workers prepare explants (single-node leafy cuttings) from the plantlets and place them on multi-cell trays. The newly prepared trays with explants are placed on conveyers and "transferred" back to the production rooms. Plantlets that have grown in the production rooms can also be "transferred" into low-temperature storage room 2 through the transfer room to slow down their growth for a few days or a few weeks in order to meet production schedules.

Two transplant growth measuring systems (Figure 1) are located in the transfer room. Each system consists of two digital cameras and a computer. By locating individual explants or a multi-cell tray with transplants in the system, the length, the width and the area of leaves of individual explants, and the maximum and height and volume of the transplant canopy can be estimated. These systems also determine the change in leaf color of explants and transplants. The quality of explants that are used for propagation and the quality of transplants that are consigned to consumers can be evaluated with these systems.

B. Production rooms

The production rooms are where the sweetpotato plantlets are grown to be propagated and/or to be consigned as transplants. There is no difference in the facilities or operational methods among these rooms. The only differences among the rooms are in the degree of cleanness or sterilization and the numbers of basic modules that they contain. The degree of cleanness is highest in the micropropagation room and lowest in the production room 2. Plantlets can be moved from the rooms with a higher degree of cleanness to one with a lower degree of cleanness, but not in the opposite direction. Two basic modules are installed in the micropropagation room and four are installed in each of production rooms 1 and 2. All of the rooms have a tray transporting system and an irrigation system.

C. Basic Module

The basic module is a bookshelf-like structure located in each production room in which plantlets in multi-cell trays are placed and cultured (Figure 2). Each basic module has a lighting unit, an air conditioning unit, a control unit and seven shelves (2670 mm x 685 mm). The vertical spacing between shelves is 520 mm. Eight multi-cell trays (approx. 300 mm x 600 mm) can be accommodated on each shelf. Each shelf is lighted with sixteen 32-W fluorescent lamps and three 16-W fluorescent lamps. The output of each lamp (excluding the 16-W lamps) can be controlled with pre-decided lighting patterns to achieve the desired PPF uniformly distributed at the surface level of empty trays. The PPF of each shelf is automatically checked once a day to find any malfunctioning lamps using a PPF sensor installed on a tray-transporting unit (described later). The PPF on the empty tray surface of each shelf can be

Figure 2. A tray transporter installed between the basic modules in each production room.

controlled between 140 to 300 μmol m^{-2} s^{-1} when the lamps are turned on.

The air conditioning unit consists of three home-use air conditioners, three mixing fans and a humidifier that are installed on the top of each module. These devices are operated with pre-decided operating patterns designed to minimize energy consumption for eliminating the heat produced by the basic module. The conditioned air is distributed by the mixing fans to each shelf through an air duct and through holes in the backside panel of the shelves. Air temperature of the basic module can be controlled in the range between 22 and 30 °C, relative humidity can be controlled up to 80 %, and air current speed can be controlled in the range between 0 and 1 m s^{-1}.

A Local Operating Network (LON) unit, which is a remote control/monitoring system, independently operates the devices of each basic module. The devices of each basic module such as air conditioners, fans, lamps, a humidifier and environmental measuring devices are operated by selecting an operation pattern from ten patterns memorized in the LON unit. Maximum, minimum and averaged values of environmental parameters of each basic module and important incidents such as errors and warnings are reported to the gateway computer. These decentralized functions of each module give a great amount of freedom for future scale-ups of this closed system.

During the dark period, if necessary, the air of each room circulates through a filtering system to remove dust, pathogens, insects and debris. An air sampler is installed in the air circulation line for monitoring microbes. The air pressure of the air circulation line is monitored to know when to change the filters.

D. Tray transporting system

Since workers are not permitted to enter the production area for handling the trays with plantlets, the transportation of trays needs to be automated. A tray transporting system is installed in each room of the production area. This system consists of a tray transporter, conveyers, and a double-door system.

The tray transporter (Figure 2) was developed by modifying a transporter that is used for inventory management in warehouses. It is mainly for transporting trays with plantlets from one location to another in the same room and/or from a room of the production area to the transfer room. The tray transporter has another function. It carries sensors and a digital camera for measuring environmental factors at any location in the basic module and for capturing continuous images during its operation.

A conveying system is installed between the transfer room and each room of the production area. It is used for transferring trays both ways. A double-door system is installed on the conveyer to separate each room of the production area from the transfer room. A door opens when trays are passed on the conveyer, but two of the doors never open at the same time to maximize the independence of each room.

E. Irrigation system

An automatic irrigation system (Figure 3) is installed on each tray transporter. The irrigation system was developed based upon the concepts of microprecision irrigation[13]. Only the proper amount of nutrient solution for a particular plant (or for a particular cell of a tray) is delivered from this irrigation system. Therefore no draining or recycling process is needed and a very high efficiency of water usage can be expected. It is also possible to adjust the amount of water for a certain plant (or cell).

When a command for irrigating a particular tray is given, the transporting/irrigation system moves to the designated tray. It lifts the tray, weighs it, and moves it to the irrigation device. Then, the irrigation device with 72 needle-type nozzles moves up and down for injecting nutrient solution. An irrigation pattern including injection positions and amounts of nutrient solution for each injection can be chosen out of 64 patterns. After injection, the tray is weighed again and then returned to the original location.

A device that automatically prepares nutrient solutions is located in the machine room (a non-sterilized area). This device dilutes stock solutions to pre-set concentrations and supplies them to the nutrient tank (1 l) of each irrigation system. The nutrient solution is delivered to each cell by gas pressure supplied by a tank of compressed nitrogen gas. Solenoid valves are installed in each of the 72 lines. The amount of nutrient solution injected into a cell is controlled by opening a valve for a specified time. A pan for catching nutrient solution and/or media that drip from a tray is installed at each irrigation system.

Figure 3. An automatic irrigation system installed on each tray transporter.

F. Low temperature storage room 2

Low temperature storage room 2 (2942 x 1592 x 2230 mm(H)) is located in the sterilized area. It is used for storing plantlets for a few days or a few weeks when it is desired to maintain their growth at a minimum. This room is dimly lighted and kept at a temperature in the range between –3 and 15 °C. An air conditioner (1.5 kW) and a humidifier (1.0 l h^{-1}) are installed. Two movable racks, each with four shelves, are placed in this storage room. Four trays with plantlets can be accommodated on each shelf. A specially designed lighting panel (32 W) is installed in each shelf, and the PPF of each shelf can be controlled with a light controller.

Non-sterilized area

The areas for transplant production in the non-sterilized area include a cleaning room, a preparation room, low temperature storage room 1 and a control room.

A. Cleaning room

Reused trays, supporting materials (media) and other materials are first cleaned in a washing area in this room. Then, they are packed in plastic bags and sterilized by dipping in hot water (about 70 °C) for a few hours. The water is heated mainly by a solar heater installed on the roof of this system, and a boiler is used as an auxiliary heater. Materials that have been cleaned and sterilized are transferred to the preparation room. This room has a door of the low temperature storage room 1 and the transplants are transferred to this room to be packed and shipped.

B. Preparation room

People and materials must pass through this area to enter the transfer room. It has a door to the air shower room and a pass-box. There is an autoclave in this room to sterilize the tools and materials that are used in the sterilized area.

C. Low temperature storage room 1

All transplants to be shipped must pass this area through a pass box between the transfer room. Transplants can be directly shipped out or stored in this storage room. The transplants can be stored here for anywhere from a few hours to a few weeks to adjust to shipping schedules. It is important to have a buffering capacity to adjust to sudden changes in the amounts of orders and/or scheduled or nonscheduled incidents in the production process. To provide more degrees of freedom to the production facility, it is important to have not only a greater capacity in terms of space but also the ability to hold transplants for a longer duration. Currently, sweetpotato transplants can be stored for about two weeks without any decrease in their economic value.

This room has six movable racks with lamps for dim lighting, similar to those used in low temperature storage 2. Four air conditioners (1.1 kW) and two humidifiers (1.5 l h^{-1}) are installed for controlling air temperature and relative humidity of this storage room.

D. Control room

Computer systems for the production managing system, a plant operating system and a database system are located in the control room (Figure 4).

Production managing system

The production managing system is the system that manages the whole process of this facility. It consists of four subsystems: a production planning subsystem, a production simulating subsystem, a process managing subsystem and a research supporting subsystem[14].

Figure 4. Computer systems for a production managing system, a plant operating system and a data base system located in the control room.

The production planning subsystem finds the best plan to produce and ship out the transplants as the orders are received. The plan takes into consideration the resources, such as space, labor, plant materials and time that are available. This subsystem develops a production schedule that specifies how many explants should be prepared on a given day, if a particular order is found to be within the production capacity of the system. The production simulating subsystem is a kind of virtual factory that simulates the production with several models. The growth of trays of plantlets is simulated based upon the environmental conditions where the trays are located. The concepts of decentralized and object-oriented systems[15] are used for simulating the actual production activities in this system. If the production plan is found to be feasible by the production simulating subsystem, the process managing subsystem gives various instructions related to tray transporting, irrigation, manual operations by workers and others. Each worker carries a handheld terminal (a small communicating device) to receive instructions from the subsystem. The workers only need to follow the instructions that are shown on the small monitor of the handheld terminal.

The research supporting subsystem is used to assist researchers with their experiments. The researchers can manage their plantlets independently from the regular production processes with the support of this subsystem.

Plant operating system

The plant operating system is the system that actually operates the facilities and monitors the actions of the facilities. It consists of a few subsystems: a facility operating subsystem, a plant monitoring subsystem and an alarm subsystem. The facility operating subsystem gives orders of particular actions to the facilities with the commands from the process managing subsystem of the production managing system. The operation of each facility and the results by the action are monitored by the plant monitoring subsystem. They can be monitored on the computer monitors. If problems are detected by the plant monitoring subsystem, the alarm subsystem takes several steps to protect the facilities. For major alarms, this subsystem sends a message to the cellular phones of the management staff, and thus the staff can check the facilities and reset them. The history of alarms and warnings are stored in a database system.

Database system

Data related to the growth of transplants and the plant operation (which are produced and/or retrieved by the production managing system, the plant operating system and the plant growth measuring system) are saved and managed in the computer of the database system. The data saved in this system is searchable and compatible with other software. Not only numerical data but also text and image can be saved in this database.

E. Research area

The research area has three laboratories (a biological lab, a physical lab, and a basic module lab) for research and other work that is done to support production. The biological lab is used to analyze plant samples and microorganisms in the air on a regular basis. The physical lab is used to examine samples of the nutrient solutions and supporting materials. The basic module lab is used for examining environmental details and conducting other experiments in the basic module.

CONCLUDING REMARKS

Using the closed system at Chiba University, the authors and their research group conduct various studies related to transplant production. Energy and mass analyses in the production area are important for improving system performance. These analyses can help to decrease energy consumption and production costs. From preliminary experiments with similar closed systems, the electricity consumed for producing a transplant for 14 days is about 0.35 MJ (= 0.1 kWh), which costs 1.5 Japanese yen[16]. Both the initial and operational costs of the closed system can be decreased by selecting facilities that have better performances and proper capacities and by optimizing their operation.

This closed system introduces many new concepts and technologies. Some of the pre-existing technologies and concepts are applied to this system, and several concepts and technologies have been newly developed for this system. Each technique or concept can be applied to other industries in agriculture, while this new concept of a closed system for transplant production itself can be utilized as a new agricultural business.

The research results from the closed system at Chiba University will create opportunities for new agricultural businesses and research areas. And these results can be utilized for further improvement of the systems and for development of small-sized systems that farmers can use for production of their own transplants. The closed transplant production system, a scaled-up system of photoautotrophic micropropagation system, can be used for producing high quality transplants of any species including woody plants. This system will maximize the production efficiency and will make it possible to produce transplants according to pre-decided schedules, which is presently difficult to achieve in agricultural production systems.

REFERENCES

1. J. Aitken-Christie, T. Kozai & M.A.L. Smith, Automation and environmental control in plant tissue culture, Kluwer Academic Publishers, Dordrecht, The Netherlands, **1995**, p.574.
2. T. Kozai, Micropropagation under photoautotrophic conditions, In: Micropropagation -technology and application-, P.C. Debergh & R.H. Zimmerman (eds.), Kluwer Academic Publishers, Dordrecht, The Netherlands, **1991**, pp. 447-469.
3. Q.T. Nguyen & T. Kozai, Environmental effects on the growth of plantlets in micropropagation, Environment Control in Biology, **1998**, 36, 59-75.
4. B.R. Jeong, K. Fujiwara & T. Kozai, Environmental control and photoautotrophic micropropagation, Horticultural Reviews, **1995**, 17, 125-172.
5. T. Kozai, K. Ohyama, F. Afreen, S. Zobayed, C. Kubota, T. Hoshi & C. Chun, Transplant production in closed systems with artificial lighting for solving global issues on environmental conservation, food, resource and energy, Proc. of ACESYS III Conference, Rutgers University, CCEA (Center for Controlled Environment Agriculture), **1999**, 31-45.
6. T. Kozai, C. Kubota, S. Zobayed, Q.T. Nguyen, F. Afreen-Zobayed & J. Heo. Developing a mass-propagation system of woody plants, In: Proc. of the 12th Toyota Conference: Challenge of Plant and Agricultural Sciences to the Crisis of Biosphere on the Earth in the 21st Century, K. Watanabe & A. Komamine (eds.), Landes Company, **1999**, pp. 293-307.
7. C. Chun & T. Kozai, Closed transplant production system at Chiba University, In: Transplant production in the 21st century, C. Kubota & C. Chun (eds.), Kluwer Academic Publishers, Dordrecht, The Netherlands, **2000**, pp. 20-27.
8. A.F.M.S. Islam, C. Chun, M. Takagaki, K. Sakami, & T. Kozai, Yield and growth of sweetpotato using plug transplants as affected by their ages and planting depths. In: Transplant production in the 21st century, C. Kubota & C. Chun (eds.), Kluwer Academic Publishers, Dordrecht, The Netherlands, **2000**, pp.149-153.
9. K. Fujiwara, T. Kozai & I. Watanabe, Development of a photoautotrophic tissue culture system for shoots and/or plantlets at rooting and acclimatization stages, Acta Horticulturae **1988**, 230, 153-158.
10. C. Kubota & T. Kozai, Growth and net photosynthetic rate of *Solanum tuberosum* in vitro under forced and natural ventilation, HortScience, **1992**, 27, 1312-1314.
11. J. Heo & T. Kozai, Development of a forced ventilation micropropagation system for enhancement of the photoautotrophic growth of sweetpotato plug plantlets cultured in vitro, Jpn. J. Trop. Agr., **1988**, 42 (Extra issue 1): 21-22.
12. S.M.A. Zobayed, C. Kubota & T. Kozai, Development of a forced ventilation micropropagation system for large-scale photoautotrophic culture and its utilization in sweet potato, In Vitro Cell. Dev. Biol.-Plant, **1999**, 35, 350-355.

13. H. Murase, Microprecision irrigation system for transplant production, In: Transplant production in the 21st century, C. Kubota & C. Chun (eds.), Kluwer Academic Publishers, Dordrecht, The Netherlands, **2000**, pp.33-37.
14. T. Hoshi, Y. Hayashi, & T. Kozai, Design concept of computerized support systems for large-scale transplant production. In: Transplant production in the 21st century, C. Kubota & C. Chun (eds.), Kluwer Academic Publishers, Dordrecht, The Netherlands, **2000**, pp.38-43.
15. T. Hoshi, Objected oriented software development support system of environmental controller in plant growth factories (in Japanese with English summary and captions), J. High Technology in Agriculture, **1992**, 3, 129-136.
16. K. Ohyama & T. Kozai, Estimating electric energy consumption and its cost in a transplant production factory with artificial lighting: A case study (in Japanese with English summary and captions). J. High Technology in Agriculture, **1998**, 10, 96-107.

Molecular Breeding of Woody Plants
N. Morohoshi and A. Komamine, editors.

PHOTOAUTOTROPHIC MICROPROPAGATION OF *RHODODENDRON*

***Carmen Valero-Aracama[1], Sayed M. A. Zobayed[2], Shyamal K. Roy[3], Chieri Kubota[1] and Toyoki Kozai[1]**

[1]*Laboratory of Environmental Control Engineering, Faculty of Horticulture, Chiba University, Matsudo, Chiba 271-8510, Japan.*

[2]*Department of Biological Sciences, University of Hull, Hull HU6 7RX, UK.*

[3]*Department of Botany, Jahangirnagar University, Savar, Dhaka, Bangladesh.*

ABSTRACT

Photoautotrophic micropropagation (no sugar in the medium under a controlled environment) has many advantages over conventional photomixotrophic micropropagation (using sugar-containing medium), such as promoted growth and development of plantlets *in vitro* and enhanced percent survival *ex vitro* with many different plant species. *Rhododendron* is an important ornamental woody species, which often shows low multiplication rates and low percentages of rooting *in vitro* and survival *ex vitro* in conventional photomixotrophic micropropagation, and which could possibly benefit from photoautotrophic culture. In the present study, photoautotrophic culture was investigated during the multiplication and preparation stages of *Rhododendron* micropropagation. Based on the results, suitable conditions for obtaining a large number of *Rhododendron* plantlets with enhanced growth included the absence of plant growth regulators in the medium, a relatively high photosynthetic photon flux, a high CO_2 concentration, two-leafed nodal cuttings as explants and full-strength medium. Normal roots were obtained in the absence of plant growth regulators, although the percentages of rooted plantlets were low. Further research is needed to enhance rooting in photoautotrophic micropropagation.

KEYWORDS

CO_2 concentration, controlled environment, explant type, IAA, IBA, medium strength, multiplication ratio, PPF, sugar, 2iP

INTRODUCTION

Photoautotrophic micropropagation, which uses sugar-free medium, has many advantages over conventional photomixotrophic micropropagation, which uses sugar-containing medium, when environmental conditions are properly controlled [1]. These advantages include enhanced photosynthesis, growth and rooting *in vitro* and increased percent survival *ex vitro*. Although most research on photoautotrophic micropropagation has focused on herbaceous species, some woody species, such as coffee (*Coffea arabusta*) [2], eucalyptus (*Eucalyptus camaldulensis*) [3], acacia (*Acacia mangium*) [4] and mangosteen (*Garcinia mangostana*) [5] have also benefited from photoautotrophic culture.

The purpose of the present study was to investigate the photoautotrophic micropropagation of *Rhododendron*, an important ornamental woody species, which often shows low multiplication ratios and low percentages of rooting *in vitro* and survival *ex vitro* in conventional photomixotrophic micropropagation.

In conventional *Rhododendron* micropropagation, single shoots are cultured in a medium supplied with 2iP and IAA to induce multiple shoots during the multiplication stage. During the preparation stage, quarter-strength medium is used, supplied with IBA to induce root formation and to prepare plantlets for *ex vitro* conditions [6]. The present study focused on the multiplication and preparation stages of micropropagation and consisted of three experiments. Expt. 1 investigated the effects of the presence or absence of plant growth regulators (PGR) conventionally used at the multiplication stage, and explant type, on the photoautotrophic growth and multiplication ratio of plantlets. Expt. 2 investigated the viability of two-leafed nodal cuttings as explants and the effects of photosynthetic photon flux (PPF) and CO_2 concentration on the photoautotrophic growth of plantlets. Expt. 3 investigated the effects of medium strength and the presence or absence of PGR conventionally used at the preparation stage on photoautotrophic growth and rooting of plantlets.

MATERIALS & METHODS

Plant materials & description of treatments

Explants obtained from *Rhododendron* "Hatsuyuki" plantlets were used in the three experiments. Culture conditions for photoautotrophic and photomixotrophic treatments in the three experiments are shown in Table 1.

Experiment 1

Shoot tips and nodal cuttings obtained from plantlets cultured photomixotrophically, each with 4-5 leaves, 25 mg fresh mass and 9.7 mg dry mass were used as explants. The presence (P) or absence (A) of PGR (5.0 mg L^{-1} 2iP+1.0 mg L^{-1} IAA) and explant type (shoot tip (S) or nodal cutting (N)) were examined after 40 days of culture in four photoautotrophic treatments, which were compared with one conventional photomixotrophic treatment (Control), with PGR in the medium and using shoot-tip explants. the number of leaves per plantlet with leaf blade length greater than 2 mm, were counted on day 40, to estimate the multiplication ratios (number of harvestable explants (HE) per plantlet). Dry mass per plantlet was measured and dry mass per HE was estimated by dividing dry mass by the multiplication ratio of the plantlets. The experiment was a factorial, completely randomized design, with 5 replications per treatment. Each vessel, containing two shoot tips and two nodal cuttings, had one replication for each explant type.

Experiment 2

Two-leafed nodal cuttings obtained from plantlets cultured photomixotrophically, each with 20 mg fresh mass and 4 mg dry mass were used as initial explants. PPF (50, 100 and 150 μmol m^{-2} s^{-1}) and CO_2 concentration (600 (L) and 2200 (H) μmol mol^{-1}) were investigated after 40 days of culture in six photoautotrophic treatments, which were compared with one conventional photomixotrophic treatment (Control), under 50 μmol m^{-2} s^{-1} PPF and 600 μmol mol^{-1} CO_2 concentration. CO_2 concentrations inside the vessels were measured on day 39 using a gas chromatograph (GC-9A, Shimadzu Co., Kyoto, Japan). On day 40, fresh mass, dry mass and leaf area per plantlet were measured, and multiplication ratios were estimated (two-leafed nodal cuttings were considered HE). The experiment was a factorial, completely randomized design, with 5 replications per treatment. Each vessel, containing 4 explants, was one replication.

Experiment 3

Two-leafed nodal cuttings obtained from plantlets cultured photoautotrophically for 40 days, each with 30 mg fresh mass and 12 mg dry mass were used as initial explants. Medium-strength (quarter-strength (Q) or full-strength (F) medium) and presence (P) or absence (A) of PGR (5 mg L^{-1} IBA) were examined after 72 days of culture in four photoautotrophic treatments. On day 72 dry mass was measured, the percentages of dry matter and rooted plantlets were estimated and the relative growth rate (RGR) was estimated from day 0 until day 72, on a plantlet dry-mass basis. CO_2 concentrations inside and outside the vessels were measured on day 71 to estimate the net photosynthetic rate (NPR) per dry mass, using the method described by Fujiwara et al. (1987) [8]. The experiment was a factorial, completely randomized design, with 6 replications per treatment. Each vessel, containing 4 explants, was one replication.

Data analysis

In all three experiments, ANOVA (analysis of variance) was used to evaluate the effects of experimental factors on photoautotrophic treatments, and the Duncan's multiple range test was used to determine significance among photoautotrophic treatments at $P \leqq 0.05$. In Expts. 1 and 2, comparisons with the Control were evaluated by a t-test at $P \leqq 0.01$ and $P \leqq 0.05$.

Table 1. Experimental conditions for *Rhododendron* micropropagation for experiments 1, 2 and 3

Micropropagation conditions	Photoautotrophic	Photomixotrophic
Culture Room		
PPF[z] (μmol m^{-2} s^{-1})	100 (Expts. 1 and 3) 50, 100 or 150 (Expt. 2)	50 (Expts. 1 and 2)
CO_2 concentration (μmol mol^{-1})	1500 (Expt. 1) 600 or 2200 (Expt. 2) 2200 (Expt. 3)	400 (Expt. 1) 600 (Expt. 2)
Photo-/dark period (h d^{-1})	16/8	16/8
Light source	white fluorescent lamps	white fluorescent lamps
Relative humidity (%)	85±10	85±10
Air temperature (°C) (photo-/dark period)	26/24	26/24
Vessels and Medium		
Vessel volume (L)	0.37	0.37
Basal medium (50 mL/vessel)	A.R.M.[y] without vitamins	A.R.M. with vitamins
Supporting material	Florialite[x]	agar (7 g L^{-1})
Sucrose concentration (g L^{-1})	0	30
No. air exchanges of the vessels (h^{-1})	4.5	0.1
No. explants (/vessel)	4	4

[z]Photosynthetic photon flux measured on the empty shelf
[y]Anderson rhododendron medium (1978) [7]
[x]Vermiculite and cellulose fiber mixture, Nisshinbo Industries Inc., Japan

RESULTS & DISCUSSION

Effects of presence or absence of PGR and explant type on multiplication ratio and growth of plantlets (Expt. 1)

Under photoautotrophic conditions the explant leaves play an important role in the subsequent growth of propagules because they produce carbohydrates by photosynthesis [9]. In the present study, the types of explants considered harvestable were two-leafed nodal cuttings for nodal-cutting treatments, and axillary shoots with at least two leaves for shoot-tip treatments. In the AN treatment, a significantly greater multiplication ratio was obtained than in the other treatments and explant dry mass was not significantly different from that in the Control (Table 2). Although a greater dry mass per HE was obtained in the AS treatment than in the other treatments, the multiplication ratio was low in the AS treatment, a condition that is not preferred for the multiplication stage. The use of photoautotrophic micropropagation, including high CO_2 concentration, relatively high PPF, high ventilation rates of the vessels and porous supporting materials, enhanced the multiplication ratio and growth of *Rhododendron* using medium without PGR and nodal cuttings as explants.

Table 2. Effects of presence (P) or absence (A) of 2iP + IAA and explant type: shoot tip (S) or nodal cutting (N), on the multiplication ratio and dry mass per HE and per plantlet on day 40 (Expt. 1)

Treatment code	Selected HE	Multiplication ratio	Dry mass / HE (mg)	Dry mass / plantlet (mg)
PS	[z]AS	[x]2.4±0.3b^{NS}	22±3.0b**	47±5.9ab**
PN	[y]NC	2.1±0.8b^{NS}	12±1.9b^{NS}	21±1.8c^{NS}
AS	AS	1.1±0.1b**	57±10.3a**	61±10.3a**
AN	NC	6.2±0.7a**	6±1.2b^{NS}	34±2.9bc**
Control	AS	2.8±0.4	9±2.5	19±2.0
[w]ANOVA				
PGR (X)		**	*	*
Expl. type (Y)		**	**	**
X × Y		**	**	NS

[z]Axillary shoots. [y]Two-leafed nodal cuttings. [x]Means ± SE in the same column followed by the same letter are non-significantly different by the Duncan Multiple Range Test at $P \leqq 0.05$. ^{NS}, *, **Non-significantly or significantly different from Control by *t*-test at $P \leqq 0.05$ or $P \leqq 0.01$, respectively. [w]Significance among the photoautotrophic treatments

Table 3. Effects of PPF (150, 100 and 50 $\mu mol\ m^{-2}\ s^{-1}$: 150, 100 and 50, respectively) and CO_2 concentration (2200 and 600 $\mu mol\ mol^{-1}$: H and L, respectively) on the growth of *Rhododendron* plantlets on day 40 (Expt. 2)

Treatment Code	Fresh mass / Plantlet (mg)	Dry mass / Plantlet (mg)	Leaf Area / Plantlet (mm^2)	Multiplication Ratio
150H	[z]29±2.0bc^{NS}	10±0.7bc*	70±6.8b**	2.2±0.2b^{NS}
100H	41±4.3a^{NS}	14±1.4a**	117±13.4a^{NS}	3.3±0.4a*
50H	30±4.2abc^{NS}	10±1.3abc*	100±16.4ab^{NS}	2.2±0.2b^{NS}
150L	24±2.2bc^{NS}	8±0.6bc^{NS}	67±5.2b*	2.2±0.2b^{NS}
100L	32±3.4ab^{NS}	10±1.1ab*	89±11.5ab^{NS}	2.2±0.3b^{NS}
50L	20±1.8c*	7±0.7c^{NS}	64±6.4b*	1.5±0.2b^{NS}
Control	39±6.3	7±1.1	92±12.7	2.1±0.4
[y]ANOVA				
PPF (X)	**	**	**	**
[CO_2] (Y)	**	**	*	**
X × Y	NS	NS	NS	NS

[z]Mean ± SE in the same column followed by the same letter are non-significantly different by the Duncan Multiple Range Test at $P \leqq 0.05$. $^{NS,\ *,\ **}$Non-significantly or significantly different from Control by *t*-test at $P \leqq 0.05$ and $P \leqq 0.01$, respectively. [y]Significance among the photoautotrophic treatments

Effects of CO_2 concentration and PPF on growth of plantlets using two-leafed nodal cuttings as explants (Expt. 2)

PPF and CO_2 concentration significantly affected the multiplication ratio and growth of plantlets (Table 3). Among photoautotrophic treatments, plantlets in the 100H treatment had a greater fresh mass and leaf area, which were not significantly different from those in the Control, and a greater dry mass and multiplication ratio, which were greater than those in the Control. These results are in accordance with those obtained with other species, such as strawberry, raspberry, asparagus [10] and *Rosa* [11], in which growth and development of plantlets were promoted under high PPF and high CO_2 concentration. Under photoautotrophic conditions, two-leafed nodal cuttings were usable explants for the multiplication stage, where plantlet growth was enhanced with the combination of 100 μmol m^{-2} s^{-1} PPF and high CO_2 concentration.

Effects of medium strength and presence or absence of PGR on growth and rooting of plantlets (Expt. 3)

A greater dry mass was obtained using full- rather than quarter-strength medium, especially in the absence of PGR (Fig. 1). The percent dry matter was not affected by medium-strength and was significantly greater in the absence than in the presence of PGR (QA and FA treatments). RGR was greater in full- than in quarter-strength medium, especially in the absence of PGR (FA treatment). NPR per dry mass was higher in full- rather than quarter-strength medium and in the absence than in the presence of PGR. The percentage of rooted plantlets was greater in the QP treatment followed by the FP treatment; however, approximately half of the rooted plantlets cultured in the presence of PGR were considered abnormal (Fig. 1).

Inorganic nutrient composition and concentrations have been optimized for

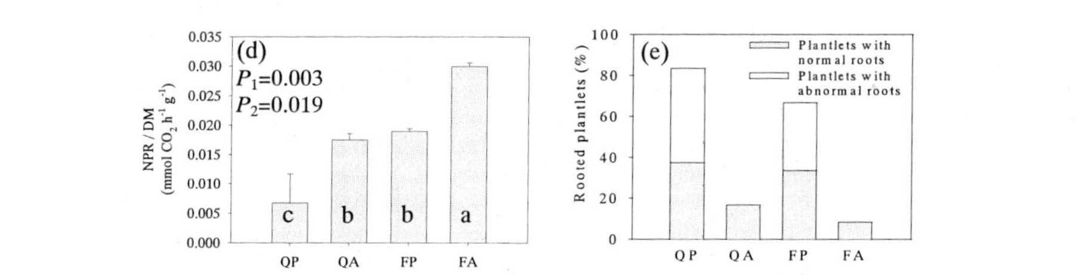

Figure 1. Plantlet dry-mass (a) and percent dry matter (b) on day 72, RGR (c) from days 0 to 72, NPR per plantlet dry-mass (d) on day 71 and percent of rooted plantlets (e) on day 72, in quarter (Q) or full (F) strength medium and in presence (P) or absence (A) of IBA (Expt. 3). P_1 and P_2 are P-values from ANOVA for medium strength and PGR variables, respectively. Letters inside the bars are comparisons by Duncan's Multiple Range Test at $P\leqq0.05$

heterotrophic micropropagation. Therefore, they might not be optimal for photoautotrophic micropropagation [12]. For the preparation stage, quarter-strength medium is used for conventional micropropagation. However, full- rather than quarter-strength medium promoted growth of *Rhododendron* plantlets in photoautotrophic micropropagation. PGR should be excluded in photoautotrophic micropropagation because in the absence rather than the presence of PGR, growth was promoted and only normal roots were obtained. Nevertheless, further research using environmental control for increasing the percentages of rooted plantlets is needed to improve *Rhododendron* micropropagation.

CONCLUSIONS

Photoautotrophic culture is an effective method for *Rhododendron* micropropagation because it enhanced the growth, the multiplication ratio and the net photosynthetic rate of plantlets. Growth of plantlets was enhanced in 100 μmol m^{-2} s^{-1} PPF and relatively high CO_2 concentration. Under such conditions, two-leafed nodal cuttings were useful explants for the multiplication stage when PGR were excluded from the medium. The same culture conditions, using full- rather than quarter-strength medium were preferred to enhance the growth of plantlets during the preparation stage. However, the absence of PGR produced low percentages of rooted plantlets. Further research on *Rhododendron in vitro* rooting under photoautotrophic conditions is needed.

REFERENCES

1. T. Kozai, Micropropagation under photoautotrophic conditions, In: *Micropropagation: Technology and Application*, P. Debergh & R. H. Zimmerman (eds.), Kluwer Academic Publishers, Dordrecht, The Netherlands, **1990**, pp. 449-471.
2. T. Kozai, Q.T. Nguyen, C. Kubota et al., Growth promotion of coffee (*Coffea arabusta*) plantlets *in vitro* by use of fibrous supports containing no sugar and culture vessels with high number of air exchanges, *Jap. J. Trop. Agr.*, **1998**, 42 (Extra issue 1), 27-28.
3. C. Kirdmanee, Y. Kitaya & T. Kozai, Effects of CO_2 enrichment and supporting material on growth, photosynthesis and water potential of *Eucalyptus* shoots/plantlets cultured photoautotrophically in vitro. *Environmental Control in Biology*, **1995**, 33 (2), 133-141.
4. M. Imelda, C. Kubota, T. Kozai et al., Photoautotrophic Micropropagation of Acacia Plantlets. *Jap. J. Trop. Agr.*, **1998**, 42(Extra issue), 29-30.
5. C. Kubota, T. Tajuddin & T. Kozai. Photoautotrophic micropropagation of mangosteen plantlets. *Jap. J. Trop. Agr.*, **1998**; 42 (Extra issue 1), 31-32.
6. W.C. Anderson, A revised tissue cultured medium for shoot multiplication of rhododendron, *J. Am. Soc. Hort. Sci.*, **1984**, 109, 343-347.
7. W.C. Anderson, Tissue culture propagation of rhododendrons, *In Vitro*, **1978**, 14: pp 334 (Abstr).
8. K. Fujiwara, T. Kozai & I. Watanabe, Fundamental studies on environments in plant tissue culture vessels (III). Measurements of carbon dioxide gas concentration in closed vessels containing tissue cultured plantlets and estimates of net photosynthetic rates of the plantlets, *J. Agr. Meteorol.*, **1987**, 43, 21-30 (in Japanese with English abstract).
9. Q.T. Nguyen & T. Kozai, Environmental effects on the growth of plantlets in micropropagation. *Environ. Control in Biol.*, **1998,** 36 (2), 59-75.
10. Y. Desjardins, F. Laforge, C. Luissier and A. Gosselin, Effect of CO_2 enrichment and high photosynthetic photon flux on the development of autotrophy and growth of tissue-cultured strawberry, raspberry and asparagus plants. *Acta Hort.*, **1988**, 230: 45-47.
11. Y De, A. Gosselin and Y. Desjardins, Carbon metabolism of micropropagated *Rosa multiflora* L., *In Vitro Cell Dev. Biol.*, **1993**, 27, 57-63.
12. T. Kozai & Y. Iwanami, Effects of CO_2 enrichment and sucrose concentration under high photon flux on plantlet growth of carnation (*Dianthus caryophyllus* L.) in tissue culture during the preparation stage, *J. Jap. Soc. Hort. Sci.*, **1988**, 57, 279-288.

Index of Authors